Financial Statistics and Mathematical Finance

Financial Statistics and Mathematical Finance

Methods, Models and Applications

Ansgar Steland

Institute for Statistics and Economics
RWTH Aachen University, Germany

WILEY

A John Wiley & Sons, Ltd., Publication

This edition first published 2012

Registered office
John Wiley & Sons Ltd, The Atrium, Southern Gate, Chichester, West Sussex, PO19 8SQ, United Kingdom

For details of our global editorial offices, for customer services and for information about how to apply for permission to reuse the copyright material in this book please see our website at www.wiley.com.

Library of Congress Cataloging-in-Publication Data

Steland, Ansgar.
Financial statistics and mathematical finance : methods, models and applications / Ansgar Steland.
Includes bibliographical references and index.
ISBN 978-0-470-71058-6
1. Business mathematics. 2. Calculus. I. Title.
HF5691.S6585 2012
332.01'5195–dc23

2012007001

A catalogue record for this book is available from the British Library.

ISBN: 978-0-470-71058-6

Set in 10/12pt Times by Thomson Digital, Noida, India.
Printed and bound in Singapore by Markono Print Media Pte Ltd.

Contents

Preface

This textbook intends to provide a careful and comprehensive introduction to some of the most important mathematical topics required for a thorough understanding of financial markets and the quantitative methods used there. For this reason, the book covers mathematical finance in the narrow sense, that is, arbitrage theory for pricing contingent claims such as options and the related mathematical machinery, as well as statistical models and methods to analyze data from financial markets. These areas evolved more or less separate from each other and the lack of material that covers both was a major motivation for me to work out the present textbook. Thus, I wrote a book that I would have liked when taking up the subject. It addresses master and Ph.D. students as well as researchers and practitioners interested in a comprehensive presentation of both areas, although many chapters can also be studied by Bachelor students who have passed introductory courses in probability calculus and statistics. Apart from a couple of exceptions, all results are proved in detail, although usually not in their most general form. Given the plethora of notions, concepts, models and methods and the resulting inherent complexity, particularly those coming to the subject for the first time can acquire a thorough understanding more quickly, if they can easily follow the derivations and calculations. For this reason, the mathematical formalism and notation is kept as elementary as possible. Each chapter closes with notes and comments on selected references, which may complement the presented material or are good starting points for further studies.

Chapter 1 starts with a basic introduction to important notions: financial instruments such as options and derivatives and related elementary methods. However, derivations are usually not given in order to focus on ideas, principles and basic results. It sets the scene for the following chapters and introduces the required financial slang. Cash flows, discounting and the term structure of interest rates are studied at an elementary level. The *return* over a given period of time, for assets usually a day, represents the most important economic object of interest in finance, as prices can be reconstructed from returns and investments are judged by comparing their return. Statistical measures for their location, dispersion and skewness have important economic interpretations, and the relevant statistical approaches to estimate them are carefully introduced. Measuring the risk associated with an investment requires being aware of the properties of related statistical estimates. For example, *volatility* is primarily related to the standard deviation and *value-at-risk*, by definition, requires the study of quantiles and their statistical estimation. The first chapter closes with a primer on option pricing, which introduces the most important notions of the field of mathematical finance in the narrow sense, namely the *principle of no-arbitrage*, the *principle of risk-neutral pricing* and the relation of those notions to probability calculus, particularly to the existence of an *equivalent martingale measure*. Indeed, these basic concepts and a couple of fundamental insights can be understood by studying them in the most elementary form or simply by examples.

Chapter 2 then discusses arbitrage theory and the pricing of contingent claims within a one-period model. At time 0 one sets up a portfolio and at time 1 we look at the result. Within this simple framework, the basic results discussed in Chapter 1 are treated with mathematical rigor and extended from a finite probability space, where only a finite number of scenarios can occur, to a general underlying probability space that models the real financial market. Mathematical separation theorems, which tell us how one can separate a given point from convex sets, are applied in order to establish the equivalence of the exclusion of arbitrage opportunities and the existence of an equivalent martingale measure. For this reason, those separation theorems are explicitly proved. The construction of equivalent martingale measures based on the Esscher transform is discussed as well.

Chapter 3 provides a careful introduction to stochastic processes in discrete time (time series), covering martingales, martingale differences, linear processes, ARMA and GARCH processes as well as long-memory series. The notion of a *martingale* is fundamental for mathematical finance, as one of the key results asserts that in any financial market that excludes arbitrage, there exists a probability measure such that the discounted price series of a risky asset forms a martingale and the pricing of contingent claims can be done by risk-neutral pricing under that measure. These key insights allow us to apply the elaborated mathematical theory of martingales. However, the treatment in Chapter 3 is restricted to the most important findings of that theory, which are really used later. Taking first-order differences of a martingale leads naturally to *martingale difference sequences*, which form whitenoise processes and are a common replacement for the unrealistic i.i.d. error terms in stochastic models for financial data and, more generally, economic data. A key empirical insight of the statistical analysis of financial return series is that they can often be assumed to be uncorrelated, but they are usually not independent. However, other series may exhibit substantial serial dependence that has to be taken into account. Appropriate parametric classes of time-series models are ARMA processes, which belong to the more general and infinite-dimensional class of linear processes. Basic approaches to estimate autocovariance functions and the parameters of ARMA models are discussed. Many financial series exhibit the phenomenon of conditional heteroscedasticity, which has given rise to the class of (G)ARCH models. Lastly, fractional differences and long-memory processes are introduced.

Chapter 4 discusses in detail arbitrage theory in a discrete-time multiperiod model. Here, trading is allowed at a finite number of time points and at each time point the trading strategy can be updated using all available information on market prices. Using the martingale theory in discrete time studied in Chapter 3, it allows us to investigate the pricing of options and other derivatives on arbitrage-free financial markets. The Cox–Ross–Rubinstein binomial model is studied in greater detail, since it is a standard tool in practice and also provides the basis to derive the famous Black–Scholes pricing formula for a European call. In addition, the pricing of American claims is studied, which requires some more advanced results from the theory of optimal stopping.

Chapter 5 introduces the reader to stochastic processes in continuous time. Brownian motion will be the random source that governs the price processes of our financial market model in continuous time. Nevertheless, to keep the chapter concise, the presentation of Brownian motion is limited to its definition and the most important properties. Brownian motion has puzzling properties such as continuous paths that are nowhere differentiable or of bounded variation. Advanced models also incorporate fractional Brownian motion and Lévy processes, respectively. Lévy processes inherit independent increments but allow for non-normal distributions of those increments including heavy tails and jump. Fractional Brownian motion is a Gaussian process as is Brownian motion, but it allows for long-range dependent increments where temporal correlations die out very slowly.

Chapter 6 treats the theory of stochastic integration. Assuming that the reader is familiar with integration in the sense of Riemann or Lebesgue, we start with a discussion of stochastic

Riemann–Stieltjes (RS) integrals, a straightforward generalization of the Riemann integral. The related calculus is relatively easy and provides a good preparation for the Itô integral. It is also worth mentioning that the stochastic RS-integral definitely suffices to study many issues arising in statistics. However, the problems arising in mathematical finance cannot be treated without the Itô integral. The key observation is that the change of the value of position $x(t) = x_t$ in a stock at time t over the period $[t, t + \delta]$ is, of course, given by $x_t \delta P_t$, where $\delta P_t = P_{t+\delta} - P_t$. Aggregating those changes over n successive time intervals $[i\delta, (i + 1)\delta]$, $i = 0, \dots, n - 1$, in order to determine the terminal value, results in the sum $\sum_{i=0}^{n-1} x(i\delta)\delta P_{i\delta}$. Now 'taking the limit $\delta \to 0$' leads to an integral $\int x_s \mathrm{d} P_s$ with respect to the stock price, which cannot be defined in the Stieltjes sense, if the stock price is not of bounded variation. Here the Itô integral comes into play. A rich class of processes are Itô processes and the famous Itô formula asserts that smooth functions of Itô processes again yield Itô processes, whose representation as an Itô process can be explicitly calculated. Further, ergodic diffusion processes as an important class of Itô processes are introduced as well as Euler's numerical approximation scheme, which also provides the common basis for statistical estimation and inference of discretely sampled ergodic diffusions.

Chapter 7 presents the Black–Scholes model, the mathematically idealized model to price derivatives which is still the benchmark continuous-time model in practice. Here one may either invest in a risky stock or deposit money in a bank account that pays a fixed interest. The Itô calculus of Chapter 6 provides the theoretical basis to develop the mathematical arbitrage theory in continuous time. The classic Black–Scholes model assumes that the volatility of the stock price is constant with respect to time, which is too restrictive in practice. Thus, we briefly discuss the required changes when the volatility is time dependent but deterministic. Finally, the generalized Black–Scholes model allows the interest rate of the 'risk-less' instrument to be random as well as dependent on time, thus covering the realistic situation that money not invested in stocks is used to buy, for example, AAA-rated government bonds.

Chapter 8 studies the asymptotic limit theory for discrete-time processes as required to construct and investigate present-day methods for decision making; that is, procedures for estimation, inference as well as model checking, using financial data in the widest sense (returns, indexes, prices, risk measures, etc.). The limit theorems, partly presented along the way when needed to develop methodologies, cover laws of large numbers and central limit theorems for martingale differences, linear processes as well as mixing processes. The methods discussed in greater detail cover the multiple linear regression with stochastic regressors, nonparametric density estimation, nonparametric regression and the estimation of autocovariances and the long-run variance. Those statistical tools are ubiquitous in the analysis of financial data.

Chapter 9 discusses some selected topics. Copulas have become an important tool for modeling high-dimensional distributions with powerful as well as dangerous applications in the pricing of financial instruments related to credits and defaults. As a matter of fact, these played an unlucky role in the 2008 financial crisis when a simplistic pricing model was applied to large-scale pricing of credit default obligations. For this reason, some of the major developments leading to the crisis are briefly reviewed, revealing the inherent complexity of financial markets as well as the need for sophisticated mathematical models and their application. Local polynomial estimation is studied in greater detail, since it has important applications to many problems arising in finance such as the estimation of risk-neutral densities conditional volatility or discretely observed diffusion processes. The asymptotic normality can be based on a powerful *reduction principle*: A (joint) smoothing central limit theorem for the innovation process $\{\epsilon_t\}$ and a derived process involving the regressors automatically implies the asymptotic normality of the local linear estimator. The testing for and detecting of change-points (structural breaks) have become active fields of current theoretical as well as applied research. Chapter 9 thus closes with a brief discussion of change-point analysis and detection with a certain focus on the detection of changes in the degree of integration.

This book features an accompanying website http://fsmf.stochastik.rwth-aachen.de

Ansgar Steland
RWTH Aachen University
Germany

Acknowledgements

I thank the following students and assistants who carefully read parts of various drafts leading up to this book: Martin Euskirchen, who also typed some chapters, Dr. Wolfgang Herff and M. Sc. Annabel Prause. In particular, I acknowledge the assistance of Dr. Mohammed Abujarad, who proof-read the final version with great care and found many typos, mistakes and made several suggestions concerning the presentation of results. Lastly, I acknowledge the patience of my family and especially my children.

1

Elementary financial calculus

1.1 Motivating examples

Example 1.1.1 *Suppose a pension fund collecting contributions from workers intends to invest a certain fraction of the fund in a certain exchange-traded stock instead of buying treasury bonds. Whereas a bond yields a fixed interest known in advance, the return of a stock is volatile and uncertain. It may substantially exceed a bond's interest, but the pension fund is also exposed to the downside risk that the stock price goes down resulting in a loss. For the pension fund it is important to know what return can be expected from the investment and which risk is associated with the investment. It would also be useful to know the amount of the invested money that is under risk. In practice, investors invest their money in a portfolio of risky assets. Then the question arises: what can be said about the relationship? In modern finance, returns are modeled by random variables that have a distribution. Thus, we have to clarify how the return distribution and its mathematical properties are related to the economic notions* expected return, volatility, *and how one can define appropriate risk measures. Further, the question arises how one can estimate these quantities from historic time series.*

Example 1.1.2 *In order to limit the loss due to the risky stock investment, the pension fund could ask a bank for a contract that pays the difference between a stop loss quote, L, and stock price, if that difference is positive when exercising the contract. Such financial instruments are called options. What is the fair price of such an option? And how can a bank initiate trades, which compensate for the risk exposure when selling the option?*

Example 1.1.3 *Suppose a steel producer agrees with a car manufacturer to deliver steel for the production of 10 000 cars in one year. The steel production starts in one year and requires a large amount of oil. In order to calculate costs, the producer wants to fix the oil price at, say, K dollars in advance. One approach is to enter a contract that pays the difference between the oil price and K at the delivery date, if that difference is positive. Such contracts are named*

Financial Statistics and Mathematical Finance: Methods, Models and Applications, First Edition. Ansgar Steland.

call options. Again, the question arises what is the fair price of such an agreement. Another possibility is to agree on a future/forward contract.

Example 1.1.4 *To be more specific and to simplify the exposition, let us assume that the steel producer needs 1 barrel whose current price at time $t = 0$ is $S_0 = 100$. To fix that price, he buys a call option with delivery price $K = 100$. The fixed interest rate is 1%. Further, suppose that the oil price, S_1, in one year at time $t = 1$ is distributed according to a two-point distribution,*

$$P(S_1 = 110) = 0.6, \qquad P(S_1 = 90) = 0.4.$$

If $S_1 = 110$ one exercises the option right and the deal yields a profit of $G = 10$. Otherwise, the option has no value. Thus, the expected profit is given by

$$E(G) = 10 \cdot 0.6 = 6.$$

Because for the buyer of the option the deal has a non-negative profit and yields a positive profit with positive probability, he or she has to pay a premium to the bank selling the option. Should the bank offer the option for the expected profit 6? Surprisingly, the answer is no. Indeed, an oil dealer can offer the option for a lower price, namely $x = 5.45$ without making a loss. The dealer buys half of the oil when entering the contract at $t = 0$ for the current price of 50 and the rest when the contract is settled. His calculation is as follows. He finances the deal by the premium x and a credit. At $t = 0$ his portfolio consists of a position in the money market, $x - 50$, and 0.5 units of oil. Let us anticipate that $x < 50$. Then at $t = 1$ the dealer has to pay back $1.01 \cdot |x - 50|$ to the bank. We shall now consider separately the cases of an increase or decreases of the oil price. If the oil price increases, the value of the oil increases to $0.5 \cdot 110 = 55$ and he receives 100 from the steel producer. He has to fix the premium x such that the net income equals the price he has to pay for the remaining oil. This means, he solves the equation

$$100 + 1.01 \cdot (x - 50) = 55$$

yielding $x = 5.445545 \approx 5.45$. Now consider the case that the oil price decreases to 90. In this case the steel producer does not exercise the option but buys the oil at the spot market. The oil dealer has to pay back the credit, sells his oil at the lower price, which results in a loss of 5. The premium x should ensure that his net balance is 0. This means, the equation

$$0.5 \cdot 90 + 1.01(x - 50) = 0$$

should hold. Solving for x again yields $x = 5.445545$. Notice that both equations yield the same solution x such that the premium is not random.

1.2 Cashflows, interest rates, prices and returns

Let us now introduce some basic notions and formulas. To any financial investment initiated at $t = t_0$ with time horizon T is attached a sequence of payments settled on a bank account that describe the investment from a mathematical point of view. Our standard notation is as follows: We denote the time points of the payments by $0 = t_0 < t_1 < \cdots < t_n = T$ and the associated payments by $X_1, \ldots, X_T$. Our sign convention will be as follows: Positive payments, $X_i > 0$, are deposits increasing the investor's bank account, whereas negative payments, $X_i < 0$, are charges.

From an economic point of view, there is a huge difference between a payment today or in the future. Thus, to compare payments, they either have to refer to the same time point t^* or one has to take into account the effects of interest. As a result, to compare investments one has to cumulate the payments discounted or accumulated to a common time point t^*. If all payments are discounted to $t^* = t_0$ and then cumulated, the resulting quantity is called the **present value**. Alternatively, one can accumulate all payments to $t^* = T$.

In practice, one has to specify how to determine times and how to measure the economic distance between two time points t_1 and t_2. It is common practice to measure the time as a multiple of a year. At this point, suppose that the dates are given using the day-month-year convention, i.e. $t = (d, m, y)$. In what follows, we denote the economic time distance between two dates t_1 and t_2 by $\tau(t_1, t_2)$. Here are some market conventions for the calculation of $\tau(t_1, t_2)$.

(i) Actual/365: Each year has 365 days and the actual number of days is used.

(ii) Actual/360: Each year has 360 days and the actual number of days is used.

(iii) 30/360: Each month has 30 days, a year 360 days.

In the following we assume that all times have been transformed using such a convention.

If the fixed interest rate is r per annum, interest is paid during the period without compound interest, the accumulated value of payments $X_1, \ldots, X_n$ at dates $t_1, \ldots, t_n$ is given by

$$V_T = \sum_{i=0}^{n} X_i(1 + \tau(t_i, 1)r).$$

The present value at $t = 0$ is calculated using the formula

$$V_0 = \sum_{i=0}^{n} X_i D(0, t_i), \qquad \text{with} \quad D(0, t_i) = \frac{1 + \tau(t_i, T)r}{1 + rT}.$$

Here $D(0, t_i)$ denotes the discount factor taking into account that the payment X_i takes place at t_i.

Often, interest is paid at certain equidistant time points, e.g. quarterly or monthly. When decomposing the year into m periods and applying the interest rate r/m to each of them, an investment of one unit of currency grows during k periods to

$$1 + \frac{r}{m}k.$$

When compound interest is taken into account, the value is

$$(1 + r/m)^k.$$

For $k = m \to \infty$ that discrete interest converges to continuous compounding

$$\lim_{m \to \infty} (1 + r/m)^m = \mathrm{e}^r.$$

Thus, the accumulation factor for an investment lasting for $t \in (0, \infty)$ years, i.e. corresponding to tm periods, equals

$$\lim_{m\to\infty} (1 + r/m)^{mt} = e^{rt}.$$

Let us now assume that the interest rate $r = r(t)$ is a function of t, such that for $r(t) > 0$, $t > 0$, the bank account, $S_0(t)$, increases continuously. There are two approaches to relate these quantities. Either start from a model or formula for $S_0(t)$ or start with $r(t)$. Let us first suppose that $S_0(t)$ is given. The annualized relative growth during the time interval $[t, t + h]$ is given by

$$\frac{1}{h}\frac{S_0(t+h) - S_0(t)}{S_0(t)}.$$

Definition 1.2.1 *Suppose that the bank account $S_0(t)$ is a differentiable function. Then the quantity*

$$r(t) = \lim_{h\downarrow 0} \frac{1}{h}\frac{S_0(t+h) - S_0(t)}{S_0(t)},$$

is well defined and is called **instantaneous (spot) rate** *or simply* **short rate**.

We have the relationship

$$r(t) = \frac{S_0'(t)}{S_0(t)} \quad \Leftrightarrow \quad S_0'(t) = r(t)S_0(t).$$

As a differential:

$$dB(t) = r(t)B(t)dt.$$

It is known that this ordinary differential equation has the general solution $S_0(t) = C\exp(\int_0^t r(s)\,ds)$, $C \in \mathbb{R}$. For our example the special solution

$$S_0(t) = \exp\left(\int_0^t r(s)\,ds\right) \tag{1.1}$$

with starting value $S_0(0) = 1$ matters. In the special case $r(t) = r$ for all t, we obtain $S_0(t) = e^{rt}$ as above.

Often, one starts with a model for the short rate. Then we define the bank account via Equation (1.1).

Definition 1.2.2 (BANK ACCOUNT)
A bank account with a unit deposit and continuous compounding according to the spot rate $r(t)$ is given by

$$S_0(t) = \exp\left(\int_0^t r(s)\,ds\right), \qquad t \geq 0.$$

When depositing x units of currency into the bank account, the time t value is $xS_0(t)$. Vice versa, for an accumulated value of 1 unit of currency at time T, one has to deposit $x = 1/S_0(T)$

at time $t = 0$. The value of $x = 1/S_0(T)$ at an arbitrary time point $t \in [0, T]$ is

$$xS_0(t) = \frac{S_0(t)}{S_0(T)}.$$

This means that the value at time $t = 0$ of a unit payment at the time horizon T is given by $S_0(t)/S_0(T)$.

Definition 1.2.3 *The* **discount factor** *between two time points $t \leq T$ is the amount at time t that is equivalent to a unit payment at time T and can be invested riskless at the bank. It is denoted by*

$$D(t, T) = \frac{S_0(t)}{S_0(T)} = \exp\left(-\int_t^T r(s)\,\mathrm{d}s\right).$$

1.2.1 Bonds and the term structure of interest rates

The basic insights of the above discussion can be directly used to price bonds and understand the term structure of interest rates.

A **zero coupon bond** pays a fixed amount of money, the **face value** or **principal** X at a fixed future time point called **maturity**. Such a bond is also referred to as a **discount bond** or **zero coupon bond**. Here and in what follows, we assume that the bond is issued by a government such that we can ignore default risk. Measuring time in years and assuming that the interest rate r applies in each year, we have learned that the present value of the payment X equals

$$P_n(X) = \frac{X}{(1+r)^n}.$$

Notice that this simple formula determines a 1-to-1 correspondence between the bond price and the interest rate. The interest rate r is the **discount rate** or **spot interest rate** for time to maturity n; *spot* rate, since that rate applies to a contract agreed on today.

Let us now consider a coupon bearing bond that pays coupons $C_1, \dots, C_k$ at times $t_1, \dots, t_k$ and the face value X at the maturity date T. This series of payments is equivalent to $k + 1$ zero coupon bonds with face values $C_1, \dots, C_k, X$ and maturity dates $t_1, \dots, t_k, T$. Thus, its price is given by the **bond price equation**

$$P(t) = \sum_{i=1}^{k} C_i P(t, t_i) + XP(t, T),$$

or equivalently

$$P(t) = \sum_{i=1}^{k} C_i P(t, t + \tau_i) + XP(t, T),$$

if $\tau_j = t_j - t$ denotes the time to maturity of the jth bond. It follows that the price of the bond can be determined by the curve $\tau \mapsto P(t, t + \tau)$ that assigns to each maturity τ the time t price for a zero coupon bond with unit principal t. It is called the **term structure of interest rates**.

There is a second approach to describe the term structure of interest rates. Let $P(t, t+m)$ denote the price at time t of a zero coupon bond paying the principal $X = 1$ at the maturity date $t+m$. Given the yearly spot rate $r(t, t+m)$ applying to a payment in m years, its price is given by

$$P(t, t+m) = \frac{1}{(1 + r(t, t+m))^m}.$$

If the coupon corresponding to the interest rate $r(t, t+m)$ is paid at n equidistant time points with continuous compounding, the formula

$$P(t, t+m) = \frac{1}{(1 + r(t, t+m)/n)^{nm}}$$

applies, which converges to the formula for continuously compounding

$$P(t, t+m) = \mathrm{e}^{-r(t,t+m)m} \quad \Leftrightarrow \quad P(t, T) = \mathrm{e}^{-r(t,T)(T-t)},$$

using the substitution $T = t + m$. The continuously compounded interest rate $r(t, T)$ is also called **yield** and the function

$$t \mapsto r(t, T)$$

the **yield curve**.

Finally, one can also capture the term structure of interest rates by the **instantaneous forward rate** at time t for the maturity date T defined by

$$f(t, T) = \frac{-\frac{\partial}{\partial T} P(t, T)}{P(t, T)} = -\frac{\partial}{\partial T} \log P(t, T).$$

Here it is assumed that the bond price $P(t, T)$ is differentiable with respect to maturity. It then follows that

$$P(t, T) = \exp\left(-\int_0^\tau f(t, t+s)\,\mathrm{d}s \right), \; r(t, t+\tau) = -\frac{1}{\tau} \int_0^\tau f(t, t+s)\,\mathrm{d}s.$$

1.2.2 Asset returns

For fixed-income investments such as treasury bonds the value of the investment can be calculated in advance, since the interest rate is known. By contrast, for assets such as exchange-traded stocks the interest rates, i.e. returns, are calculated from the quotes that reflect the market prices.

Let S_t be the price of a stock at time t. Since such prices are quoted at certain (equidistant) time points, it is common to agree that the time index attains values in the discrete set of natural numbers, $\mathbb{N}$. If an investor holds one share of the stock during the time interval from time $t-1$ to t, the asset price changes to

$$S_t = S_{t-1}(1 + R_t),$$

where

$$R_t = \frac{S_t - S_{t-1}}{S_{t-1}} = \frac{S_t}{S_{t-1}} - 1$$

is called the simple net return and

$$1 + R_t = \frac{S_t}{S_{t-1}}$$

are the gross returns. How are asset returns aggregated over time? Suppose an investor holds a share between s and $t = s + k$, i.e. over k periods, $s, t, k \in \mathbb{N}$ (or more generally $s, t, k \in [0, \infty)$). Define the k-period return

$$R_t(k) = \frac{S_t - S_s}{S_s} = \frac{S_t}{S_s} - 1.$$

One easily checks the following relationship between the simple returns $R_{s+1}, \ldots, R_t$ and the k-period return:

$$1 + R_t(k) = \frac{S_t}{S_s} = \prod_{i=s+1}^{t} \frac{S_i}{S_{i-1}} = \prod_{i=s+1}^{t} (1 + R_i).$$

When an asset is held for k years, the annualized average return (effective return) is given by the geometric mean

$$R_{t,k} = \left[\prod_{i=0}^{k-1} (1 + R_{t+i}) \right]^{1/k} - 1.$$

A fixed-income investment with a annualized interest rate of $R_{t,k}$ yields the same accumulated value. Note that

$$R_{t,k} = \exp \left[\frac{1}{k} \sum_{i=0}^{k-1} \log(1 + R_{t+i}) \right] - 1. \tag{1.2}$$

The natural logarithm of the gross returns,

$$r_t = \log(1 + R_t) = \log \frac{S_t}{S_{t-1}}$$

is called log return. Using Equation (1.2) we see that the k-period log return for the period from s to $t = s + k$ can be calculated as

$$r_t(k) = \log(1 + R_t(k)) = \sum_{i=s+1}^{t} \log(1 + R_i) = \sum_{i=s+1}^{t} r_i.$$

Thus, in contrast to the returns R_t the log returns possess the pleasant property of additivity w.r.t. time aggregation.

Using these definitions we obtain the following fundamental multiplicative decomposition of an asset price:

$$S_t = S_0 \prod_{i=1}^{t} (1 + R_i) = S_0 \prod_{i=1}^{t} \exp(r_i).$$

1.2.3 Some basic models for asset prices

When a security is listed on a stock exchange, there exists no quote before that time. Let us denote the sequence of price quotes, often the daily closing prices, by $S_0, S_1, \ldots$. Since $S_0 > 0$ denotes the first quote, it is often regarded as a constant. If one wants to avoid possible effects of the initial price, one puts formally $S_0 = 0$.

A first approach for a stochastic model is to assume that the price differences are given by

$$\Delta + u_n, \qquad n = 1, 2, \ldots$$

with a deterministic, i.e. nonrandom, constant $\Delta \in \mathbb{R}$ and i.i.d. random variables u_n, $n \in \mathbb{N}$, with common distribution function F such that

$$E(u_n) = 0, \qquad \text{Var}\,(u_n) = \sigma^2 \in (0, \infty), \qquad \forall n \in \mathbb{N}.$$

In the present context, it is common to name the u_n innovations. When referring to the sequence of innovations, we shall frequently write $\{u_n : n \in \mathbb{N}_0\}$ or, for brevity of notation, $\{u_n\}$ if the index set is obvious. The above model for the differences implies that the price process is given by

$$S_t = S_0 + \sum_{i=1}^{t} (\Delta + u_i) = S_0 + t\Delta + \sum_{i=1}^{t} u_i, \qquad t = 0, 1, \ldots$$

where we put $u_0 = 0$ and agree on the convention that $\sum_{i=1}^{0} a_i = 0$ for any sequence $\{a_n\}$. S_t is called (arithmetic) **random walk** and **random walk with drift** if $\Delta \neq 0$. Obviously

$$E(S_t) = S_0 + \Delta\, t$$

and

$$\text{Var}\,(S_t) = t\sigma^2.$$

This particular model for an asset price dates back to the work of Bachelier (1900).

An alternative approach is based on the log returns. Let us denote

$$R_i := \log(S_i / S_{i-1}), \qquad i \geq 1.$$

Then

$$S_t = S_0 \prod_{i=1}^{t} S_i / S_{i-1} = S_0 \prod_{i=1}^{t} \exp(R_i).$$

The associated log price process is then given by

$$\log S_t = \log S_0 + \sum_{i=1}^{t} R_i, \qquad t = 0, 1, \ldots,$$

which is again a random walk.

A classic distributional assumption for the log returns $\{R_n\}$ is the normal one,

$$R_i \overset{i.i.d.}{\sim} N(\mu, \sigma^2)$$

with $\mu \in \mathbb{R}$ and $\sigma^2 > 0$. As a consequence, the log prices are normally distributed as well,

$$\log(S_t) = \log(S_0) + \sum_{i=1}^{t} R_i \sim N(\log(S_0) + t\mu, t\sigma^2).$$

Thus, S_t follows a lognormal distribution. Let us summarize some basic facts about that distribution:

A random variable X follows a **lognormal distribution** with parameters $\mu \in \mathbb{R}$ (**drift**) and $\sigma > 0$ (**volatility**) if $Y = \log(X) \sim N(\mu, \sigma^2)$. X takes on values in the interval $(0, \infty)$ and

$$P(\log X \leq y) = \frac{1}{\sqrt{2\pi}\sigma} \int_{-\infty}^{y} \mathrm{e}^{-(t-\mu)^2/2\sigma^2} \, \mathrm{d}t, \qquad y \in (0, \infty).$$

The change of variable $u = \mathrm{e}^t$ leads to

$$P(X \leq \mathrm{e}^y) = P(\log X \leq y) = \int_{-\infty}^{\mathrm{e}^y} \frac{1}{\sqrt{2\pi}\sigma u} \mathrm{e}^{-(\log u - \mu)^2/2\sigma^2} \, \mathrm{d}u.$$

By evaluating the right-hand side at $y = \log x$, we see that the density $f(x)$ of X is given by

$$f(x) = \frac{1}{\sqrt{2\pi} x \sigma} \mathrm{e}^{-(\log x - \mu)^2/2\sigma^2} \mathbf{1}(x > 0), \qquad x \in \mathbb{R}. \tag{1.3}$$

Now it is easy to verify that mean and variance of X are given by

$$E(X) = \mathrm{e}^{\mu + \sigma^2/2} \quad \text{and} \quad \mathrm{Var}\,(X) = \mathrm{e}^{2\mu + \sigma^2}(\mathrm{e}^{\sigma^2} - 1).$$

In order to model distributions that put more mass to extreme values than the standard normal distribution, one often uses the ***t*-distribution with *n* degrees of freedom** defined via the density function

$$f(x) = \frac{1}{n\pi} \frac{\Gamma((n+1)/2)}{\Gamma(n/2)} \left(1 + \frac{x^2}{n}\right)^{-\frac{n+1}{2}}, \qquad x \in \mathbb{R},$$

which is parametrized by $n \in \mathbb{N}$. By symmetry, its expectation is zero and the variance turns out to be $n/(n-2)$, if $n > 2$.

Several questions arise: Which of the above two models holds true or provides a better approximation to reality? Are returns and log returns, respectively, normally distributed? Are asset returns symmetrically distributed? How can we estimate important distributional parameters such as μ, σ^2 or the skewness? Does the assumption of independent returns apply to real returns? Do price processes follow random walk models at all? What is the effect of changes of economic conditions on the distribution of returns? Can we test or detect such effects? How can we model the stochastic relationship between the return series of, say, m securities?

There is some evidence that some financial variables have much heavier tails than a normal distribution.

A random variable X has a **stable distribution** or is **stable**, if X has a **domain of attraction**. The latter means that there exist i.i.d. random variables $\{\xi_n\}$ and sequences $\{\sigma_n\} \subset (0, \infty)$ and $\{\mu_n\} \subset \mathbb{R}$, such that

$$\frac{1}{\sigma_n} \sum_{i=1}^{n} \xi_i + \mu_n \overset{d}{\to} X,$$

as $n \to \infty$. The classic central limit theorem tells us that the $X \sim N(\mu, \sigma^2)$ is stable. By the Lévy–Khintchine formula, the characteristic function

$$\varphi(\theta) = E(e^{\imath\theta X}), \qquad \theta \in \mathbb{R},$$

where $\imath^2 = -1$, of a stable random variable X has the representation

$$\varphi(\theta) = \begin{cases} \exp\left\{\imath\mu\theta - \sigma^\alpha|\theta|^\alpha\left(1 - \imath\beta(\operatorname{sgn}(\theta))\tan\frac{\pi\alpha}{2}\right)\right\}, & \alpha \neq 1, \\ \exp\left\{\imath\mu\theta - \sigma|\theta|\left(1 + \imath\beta\frac{2}{\pi}(\operatorname{sgn}(\theta))\log|\theta|\right)\right\}, & \alpha = 1, \end{cases}$$

where $0 < \alpha \leq 2$ is the **stability (characteristic) exponent**, $-1 < \beta < 1$ the **skewness parameter**, $\sigma > 0$ the **scale parameter** and $\mu \in \mathbb{R}$ the **location parameter**. For $\alpha = 2$ one obtains the normal distribution $N(\mu, \sigma^2)$, since then $\varphi(\theta) = \exp(\imath\mu\theta - \sigma^2\theta^2)$. The tails of a standard normal distribution decay exponentially fast,

$$P(|X| > x) \sim \sqrt{\frac{2}{\pi}} \frac{e^{-x^2/2}}{x}, \qquad x \to \infty \qquad (X \sim N(0, 1)).$$

By contrast, the **tails** of a stable random variable X with characteristic exponent $0 < \alpha < 2$ decay as $x^{-\alpha}$, since

$$\lim_{x\to\infty} x^\alpha P(X > x) = C_\alpha \frac{1+\beta}{2} \sigma^\alpha \tag{1.4}$$

and

$$\lim_{x\to\infty} x^\alpha P(X < -x) = C_\alpha \frac{1-\beta}{2} \sigma^\alpha, \tag{1.5}$$

where $C_\alpha = \left(\int_0^\infty x^{-\alpha} \sin(x)\,dx\right)^{-1}$.

Stable distributions appear as a special case of infinitely divisible distributions. A random variable (or random vector) X and its distribution are called **infinitely divisible**, if for every $n \in \mathbb{N}$ there exist independent and identically distributed random variables $X_{n1}, \ldots, X_{nn}$ such that

$$X \overset{d}{=} X_{n1} + \cdots + X_{nn}.$$

Those infinitely divisible distributions are exactly the distributions that can appear as limits of the distributions of sums $\sum_{k=1}^{n} X_{nk}$ of such arrays of row-wise i.i.d. random variables. Let X be a d-dimensional random vector and again let $\varphi(\theta) = E(\exp(\imath\theta' X))$, $\theta \in \mathbb{R}^d$, be its characteristic function. Then, the **Lévy–Khintchine formula** asserts that

$$\varphi(\theta) = \exp\left\{\imath\theta' b - \frac{1}{2}\theta' C\theta + \int_{\mathbb{R}^d} \left(e^{\imath\theta' x} - 1 - \imath\theta' h(x)\right) d\nu(x)\right\}, \tag{1.6}$$

where

$$h(x) = x\mathbf{1}(|x| \leq 1), \qquad x \in \mathbb{R}^d,$$

is a *truncation function*, $b \in \mathbb{R}^d$ and C a symmetric and non-negative definite $(d \times d)$-matrix and ν a **Lévy measure**, that is a positive measure on the Borel sets of $\mathbb{R}^d$ such that $\nu(\{0\}) = 0$ and

$$\int_{\mathbb{R}^d} \left(|x|^2 \wedge 1\right) \, \mathrm{d}\nu(x) < \infty.$$

As a consequence, $\varphi(\theta)$ is characterized by the **triplet** (b, C, ν).

The characteristics of the normal distribution $N(\mu, \sigma^2)$ are $(b, C, \nu) = (\mu, \sigma^2, 0)$, of course. For a Poisson distribution with intensity λ, the characteristic function is

$$\varphi(\theta) = \exp(\lambda(\mathrm{e}^{\imath\theta} - 1)),$$

which results, if we put $b = \lambda$, $C = 0$ and ν the one-point measure that assigns mass λ to the single point 1.

1.3 Elementary statistical analysis of returns

We have seen that price processes can be build from returns R_t that are modeled as random variables. For simplicity of our exposition, let us assume that $R_1, \ldots, R_T$ are independent and identically distributed. To simplify notation, let R denote a generic return, i.e. $R \stackrel{d}{=} R_1$ which means that for any event A we have $P(R \in A) = P(R_1 \in A)$.

But before focusing on returns, let us briefly review the most basic probabilistic quantities to which we will refer frequently in the following for an arbitrary random variable X. In general, the distribution of a random variable is uniquely determined by its **distribution function (d.f.)**

$$F_X(x) = P(X \leq x), \qquad x \in \mathbb{R}.$$

If $f : \mathbb{R} \to \mathbb{R}$ is a density, i.e. non-negative function with $\int f(x)\,\mathrm{d}x = 1$, then the d.f. $F(x)$ can be calculated by

$$F(x) = \int_{-\infty}^{x} f(t)\,\mathrm{d}t, \qquad x \in \mathbb{R}.$$

A random variable X that attains a density function f is called a **continuous random variable**. Usually, it is assumed that returns are continuous random variables in that sense.

The first moment is defined by $\mu = E(X)$ and can be calculated for a continuous random variable via

$$\mu = E(X) = \int_{-\infty}^{\infty} x f(x)\,\mathrm{d}x.$$

$E(X)$ is also called the **expectation** or **mean** of X. If X is a **discrete random variable**, that is X takes values in some discrete set $\{x_1, x_2, \cdots\}$ of possible values with corresponding probabilities $p_1, p_2, \ldots$ such that

$$P(X = x_i) = p_i, \qquad i = 1, 2, \ldots,$$

then

$$E(X) = \sum_{i=1}^{\infty} x_i p_i.$$

More generally, the **kth moment** of X is defined as $E(X^k)$ and $E|X|^k$ is referred to as the **kth absolute moment**. Assumptions on the existence of higher moments control the probability of **outliers**, that is extreme values. Indeed, by virtue of Markov's inequality, the probability that X takes values larger than $c > 0$ in absolute value decays faster for increasing c, if higher moments exist, since

$$P(|X| > c) \leq \frac{E|X|^k}{c^k}.$$

Compare this inequality with the formulas (1.4) and (1.5) for the special class of stable distributions. As extreme values (outliers) of daily returns, usually negative ones, correspond to unexpected high-impact news such as a crash, the behavior of the **tail probabilities** $P(X < -c)$ and $P(X > c)$, $c > 0$, are of substantial interest, and moment assumptions automatically constrain them.

Suppose we are given a random sample $X_1, \ldots, X_T$ of sample size T. The **empirical distribution function** of the sample $X_1, \ldots, X_T$ is defined as

$$F_T(x) = \frac{1}{T} \sum_{t=1}^{T} \mathbf{1}(X_t \leq x), \qquad x \in \mathbb{R}.$$

Notice that $F_T(x)$ is the fraction of observations that are less or equal than x.

For a distribution function F let

$$F^{-1}(y) = \inf\{x : F(x) \geq y\}$$

denote the left-continuous inverse called **quantile function**. Applying that definition to the empirical distribution function yields the **sample quantile function**

$$F_T^{-1}(p) = \inf\{x : F_T(x) \geq p\} = X_{(\lceil np \rceil)}, \qquad p \in (0, 1).$$

For a fixed p, $F_T^{-1}(p)$ is called the **sample p-quantile** or **empirical p-quantile**. Here $X_{(1)} \leq \cdots \leq X_{(T)}$ denotes the **order statistic** and $\lceil x \rceil$ is the smallest integer larger or equal to x. Notice that $X_{(\lceil np \rceil)} = X_{(\lfloor np \rfloor + 1)}$ where $\lfloor x \rfloor$ is the floor function, i.e. the largest integer that is less than or equal to x. Quantiles play an important role in characterizing a distribution. The sample 0.5-quantile is called the **median** and is also denoted by x_{med}. Together with the 0.25- and 0.75-quantiles,

$$Q_1 = F_T^{-1}(0.25), \qquad Q_3 = F_T^{-1}(0.75),$$

called **quartiles**, we get a picture where the lower (upper) fourth and the central 50% of the data are located. Augmenting these three statistics with the minimum and maximum defining the range of the data set, we obtain the so-called **five-point summary** $x_{\min}, Q_1, x_{\text{med}}, Q_3, x_{\max}$. Those five numbers already provide an informative view on the distribution of the data. The **boxplot (box and whiskers plot)** is a convenient graphical representation by a box symbolizing the central half of the data between Q_1 and Q_3 and straight lines connecting $x_{\min}$ and Q_1 as well as Q_3 and $x_{\max}$. It is also common to replace $(x_{\min}, x_{\max})$ by the quantiles $(F_T^{-1}(p), F_T^{-1}(1-p))$. Typical values for p are $p = 0.01, 0.05$ and 0.1.

Sample quantiles are asymptotically normal under fairly general conditions. Let $p \in (0, 1)$ and denote by $x_p = F^{-1}(p)$ the theoretical p-quantile. If F attains a density that is positive in a neighborhood of x_p, then

$$\sqrt{T}(F_T^{-1}(p) - x_p) \stackrel{d}{\to} N(0, p(1-p)/f(x_p)^2), \tag{1.7}$$

as $T \to \infty$. The problem arises that the asymptotic variance depends on the unknown density, which has to be estimated by some appropriate estimator $\widehat{f}_T$. We shall discuss this issue in Section 1.3.4 and anticipate that such an estimator can be defined having nice mathematical properties under fairly weak regularity conditions that do not impose a constraint on the shape of the density f, which is of particular importance when analyzing financial data such as returns. Based on the large sample result (1.7), which still holds true when plugging in a consistent estimator, it is straightforward to construct the confidence interval for x_p,

$$\left[F_T^{-1}(p) - z_{1-\alpha/2} \frac{\sqrt{p(1-p)}}{\widehat{f}_T(x_p)}, F_T^{-1}(p) + z_{1-\alpha/2} \frac{\sqrt{p(1-p)}}{\widehat{f}_T(x_p)} \right],$$

which attains the coverage probability $1 - \alpha$, if $T \to \infty$, where $z_{1-\alpha/2}$ denotes the $(1 - \alpha/2)$-quantile of the standard normal distribution. We discuss the derivation of such confidence intervals in greater detail in the next section.

1.3.1 Measuring location

Measures of locations are usually defined in terms of moments or quantiles. The expectation is the most commonly used measure of location of a random variable.

Returning to our problem to analyze financial returns, the problem arises that the distribution of the returns is unknown to us. But then the mean return $\mu = E(R)$ is unknown as well. The best we can do is to use statistical estimators, i.e. functions of the data $R_1, \ldots, R_T$, which output a value that is regarded as a good estimate for μ. A standard approach to obtain such estimators for quantities that are defined in terms of expectations is to replace the averaging with respect to the distribution by averaging with respect to the so-called **empirical probability measure** that attaches equal mass $1/T$ to the values $R_1, \ldots, R_T$. The expectation with respect to that discrete distribution is simply the **arithmetic mean**

$$\overline{R} = \overline{R}_T = \frac{1}{T} \sum_{t=1}^{T} R_t.$$

It is easy to check that $E(\overline{R}_T) = E(R_1) = \mu$, and this calculation holds true whatever the value μ attains. In statistics, an estimator satisfying that property is called an **unbiased estimator**. It tells us that, averaged over all possible scenarios ω corresponding to all possible values

$r = R(\omega)$ for the return and weighted with the corresponding probabilities, the estimator estimates the right value, namely μ.

Suppose we have observed T daily log returns $R_1, \ldots, R_T$ and aim at testing the hypothesis that their common mean $\mu = E(R_1)$ equals some specified value μ_0. The corresponding two-sided statistical testing problem is then given by

$$H_0 : \mu = \mu_0 \qquad \text{versus} \qquad H_1 : \mu \neq \mu_0.$$

Assuming that the returns are i.i.d. and follow a normal law suggest using the t-test that is based on the test statistic

$$Z = \sqrt{T}\frac{\overline{R}_T - \mu_0}{S_T} \tag{1.8}$$

with

$$S_T = \sqrt{\frac{1}{T-1}\sum_{t=1}^{T}(R_t - \overline{R}_T)^2};$$

the statistic S_T will be discussed in greater detail in the next subsection. Under the null hypothesis H_0, the statistic Z follows a t-distribution with $df = T - 1$ degrees of freedom. Consequently, we may reject H_0 at a significance level of $\alpha \in (0, 1)$, if

$$|T| > t(df)_{1-\alpha/2},$$

where $t(df)_{1-\alpha/2}$ denotes the $(1 - \alpha/2)$–quantile of the $t(df)$-distribution.

If the log returns are non-normal, one can often rely on the central limit theorem which asserts that the statistic Z is asymptotically normal. Hence, the null hypothesis is then rejected, if $|Z| > z_{1-\alpha/2}$.

Example 1.3.1 *For the FTSE log returns illustrated in Figure 1.1, one gets* $z = 2.340558$, *which exceeds the critical value* 1.959964 *corresponding to the* 5% *significance level, indicating that the mean log return differs from zero and is actually positive. However, this assertion is not valid on the* 1% *significance level.*

Often, one is also interested to provide interval estimates for the mean. Again assuming i.i.d. normal returns, a **confidence interval** for the mean with coverage probability $1 - \alpha$, is an interval $[L, U]$ where $L = L(R_1, \ldots, R_T)$ and $U = U(R_1, \ldots, R_T)$ are functions of the sample such that

$$P(L \leq \mu \leq U) = 1 - \alpha$$

for any $\mu \in \mathbb{R}$. Such a confidence interval is given by

$$L = \overline{R}_T - t(df)_{1-\alpha/2}\frac{S_T}{\sqrt{T}}, \qquad U = \overline{R}_T + t(df)_{1-\alpha/2}\frac{S_T}{\sqrt{T}},$$

where, as above, $df = T - 1$. This can be easily established by noting that the event $L \leq \mu \leq U$ is equivalent to

$$-t(df)_{1-\alpha/2} \leq \sqrt{T}(\overline{R}_T - \mu)/S_T \leq t(df)_{1-\alpha/2}.$$

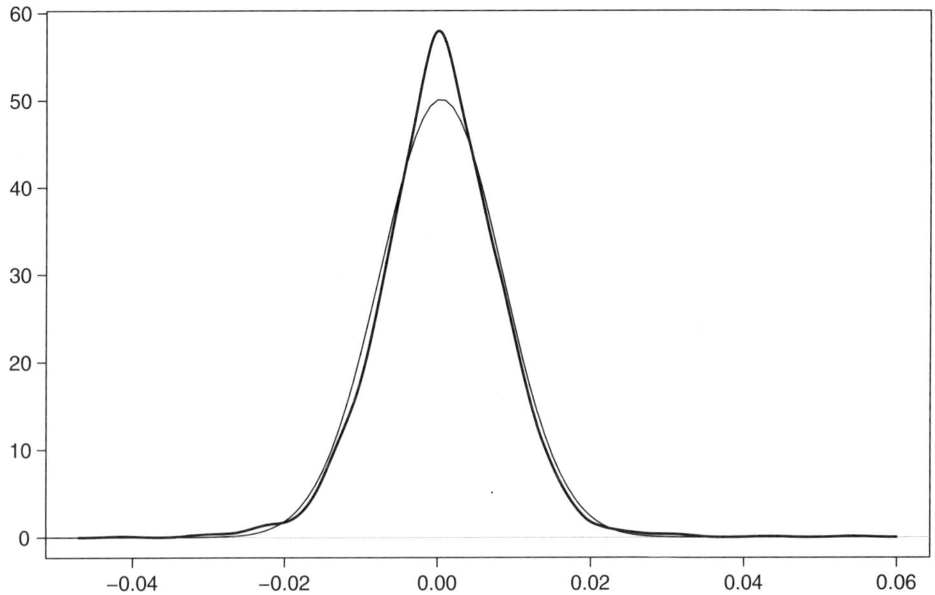

Figure 1.1 Kernel density estimate of the FTSE daily log returns with cross-validated bandwidth choice.

But the latter event occurs with probability $1 - \alpha$, since

$$\sqrt{T}(\overline{R}_T - \mu)/S_T \sim t(df).$$

However, usually daily returns are not normal but affected by **stylized facts** such as asymmetry, peakedness (more mass around zero) and heavier tails than under a normal law. This can be easily seen from Figure 1.1. The famous central limit theorem asserts that the statistic Z defined in Equation (1.8) is asymptotically standard normal, as long as the returns are i.i.d. with existing fourth moment.[1] Consequently, a valid asymptotic test is given by the decision rule

$$\text{reject } H_0 \text{ if } |Z| > z_{1-\alpha/2},$$

where $z_{1-\alpha/2} = \Phi^{-1}(1 - \alpha/2)$ denotes the $(1 - \alpha/2)$-quantile of the $N(0, 1)$-distribution. In the same vein, an asymptotic confidence interval for μ is obtained by replacing the quantiles of the $t(df)$-distribution in the formulas for L and U by the respective quantiles of the standard normal law.

Similarly, one may construct an asymptotic confidence interval for μ based on the central limit theorem. In this case, the probability of the event

$$-z_{1-\alpha/2} \leq \sqrt{T}(\overline{R}_T - \mu)/S_T \leq z_{1-\alpha/2},$$

[1] This result even remains true under the substantially weaker assumption that the log returns are a stationary martingale difference sequence.

which is equivalent to the event

$$L' = \overline{R}_T - z_{1-\alpha/2}\frac{S_T}{\sqrt{T}} \leq \mu \leq \overline{R}_T + z_{1-\alpha/2}\frac{S_T}{\sqrt{T}} = U',$$

converges to $1-\alpha$, as $T \to \infty$. Thus, a confidence interval with **asymptotic coverage probability** $1-\alpha$ is given by the random interval $[L', U']$.

Example 1.3.2 *For the FTSE log returns one calculates the asymptotic* 95% *confidence interval* $[l, u] = [0.0000702, 0.000793]$ *for the mean log return.*

It is a general insight, supported by many empirical studies, that the statistical analysis of financial returns should not be based on procedures assuming the classic assumptions of normality and independent observations, since those assumptions are usually violated. Therefore, large sample theory forms the mathematical core for inferential procedures in finance.

1.3.2 Measuring dispersion and risk

The mean $\mu = E(R_t)$ tells us where the distribution is located; it is a measure for the center of the distribution. Then we can determine for each return R_t its distance $|R_t - \mu|$ from the mean. The mean squared distance,

$$\sigma^2 = \text{Var}(R) = E(R-\mu)^2 = E(R^2) - \mu^2$$

is called the **variance** of R. Its square root,

$$\sigma = \sigma_R = \sqrt{\text{Var}(R)}$$

is called the **standard deviation**. Variance and standard deviation can be defined for any random variable X with existing second moment. If X and Y are independent random variables with $EX^2 < \infty$ and $EY^2 < \infty$, then

$$\text{Var}(X+Y) = \text{Var}(X) + \text{Var}(Y)$$

yielding $\sigma_{X+Y} = \sqrt{\text{Var}(X+Y)} = \sqrt{\sigma_X^2 + \sigma_Y^2}$, whereas in the general case

$$\text{Var}(X+Y) = \text{Var}(X) + \text{Var}(Y) + 2\,\text{Cov}(X,Y).$$

Here

$$\text{Cov}(X,Y) = E(X-EX)(Y-EY)$$

is called the **covariance of** X **and** Y.

When considering daily (log) returns, σ is also frequently called **(actual) volatility**. When volatility of returns is addressed, it is important to be aware of the corresponding unit of time, e.g. yearly, monthly or daily. The **annualized volatility** σ_{an} is the standard deviation of the yearly return, whereas **generalized volatility** addresses the volatility corresponding to the time horizon τ (in years) given by

$$\sigma_{\text{an}}\sqrt{\tau}.$$

Notice that the formula coincides with the standard deviation of the return $R(\tau)$ corresponding to the time period τ, if τ is an integer and the yearly log returns are identically distributed and uncorrelated, since then the additivity of log returns gives $R(\tau) = \sum_{t=1}^{\tau} R'_t$ where $R'_1, \ldots, R'_\tau$ denote the τ yearly log returns. But then

$$\sigma_{R(\tau)} = \sqrt{\sum_{t=1}^{\tau} \text{Var}\,(R'_t)} = \sqrt{\tau}\sigma',$$

where σ' is the volatility of the yearly returns R'_t. However, usually the annualized volatility is determined from the actual volatility of the daily log returns. Since there are 252 trading days in a year, annualized volatility σ_{an} and actual volatility σ are related by

$$\sigma_{\text{an}} = \sigma\sqrt{252}.$$

The monthly volatility is then given by $\sigma_{\text{m}} = \sigma\sqrt{252/12}$.

Estimation of the variance and standard deviation is usually based on the plug-in principle already explained in the previous subsection. Given a sample $R_1, \ldots, R_T$ of returns, it naturally leads to the the **empirical variance** or **sample variance**

$$V_T^2 = \frac{1}{T}\sum_{t=1}^{T}(X_t - \overline{R}_T)^2.$$

A tedious calculation shows that $E(V_T^2) = \frac{T-1}{T}\sigma^2$, i.e. V_T^2 is not an unbiased estimator of the variance. Thus, in practice the estimator

$$S_T^2 = \frac{1}{T-1}\sum_{t=1}^{T}(R_t - \overline{R}_T)^2$$

is used. The corresponding estimator for the standard deviation is the square root, $S_T = \sqrt{S_T^2}$, of that expression. Estimates of the various volatilities discussed above can be obtained by substituting σ by S_T. For example, if the R_ts are daily log returns, annualized volatility is estimated by $S_T\sqrt{252}$.

1.3.2.1 Value-at-risk

Another risk measure that has been become the de-facto standard in the financial industry is value-at-risk. Recall that the profit or loss (P&L) of any investment during a time period $[0, h]$ is uncertain and therefore represents a risk exposure, namely to suffer a loss. Roughly speaking, value-at-risk is a risk measure that represents the smallest loss we are exposed to with probability α. Here the risk probability α is chosen by us; common values are 1% and 5%. Let V_t denote the **marked-to-market** value of a long position at time t, i.e. the value is based on the current market value. Then the profit is $\Delta V = V_{t+h} - V_t$, where negative values are losses. Now let us consider the loss $L = -\Delta V$ and let v be the fixed value satisfying

$$P(L > v) = \alpha.$$

This means, with a probability of α we suffer a loss exceeding v. That number v (a loss) is called **value-at-risk (VaR)** at the probability level α and denoted by VaR or VaR_α. Roughly speaking, it is the *smallest* loss among the largest losses occurring with probability α. By definition,

$$\mathrm{VaR}_\alpha = F_L^{-1}(1-\alpha),$$

where F_L^{-1} denotes the quantile function associated to the loss distribution. That means, value-at-risk is the $(1-\alpha)$-quantile of the loss distribution. Notice that value-at-risk can be also defined by the α-quantile of the P&L distribution,

$$\mathrm{VaR}_\alpha = -F_{\Delta V}^{-1}(\alpha).$$

Often, VaR is calculated on a daily basis. If the daily 1% value-at-risk of a position is 100 000, the probability that the value of the position will fall below −100 000 is 1%; with probability 1% we suffer a loss being larger than 100 000.

Since VaR is defined as a quantile, we may estimate it by the corresponding sample quantiles. If $L_1, \ldots, L_T$ are i.i.d. losses corresponding to the time horizon h,

$$\widehat{\mathrm{VaR}}_\alpha = L_{\lceil n(1-\alpha)\rceil}.$$

Statistical tests and the calculation of confidence intervals can therefore be based on the large sample theory of quantiles discussed above. In the same vein, the asymptotic confidence intervals carry over to confidence intervals for value-at-risk.

1.3.2.2 Expected shortfall, lower partial moments and coherent risk measures

VaR gives the smallest loss among the largest losses occurring with probability α. It is natural to average those losses, that is to consider the conditional expectation of the profit or loss L over a given period of time

$$S_\alpha(L) = E(L|L \le \mathrm{VaR}_\alpha)$$

is called the **expected shortfall** or **conditional value-at-risk**. One can show that

$$S_\alpha(X) = -\frac{1}{\alpha}\int_0^\alpha F_{\Delta V}^{-1}(x)\,\mathrm{d}x.$$

For this reason, $S_\alpha(X)$ is also called the **average value-at-risk**.

Clearly, we do not worry about realizations l of L with $l > E(L)$, but are concerned about the **downside risk**, that is losses below the expectation $E(L)$ of the position. If L is symmetrically distributed, then $P(L < E(L)) = P(L > E(L))$ and the variance or standard deviation provide meaningful measures for the downside risk. But especially for asymmetric distributions it makes sense to consider the **semivariance** defined as

$$E\big(\min(0, L - EL)^2\big).$$

Often there exists a benchmark profit b to which a portfolio is compared. If the portfolio does not outperform the given benchmark b, that is if $L \le b$, then $b - L$ is the loss we suffer when we have a long position in the portfolio. The mth moment of the corresponding random variable $(b-L)\mathbf{1}(L \le b)$,

$$LP_m(L) = E\left((b-L)^m \mathbf{1}(L \le b)\right)$$

is called the **lower partial moment of the order** m, provided it exists. Notice that

$$LP_0(L) = P(L \leq b)$$

is the probability that the portfolio does not outperform the benchmark, and $LP_1(L)$ is the expected underperformance.

All the quantities discussed above assign a real number to a random variable interpreted as the loss of a portfolio or position over some fixed period of time, and that number is interpreted as a quantitative measure of the risk. The question arises which properties (axioms) such a risk measure should satisfy. Generally, a **risk measure** or **risk functional** ρ is a function defined on a sufficiently rich set $\mathcal{A}$ of random variables (random payment profiles) taking values in the real numbers. Given such a risk measure ρ, we may distinguish risky payments with non-negative risks and **acceptable payments** with negative risks.

A risk measure $\rho : \mathcal{A} \to \mathbb{R}$ is called **coherent**, if it satisfies the following four axioms:

(i) $X \leq Y$ implies that $\rho(X) \leq \rho(Y)$ for all $X, Y \in \mathcal{A}$ (monotonicity).

(ii) $\rho(X + Y) \leq \rho(X) + \rho(Y)$ for all $X, Y \in \mathcal{A}$ (subadditivity).

(iii) $\rho(aX) = a\rho(X)$ for $a > 0$ (positive homogeneity).

(iv) $\rho(X + a) = \rho(X) - a$ for any $X \in \mathcal{A}$ and $a \in \mathbb{R}$ (translational invariance).

Sometimes, a further axiom is considered

(v) If $X \stackrel{d}{=} X'$, then $\rho(X) = \rho(X')$ (distributional invariance).

Axiom (i) requires that the risk of a position increases, if the random payment profile increases for all states $\omega \in \Omega$. The second axiom addresses an important aspect of risk management: Risks associated to two positions may cancel when aggregating them. The standard deviation $\sigma(X) = \sqrt{\text{Var}\,(X)}$ satisfies axiom (ii) and (iii). To see (ii), use the inequality

$$\text{Cov}(X, Y) \leq \sigma(X)\sigma(Y)$$

to obtain

$$\begin{aligned}
\sigma(X+Y) &= 2\sqrt{\text{Var}\left(\frac{X}{2} + \frac{Y}{2}\right)} \\
&\leq 2\sqrt{\left(\frac{1}{2}\right)^2 \sigma_X^2 + \left(\frac{1}{2}\right)^2 \sigma_Y^2 + \sigma_X \sigma_Y} \\
&= 2\sqrt{\left(\frac{\sigma_X}{2} + \frac{\sigma_Y}{2}\right)^2} \\
&= \sigma_X + \sigma_Y
\end{aligned}$$

for *any* pair (X, Y) of random variables with existing second moments and arbitrary correlation. This also implies that in the Gaussian world value-at-risk also satisfied axiom (ii). This can be seen as follows. Notice that value-at-risk for a random P&L $X \sim N(\mu, \sigma^2)$ is given by

$$\text{VaR}_\alpha(X) = \mu + \Phi^{-1}(\alpha)\sigma.$$

Further, for a random vector (X, Y) distributed according to a bivariate normal distribution with marginals $N(\mu_X, \sigma_X^2)$, $N(\mu_Y, \sigma_Y^2)$ and covariance γ the sum $X + Y$ is again Gaussian,

$$X + Y \sim N(\mu_X + \mu_Y, \sigma_{X+Y}^2), \qquad \sigma_{X+Y}^2 = \sigma_X^2 + \sigma_Y^2 + 2\gamma,$$

such that the α-quantile of $X + Y$ is

$$\mathrm{VaR}_\alpha(X + Y) = \mu_X + \mu_Y + \Phi^{-1}(\alpha)\sigma_{X+Y}.$$

Hence, $\sigma_{X+Y} \leq \sigma_X + \sigma_Y$ immediately implies

$$\mathrm{VaR}_\alpha(X + Y) \leq \mathrm{VaR}_\alpha(X) + \mathrm{VaR}_\alpha(Y).$$

However, for general distributions axiom (ii) can be violated, such that in general value-at-risk is not a coherent risk measure, which is probably the main criticism against value-at-risk.

Axiom (iii) is a scaling property, which allows us to compare risks expressed in different currencies, for example. Finally, the fourth axiom means that when adding a fixed payment to the position, in order to compensate losses and reduce the risk in this way, the risk measure is also reduced by exactly that amount, and, by contrast, withdrawing cash increases the risk. Then $\rho(X)$ can be interpreted as the amount of capital needed to eliminate the risk and transform a position into a acceptable payment. Obviously, that axiom is not satisfied by the standard deviation.

One can show that the expected shortfall satisfies all axioms and is therefore a coherent risk measure. More generally, any risk measure allowing a representation

$$\rho(X) = \sup_{P \in \mathcal{P}} E_P(-X),$$

where $\mathcal{P}$ is a set of probability measures and E_P indicates that the expectation is calculated under P, can be shown to be a coherent risk measure. For $S_\alpha(X)$ the set $\mathcal{P}$ is given by all densities that are bounded by $1/\alpha$.

1.3.3 Measuring skewness and kurtosis

The most common approach to measure skewness, i.e. departures from symmetry, is to consider the third standardized moment,

$$\mu_3^* = E\left(\frac{R_1 - \mu}{\sigma}\right)^3,$$

where $\mu = E(R_1)$ and $\sigma^2 = \mathrm{Var}\,(R_1)$. Notice that $\mu_3^* = 0$, if $R_1 - \mu \stackrel{d}{=} \mu - R_1$.[2]

Given a sample $R_1, \ldots, R_T$, one uses the estimator

$$\widehat{\mu}_3^* = \frac{1}{T}\sum_{t=1}^{T}\left(\frac{R_t - \overline{R}_T}{S_T}\right)^3.$$

[2] If $X \stackrel{d}{=} -X$ and f is a function with $f(-x) = -f(x)$ and $Ef(X) \in \mathbb{R}$, then $Ef(X) = Ef(-X) = -Ef(X)$, which implies $Ef(X) = 0$.

The statistic $\widehat{\mu}_3^*$ is very sensitive with respect to outliers. An alternative measure based on quantiles is to compare the distance between the 0.75-quantile and the median and the distance between the median and the 0.25-quantile, expressed as a fraction of the maximum value, i.e.

$$\gamma = \frac{[F^{-1}(0.75) - F^{-1}(0.5)] - [F^{-1}(0.5) - F^{-1}(0.25)]}{F^{-1}(0.75) - F^{-1}(0.25)}.$$

The corresponding estimator based on $R_1, \dots R_T$ is

$$\widehat{\gamma}_T = \frac{[Q_3 - x_{\text{med}}] - [x_{\text{med}} - Q_1]}{Q_3 - Q_1}.$$

Since sample quantiles, particularly Q_1, Q_3 and x_{med} are more robust than an arithmetic mean, $\widehat{\gamma}_T$ provides a reliable measure of skewness even for data sets from distributions with heavy tails.

A common approach to measure deviations from the shape of the Gaussian density is based on the fourth standardized moment,

$$\mu_4^* = E\left(\frac{R_1 - \mu}{\sigma}\right)^4,$$

also called **kurtosis**. Since for a normal distribution one obtains $\mu_4^* = 3$, it is common to consider the **excess kurtosis**,

$$\kappa = \mu_4^* - 3.$$

Distributions such as the normal one with an excess kurtosis equal to 0 are called **mesokurtic**. The standard interpretations when $\kappa \neq 0$ are as follows. A distribution with $\kappa > 0$ is called **leptokurtic**. It has a more pronounced peak compared to the normal law and lighter tails. A distribution with $\kappa < 0$ is called **platykurtic**. Such distributions have a flatter peak and heavier tails than the Gaussian density. Kurtosis and excess kurtosis are estimated by their sample analogs

$$\widehat{\mu}_4^* = \frac{1}{T}\sum_{t=1}^{T}\left(\frac{R_t - \overline{R}_T}{S_t}\right)^4$$

and

$$\widehat{\kappa}_T = \widehat{\mu}_4^* - 3,$$

respectively.

1.3.4 Estimation of the distribution

We have already discussed that financial returns for shorter time horizons tend to violate properties of the normal distribution. Taking for granted that the return distribution attains a density function[3] f in the sense that the distribution function

[3] For some financial instruments that assumption is violated, since there are trading periods where the price remains constant such that the return is 0.

$F(x) = P(R_1 \le x)$ can be represented as

$$F(x) = \int_{-\infty}^{x} f(t)\,\mathrm{d}t, \qquad x \in \mathbb{R},$$

the question arises how we can estimate the density f. Noticing that $f(x) = F'(x)$, we may approximate $f(x)$ by the difference ratio

$$f(x) \approx \frac{F(x+h) - F(x-h)}{2h},$$

for small $h > 0$. A natural approach is to estimate the right-hand side by plugging in the empirical distribution function $F_T(x) = T^{-1} \sum_{t=1}^{T} \mathbf{1}(R_t \le x)$ of T historical returns $R_1, \ldots, R_T$ and to regard the resulting expression as an estimate for $f(x)$. Noting that

$$\mathbf{1}(R_t \le x + h) - \mathbf{1}(R_t \le x - h) = \mathbf{1}(x - h < R_t \le x + h),$$

this idea leads to the estimator

$$x \mapsto \frac{1}{Th} \sum_{t=1}^{T} \frac{1}{2} \mathbf{1}\left(-1 < \frac{R_t - x}{h} \le 1 \right), \qquad x \in \mathbb{R}.$$

Each of the T summands corresponds to the density $K_0(z) = \frac{1}{2}\mathbf{1}(-1 < |z| \le 1)$, $z \in \mathbb{R}$, of the uniform distribution on $(-1, 1]$ evaluated at the points $(x - R_t)/h$, $t = 1, \ldots, T$. Obviously, as a function of x the above density estimator is discontinuous, which results in many spurious jumps. If we replace the discontinuous density K_0 by other density functions, we arrive at the **Rosenblatt–Parzen** kernel density estimator

$$\widehat{f}_{Th}(x) = \frac{1}{Th} \sum_{t=1}^{T} K([R_t - x]/h), \qquad x \in \mathbb{R}.$$

The parameter h is called the **bandwidth**. It has a strong influence on the resulting estimator. If h is chosen too small, there will be many spurious artifacts such as local extrema in the graph, whereas too large values for the bandwidth lead to oversmoothing. K, called the **smoothing kernel**, is usually chosen as an arbitrary unimodal density function with finite second moment that is symmetric around zero. Table 1.1 lists some smoothing kernels frequently used in practice.

Table 1.1 Some commonly used smoothing kernels for nonparametric density estimation.

Kernel	Definition
Triangular	$(1 - \lvert x\rvert)\mathbf{1}(\lvert x\rvert \le 1)$
Cosine	$(\pi/4)\cos(x\pi/2)$
Gaussian	$(2\pi)^{-1}\exp(-x^2/2)$
Epanechnikov	$(3/4)(1 - x^2)\mathbf{1}(\lvert x\rvert \le 1)$
Biweight	$(15/16)(1 - x^2)^2\mathbf{1}(\lvert x\rvert \le 1)$
Silverman	$(1/2)\exp(-\lvert x\rvert/\sqrt{2})\sin(\lvert x\rvert/\sqrt{3} + \pi/4)$

Notice that the estimator $\widehat{f}_{Th}(x)$ allows the following nice interpretation: If K is a density that is symmetric around 0 with unit variance, then

$$x \mapsto \frac{1}{h} K\left(\frac{x-m}{h}\right), \qquad x \in \mathbb{R},$$

is a density with mean m and standard deviation h for any fixed $m \in \mathbb{R}$ and $h > 0$. Consequently, $\widehat{f}_{Th}(x)$ averages those T densities $x \mapsto h^{-1}K([x - R_t]/h)$, $t = 1, \ldots, T$, associated to the observed values.

It is worth discussing some further basic properties of the kernel density estimator, in order to understand why it estimates any underlying density under fairly general conditions. Another issue we have to discuss is the question how to select the smoothing kernel and the bandwidth. First, notice that it is easy to check that $\widehat{f}_{Th}(x)$ indeed is a density function, if K has that property. Further, $\widehat{f}_{Th}$ inherits its smoothness properties from K. In particular, we may estimate $f'(x)$ by $\widehat{f}'_{Th}(x)$. Provided the returns $R_1, \ldots, R_T$ form an i.i.d. sample, we obtain

$$E(\widehat{f}_{Th}(x)) = \int \frac{1}{h} K\left(\frac{z-x}{h}\right) f(z)\,\mathrm{d}z = (K_h \star f)(x),$$

where $K_h(z) = h^{-1}K(z/h)$ is the rescaled kernel and $\star$ denotes the convolution operator. It follows that the Parzen–Rosenblatt estimator is not an unbiased estimator for f; its bias equals

$$b_h(x) = E(\widehat{f}_{Th}(x)) - f(x) = (K_h \star f)(x) - f(x).$$

However, Bochner's lemma, cf. Lemma A.2.1, implies that the convolution $(K_h \star f)(x)$ converges to $f(x)$, as $h \to 0$. Thus, the bandwidth should be chosen as a decreasing function of the sample size T. Under the i.i.d. assumption, it is easy to verify that the variance equals

$$\sigma^2_{Th}(x) = \operatorname{Var}(\widehat{f}_{Th}(x)) = \frac{1}{Th}\left[(K_h^2 \star f)(x) - (K_h \star f)^2(x)\right],$$

where $K_h^2(z) = h^{-1}K^2(z/h)$, $z \in \mathbb{R}$. Again, Bochner's lemma implies that the expression in brackets converges to finite constant, such that the variance of $\widehat{f}_{Th}$ is of the order $1/Th$ and tends to 0, if $Th \to \infty$. Let us consider the mean squared error (MSE),

$$\mathrm{MSE}(\widehat{f}_{Th}(x); f(x)) = E(\widehat{f}_{Th}(x) - f(x))^2,$$

which can be decomposed into its two additive components, the variance $\sigma^2_{Th}(x)$ and the squared bias $b_h^2(x)$,

$$\mathrm{MSE}(\widehat{f}_{Th}(x); f(x)) = \sigma^2_{Th}(x) + b_h^2(x).$$

We see that the MSE converges to zero for any bandwidth choice satisfying

$$h \to 0 \qquad \text{and} \qquad Th \to \infty.$$

To get further insights, we need the following notion. A kernel K is called the **kernel of the order** r, if

$$\int z^j K(z)\,\mathrm{d}z = \begin{cases} 1, & j = 0, \\ 0, & j = 1, \ldots, r-1, \\ c \neq 0, & j = r. \end{cases}$$

For example, the kernel $K(x) = \left(\frac{9}{8} - \frac{15}{8}x^2\right) \mathbf{1}(|x| \leq 1)$, $x \in \mathbb{R}$, is a kernel of order 4. Let us assume that the underlying density f is r times differentiable. Then one can easily establish the expansions

$$E(\widehat{f}_{Th}(x)) = f(x) + h^r f^{(r)}(x) \frac{(-1)^r}{r!} \int u^r K(u)\,\mathrm{d}u + o(h^r),$$

$$\mathrm{Var}\,(\widehat{f}_{Th}(x)) = \frac{1}{Th} f(x) \int K^2(z)\,\mathrm{d}z + o(1/Th),$$

which yield the following expansion for the MSE

$$\mathrm{MSE}(\widehat{f}_{Th}(x); f(x)) = \frac{f(x)R(K)}{Th} + h^{2r}[f^{(r)}(x)]^2 M_r^2 + o(h^{2r} + 1/Th),$$

where

$$M_r = \frac{(-1)^r}{r!} \int u^r K(u)\,\mathrm{d}u$$

and

$$R(g) = \int g^2(x)\,\mathrm{d}x$$

measures the roughness of a L_2 function g. These expansion show that higher-order kernels reduce the order of the bias, which is now $O(h^{2r})$.

A bandwidth choice is called **local asymptotically optimal bandwidth**, if it minimizes the dominating terms of the above expansion represented by the function

$$h \mapsto \frac{f(x)R(K)}{Th} + h^{2r}[f^{(r)}(x)]^2 M_r^2, \qquad h > 0.$$

It is easy to see that the optimal bandwidth is given by

$$h^*(x) = h_T^*(x) = \left(\frac{f(x)R(K)}{2rM_r^2[f^{(r)}(x)]^2 T} \right)^{1/(2r+1)}.$$

In particular, we see that for a second-order kernel the optimal bandwidth is of the order $O(T^{-1/5})$. Notice that this approach leads to a local bandwidth choice. In order to use that approach in practice, one needs pilot estimators of the density $f(x)$ and the derivative $f^{(r)}(x)$.

However, more common are global approaches based on the **integrated mean squared error (IMSE)**

$$\mathrm{IMSE}(\widehat{f}_{Th}; f) = \int \mathrm{MSE}(\widehat{f}_{Th}(x); f(x))\,\mathrm{d}x = \int E\left(\widehat{f}_{Th}(x) - f(x)\right)^2 \mathrm{d}x.$$

For $r = 2$ one obtains the expansion

$$\mathrm{IMSE}(\widehat{f}_{Th}; f) = \frac{R(K)}{Th} + \frac{1}{4}h^4 M_2^2 \int [f^{(2)}(x)]^2 \, \mathrm{d}x + o(h^4 + 1/Th).$$

Neglecting the remainder yields the **asymptotic integrated mean squared error (AMISE)**,

$$\mathrm{AMISE}(h) = \frac{R(K)}{Th} + \frac{1}{4}h^4 M_2^2 \int [f^{(2)}(x)]^2 \, \mathrm{d}x,$$

which we now study as a function of the bandwidth h. The optimal bandwidth h_{opt} that minimizes the AMISE and is easily shown to be

$$h_{\mathrm{opt}} = C_0 T^{-1/5},$$

where

$$C_0 = M_2^{-2/5} R(K)^{1/5} \left[\int f^{(2)}(x)]^2 \, \mathrm{d}x \right]^{-1/5}.$$

Unfortunately, the constant C_0 is unknown. The **normal reference rule-of-thumb** determines the constant for the standard normal distribution with mean zero and variance σ^2 as a reference model. When also using a normal kernel for smoothing, we obtain the optimal bandwidth

$$h_{\mathrm{opt}}^* = (4\pi)^{-1/10} \left[(3/8)\pi^{-1/2} \right]^{-1/5} \sigma \cdot T^{-1/5} \approx 1.06\sigma T^{-1/5}.$$

This choice if often used in practice with σ estimated by the sample standard deviation of the data.

Clearly, an undesirable feature of the above approach is that the method is tuned to a fixed reference distribution, as it tries to estimate the asymptotically optimal bandwidth in this case, although the kernel density aims at estimating an arbitrary (smooth) density. Thus, fully automatic procedures that do not make such restrictions are usually applied. Widespread approaches are unbiased and biased least-squares cross-validation, which we shall briefly discuss here.

Least squares unbiased cross-validation minimizes a nonparametric estimator of the integrated squared error and therefore provides an optimal bandwidth tailord to all x in the support instead of fixing some x. Since

$$\int \left[\widehat{f}_{Th} - f(x) \right]^2 \mathrm{d}x = \int \widehat{f}_{Th}^2(x) \, \mathrm{d}x - 2 \int \widehat{f}_{Th}(x) f(x) \, \mathrm{d}x + \int f(x)^2 \, \mathrm{d}x,$$

minimizing the IMSE is equivalent to minimizing the first two terms on the right-hand side. Observe that

$$\int \widehat{f}_{Th}(x) f(x) \, \mathrm{d}x = E_R(\widehat{f}_{Th}(R)),$$

if $R \sim f$ is independent from $R_1, \ldots, R_T$ and E_R denotes the expectation with respect to R. Thus, we may estimate $E_R(\widehat{f}_{Th}(R))$ by

$$\widehat{f}_{T,-i} = \frac{1}{(T-1)h} \sum_{t=1, t \neq i}^{T} K\left(\frac{R_t - R_i}{h}\right).$$

That estimate is called the **leave-one-out estimate of** $f(X_i)$. The first term is estimated by plugging in the kernel density estimate,

$$\begin{aligned} \int \widehat{f}^2_{Th}(x)\,\mathrm{d}x &= \frac{1}{T^2h^2} \sum_{t=1}^{T}\sum_{s=1}^{T} \int K\left(\frac{R_t - x}{h}\right) K\left(\frac{R_s - x}{h}\right) \mathrm{d}x \\ &= \frac{1}{T^2h} \sum_{t=1}^{T}\sum_{s=1}^{T} (K \star K)\left(\frac{R_t - R_s}{h}\right). \end{aligned}$$

Least squares cross-validation uses these estimators and minimizes the objective function

$$\mathrm{UCV}(h) = \frac{1}{T^2h} \sum_{t=1}^{T}\sum_{s=1}^{T} (K \star K)\left(\frac{R_t - R_s}{h}\right) - \frac{2}{T(T-1)h} \sum_{s=1}^{T} \sum_{t=1, t \neq s}^{T} K\left(\frac{R_t - R_s}{h}\right),$$

which has to be done numerically. Thus, the expectation of both terms yielding UCV(h) match the first two terms of the IMSE. One can show that, asymptotically, minimizing CV(h) is indeed equivalent to minimizing

$$B_1 h^4 + \frac{R(K)}{Th},$$

where

$$B_1 = \frac{M_2^2}{4} \left\{ \int [f^{(2)}(x)]^2\,\mathrm{d}x \right\}.$$

From here it is easy to see that the minimizer of the last display coincides with the minimizer of the IMSE. Moreover, one can even show that

$$\frac{h_{\mathrm{LCV}} - h_{\mathrm{opt}}}{h_{\mathrm{opt}}} \to 0,$$

as $T \to \infty$, in probability, a strong justification of the method.

Biased least-squares cross-validation minimizes another estimate of the asymptotic mean squared error (AMISE). Recall that

$$\mathrm{AMISE}(h) = \frac{R(K)}{Th} + \frac{1}{4} K_2^2 h^4 R(f'').$$

The optimal bandwidth is given by

$$h_0 = \left(\frac{R(K)}{M_2^2 T R(f'')} \right)^{1/5}.$$

A natural estimate for the only unknown quantity $R(f'')$ is $R(\widehat{f}''_T)$, where $\widehat{f}''_T$ is the second derivative of the kernel estimator $\widehat{f}_T$, but it turns out that

$$E(R(\widehat{f}''_T)) = R(f'') + \frac{R(K'')}{Th^5} + O(h^2).$$

One can do better by estimating the positive bias. This leads to the estimator $R(\widehat{f}''_T) - \frac{R(K'')}{Th^5}$. Noticing that

$$R(\widehat{f}''_T) = \frac{R(K'')}{Th^5} + \frac{2}{T^2h^5} \sum_{1 \le s < t \le T} \phi\left(\frac{X_t - X_s}{h}\right),$$

where

$$\phi(x) = \int K''(u)K''(u+x)\,\mathrm{d}u, \qquad x \in \mathbb{R},$$

This leads to the biased cross-validation function

$$\mathrm{BCV}(h) = \frac{R(K)}{Th} + \frac{K_2^2}{2T^2h} \sum_{1 \le s < t \le T} \phi\left(\frac{X_t - X_s}{h}\right),$$

which is then minimized.

Figure 1.1 illustrates the kernel density estimator for the daily log returns of the FTSE from 1991 to 1998. The bandwidth is selected by the biased least-squares cross-validation method.

1.3.5 Testing for normality

Asset returns are often non-normal, particularly returns corresponding to small time lags such as daily or intraday returns. In order to check the hypothesis that the returns are normal, many statistical tests have been proposed in the literature. At this point, we shall discuss those tests that are most widely used in practice.

Let $R_1, \ldots, R_T$ be an i.i.d. sample of returns with common d.f. F. We aim at testing the null hypothesis that F is a normal distribution,

$$H_0 : F \in \{\Phi_{\mu,\sigma^2} : \mu \in \mathbb{R}, \sigma^2 > 0\}$$

against the alternative hypothesis

$$H_1 : F \notin \{\Phi_{(\mu,\sigma^2)} : \mu \in \mathbb{R}, \sigma^2 > 0\}.$$

Notice that H_1 means that for all $\mu \in \mathbb{R}$ and $\sigma^2 > 0$ there exists at least one $x \in \mathbb{R}$ such that $F(x) \neq \Phi_{(\mu,\sigma^2)}(x)$.

The **Jarque and Bera test** is given by

$$J_T = T\left(\frac{\widehat{\mu}_3^2}{6} + \frac{(\widehat{\mu}_4 - 3)^2}{24}\right),$$

where $\widehat{\mu}_3$ is the sample skewness and $\widehat{\mu}_4$ the sample kurtosis. Since J_T is asymptotically $\chi^2(2)$-distributed, as $T \to \infty$, one rejects H_0, if $J_T > \chi^2(2)_{1-\alpha}$. However, the test should be used only for large data sets. Notice that the Jarque and Bera test measures the departure of the sample skewness and kurtosis from their theoretical values under the null hypothesis.

Another class of tests is based on the following idea. If the null hypothesis is true, we estimate the parameters μ and σ^2 by their sample analogs $\widehat{\mu}_T$ and S_T^2. The corresponding estimate of the distribution function is then $\Phi_{(\widehat{\mu}_T, S_T^2)}(x)$. If the alternative hypothesis is true, we may rely on the empirical distribution function, i.e.

$$\widehat{F}_T(x) = \frac{1}{T}\sum_{t=1}^{T} \mathbf{1}(R_t \leq x), \qquad x \in \mathbb{R},$$

which provides a consistent estimator of $F(x)$ without assuming any specific shape of the distribution. Now we can compare those two estimates by calculating the maximum deviation. This motivates the **Lilliefors test statistic**

$$L = \sup_{t \in \mathbb{R}} |\widehat{F}_T(t) - \Phi_{(\widehat{\mu}_T, S_T^2)}(t)|.$$

The asymptotic distribution of L is none of the standard distributions that have appeared so far. To conduct the test on the 5% significance level, one compares L with the critical value $0.805/\sqrt{T}$. However, the test is implemented in standard statistical software.

Sometimes, one wants to test the simple null hypothesis $H_0 : F = \Phi_{(\mu_0, \sigma_0^2)}$ against the alternative hypothesis $H_1 : F \neq \Phi_{(\mu_0, \sigma_0^2)}$ for some known constants $\mu_0 \in \mathbb{R}$ and $\sigma_0^2 > 0$. In this case, one may calculate

$$KS = \sup_{t \in \mathbb{R}} |\widehat{F}_T(t) - \Phi_{(\mu_0, \sigma_0^2)}(t)|.$$

That test is called the **Kolmogorov–Smirnov test**.

1.4 Financial instruments

Before proceeding, we shall introduce some financial slang and basic financial instruments. From an economic point of view, a trade is an agreement between two parties, a buyer and a seller, to buy or sell a certain amount of an asset at a certain date. The buyer attains a **long position** in the asset and the seller a **short position**. Associated to each trade are payments. For a given party we agree on the following sign convention: If the party receives a payment, it gets a positive sign. If the party has to pay the amount, we assign a negative sign.

1.4.1 Contingent claims

The payments of many financial instruments depend on other instruments or variables, often securities such as stocks, stock indices, oil, energy prices, or commodities, which are then called the **underlying** of such an instrument. It is even possible to buy financial instruments whose payment depends on quantities such as the weather.

Derivatives and futures are used for hedging risks associated with the production and distribution of goods and services in the real economy and, indeed, they are needed for those purposes. But they are also used a lot for pure speculation. To some extent speculators are needed as counterparties for hedges, but some markets are dominated by excessive speculation leading to substantial bubbles. For example, the unethical speculation in agricultural commodities since 2005, when volatility increased due to extreme weather incidents and increasing demand, is regarded as a substantial factor for record highs of food prices in developing countries leading to social instability and starvation.

A financial instrument whose payoff depends on another quantity is called a **contingent claim**. We shall give a mathematical definition later. If the underlying is a security such as an exchange-traded stock, it is called **derivative asset**. In what follows, we introduce the most important derivatives and related instruments and contracts.

1.4.2 Spot contracts and forwards

Definition 1.4.1 *A* **spot contract** *is an agreement to buy or sell an asset at the same day at a certain price called spot price that we shall denote by S_t. In the following, we shall assume that $t = 0$ stands for the time when a trade is initiated and T denotes the time horizon when the trade is settled. By contrast,* **forward** *contracts are agreements to buy or sell an asset at a future time at a price that is fixed when the parties agree on the contract, i.e. today. A forward allows the holder of the long position to buy the asset at a future time point T, the delivery date, at a fixed delivery price K, which coincides with the forward price F. The payoff of a long forward contract is $S_T - K$ and $K - S_T$ for a short position.*

The markets where spot contracts are traded are called spot markets. Forwards are traded over-the-counter (OTC), usually between financial institutions such as banks and their clients, e.g. an enterprise or private investor. There are no cash payments in $t = 0$. A forward is settled at the delivery date T when the seller has to deliver the asset to the buyer. However, often the parties agree on cash settlement. If the price at delivery, S_T, is higher than the delivery price K, the holder of a long position receives the payment $S_T - K$ and makes a profit. That additional payment hast the effect that he buys the asset for the forward price $F = K$, since $-S_T + (S_T - K) = -K$. But if the price is lower, he has to pay the difference to the seller. Again due to this additional payment, the net price of buying the asset is the delivery price.

1.4.3 Futures contracts

Definition 1.4.2 *Futures are standardized forward contracts usually traded on an exchange.*

For instance, the NYMEX light sweet crude oil futures is a contract on the physical delivery of 1000 barrel during a specified month. Standardization and handling by exchanges allows market participants to actively trade the contracts. Thus, in contrast to forwards, which can be highly specialized nontradeable agreements, futures can be very liquid financial instruments. The exchange specifies in detail the asset, how many units will be delivered under one contract (the contract size), the delivery date and how and where the asset will be delivered. For many contracts physical delivery is not possible or inconvenient and cash settlement applies. Here an equivalent cash payment between the parties is initiated. A futures contract can be bought and sold at any time point until its delivery date. The corresponding price is the futures price. At each trading day a settlement price is quoted, usually the closing price immediately before the end of trading day. The settlement price is used to determine the margins that are required from any investor. The investor has to deposit funds in a margin account. When entering a contract, the so-called initial margin has to be paid. At each trading day the account is marked to market to adjust the possible losses and gains. When the futures price rises, the holder of a long position makes a profit that is exactly the loss of the holder of the short position. The broker of an investor who is short reduces the margin account by the loss and the exchange transfers the money to the broker of the counter party where it increases the margin account. This is called daily settlement. If the margin account falls below the maintenance margin, the investor receives a margin call to deposit further funds. Otherwise the broker will close-out the position, i.e. neutralizing the existing contract.

1.4.4 Options

Options are agreements that give the holder of a long position the right, but not the obligation, to buy or sell the underlying at a fixed price in the future under certain conditions. There are a vast number of options traded nowadays; the most basic options are described in the following definition.

Definition 1.4.3 (EUROPEAN CALL/PUT OPTION, BASIS PRICE, EXPIRATION DATE)
A **European call option** *gives the holder the right to buy the underlying at a specified price, the* **strike price** *or* **basis price** *K at a fixed time point T called* **maturity** *or* **expiration date**. *The holder of a* **European put option** *has the right to sell the underlying for the strike price at maturity. If S_t stands for spot price of the underlying at time $t \in [0, T]$, we will denote the price (fair value) of a European call at time t by $C_e(S_t, K, t, T)$. Our notation for the price of a put will be $P_e(S_t, K, t, T)$. $T - t$ is called the* **time-to-maturity**.

Often, cash settlement applies. This means, the buyer does not get the underlying but the equivalent amount of money he would realize as a profit when buying the underlying for the strike price and selling it on the market. Denote by $C(S_t, K, t, T)$ the price of such an option at time t. At time T it coincides with the payoff given by

$$s \mapsto C(s, K, T, T), \qquad s \in [0, \infty).$$

The holder of a European call exercises the option, if $S_T > K$. The profit is $S_T - K$. Thus,

$$C_e = C_e(S_T, K, T, T) = \begin{cases} S_T - K, & S_T > K, \\ 0, & S_T \leq K, \end{cases}$$

which can be written in the form

$$C_e = \max(0, S_T - K) = (S_T - K)^+.$$

Similarly, for a European put option we have

$$P_e = P_e(S_T, K, T, T) = \max(0, K - S_T) = (K - S_T)^+.$$

The **internal value** of an option is its positive cashflow when one would exercise it. For a European call it is given by $(S_t - K)\mathbf{1}(S_t > K)$ and for a put equals $(K - S_t)\mathbf{1}(S_t < K)$. An option is **in the money**, when its internal value is positive ($S_t > K$ for a call, $S_t < K$ for a put), and it is called **out of the money** if the internal value is 0. ($S_t < K$ for a call, $S_t > K$ for a put). The ratio K/S is called **moneyness**.

Example 1.4.4 (Portfolio Insurance)
European put options can be used to solve the problem discussed in Exercise 1.1.1. Suppose the pension funds intends to buy a portfolio of stocks, frequently called basket of stocks, whose current price is $S_t = 110$. *Further, assume that the pension fund can buy a European put option on that basket. If the pension fund is willing to take a (downside) risk of at most* 10 *units of currency, a put with strike* 100 *has to be chosen.*

The portfolio of the pension fund consists of the basket and one put option. Consider its value at maturity T. *If* $S_T > 100$, *the put option is out of the money, i.e. its value is* 0, *such that the porfolio's value is* S_T. *In the case* $S_T \leq 100$, *the payoff of the put option is* $100 - S_T$ *such that the porfolio's value is* $V_T = S_T + (100 - S_T) = 100$. *It follows that the loss can not exceed* 10 *units of currency.*

1.4.5 Barrier options

The value of a barrier call option depends on whether the price of the underlying touches a certain value called barrier. Knock-out options die if the barrier is reached, whereas knock-in options are activated in this case.

Definition 1.4.5 *A European barrier option with expiration date* T, *barrier* B, $B < S_0$ *and* $B < K$, *and strike price* K *gives the option holder the right to buy the underlying at time* T, *if*

$$S_t > B \text{ for all } 0 \leq t \leq T \quad \text{(down-and-out)}$$

and

$$S_t < B \text{ for all } 0 \leq t \leq T \quad \text{(up-and-out)},$$

respectively. For a knock-in option the right is activated when

$$S_t \leq B \text{ for some } t \in [0, T] \quad \text{(down-and-in)}$$

or

$$S_t \geq B \text{ for some } t \in [0, T] \quad \text{(up-and-in)}.$$

American-style options allow buying the underlying at an arbitrary time point provided they are activated.

Barrier options are examples of **path-dependent options** whose payoff and value depends on the price trajectory S_t, $0 \leq t \leq T$, during the lifetime of the contract.

Definition 1.4.6 *An American* **average price call options** *is given by the payoff profile*

$$\max(0, \overline{S_t} - K), \quad t = 1, \dots, T,$$

where K *stand for the exercise price and*

$$\overline{S_t} = \frac{1}{t} \sum_{i=1}^{t} S_i, \quad t = 1, \dots, T,$$

denotes the average price. American **strike call options** *have the payoff profile* $\max(0, S_t - \overline{S_t})$, $t = 1, \ldots, T$. *Their exercise price is determined when the option is exercised. The corresponding European-style options are given by the payoffs* $\max(0, \overline{S_T} - K)$ *and* $\max(0, S_T - \overline{S_T})$ *at maturity, respectively.*

1.4.6 Financial engineering

By combining financial instruments, particularly derivatives, one can implement interesting payoff profiles. For example, a **long straddle** consists of a long position in a European call option and a long position in a European put with the same underlying and the same maturity, both in the money. For large increases of the stock price, the long positions provides a profit, whereas for large decreases of the stock price the put earns the money. In this way, one can create a position that makes a profit if the stock price changes, independent of the direction.

Basically, we shall see that the fair price π of a derivative or a contingent claims can be calculated by an expectation $E^*(C^*)$ of the discounted payoff C^* of the derivative under a certain probability measure. This automatically also allows us to price portfolios of contingent claims. Suppose such a portfolio consists of n positions given by the discounted payoffs $C_1^*, \ldots, C_n^*$ of each claim and the numbers of contracts $x_1, \ldots, x_n$ we held. Since expectations are linear, the fair price of the portfolio is

$$E^* \left(\sum_{i=1}^{n} x_i C_i^* \right) = \sum_{i=1}^{n} x_i \pi_i,$$

where $\pi_i = E^*(C_i^*)$ is the fair price of the ith claim.

In financial engineering, artificial portfolios of derivatives are often constructed in order to generated certain payoff profiles, for example in order to simultaneously hedge risks and generate opportunities for a profit, or as a complex financial product for customers. If a given payoff profile, Z, can be constructed by a portfolio such that $Z = \sum_{i=1}^{n} x_i C_i$, then the above formula allows us to determine the fair price of such a complex product. What makes such products challenging and risky is the fact that the underlying instruments $C_1, \ldots, C_n$ may have quite different risk exposures to risk factors such as interest rates, price changes of the underlying, volatility changes of the underlying or the risk that the issuer of the instrument defaults. Furthermore, the underlying portfolio is often unknown to the customer, which hinder his or her evaluation of the risk associated to such a product.

1.5 A primer on option pricing

This section is devoted to an introduction to some basic ideas and principles that lead to a powerful and elegant theory of option pricing. It is a matter of fact that they can be explained and understood in the simplest framework of a financial market with one asset and one European call option. We will obtain first convincing answers to the question on how to determine a fair price for a contingent claim, but simultaneously these answers give rise to various questions on how to extend them to more general and realistic frameworks.

1.5.1 The no-arbitrage principle

The no-arbitrage principle says that on an idealized financial market the prices do not allow for a riskless profit, i.e. there is no *free lunch*. Such arbitrage opportunities can arise if, for example, the prices in New York are higher than in London or the price of a bond is less than the fair value of its future payments. For what follows, we use the following mathematical definition.

Definition 1.5.1 *An* **arbitrage opportunity** *is a transaction yielding a random payment* X_1 *in* $t = 1$ *with initial value* x_0 *in* $t = 0$ *such that*

$$x_0 \leq 0 \qquad \textit{(no costs)}$$

and

$$X_1 \geq 0 \; P - a.s., \qquad \textit{and} \qquad P(X_1 > 0) > 0.$$

Example 1.5.2 *Let us apply the no-arbitrage principle to determine the fair value* F_0 *of a forward contract, i.e. the arbitrage-free price that applies at time* $t = 0$*. We claim that there is a unique no-arbitrage forward price, namely* $F_0 = S_0 e^{rT}$*, when assuming continuous compounding. Assume* $F_0 > S_0 e^{rT}$*. In this case, the seller can make a riskless profit by borrowing* S_0 *at time zero and buying the underlying. At maturity, he sells the underlying at the delivery price* K*, pays back* $S_0 e^{rT}$ *and earns* $F_0 - S_0 e^{rT} > 0$*. If* $F_0 < S_0 e^{rT}$*, the buyer sells the underlying aud puts the money to the bank. At maturity he receives* $S_0 e^{rT}$ *and pays* F_0 *for the underlying, leaving a profit* $S_0 e^{rT} - F_0$*. It is interesting and important to note that the forward price does not depend on the price of the underlying at maturity.*

The no-arbitrage principle also immediately leads to a simple formula that relates the price of an European call and European put. The idea is to set up a portfolio that leads to the same payoff as an European call option with maturity T and strike price K. If we buy a stock and sell a zero bond with nominal K, the value at time T is $S_T - K$. If we add a put to the portfolio, its value at maturity is zero, if $S_T > K$, but $K - S_T$, if $S_T \leq K$. It follows that the value of the portfolio is 0, if $S_T \leq K$, but $S_T - K$, if $S_T > K$. Its value at time 0 is

$$\pi(P_e) - K e^{-rT} + S_0$$

and must be equal to the fair price of the call, which establishes the **put-call parity**

$$\pi(C_e) = \pi(P_e) - K e^{-rT} + S_0.$$

The existence of arbitrage opportunities, which is ruled out by the no-arbitrage principle, means that the current prices of financial instruments are not balanced with their future payments. Many economists argue that on real financial markets arbitrage can at best exist temporarily, since they are discovered by market participants that then enter trades that quickly remove the arbitrage opportunity. If, for instance, the price of a financial instrument is too low and provides a free lunch, speculators will enter long positions such that its price will rise until the riskless profit disappears. We shall see that the no-arbitrage principle is a powerful and simple approach to determine fair prices.

1.5.2 Risk-neutral evaluation

The evaluation of a random (future) payment X depends on the preferences that can be expressed via a probability measure on the underlying measure space. The crucial question is whether a fixed payment, i.e. the case $X(\omega) = x_0$, for all $\omega \in \Omega$ and some fixed x_0, is preferred to a risky payment that offers the chance that the event $\{X > x_0\}$ occurs, but usually at the risk that the event $\{X < x_0\}$ may occur as well.

For simplicity of exposition, let us assume that the uncertainty about the future payment is measured in terms of the volatility, i.e. the square root of the variance. Given two investment opportunities with equal means, a **risk-averse** investor prefers the alternative with the smaller variance. By contrast, if the investor is **risk neutral**, he has no preference at all, since he ignores the variance.

In a risk-neutral world of risk-neutral investors everybody just looks at the mean. Let us denote the probability measure corresponding to this risk-neural world by P^*. Under P^* a stock is preferred to a riskless investment, if and only if its expected return is higher than the riskless return earned on a bank account. Denote the stock's price at time t by S_t and denote its random return by R. We assume that the price S_0 at $t = 0$ is a constant S_0 known to us. Then the random price at $t = 1$ is given by

$$S_1 = S_0(1 + R).$$

In a risk-neutral world the value of that payment is given by

$$E^*(S_1) = S_0(1 + E^*(R)).$$

Here and throughout, the symbol E^* means that the expectation is calculated under the probability measure P^*. If we deposit the initial capital S_0 in a bank account, we obtain $S_0(1 + r)$. The principle of no-arbitrage implies that $E^*(S_1)$ and $S_0(1 + r)$ must coincide, i.e.

$$E^*(S_1) = S_0(1 + r) \qquad \Leftrightarrow \qquad E^*\left(\frac{S_1}{1 + r}\right) = S_0.$$

As a consequence, under risk-neutral pricing the (fair) price of the stock can be calculated as an expectation under the probability measure P^*. Can we calculate P^* from the above equation?

To get first insights, we shall study a very simple one-period model for a financial market consisting of one stock and one European call option on that stock. To make the model as simple as possible, let us assume a binomial model for the stock price where the price can either go up or go down. In this case, we may choose the sample space $\Omega = \{+, -\}$ to represent the possible future states of our financial market, equipped with the power set sigma field. The real probability measure P is uniquely determined by $P(\{+\}) = p$, $p \in (0, 1)$. Notice that we exclude the trivial cases $p = 0$ and $p = 1$. We model the stock price by

$$S_1(\omega) = \begin{cases} S_0 u, & \omega = +, \\ S_0 d, & \omega = -, \end{cases}$$

with constants u (up factor) and d (down factor) satisfying $0 < d < 1 + r < u$. The European call is given by its payoff

$$C_e = \begin{cases} S_1 - K, & S_1 > K, \\ 0, & S_1 \leq K. \end{cases}$$

To avoid trivialities, we shall assume that the strike price K ensures that $S_0 d < K < S_1 u$.

In the above simple model the risk-neutral probability measure P^* is uniquely determined by $p^* = P^*(\{+\})$. The risk-neutral pricing formula $E^*(S_1) = S_0(1+r)$ is now equivalent to the equation

$$p^* S_0 u + (1 - p^*) S_0 d = S_0(1+r),$$

which has the unique solution

$$p^* = \frac{1+r-d}{u-d}.$$

This means, given the model parameters r, d and u we can determine P^*. Relying on the principle of risk-neutral pricing, the fair value of any random payment X_1 at time $t = 1$ can be calculated by

$$\pi(X_1) = E^*(X_1/(1+r)).$$

In particular, for a European call option on a stock we obtain

$$\pi(C_e) = p^* \frac{S_0 u - K}{1+r}.$$

Example 1.5.3 *Recall Example 1.1.3 and Example 1.1.4, where the oil price was assumed to either go up by* 10% *or go down by* 10%. *This means that we have* $u = 1.1$ *and* $d = 0.9$. *The riskless rate was* $r = 0.01$. *Hence, the risk-neutral probability measure* P^* *is given by*

$$p^* = \frac{1+r-d}{u-d} = \frac{1.01 - 0.9}{0.2} = 0.55,$$

yielding the risk-neutral option price

$$E^*(C_e/(1+r)) = \frac{10}{1.01} 0.55 = 5.445545.$$

This is exactly the lower price limit calculated by the oil trader.

Let us slightly generalize our model to allow for a trinomial model for the stock price. We put $\Omega = \{+, \circ, -\}$ and assume that, given three factors $d < m < u$, the stock price at time $t = 1$ satisfies

$$S_1(\omega) = \begin{cases} S_0 u, & \omega = +, \\ S_0 m, & \omega = \circ, \\ S_0 d, & \omega = -. \end{cases}$$

The risk-neutral probability measure P^* is now determined by $p_1^*, p_2^*, p_3^* \in [0, 1]$ such that $p_1^* + p_2^* + p_3^* = 1$. In this model, the pricing formula $E^*(S_1) = S_0(1+r)$ leads to

$$p_1^* u + p_2^* m + (1 - p_1^* - p_2^*) d = (1+r) \quad \Leftrightarrow \quad p_1^*(u-d) + p_2^*(m-d) = (1+r) - d.$$

This equation has infinite solutions. The special solution corresponding to $p_2^* = 0$ is the solution of the binomial model. In general, the solutions can be parameterized by p_2^* yielding

$$p_1^* = \frac{1+r-d+p_2^*(m-d)}{u-d}, \qquad p_2^* \in [0, 1],\ p_3^* = 1 - p_1^* - p_2^*.$$

It follows that pricing using the risk-neutral approach is not unique; there are infinitely many prices.

Exercise 1.5.4 *Determine all risk-neutral probability measures. Which conditions on d, m, u and r are required?*

1.5.3 Hedging and replication

Options are usually written by banks that are interested in hedging the risk of such a deal. Again, we consider a European option C_e on a stock S_1 that follows a binomial model. By introducing the notations $S_1(-)$, $S_1(+)$ and $C_e(-)$, $C_e(+)$, we shall see that the formulas we are going to derive hold for general options as well. The question arises whether it is possible to set up a portfolio that neutralizes any risk from the option deal. If we had a portfolio that exactly reproduces the option, we could buy that portfolio to neutralize the financial effect of selling the option to a customer. So, let us assume the bank holds a portfolio (θ_0, θ_1), where θ_0 is the amount of cash deposited in the bank account and θ_1 stands for the shares. Denote the value of the portfolio at time t by V_t. The portfolio neutralizes the option if it has the same value at $t = 0$ and $t = 1$. Obviously,

$$V_0 = \theta_0 + \theta_1 S_0,$$

and

$$V_1(\omega) = \begin{cases} \theta_0(1+r) + \theta_1 S_0 u, & \omega = +, \\ \theta_0(1+r) + \theta_1 S_0 d, & \omega = -. \end{cases}$$

The value W_0 of the option at time 0 is its price $\pi(C_e)$, and at time 1

$$W_1(\omega) = \begin{cases} S_0 u - K, & \omega = +, \\ 0, & \omega = -. \end{cases}$$

The portfolio replicates the option if $V_t(\omega) = W_t(\omega)$ holds true for all $\omega \in \Omega$ and all $t \in \{0, 1\}$. This leads to the equations

$$V_0 = \pi(C_e) \tag{1.9}$$

and

$$\theta_0(1+r) + \theta_1 S_0 u = S_0 u - K \tag{1.10}$$

$$\theta_0(1+r) + \theta_1 S_0 d = 0 \tag{1.11}$$

Substitute $\theta_0(1+r) = -\theta_1 S_0 d$ (Equation (1.11)) into Equation (1.10) to obtain

$$-\theta_1 S_0 d + \theta_1 S_0 u = S_0 u - K \Leftrightarrow \theta_1(S_0 u - S_0 d) = S_0 u - K \tag{1.12}$$

$$\Leftrightarrow \theta_1(S_0 u - S_0 d) = C_e(+) - C_e(-). \tag{1.13}$$

Thus, noting that $S_0 u - S_0 d = S_1(+) - S_1(-)$, we arrive at

$$\theta_1 = \frac{C_e(+) - C_e(-)}{S_1(+) - S_1(-)}.$$

This ratio, the number of shares needed to replicate (exactly!) the option, is called the **hedge ratio**. In Example 1.1.4 the hedge ratio is $\theta_1 = 10/20 = 1/2$. Indeed, the oil trader bought half of the oil at time $t = 0$, i.e. he constructed the hedge portfolio. For θ_0 we obtain the formula

$$\theta_0 = C_e(-) - \frac{C_e(+) - C_e(-)}{u - d}\frac{d}{1 + r}.$$

For our example, we obtain $\theta_0 = 0 - \frac{10}{1.1-0.9} \cdot \frac{0.9}{1.01} \approx -44.55$. This means, the oil trader borrows the amount 44.554 from the bank. Since he receives the premium 5.45, he can buy the oil to hedge the option. The initial costs for the hedge, the **replication costs**, are $V_0 = \theta_0 + \theta_1 S_0$. These replication costs should be equal to the fair price of the option.

Exercise 1.5.5 *Show that* $V_0 = E^*\left(\frac{C_e}{1+r}\right)$, *if* P^* *is the probability measure given by* $p^* = \frac{1+r-d}{u-d}$.

1.5.4 Nonexistence of a risk-neutral measure

Consider a financial market with two stocks following a binomial model with up factors u_1, u_2 and down factors d_1, d_2. Risk-neutral evaluation now leads us to two equations, namely

$$p^*u_1 + (1 - p^*)d_1 = 1 + r,$$
$$p^*u_2 + (1 - p^*)d_2 = 1 + r,$$

for the free parameter p^*. Depending on the parameters r, d_1, d_2, u_1, u_2, there may be no solution. Consequently, there may be no risk-neutral probability measure at all.

1.5.5 The Black–Scholes pricing formula

We shall now discuss the famous Black–Scholes option pricing formula, although we have to anticipate some results derived later in this book.

Suppose we have a risk-neutral pricing measure P^* at our disposal and consider a European call option on a stock with price S_t and strike K. The payoff at maturity T is $C = \max(S_T - K)$. Suppose that a fixed interest is paid in each period and let us express the corresponding discount factor in the form e^{-r} for some $r > 0$. Then the discounted payoff is $C^* = e^{-rT}\max(S_T - K, 0)$. In a risk-neutral world, we must have $E^*(C) = C_0$, where C_0 denotes the fair price at time $t = 0$ of the random payment C, or, equivalently,

$$C_0 = E^*(C^*) = e^{-rT}E^*(\max(S_T - K, 0)).$$

This means, we may calculate the fair price of the European call option by evaluating the expression on the right-hand side, which requires determination of the distribution of S_T under P^*.

The famous Black–Scholes model assumes that under the real probability measure log prices are normally distributed, say, with drift parameter $\mu \in \mathbb{R}$ and volatility $\sigma > 0$. Then it turns out that under P^* the log price S_T at maturity follows a lognormal distribution with

$$\text{drift } \log S + (r - \sigma^2/2)T \quad \text{and volatility } \sigma\sqrt{T}.$$

Here $S = S_0$ denotes today's stock price, which is the basis to determine the fair price of the option.

We will apply the following result: Suppose that X follows a lognormal distribution with parameters $m \in \mathbb{R}$ and $s > 0$. Then

$$E(X-K)^+ = \mathrm{e}^{m+s^2/2}\Phi\left(\frac{m-\log K}{s}+s\right) - K\Phi\left(\frac{m-\log K}{s}\right). \tag{1.14}$$

We will give a sketch of the derivation and encourage the reader to work out the details. To check that nice result, first notice that for $x \geq 0$ we have $(X-K)^+ \geq x \Leftrightarrow X-K \geq x$. Hence, denoting the density of X by $f(x)$,

$$\begin{aligned} E(X-K)^+ &= \int_0^\infty P(X \geq K+x)\,\mathrm{d}x \\ &= \int_0^\infty \int_{K+x}^\infty f(t)\,\mathrm{d}t\,\mathrm{d}x \\ &= \int_K^\infty \int_x^\infty f(t)\,\mathrm{d}t\,\mathrm{d}z, \end{aligned}$$

where we made the change of variable $z = x + K$. If we plug in Equation (1.3), the formula for the density of a lognormal distribution, we arrive at

$$E(X-K)^+ = \int_K^\infty \int_x^\infty \frac{1}{\sqrt{2\pi}st}\mathrm{e}^{-(\log t - m)^2/2s^2}\,\mathrm{d}t\,\mathrm{d}x.$$

Substituting $z = (\log t - m)/s$, such that $\mathrm{d}z = \mathrm{d}t/st$, leads to the integral

$$\int_K^\infty \int_{(\log x - m)/s}^\infty \varphi(z)\,\mathrm{d}z\,\mathrm{d}x,$$

where $\varphi(x) = 1/\sqrt{2\pi}\mathrm{e}^{-x^2/2}$ denotes the density of the standard normal distribution. Apply the integration by parts rule $\int uv' = uv| - \int u'v$ with $u(x) = \int_{(\log x - m)/s}^\infty \varphi(z)\,\mathrm{d}z$ and $v'(x) = 1$ to obtain that

$$E(X-K)^+ = \int_K^\infty \varphi\left(\frac{\log x - m}{s}\right)\mathrm{d}x - K\Phi\left(\frac{m-\log K}{s}\right),$$

where $\Phi(x) = \int_{-\infty}^x \varphi(t)\,\mathrm{d}t$ denotes the d.f. of the standard normal distribution. Finally, using the substitution $z = \log x$ one easily verifies Equation (1.14).

Now let us apply formula (1.14) with $m = \log S_0 + (r - \sigma^2/2)T$ and $s = \sigma\sqrt{T}$:

$$\begin{aligned} E^*(S_T-K)^+ = \mathrm{e}^{\log S + rT}\Phi&\left(\frac{\log(S/K)+(r-\sigma^2/2)T}{\sigma\sqrt{T}} + \sigma\sqrt{T}\right) \\ - K\Phi&\left(\frac{\log(S/K)+(r-\sigma^2/2)T}{\sigma\sqrt{T}}\right). \end{aligned}$$

Therefore, the fair price of a European call option is given by

$$\pi(C_e) = E^*(C^*) = \mathrm{e}^{-rT}E^*(S_T-K)^+ = S_0\Phi(d_1) - K\Phi(d_2)\mathrm{e}^{-rT}, \tag{1.15}$$

where

$$d_1 = \frac{\log(S/K) + (r + \sigma^2/2)T}{\sigma\sqrt{T}},$$
$$d_2 = \frac{\log(S/K) + (r - \sigma^2/2)T}{\sigma\sqrt{T}}.$$

It turns out that, by virtue of the call-put parity the price of an European put option is then given by

$$\pi(C_p) = \pi(C_e) + Ke^{-rT} - S.$$

Further, in order to obtain the time t value of such options with time to maturity $\tau = T - t$, one only has to replace T by τ and let S denote the time t price of the underlying.

1.5.6 The Greeks

The Black–Scholes price formula explicitly shows on which quantities the fair arbitrage-free price of a European call option depends: Besides the option parameters K, T and the initial price S_0, which are fixed in the contract, the formula depends on the risk-free interest rate, r, and the volatility σ of the the log stock price. For risk management it is essential to know how sensitive a position is with respect to those quantities. If, for example, the volatility of the underlying increases, this will affect immediately the value of a position in a European option.

1.5.6.1 First-order Greeks

We shall now introduce the first-order greeks by referring to a European call priced within the Black–Scholes model. However, these definitions apply to any derivative.

In order to allow easy interpretation, we would like to define the sensitivity with respect to the stock price as the rate of the option's price V if the stock price changes by one unit of currency, i.e. as the ratio $\frac{\Delta\pi(C_e)}{\Delta S}$. Having an explicit formula for $\pi(C_e)$, obviously a differentiable function of S, T, σ and r, which are now regarded as variables, we can provide a rigorous definition of the sensitivity with respect to the changes of the stock price, called **Delta**, in terms of the partial derivative

$$\Delta = \frac{\partial\pi(C_e)}{\partial S}.$$

In the same vein, we may introduce the sensitivity with respect to a change of the expiration date T, which is called **Theta**,

$$\Theta = \frac{\partial\pi(C_e)}{\partial T}.$$

The parameter **Vega** (or **Kappa**) measures the rate of the option's price with respect to changes of the volatility and is defined as the corresponding partial derivative

$$\nu = \frac{\partial\pi(C_e)}{\partial\sigma}.$$

Table 1.2 Greeks for European options

Greek	Call Option	Put Option
$\Delta = \frac{\partial \pi(C_e)}{\partial S}$	$\Phi(d_1)$	$\Phi(d_1) - 1$
$\Theta = \frac{\partial \pi(C_e)}{\partial T}$	$-\frac{S\varphi(d_1)\sigma}{2\sqrt{T}} - rKe^{-rT}\Phi(d_2)$	$-\frac{S\varphi(d_1)\sigma}{2\sqrt{T}} + rKe^{-rT}\Phi(-d_2)$
$\nu = \frac{\partial \pi(C_e)}{\partial \sigma}$	$S\varphi(d_1)\sqrt{T}$	$S\varphi(d_1)\sqrt{T}$
$\rho = \frac{\partial \pi(C_e)}{\partial r}$	$KTe^{-rT}\Phi(d_2)$	$-KTe^{-rT}\Phi(-d_2)$
$\Gamma = \frac{\partial^2 \pi(C_e)}{\partial S^2}$	$\frac{\varphi(d_1)}{S\sigma\sqrt{T}}$	$\frac{\varphi(d_1)}{S\sigma\sqrt{T}}$

Calculations assume the Black-Scholes model

Finally, **Rho** is the standard notation for the sensitivity with respect to changes of the interest rate and formally given by

$$\rho = \frac{\partial \pi(C_e)}{\partial r}.$$

Table 1.2 lists the resulting formulas assuming the Black–Scholes model.

The first-order greeks allow us to approximate the option's price by a linear function. For example, if the price of the underlying changes from S to $\widetilde{S}$, knowing $\Delta = \frac{\partial \pi(C_e)}{\partial S}$ provides the approximation

$$\pi \approx \pi(C_e) + \frac{\partial \pi(C_e)}{\partial S}(\widetilde{S} - S),$$

which is accurate if $|\widetilde{S} - S|$ is small.

It is important to note that these partial derivatives are still functions of the remaining variables. Hence, their values depend on the values of those variables, the model parameters. If more than one parameter changes, it can not be seen from a single sensitivity measure how the option price reacts.

Observing that, given the strike price K, the variables S, T, r, σ determine $\pi(C_e)$, it is clear that the above greeks form the gradient

$$\frac{\partial \pi(C_e)}{\partial \vartheta} = \left(\frac{\partial \pi(C_e)}{\partial S}, \frac{\partial \pi(C_e)}{\partial T}, \frac{\partial \pi(C_e)}{\partial \sigma}, \frac{\partial \pi(C_e)}{\partial r} \right)' = (\Delta, \Theta, \nu, \rho)',$$

where $\vartheta = (S, T, \sigma, r)'$. The corresponding linear approximation following from Taylor's theorem is then given by

$$\pi \approx \pi(C_e) + \frac{\partial \pi(C_e)}{\partial \vartheta}(\widetilde{\vartheta} - \vartheta) = \pi(C_e) + \Delta(\widetilde{S} - S) + \Theta(\widetilde{T} - T) + \nu(\widetilde{\sigma} - \sigma) + \rho(\widetilde{r}, - r),$$

if the parameters change from ϑ to $\widetilde{\vartheta} = (\widetilde{S}, \widetilde{T}, \widetilde{\sigma}, \widetilde{r})'$.

1.5.6.2 Second-order Greeks

The first-order greeks correspond to first-order partial derivatives yielding linear approximations of the option's price. The next step is to take into account second-order partial derivatives as well, which lead to quadratic approximations.

Of particular concern is the dependence of the option price on the price of the underlying. The second-order partial derivative

$$\Gamma = \frac{\partial^2 \pi(C_e)}{\partial S^2}$$

is called **Gamma**.

1.5.7 Calibration, implied volatility and the smile

In order to price options with the Black–Scholes pricing formula, one has to specify the interest rate r. Usually, one takes the yield of a treasury bill with a short maturity. Further, one needs to determine in some way the volatility σ, which is not directly observable. Basically, there are two approaches. The *statistical approach* is to estimate σ from historical data as discussed in Section 1.3.2. Another approach frequently applied in finance is **calibration**, which means that an unknown parameter of a formula for some quantity is determined (calibrated) by matching the formula with real market data for that quantity. This has the advantage that the model reproduces current market data and is therefore often preferred by traders, analysts and bankers, since they tend to mistrust models and methods that seem to contradict markets.

In the case of option pricing by the Black–Scholes formula one calibrates the model by matching the prices predicted by the Black–Scholes formula with real market prices for options by varying the free parameter σ. Notice that equating Equation (1.15) to a actual price leads to a nonlinear equation for σ. The matching is done for a fixed strike price K and a fixed time to maturity $T - t$. The volatility σ determined in this way is called the **implied volatility**.

In theory, the volatility σ of the underlying asset is constant across strike prices and maturities. However, when determining the implied volatility for different values of K and T, one observes a dependence on those parameters. Sometimes the volatility is a decreasing function of K, a phenomenon called **volatility skew**. In other cases, particularly for options on foreign currencies, the volatility is lower for at-the-money options and gets larger as the option moves into the money or out of the money. This effect is called **volatility smile**. The dependence on K is usually parametrized by the **moneyness** or **strike ratio**, S/K. If one calculates the implied volatility over a two-dimensional grid of values for the strike K (or K/S) and the maturity T, one obtains a two-dimensional curve called the **volatility surface**. Figure 1.2 shows a volatility surface for SIEMENS AG.

1.5.8 Option prices and the risk-neutral density

There is an interesting and important relationship between option prices and the probability density of the risk-neutral probability measure used for pricing. The validity of this relationship is not restricted to the Black–Scholes model, but is an intrinsic structural properties of a financial market. It can be used to infer the risk-neutral probability from option prices.

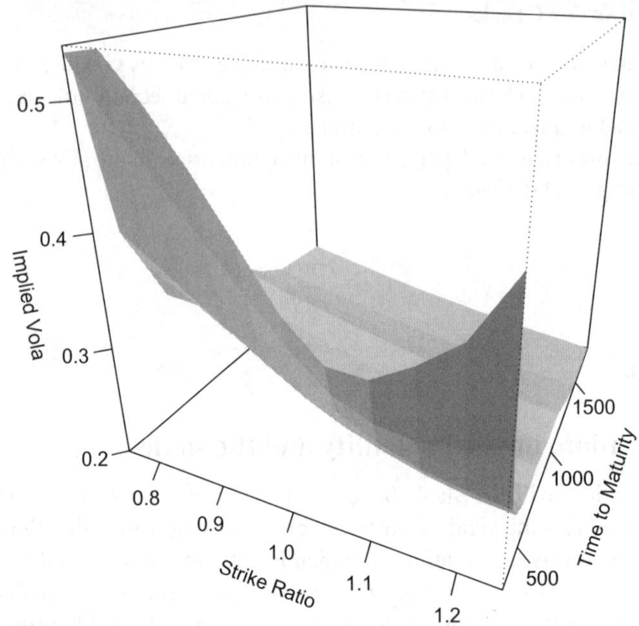

Figure 1.2 Volatility surface at November 4th of European call options on SIEMENS AG for maturities ranging from November 2011 to December 2015. Time to maturity measured in days, data taken from DATASTREAM.

Recall the starting point of our derivation of the Black–Scholes formula, namely the equation

$$C_e(K) = \mathrm{e}^{-rT} E^*(S_T - K)^+, \tag{1.16}$$

which we now study as a function of the strike price K. We also denote the risk-neutral price by $C_e(K)$ to indicate that we do not refer to the Black–Scholes formula. At this point, it is only assumed that there exists a risk-neutral measure P^* used to price random future payments. Let us also assume that the terminal stock stock price S_T attains a probability density under the risk-neutral probability measure P^*, which we will denote by $\varphi_T^*(x)$. This means,

$$P^*(S_T \le x) = \int_{-\infty}^{x} \varphi_T^*(u)\,\mathrm{d}u, \qquad x \in \mathbb{R}.$$

Then we may rewrite Equation (1.16) as

$$C_e(K) = \mathrm{e}^{-rT} \int_{-\infty}^{\infty} (x - K)^+ \varphi_T^*(x)\,\mathrm{d}x = \mathrm{e}^{-rT} \int_{K}^{\infty} \varphi_T^*(x)\,\mathrm{d}x.$$

Apply the formula

$$\frac{d}{\mathrm{d}t} \int_{a(t)}^{b(t)} f(x,t)\,\mathrm{d}x = f(b(t),t)b'(t) - f(a(t),t)a'(t) + \int_{a(t)}^{b(t)} \frac{\partial f(x,t)}{\partial t}\,\mathrm{d}x$$

to obtain that the first derivative of the risk-neutral price of a European call satisfies

$$\frac{\partial C_e(K)}{\partial K} = -\int_K^\infty \varphi_T^*(x)\,\mathrm{d}x$$

and the second derivative

$$\frac{\partial^2 C_e(K)}{\partial K^2} = \varphi_T^*(K).$$

As a consequence, one may determine the risk-neutral probability measure by analyzing option prices for different strike prices K. Since any probability density function is non-negative, we also see that the option prices is a *convex* function of the strike price.

1.6 Notes and further reading

A popular text on options, futures and other derivatives avoiding mathematics is the comprehensive book of Hull (2009). It explains in great detail and accompanied by many examples the economic reasoning behind such financial instruments and how the corresponding markets operate, provides basic formulas for the valuation of such financial operations and sketches at an elementary level the mathematical theory behind it. We also refer to the introductions to mathematical finance of Baird (1992), Pliska (1997) and Buchanan (2006), which focus more or less on the discrete-time setting and finite probability spaces, respectively. For the theory of coherent risk measure we refer to the seminal work Artzner et al. (1999), the recent monograph Pflug and Römisch (2007) and the discussion Embrechts et al. (2002) of dependence measures and their properties. There are various text books on the general theory of statistics including estimation, optimal hypothesis testing and confidence intervals, for example Lehmann and Romano (2005) or Shao (2003). Financial statistics is discussed in Lai and Xing (2008). More on kernel smoothing methods and their properties can be found in the monographs Silverman (1986), Härdle (1990), Fan and Gijbels (1996) and Wand and Jones (1995). The problem how to select the bandwidth one may additional consult Scott and Terrell (1987) and Savchuk et al. (2010). For a recent approach using singular spectrum analyses, we refer to Golyandina et al. (2011).

References

Artzner P., Delbaen F., Eber J.M. and Heath D. (1999) Coherent measures of risk. *Math. Finance* **9**(3), 203–228.

Bachelier L. (1900) Théorie de la spéculation. *Annales Scientifiques de l' É.N.S.* 3e **17**, 21–86.

Baird A.J. (1992) *Option Market Making: Trading and Risk Analysis for the Financial and Commodity Option Markets*. Wiley Finance, Wiley & Sons.

Buchanan J. (2006) *An Undergraduate Introduction to Mathematical Finance*. World Scientific, Singapore.

Embrechts P., McNeil A.J. and Straumann D. (2002) Correlation and dependence in risk management: properties and pitfalls *Risk Management: Value at Risk and Beyond (Cambridge, 1998)*. Cambridge Univ. Press Cambridge pp. 176–223.

Fan J. and Gijbels I. (1996) *Local Polynomial Modelling and its Applications*. vol. 66 of *Monographs on Statistics and Applied Probability*. Chapman & Hall, London.

Golyandina N., Pepelyshev A. and Steland A. (2011) New approaches to nonparametric density estimation and selection of smoothing parameters. **56**(7), 2206–2218.

Härdle W. (1990) *Applied Nonparametric Regression*. vol. 19 of *Econometric Society Monographs*. Cambridge University Press, Cambridge.

Hull J.C. (2009) *Options, Futures, and Other Derivatives*. 7th edn. Pearson Prentice Hall.

Lai T.L. and Xing H. (2008) *Statistical Models and Methods for Financial Markets*. Springer Texts in Statistics. Springer, New York.

Lehmann E.L. and Romano J.P. (2005) *Testing Statistical Hypotheses*. Springer Texts in Statistics third edn. Springer, New York.

Pflug G.C. and Römisch W. (2007) *Modeling, Measuring and Managing Risk*. World Scientific Publishing Co. Pte. Ltd., Hackensack, NJ.

Pliska S. (1997) *Introduction to Mathematical Finance*. Blackwell Publishing, Oxford.

Savchuk O.Y., Hart J.D. and Sheather S.J. (2010) Indirect cross-validation for density estimation. *J. Amer. Statist. Assoc.* **105**(489), 415–423. With supplementary material available online.

Scott D.W. and Terrell G.R. (1987) Biased and unbiased cross-validation in density estimation. *J. Amer. Statist. Assoc.* **82**(400), 1131–1146.

Shao J. (2003) *Mathematical Statistics*. Springer Texts in Statistics 2nd edn. Springer-Verlag, New York.

Silverman B.W. (1986) *Density Estimation for Statistics and Data Analysis*. Monographs on Statistics and Applied Probability. Chapman & Hall, London.

Wand M.P. and Jones M.C. (1995) *Kernel Smoothing*. vol. 60 of *Monographs on Statistics and Applied Probability*. Chapman and Hall Ltd., London.

2

Arbitrage theory for the one-period model

This chapter is devoted to a detailed study of the pricing of options and, more generally, arbitrary random payments that are called contingent claims. To keep the theory simple and clean but general enough to obtain valuable insights, we confine our study to a one-period model. Further, for simplicity of proofs, some results are given for a finite sample space that implies that assets can attain only a finite number of values.

Having observed that the principle of risk-neutral valuation yields an elegant and simple solution to the option pricing problem, we shall elaborate conditions for the existence and uniqueness of pricing measures and discuss how they can be calculated. When such a pricing measure exists, the question arises whether the corresponding prices really preclude arbitrage. Another issue of interest is to which extent contingent claims can be hedged, i.e. replicated, at all.

2.1 Definitions and preliminaries

We consider a financial market with one riskless investment opportunity, e.g., a bond or bank account, and d risky assets. In practice, the risky assets are usually exchange-traded stocks and we will frequently refer to the assets as stocks, but the theory applies to investment funds as well. The only requirement is that prices are random. The one-period model assumes that there are two time points, $t = 0$ and $t = 1$, where the investor can trade and consume, respectively. At $t = 0$ the investor sets up his portfolio, and at $t = 1$ the portfolio value is determined by closing all positions, at least virtually.

Let us now introduce the details of the model and some notations. r denotes the fixed and deterministic riskless interest rate for the bond and bank account, respectively. We assume that r applies to both deposits and loans. Let x denote the balance at $t = 0$ and fix the interpretation of the sign of x as follows. If $x > 0$, we are given a deposit, whereas $x < 0$ indicates that the

Financial Statistics and Mathematical Finance: Methods, Models and Applications, First Edition. Ansgar Steland.

investor has the liability $|x|$ due to a loan or credit. The value of the bank account at time $t = 1$ is then given by

$$v = (1 + r)x.$$

If the riskless investment opportunity is given by a bond with initial price $S_{00} \in (0, \infty)$, its value at time $t = 1$ is

$$S_{10} = (1 + r)S_{00}.$$

To simplify the exposition, we will assume $S_{00} = 1$ corresponding to a bank account.

Let us denote by $S_{01}, \ldots, S_{0d}$ the given prices of the d risky assets. The underlying probability space is denoted by $(\Omega, \mathcal{F}, P)$. We interpret an outcome $\omega \in \Omega$ as a possible scenario for the market at time $t = 1$. P stands for the true probability measure of the real world that assigns measurable sets $A \in \mathcal{F}$ their probability of occurence, $P(A)$.

The risky and unknown market prices at the end of the period are modeled by random variables

$$S_{10}, \ldots, S_{1d} : (\Omega, \mathcal{F}, P) \to (\mathbb{R}, \mathcal{B})$$

with $S_{10} = (1 + r)S_{00}$. The random vector $S_0 = (S_{00}, \ldots, S_{0d}) \in \mathbb{R}^{d+1}$ is called a **price vector**.

Definition 2.1.1

(i) Let $S_0 = (S_{00}, \ldots, S_{0d})'$ be a known price vector, i.e. $S_{0i} > 0$ for $i = 0, \ldots, d$ are known initial prices, and $S_1 = (S_{10}, \ldots, S_{1d})'$ be a random vector such that $S_{1i} : (\Omega, \mathcal{F}, P) \to (\mathbb{R}, \mathcal{B})$, $i = 1, \ldots, d$, are positive random variables. Then $\{S_t : t = 0, 1\}$ is called a **price process**.

(ii) $\{S_t^* : t \in 0, 1\}$ with $S_0^* = S_0$ and $S_1^* = (S_{10}^*, \ldots, S_{1d}^*)'$ where $S_{1i}^* = \frac{S_{1i}}{1+r}$ for $i = 0, \ldots, d$, is called a **discounted price process**.

Mathematically, our financial market is now defined by $((\Omega, \mathcal{F}, P), \{S_t\})$.

Definition 2.1.2 (PORTFOLIO)
A vector $\varphi = (\varphi_0, \ldots, \varphi_d)' \in \mathbb{R}^{d+1}$ is called a **portfolio**, *$(\varphi_1, \ldots, \varphi_d)'$ is named a* **portfolio in the (risky) assets**.

Let us agree on the following sign convention: $\varphi_i > 0$ signifies a long position and $\varphi_i < 0$ a short position.

Definition 2.1.3

(i) For any portfolio $\varphi = (\varphi_0, \ldots, \varphi_d)' \in \mathbb{R}^{d+1}$ and each price process $\{S_t : t = 0, 1\}$ the process $\{V_t : t = 0, 1\}$ with

$$V_t = V_t(\varphi) = \varphi' S_t = \sum_{i=0}^{d} \varphi_i S_{ti}, \quad t = 0, 1,$$

is called the **value process of the portfolio** φ.

(ii) The **discounted value process** $\{V_t^* : t = 0, 1\}$ *is defined by*

$$V_t^* = V_t^*(\varphi) = \varphi' S_t^*, t = 1.$$

Clearly, V_t accounts for the value of the portfolio φ. To this end, suppose $\varphi_i \geq 0$. Buying φ_i shares of asset i costs $\varphi_i S_{1i}$. The required initial capital to establish the portfolio is

$$V_0 = \varphi_0 + \sum_{i=1}^{d} \varphi_i S_{0i}.$$

The amount $\sum_{i=1}^{d} \varphi_i S_{0i}$ is needed to buy the assets. The rest $V_0 - \sum_{i=1}^{d} \varphi_i S_{0i}$ is deposited in a bank account, if it is positive; otherwise it represents a credit required to finance the portfolio. The financial contribution of each short position, i.e. those positions with $\varphi_i < 0$, to initiate the long positions is given by $|\varphi_i S_{0i}|$.

Definition 2.1.4

(i) $\{G_t : t = 0, 1\}$ *with* $G_t = G_t(\varphi) = V_t(\varphi) - V_0(\varphi)$, $t = 0, 1$, *is called the* **gains process**.

(ii) $\{G_t^* : t = 0, 1\}$ *with* $G_t^* = G_t^*(\varphi) = V_t^*(\varphi) - V_0^*(\varphi), t = 0, 1$, *is called the* **discounted gains process**.

2.2 Linear pricing measures

We aim at investigating under which conditions portfolios providing riskless profits may exist. The surprising result is that this is closely related to the existence of a pricing measure allowing us to price payments using the principle of risk neutrality. To obtain a 1-to-1 characterization, we introduce dominant portfolios, a notion that is stronger and more pleasant for the owner, than an arbitrage opportunity.

Definition 2.2.1 *A portfolio φ is called* **dominant**, *if there exists a portfolio $\widetilde{\varphi} \neq \varphi$ such that $V_0(\varphi) = V_0(\widetilde{\varphi})$ and $V_1(\varphi) > V_1(\widetilde{\varphi})$ P-a.s.; we say φ dominates $\widetilde{\varphi}$.*

Lemma 2.2.2 *A dominant porfolio exists, if and only if there is some portfolio φ with*

$$V_0(\varphi) = 0 \text{ and } V_1(\varphi) > 0 \ P - a.s.$$

Proof.

'$\Rightarrow$': Suppose $\overline{\varphi}$ dominates $\widetilde{\varphi}$. Define $\varphi = \overline{\varphi} - \widetilde{\varphi}$. By linearity of the mapping $\varphi \mapsto V(\varphi)$, we may conclude that $V_0(\varphi) = V_0(\overline{\varphi}) - V_0(\widetilde{\varphi}) = 0$ and $V_1(\varphi) = V_1(\overline{\varphi}) - V_1(\widetilde{\varphi}) > 0$, a.s.

'$\Leftarrow$': Obviously, a portfolio φ with $V_0(\varphi) = 0$ and $V_1(\varphi) > 0$ a.s. is better than investing nothing.

Proposition 2.2.3 *Suppose that Ω is a finite set. There exists a dominant portfolio, if and only if there is some portfolio $\widetilde{\varphi}$ with*

$$V_0(\widetilde{\varphi}) < 0 \text{ and } V_1(\widetilde{\varphi}) \geq 0 \quad P-a.s.$$

Proof. We show the necessity of the condition. By Lemma 2.2.2, there exists some portfolio φ with $V_0^*(\varphi) = 0$ and $V_1^*(\varphi) > 0$, P-a.s., implying

$$G_1^*(\varphi)(\omega) = V_1^*(\varphi)(\omega) - V_0(\varphi) > 0$$

for all $\omega \in \Omega$. Thus,

$$\delta := \min_{\omega \in \Omega} G_1^*(\varphi)(\omega) > 0.$$

Next, define $\widetilde{\varphi} \in \mathbb{R}^{d+1}$ by $\widetilde{\varphi}_i = \varphi_i, i = 1, \ldots, d$, and $\widetilde{\varphi}_0 = -\sum_{i=1}^d \varphi_i S_{0i}^* - \delta$. We obtain

$$V_0^*(\widetilde{\varphi}) = \widetilde{\varphi}_0 + \sum_{i=1}^d \widetilde{\varphi}_i S_{0i}^* = -\delta < 0$$

and for any $\omega \in \Omega$

$$\begin{aligned} V_1^*(\widetilde{\varphi})(\omega) &= \widetilde{\varphi}_0 + \sum_{i=1}^d \varphi_i S_{1i}^*(\omega) \\ &= -\sum_{i=1}^d \varphi_i S_{0i}^*(\omega) - \delta + \sum_{i=1}^d \varphi_i S_{1i}^*(\omega) \\ &= -\delta + \sum_{i=1}^d \varphi_i (S_{1i}^* - S_{0i})(\omega) \\ &= -\delta + G_1^*(\widetilde{\varphi})(\omega) \geq 0. \end{aligned}$$

By definition, the existence of a dominant portfolio implies that there are two portfolios having the same price, such that one is always better, i.e. its payoff is higher with probabity 1. This could be excluded if the price can be determined as a monotone function of the payment in $t = 1$.

Definition 2.2.4

(i) *Assume that Ω is finite. $\pi = \{\pi(\omega) : \omega \in \Omega\} \subset [0, \infty)$ is called a* **(linear) pricing measure***, if for all $\varphi \in \mathbb{R}^{d+1}$*

$$V_0^*(\varphi) = \sum_{\omega \in \Omega} \pi(\omega) V_1^*(\omega).$$

(ii) For arbitrary Ω equipped with a σ field $\mathcal{F}$, a measure π is called a **pricing measure**, *if*

$$V_0^*(\varphi) = \int V_1^*(\omega) \mathrm{d}\pi(\omega) \quad \textit{for all } \varphi \in \mathbb{R}^{d+1}.$$

Suppose that we are given such a pricing measure π. It is then easy to see that it is a probability measure, i.e.

$$\sum_{\omega \in \Omega} \pi(\omega) = 1 \text{ and } \int_{\Omega} \mathrm{d}\pi = 1, \text{ respectively.}$$

Lemma 2.2.5 *Assume that Ω is finite.*

(i) For any pricing measure $\{\pi(\omega) : \omega \in \Omega\}$ we have

$$S_{0j} = \sum_{\omega} \pi(\omega) S_{1j}^*(\omega), \quad j = 1, \ldots, d. \tag{2.1}$$

(ii) If $\{\pi(\omega) : \omega \in \Omega\}$ is a probability measure such that Equation (2.1) holds true, then it is a pricing measure.

Proof.

(i) Fix $j \in \{1, \ldots, d\}$ and consider the portfolio φ given by $\varphi_i = \mathbf{1}(i = j),\ i = 0, \ldots, d$. Then $V_0(\varphi) = S_{0j}$ and $V_1^*(\varphi) = S_{1j}^*$. If $\{\pi(\omega)\}$ is a pricing measure, we may conclude that

$$S_{0j} = V_0(\varphi) = \sum_{\omega \in \Omega} \pi(\omega) V_1^*(\omega) = \sum_{\omega \in \Omega} \pi(\omega) S_{1j}^*(\omega).$$

(ii) This follows easily by observing that $\varphi \mapsto V_t(\varphi)$ is a linear mapping.

Lemma 2.2.5 tells us that the initial prices are a linear function of the possible future prices weighted by the pricing measure. That linear structure allows us to use what is known about systems of linear equations of the form $\mathbf{Ax} = \mathbf{b}$, where $\mathbf{A}$ is a $n \times p$ matrix, $\mathbf{b}$ an n-dimensional vector and $\mathbf{x}$ the unknown p-dimensional solution vector, to treat the pricing problem. The details are as follows. When $\Omega = \{\omega_1, \ldots, \omega_n\}$ is finite and $\pi = \{\pi(\omega)\}$ a pricing measure, then the linear equations

$$S_{0j}^* = \sum_{i=1}^{n} \pi_i \cdot S_{1j}^*(\omega_i), \quad j = 1, \ldots, d,$$

hold true, if we put $\pi_i = \pi(\omega_i)$, $i = 1, \ldots, n$. Vice versa, any solution $(\pi_1, \ldots \pi_n)'$ of these equations satisfying the additional constraints

$$0 \le \pi_1, \ldots, \pi_n \le 1, \quad \text{and} \quad \sum_{i=1}^{n} \pi_i = 1$$

yields a pricing measure. In matrix notation, we have

$$\mathbf{D}^*\pi = \mathbf{b}^*,$$

where

$$\mathbf{D}^* = \begin{pmatrix} 1 & \dots & 1 \\ S_{11}^*(\omega_1) & \dots & S_{11}^*(\omega_n) \\ \vdots & \vdots & \vdots \\ S_{1d}^*(\omega_1) & \dots & S_{1d}^*(\omega_n) \end{pmatrix} \quad \text{and} \quad \mathbf{b}^* = \begin{pmatrix} 1 \\ S_{01}^* \\ \vdots \\ S_{0d}^* \end{pmatrix}$$

$\mathbf{D}^*$ is called the **discounted payment matrix** or **discounted payoff matrix**.

The following theorem clarifies the relationship between pricing measures and dominant portfolios.

Theorem 2.2.6 *Assume that $\Omega = \{\omega_1, \dots, \omega_n\}$ is finite. There is a linear pricing measure if and only if there is no dominant portfolio*

Proof. The problem to find some $\pi \geq 0$ with $\mathbf{D}^*\pi = \mathbf{b}^*$ can written as the following linear program.

$$\max_{\pi \in \mathbb{R}^n} \mathbf{0}'\pi \text{ such that } \mathbf{D}^*\pi = \mathbf{b}^*, \pi \geq 0, \tag{P}$$

where $\mathbf{0} \in \mathbb{R}^n$ is the vector of zeroes and $\mathbf{D}^*$ the discounted payment matrix. Suppose there is some pricing measure π. Then (P) has a solution. The duality theorem of linear programming asserts that this is equivalent to the existence of a solution $\varphi = (\varphi_0, \dots, \varphi_d)' \in \mathbb{R}^{d+1}$ of the dual problem

$$\min_{\varphi \in \mathbb{R}^{d+1}} \varphi'\mathbf{b}^* \text{ such that } \varphi'\mathbf{D}^* \geq 0, \tag{D}$$

and the optimal values coincide. The solution φ is a portfolio with $V_0(\varphi) = V_0^*(\varphi) = \varphi'\mathbf{b}^* = 0$ and $((\mathbf{D}^*)'\varphi)_k = \varphi_0 + \sum_{i=1}^d \varphi_i S_{1i}^*(\omega_k) = V_1^*(\varphi)(\omega_k) \geq 0$ for $k = 1, \dots, n$. That is, φ is a portfolio ensuring the minimal price $V_0(\varphi) = 0$ as well as $V_1^*(\varphi) \geq 0$. Therefore, there is no portfolio with $V_0 < 0$ and $V_1 \geq 0$, cf. Proposition 2.2.3. To show sufficiency, suppose there is no dominant portfolio. By Proposition 2.2.3, there is no portfolio with $V_0(\varphi) < 0$ and $V_1(\varphi)(\omega) \geq 0$ for all $\omega \in \Omega$. The choice $\varphi = \mathbf{0} \in \mathbb{R}^{d+1}$ yields $V_0(\varphi) = V_1(\varphi) = 0$ and solves (D), since any portfolio with $V_1 \geq 0$ must satisfy $V_0 \geq 0$. The corresponding solution of (P) can be used as a pricing measure.

2.3 More on arbitrage

In Definition 1.4.1, we introduced the notion of an arbitrage opportunity. We shall now specify it to our market model and than discuss some equivalent characterizations.

Definition 2.3.1 *A portfolio* $\varphi \in \mathbb{R}^{d+1}$ *is called a* **arbitrage opportunity** *or* **arbitrage portfolio**, *if* $\varphi' S_0 \leq 0$ *(no costs),*

$$\varphi' S_1 \geq 0 \ P\text{-a.s.} \quad \text{and} \quad P(\varphi' S_1 > 0) > 0.$$

This means, at time 1 *the portfolio's payoff is non-negative, almost surely, and the probability to make a real profit is positive.*

Remark 2.3.2 *Notice that the arbitrage property of a portfolio depends on the underlying probability measure P. Moreover, that property is invariant on the set*

$$\mathcal{P} = \mathcal{P}(P) = \{\widetilde{P} : \widetilde{P} \text{ is a probability measure on } (\Omega, \mathcal{F}) \text{ with } \widetilde{P} \sim P\}.$$

The proofs of those facts are left to the reader.

Recall the following probabilistic notion: Two (probability) measures ν and μ on $(\Omega, \mathcal{F})$ are called **equivalent**, denoted by $\nu \sim \mu$, if they share the null sets, i.e.

$$\mu(A) = 0 \quad \Leftrightarrow \quad \nu(A) = 0 \quad \text{for all } A \in \mathcal{F}.$$

Definition 2.3.3 *A financial market is called* **arbitrage-free**, *if there do not exist any arbitrage opportunities. In this case, the financial market statisfies the* **no-arbitrage condition**.

Since nobody knows exactly what's going on in reality, arbitrage-freeness is a property of a mathematical model for a financial market. Such a model is given by $(\Omega, \mathcal{F})$, the price process S_t and the probability P.

The no-arbitrage condition can also be expressed in terms of the d-dimensional gains process.

Lemma 2.3.4 *A financial market is arbitrage-free if and only if for any portfolio* $\varphi = (\varphi_1, \ldots, \varphi_d)'$ *in the stocks the following implication holds true*

$$\varphi' G_1^* \geq 0 \text{ a.s.} \quad \Rightarrow \quad \varphi' G_1^* = 0 \text{ a.s.}$$

Proof.

'$\Rightarrow$': Suppose the market is arbitrage-free and choose $(\varphi_1, \ldots, \varphi_d) \in \mathbb{R}^d$ with $\varphi' G_1^* \geq 0$, a.s. We shall contradict the claim that $\varphi' G_1^* > 0$ with positive probability.

To do so, we show that in this case we can find some $\widetilde{\varphi}_0 \in \mathbb{R}$, such that $\widetilde{\varphi} = (\widetilde{\varphi}_0, \varphi_1, \ldots, \varphi_d)'$ represents an arbitrage portfolio, i.e.

$$\widetilde{\varphi}' S_0 \leq 0, \quad \widetilde{\varphi}' S_1 \geq 0 \text{ a.s.}, \quad P(\widetilde{\varphi}' S_1 > 0) > 0.$$

By assumption, we have with positive probability

$$\varphi' G_1^* = \sum_{i=1}^{d} \varphi_i (S_{1i}^* - S_{0i}) = \sum_{i=1}^{d} \varphi_i S_{1i}^* - \sum_{i=1}^{d} \varphi_i S_{0i} > 0$$

$$\overset{(1+r)}{\Rightarrow} \quad \sum_{i=1}^{d} \varphi_i S_{1i} - \sum_{i=1}^{d} \varphi_i S_{0i}(1+r) > 0.$$

Put $\widetilde{\varphi}_0 = -\sum_{i=1}^{d} \varphi_i S_{0i}$, i.e. finance the portfolio $\varphi_1, \ldots, \varphi_d$ by a credit, and let $\widetilde{\varphi} = (\widetilde{\varphi}_0, \varphi_1, \ldots, \varphi_d)$. Then

$$\varphi' S_0 = 0$$

and

$$\varphi' S_1 = \underbrace{-\sum_{i=1}^{d} \varphi_i S_{0i}(1+r)}_{\text{value of the credit}} + \sum_{i=1}^{d} \varphi_i S_{1i} > 0$$

with positive probability. Thus, $\widetilde{\varphi}$ is an arbitrage portfolio which yields the contradiction.

'$\Leftarrow$': Assume that the implication

$$\widetilde{\varphi}' G_1^* \geq 0 \quad \text{a.s.} \quad \Rightarrow \quad \widetilde{\varphi}' G_1^* = 0 \quad \text{a.s.}, \tag{2.2}$$

holds true for any portfolio $\widetilde{\varphi} = (\widetilde{\varphi}_1, \ldots, \widetilde{\varphi}_d)$ in the stocks. Let $\varphi = (\varphi_0, \ldots, \varphi_d)$ be an arbitrage portfolio, i.e.

$$\varphi' S_0 \leq 0, \quad \varphi' S_1 \geq 0 \quad \text{a.s.}, \quad P(\varphi' S_1 > 0) > 0. \tag{2.3}$$

We obtain

$$\begin{aligned} \varphi' G_1^* &= \sum_{i=1}^{d} \varphi_i S_{1i}^* - \sum_{i=1}^{d} \varphi_i S_{0i} \\ &= \sum_{i=0}^{d} \varphi_i S_{1i}^* - \underbrace{\sum_{i=0}^{d} \varphi_i S_{0i}}_{\leq 0 \text{ (a.s.)}} \\ &\geq \sum_{i=0}^{d} \varphi_i S_{1i} \cdot \frac{1}{1+r} = A. \end{aligned}$$

Since $A = \varphi' S_1/(1+r)$, Equation (2.3) yields $A \geq 0$ a.s. and $P(A > 0) > 0$. This contradicts (2.2).

2.4 Separation theorems in $\mathbb{R}^n$

In this section, we discuss some important separation theorems that will be used in the following. Such theorems assert that one can identify certain subsets by a linear function in the sense that for each x the value of that linear function determines whether or not the point belongs to the subset. We shall need a special version, but let us start with the following basic one.

Theorem 2.4.1 *Assume that $K \subset \mathbb{R}^n$ is a closed convex set with $0 \notin K$. Then there exists some $\lambda \in \mathbb{R}^n$ and a real number $a > 0$, such that*

$$\lambda' x \geq a, \quad \textit{for all } x \in K.$$

Here $a = \text{dist}(0, K)^2$.

Proof. Select $r > 0$ such that the closed ball with center 0 and radius r, i.e.

$$\overline{B(0, r)} = \{y \in \mathbb{R}^n : \|y\| \leq r\},$$

hits K. Put $M = K \cap \overline{B(0, r)} \neq \emptyset$ and note that M is compact. Therefore, the continuous mapping

$$x \mapsto \|x\|, \quad x \in \mathbb{R}^n,$$

attains its minimum on M. Denote the unique minimum by $x_0 \in M$. It follows that

$$\|x\| \geq \|x_0\| \quad \text{for all } x \in K. \tag{2.4}$$

Since K is convex, we have for all $x \in K$ and $\alpha \in [0, 1]$

$$\alpha x + (1 - \alpha)x_0 = x_0 + \alpha(x - x_0) \in K.$$

Hence, we can apply Equation (2.4) with $x = x_0 + \alpha(x - x_0)$ and obtain

$$\|x_0 + \alpha(x - x_0)\| \geq \|x_0\| \quad \Leftrightarrow \quad (x_0 + \alpha(x - x_0))'(x_0 + \alpha(x - x_0)) \geq x_0' x_0.$$

The last statement is equivalent to

$$2\alpha(x - x_0)' x_0 + \alpha^2 (x - x_0)'(x - x_0) \geq 0, \quad \alpha \in [0, 1].$$

For any $\alpha \to 0$ we obtain $(x - x_0)' x_0 \geq 0$. Thus, if we put $\lambda := x_0$, then

$$\lambda' x \geq a := \lambda' \lambda, \quad \text{for all } x \in K.$$

As a preparation we note the following simple fact.

Lemma 2.4.2 *Let $\{\lambda, \lambda_k\} \subset \mathbb{R}^n$ and $\{x, x_k\} \subset \mathbb{R}^n$ be two sequences with $\lambda_k \to \lambda$, $k \to \infty$, and $x_k \to x$, $k \to \infty$. Then*

$$\lambda_k' x_k \to \lambda' x, \quad k \to \infty.$$

Proof. Clearly, $\lambda_k \to \lambda$, implies $\|\lambda_k\| \to \|\lambda\|$, as $k \to \infty$. Therefore, $\|\lambda_k\| \leq C$ for some constant $C > 0$. The decomposition $\lambda_k' x_k - \lambda' x = \lambda_k'(x_k - x) + (\lambda_k - \lambda)' x$ yields

$$\|\lambda_k' x_k - \lambda' x\| \leq \|\lambda_k\| \|x_k - x\| + \|\lambda_k - \lambda\| \|x\|.$$

Now the result follows easily.

Theorem 2.4.3 *Let $K \subset \mathbb{R}^n$ be a nonempty and convex set with $0 \notin K$; 0 may be a boundary point. Then there is some $\lambda \in \mathbb{R}^n$, such that*

$$\lambda' x \geq 0, \quad \textit{for all } x \in K$$

with strict inequality for at least one $x \in K$.

Proof. In the case $m = \inf_{x \in K} \|x\| > 0$ the infimum m is attained for some point $0 \neq x_0 \in \overline{K}$. Since $\overline{K}$ is closed and convex, we have for any $\alpha \in [0, 1]$

$$\|x_0 + \alpha(x - x_0)\| \geq \|x_0\|, \quad \text{for all } x \in K \subset \overline{K},$$

and $\alpha \to 0$ leads to $\lambda' x \geq \|\lambda\|^2 > 0$, if we put $\lambda := x_0$.

Now assume that 0 is a boundary point of K. The basic idea is to shift the set K a little and to apply the above separation theorem. For the reader's convenience, we provide the details, which are a little subtle.

We shall first show that $\overline{K}$ is a proper subset of $\mathbb{R}^n$, i.e. $\overline{K} \neq \mathbb{R}^n$, by constructing some $y \in \mathbb{R}^n \backslash \overline{K}$. Let $\{x_1, \ldots, x_m\}$ be a maximal set of linear independent vectors of K that span K. That is, any vector of K can be represented as a linear combination of $x_1, \ldots, x_m$, although not all linear combinations may lie in K. Consider the vector

$$y = -(x_1 + \cdots + x_m).$$

Suppose $y \in \overline{K}$. Then there is some sequence $\{y_k\} \subset K$ with $y_k \to y$. Each y_k can be represented as a linear combination of $x_1, \ldots, x_m$. This means that there are real coefficients λ_{ki}, such that

$$y_k = \sum_{i=1}^m \lambda_{ki} x_i.$$

Now it follows that $\lambda_{ki} \to -1$, $k \to \infty$, since $y_k \to y$. Consequently, there is some index $k_0 \in \mathbb{N}$ with $\lambda_{k_0,i} < 0$, $i = 1, \ldots, m$. We may conclude that

$$y_{k_0} + \sum_{i=1}^m (-\lambda_{k_0,i}) x_i = 0.$$

Observe that the coefficients $1, -\lambda_{k_0,1}, \ldots, -\lambda_{k_0,m}$ are positive, but they do not necessarily sum up to 1. If we divide them by their sum, $1 - \sum_{i=1}^m \lambda_{k_0,i}$, we obtain a convex combination of $y_{k_0}, x_1, \ldots, x_m \in K$, which equals 0. This contradicts the assumption $0 \notin K$. Thus, $\overline{K} \neq \mathbb{R}^n$ follows.

Since 0 is a boundary point of K, we know that $\inf_{x \in K} \|x\| = 0$.

We may choose some sequence $\{z_k\} \subset \mathbb{R}^n$ satisfying

$$z_k \to 0, \ k \to \infty, \quad \text{and} \quad \inf_{x \in K} \|x - z_k\| > 0. \tag{2.5}$$

To construct that sequence, let x be an interior point of K and denote by $T(x) = -x$ the reflection at the origin. Then $y = T(x) \notin K$, since otherwise the line segment connecting x and y lies in K, which would imply $0 \in K$. Now choose $\varepsilon > 0$ such that $B(x, \varepsilon) \subset K$. By continuity of T, the set

$$V = T(B(x, \varepsilon)) = \{T(z) : z \in B(x, \varepsilon)\}$$

is an open neighborhood of $y = -x$ with $V \cap K = \emptyset$, if we eventually reduce ε. Thus, y is not a boundary point. We claim that all points u of the line segment connecting y and 0 satisfy $\inf_{z \in K} \|u - z\| > 0$. Indeed, if u is a point with $\inf_{z \in K} \|u - z\| = 0$, we can find some $\widetilde{u} \in K$ in a vicinity of u, such that the line passing through $\widetilde{u}$ and the origin cuts the set $B(x, \varepsilon)$. But this implies $0 \in K$. Having constructed the point y, the sequence $z_k = y/k$, $k \in \mathbb{N}$, satisfies Equation (2.5). Now the set $K_k = K - z_k$ satisfies $\inf_{x \in K_k} \|x\| = \inf_{x \in K} \|x - z_k\| > 0$ by construction. Due to Theorem 2.4.1, there is some $\lambda_k \in \mathbb{R}^n$ and some real $a_k > 0$ with

$$\lambda_k' x \geq a_k, \qquad \text{for all } x \in K_k.$$

Without loss of generality (w.l.o.g.) we can assume $\|\lambda_k\| = 1$. The sequence $\{\lambda_k\}$ lies in the compact set $S^{n-1} = \{x \in \mathbb{R}^n : \|x\| = 1\}$. Therefore, there is some convergent subsequence $\{\lambda_{k_l} : l \in \mathbb{N}\} \subset \{\lambda_k : k \in \mathbb{N}\}$ with limit λ. Since $z_l \to 0$, we may apply Lemma 2.4.2 to obtain

$$\lambda' x = (\lim_{l \to \infty} \lambda_{k_l})'(\lim_{l \to \infty} (x - z_{k_l})) \geq 0$$

for all $x \in K$.

It remains to show that λ is not perpendicular to K. Assume the contrary, i.e. $\lambda' x = 0$ for all $x \in K$. Then $K \subset U = \{y \in \mathbb{R}^n : y \perp \lambda\} \subsetneq \mathbb{R}^n$ contradicting our assumption that K is not contained in some linear subspace of $\mathbb{R}^n$. Therefore, we can find some $x_0 \in K$ with $\lambda' x_0 > 0$, which completes the proof.

It is known from linear algebra that for any linear subspace $U \subset \mathbb{R}^n$ one can find some $\lambda \in \mathbb{R}^n$ ensuring $\lambda' x = 0$ for all $x \in U$. The next theorem asserts that λ can be selected such that the corresponding linear functional $\lambda(x) = \lambda' x$ is positive on a given compact set K, provided that $U \cap K = \emptyset$. Recall the notation

$$A - B = \{x \in \mathbb{R}^n : x = a - b \text{ with } a \in A \text{ and } b \in B\},$$

and the following simple lemma.

Lemma 2.4.4 *The following assertions hold.*

(i) If A and B are convex, $A - B$ is convex.

(ii) If A is compact and B is closed, then $A - B$ is closed.

Proof. Suppose $x_n = a_n - b_n \to x$, $n \to \infty$, for $\{a_n\} \subset A$ and $\{b_n\} \subset B$. We have to show that $x \in A - B$, i.e. $x = a - b$ with $a \in A$ and $b \in B$. Since A is compact, i.e. closed and bounded, $a_{n_k} \to a$, $k \to \infty$, for a subsequence $\{a_{n_k} : k \in \mathbb{N}\}$ and some $a \in A$. This implies that $b_{n_k} = a_{n_k} - x_{n_k}$ converges to $a - x$, as $k \to \infty$. Since B is closed, the limit $b := a - x$ lies in B. But then $x = a - b \in A - B$.

Theorem 2.4.5 *Let $U \subset \mathbb{R}^n$ be a linear subspace and $\emptyset \neq K \subset \mathbb{R}^n$ a compact and convex set with $U \cap K = \emptyset$. Then there is some $\lambda \in \mathbb{R}^n$, such that*

$$\lambda' x = 0, \quad \textit{for all } x \in U,$$

and

$$\lambda' x > 0, \quad \textit{for all } x \in K.$$

Proof. $K - U$ is convex and closed. Further, $K \cap U = \emptyset$ implies $0 \notin K - U$. By Theorem 2.4.1, there is some $\lambda \in \mathbb{R}^n$ and some real $a > 0$ with

$$\lambda' x \geq a, \quad \text{for all } x \in K - U.$$

We may conclude that

$$(*) \quad \lambda'(y - x) = \lambda' y - \lambda' x \geq a, \quad \text{for all } y \in K \text{ and } x \in U.$$

Suppose $\lambda' x \neq 0$ for some $x \in U$. Since $\lambda'(kx) = k \cdot \lambda' x$ for all $k \in \mathbb{Z}$, we can find some integer k with $\lambda' y - \lambda'(kx) < a$, which contradicts $(*)$. Therefore, $\lambda' x = 0$ for all $x \in U$ yielding $\lambda' y \geq a > 0$ for all $y \in K$.

2.5 No-arbitrage and martingale measures

Recall that two probability measures P and Q defined on some measurable space $(\Omega, \mathcal{F})$ are called equivalent, denoted by $P \sim Q$, if their null sets agree, i.e. if

$$P(A) = 0 \quad \Leftrightarrow \quad Q(A) = 0$$

holds for each $A \in \mathcal{F}$. In this case, the Radon–Nikodym derivative, $\mathrm{d}P/\mathrm{d}Q$, of P w.r.t. Q exists and is strictly positive.

Definition 2.5.1 (Equivalent Martingale Measure)
Consider a financial market given by a probability space $(\Omega, \mathcal{F}, P)$, a riskless interest rate r and a risky asset $\{S_{t1} : t = 1, 2\}$.

(i) A probability measure P^* on $\mathcal{F}$ is called a **martingale measure**, *if*

$$S_{0i} = E^*(S_{1i}^*) = E^*\left(\frac{S_{1i}}{1+r}\right), \quad i = 1, \ldots, d.$$

Here and in the following E^* denotes the expectation w.r.t. to P^*.

(ii) A probability measure P^* on $(\Omega, \mathcal{F})$ is called an **equivalent martingale measure (w.r.t.** *P), if P^* is a martingale measure with $P^* \sim P$.*

(iii) Denote by

$$\mathcal{P} = \{P^* : P^* \text{ is an equivalent martingale measure}\}$$

the set of all equivalent martingale measures (w.r.t. P).

Notation 2.5.2 *In the following, we will use the notation*

$$E^*(X) = E_{P^*}(X) = \int_\Omega X \, \mathrm{d}P^*.$$

Remark 2.5.3

(i) Since $S^*_{0i} = S_{0i}$, the above condition is equivalent to

$$S^*_{0i} = E^*(S^*_{1i}), \qquad i = 1, \ldots, d.$$

This means, the d discounted price processes $\{S^*_{ti} : t = 0, 1\}$ are so-called martingales *under the measure P^*. We will study martingales in greater detail in the next chapter as they play a crucial role in mathematical finance.*

(ii) Notice that the notion of an equivalent martingale measure, and therefore the set $\mathcal{P}$, depends on the true probability measure P.

(iii) P^* is a martingale measure, if and only if the expected discounted profit under P^* when buying the share i is 0,

$$E^*(G^*_{1i}) = E^*(S^*_{1i} - S_{0i}) = 0, \quad i = 1, \ldots, d.$$

The notion of a martingale measure is crucial in mathematical finance. Recall from Chapter 1 that evaluating a random payment X by an expectation ignores the risk associated with X. Thus, it is crucial to distinguish the martingale measure P^* from the true probability measure P. Generally, the expectations $E_P(S_{1i}/(1+r))$ differ from the market prices. But a martingale measure belongs to the risk-neutral world where risk is ignored but discounted random payments are correctely priced by taking expectations unter P^*. This means, P^* attaches probabilities to the market scenarios $\omega \in \Omega$ and events $A \subset \Omega$, respectively, in such a way that the evaluation of a random payment can be made *under the assumption* that we live in a risk-neutral world.

Our next goal is one of the main results of option pricing, namely that the no-arbitrage condition is equivalent to the existence of an equivalent martingale measure P^*, i.e. $\mathcal{P} \neq \emptyset$. Since the proof for an arbitrary probability space is rather involved, we first consider the case of a finite probability space.

Thus, to this end we assume that

$$\Omega = \{\omega_1, \ldots, \omega_n\} \text{ for some } n \in \mathbb{N},$$
$$\mathcal{F} = \text{Pot}(\Omega),$$
$$P(\{\omega\}) > 0 \quad \forall\, \omega \in \Omega.$$

Notice that each random variable $Z\colon \Omega \to \mathbb{R}$ can be mapped to the vector $\big(Z(\omega_1), \ldots, Z(\omega_n)\big)' \in \mathbb{R}^n$. Vice versa, the mapping

$$\mathbb{R}^n \ni (y_1, \ldots, y_n)' \mapsto Z,\ Z(\omega_i) = y_i,\ i = 1, \ldots, n,$$

defines a random variable. Clearly, this mapping is 1-1, and thus from now on we shall identify random variables with vectors in this way and make use of matrix calculus. Recall the definition of the $((d+1) \times n)$-dimensional payment matrix

$$\mathbf{D} = \begin{bmatrix} S_{10}(\omega_1), \ldots, S_{10}(\omega_n) \\ \vdots \qquad\qquad \vdots \\ S_{1d}(\omega_1), \ldots, S_{1d}(\omega_n) \end{bmatrix}.$$

The ith row corresponds to the ith investment object, and the jth colum to the realizations under the possible market scenarios. We have the correspondence

$$\varphi' S_1 = \mathbf{D}'\varphi,$$

where, as explained above, $\varphi' S_1$ is identified with the vector $(\varphi' S_1(\omega_1), \ldots, \varphi' S_1(\omega_n))'$.

Theorem 2.5.4 (FUNDAMENTAL THEOREM OF ASSET PRICING, FINITE Ω)
Let $\Omega = \{\omega_1, \ldots, \omega_n\}$. If the financial market is arbitrage-free, then there exists a vector $\psi \in \mathbb{R}^n$ with positive entries ψ_j, such that

$$S_0 = \mathbf{D}\psi.$$

An equivalent martingale measure P^* on $\mathcal{F}$ is given by

$$P^*(\{\omega_i\}) = \frac{\psi_i}{\sum_{j=1}^n \psi_j}, \qquad i = 1, \ldots, n,$$

i.e. $E^*(S_1^*) = S_0$ holds true, and $\sum_{j=1}^n \psi_j = (1+r)^{-1}$ represents the discount factor.

Proof. Consider the set

$$U = \left\{(-\varphi' S_0, (\varphi' S_1)')' : \varphi \in \mathbb{R}^{d+1}\right\} \subset \mathbb{R} \times \mathbb{R}^n = \mathbb{R}^{n+1},$$

and notice that U is a linear subspace of $\mathbb{R}^{n+1}$. Let

$$M = \left\{(y_0, \ldots, y_n)' \in \mathbb{R}^{n+1} : y_i \geq 0,\ i = 0, \ldots, n;\ y_j > 0 \text{ for some } j = 1, \ldots, n\right\}.$$

Clearly, $0 \notin M$, and we have

$$\text{The market is arbitrage-free} \quad \Leftrightarrow \quad U \cap M = \emptyset.$$

Next consider

$$K = \left\{(y_0, \ldots, y_n)' \in M : \sum_{i=0}^n y_i = 1\right\}.$$

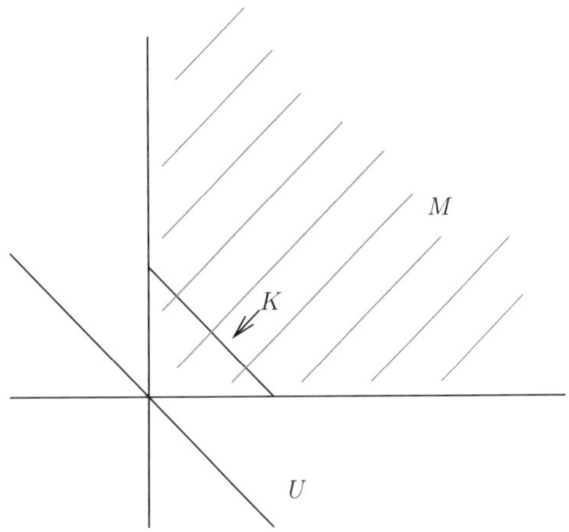

Figure 2.1 U is the linear subspace of payoff profiles $\varphi' S_1$ attainable by a portfolio φ, M corresponds to the arbitrage opportunities provided $\varphi' S_0 \le 0$, and K is a compact subset of M, which can be separated from U by virtue of Theorem 2.4.5.

K is a nonempty, convex and compact subset of M with $U \cap K = \emptyset$.
By Theorem 2.4.5, we can separate K and U, i.e.

$$\exists \ \lambda = (\lambda_0, \lambda_1, \ldots, \lambda_n) \in \mathbb{R}^{n+1},$$

such that

$$\begin{aligned} \lambda' x = 0, &\quad \forall\, x \in U, \\ \lambda' x > 0, &\quad \forall\, x \in K. \end{aligned}$$

See Figure 2.1 for an illustration.

Since $e_j = (0, \ldots, 0, 1, 0, \ldots, 0)' \in K$, we have $\lambda_j = \lambda' e_j > 0$, $j = 0, \ldots, n$. Thus, if we recall the definition of U and put $\widetilde{\lambda} = (\lambda_1, \ldots, \lambda_n)'$:

$$\begin{aligned} & -\lambda_0 \varphi' S_0 + \widetilde{\lambda}' \underbrace{\varphi' S_1}_{=\mathbf{D}'\varphi} &= 0 \qquad & \forall\, \varphi \in \mathbb{R}^{d+1} \\ \Leftrightarrow \quad & -\lambda_0 S_0' \varphi + (\mathbf{D}\widetilde{\lambda})' \varphi &= 0 \qquad & \forall\, \varphi \in \mathbb{R}^{d+1} \\ \Leftrightarrow \quad & (\mathbf{D}\widetilde{\lambda})' \varphi &= (\lambda_0 S_0)' \varphi \quad & \forall\, \varphi \in \mathbb{R}^{d+1}. \end{aligned}$$

That means

$$\mathbf{D}\widetilde{\lambda} = \lambda_0 S_0 \quad \Leftrightarrow \quad S_0 = \mathbf{D} \left(\frac{\widetilde{\lambda}}{\lambda_0} \right).$$

Now define

$$\psi := (\psi_1, \ldots, \psi_n)' := \frac{\widetilde{\lambda}}{\lambda_0} = \left(\frac{\lambda_1}{\lambda_0}, \ldots, \frac{\lambda_n}{\lambda_0} \right)' \in \mathbb{R}^n.$$

Clearly,

$$p_j^* := P^*(\{\omega_j\}) := \frac{\psi_j}{\sum_{k=1}^n \psi_k}, \quad j = 1, \ldots, n,$$

defines a probability measure P^* on $(\Omega, \mathcal{F})$ with $P^* \sim P$. Indeed, $\lambda_j > 0$ for all j implies $\psi_j > 0$ for all j.

Now choose $\overline{\varphi} \in \mathbb{R}^{d+1}$ with

$$\mathbf{D}'\overline{\varphi} = \mathbf{1} \cdot (1 + r)$$

and notice that $\overline{\varphi}$ replicates the bank account. Thus, the initial price of $\overline{\varphi}$ is 1, since otherwise we obtain an arbitrage opportunity. We obtain

$$1 = \overline{\varphi}' S_0 = \overline{\varphi}'(\mathbf{D}\psi) = \psi'\mathbf{D}'\overline{\varphi} = \psi'\mathbf{1} \cdot (1 + r) = (1 + r) \sum_{j=1}^n \psi_j.$$

Thus, $\sum_{j=1}^n \psi_j = \frac{1}{1+r}$ is the discounting factor in our model. Finally, using again the fact that $S_0 = \mathbf{D}\psi$ we obtain for $i = 1, \ldots, d$

$$\begin{aligned} E^*(S_{1i}^*) &= E^*\left(\frac{S_{1i}}{1+r}\right) \\ &= \frac{1}{1+r} \sum_{j=1}^n S_{1i}(\omega_j) p_j^* \\ &= \frac{1}{1+r} \sum_{j=1}^n D_{ij}\psi_j \Big/ \sum_{k=1}^n \psi_k \\ &= \sum_{j=1}^n D_{ij}\psi_j = (\mathbf{D}\psi)_i = S_{0i}\,. \end{aligned}$$

Consequently, $E^*(S_1^*) = S_0$, which completes the proof.

Our next goal is to extent the result to general probability spaces. As a preparation, let us recall the following simple fact.

Lemma 2.5.5 *If $E(X\mathbf{1}_{\{X<0\}}) \geq 0$, then $X \geq 0$ a.s.*

Proof. Since $X(\omega)\mathbf{1}_{\{X(\omega)<0\}} \leq 0$ for all $\omega \in \Omega$, we obtain $\int X\mathbf{1}_{\{X<0\}} \mathrm{d}P \leq 0$ by the monoticity of the measure integral. Hence, $\int X\mathbf{1}_{\{X<0\}} \mathrm{d}P = 0$. But this is equivalent to $X\mathbf{1}_{\{X<0\}} = 0$ $P-$a.s. such that $N = \{X\mathbf{1}_{\{X<0\}} \neq 0\}$ and therefore $A = \{X < 0\} \subset N$ are P-null sets. Consequently, $P(X \geq 0) = 1$.

Theorem 2.5.6 (FUNDAMENTAL THEOREM OF ASSET PRICING, GENERAL Ω) *The following assertions hold.*

(i) *If $\mathcal{P} \neq \emptyset$, the financial market is arbitrage-free.*

(ii) *If the market is arbitrage-free, then there is some equivalent martingale measure $P^* \in \mathcal{P}$, such that $\mathrm{d}P^*/\mathrm{d}P$ is bounded.*

Proof.

(i) Suppose there is some $P^* \in \mathcal{P}$. We will contradict the following
Claim: There is some arbitrage opportunity φ.
Let $\varphi \in \mathbb{R}^{d+1}$ be an arbitrary portfolio, such that $\varphi' S_1 \geq 0$ P-a.s. and $E(\varphi' S_1) > 0$. By equivalence of P and P^*, we have $E^*(\varphi' S_1) > 0$. We may conclude that

$$\sum_{i=0}^{d} \varphi_i E^* \left(\frac{S_{1i}}{1+r} \right) = E^* \left(\frac{\varphi' S_1}{1+r} \right) > 0.$$

Since P^* is a martingale measure, $S_0 = E^*(S_1/(1+r))$, which leads to positive costs,

$$\varphi' S_0 = \sum_{i=0}^{d} \varphi_i S_{0i} = \sum_{i=0}^{d} \varphi_i E^* \left(\frac{S_{1i}}{1+r} \right) > 0.$$

This means, φ cannot be an arbitrage portfolio, which is a contradiction.

(ii) Recall the definition of the d-dimensional discounted gains process of the d risky assets:

$$G_1^* = (G_{11}^*, \ldots, G_{1d}^*), \quad G_{1i}^* = S_{1i}^* - S_{0i}, \quad i = 1, \ldots, d.$$

By Lemma 2.3.4, the no-arbitrage condition is equivalent to

$$(*) \quad \forall \varphi \in \mathbb{R}^d : \sum_{i=1}^{d} \varphi_i G_{1i}^* \geq 0 \ P\text{-a.s.} \Rightarrow \sum_{i=1}^{d} \varphi_i G_{1i}^* = 0 \ P\text{-a.s.}$$

We will show that this implies the existence of an equivalent martingale measure $P^* \sim P$. Recall that $P^* \sim P$ is a martingale measure, if and only if

$$E^*(S_1^*) = S_0 \quad \Leftrightarrow \quad E^*(G_1) = 0.$$

For brevity, we shall write $G = G_1^*$ for the discounted prices at time 1 in the rest of the proof. That notation will not interfere with standard notation, since we have only to deal with the discounted prices in what follows. Let us first consider the case that G is P-integrable, i.e. $E|G| < \infty$. Define

$$\mathcal{K} = \{Q : \ Q \text{ is a probability measure with } Q \sim P \text{ and } \mathrm{d}Q/\mathrm{d}P \text{ bounded}\}.$$

Obviously, $\mathcal{K} \neq \emptyset$, since $P \in \mathcal{K}$. Now, for all $Q \in \mathcal{K}$:

$$E_Q|G| = \int |G(\omega)| \frac{\mathrm{d}Q}{\mathrm{d}P}(\omega) \, \mathrm{d}P(\omega) \leq C E_P|G| < \infty,$$

if $\mathrm{d}Q/\mathrm{d}P \le C < \infty$. Thus, G is integrable under all measures $Q \in \mathcal{K}$, too. It is straightforward to verify that $\mathcal{K}$ is a convex set. Let us now consider the corresponding set of expectations of G under $Q \in \mathcal{K}$, i.e.

$$\mathcal{E} = \{E_Q(G) : Q \in \mathcal{K}\} \subset \mathbb{R}^d.$$

There exists an equivalent martingale measure P^*, if $0 \in \mathcal{E}$. Indeed, in this case any $P^* \in \mathcal{K}$ with $E^*(G) = 0 \in \mathcal{E}$ does the job. Thus, we have to show $0 \in \mathcal{E}$. Let us contradict the

Claim: $0 \notin \mathcal{E}$

First, notice that $\mathcal{E}$ is convex. Indeed, if $x_1, x_2 \in \mathcal{E}$ and $\alpha \in [0, 1]$, then there exist $Q_1, Q_2 \in \mathcal{K}$ with $x_i = E_{Q_i}(G),\ i = 1, 2$. Then,

$$\begin{aligned}\alpha x_1 + (1-\alpha)x_2 &= \alpha \int G\,\mathrm{d}Q_1 + (1-\alpha)\int G\,\mathrm{d}Q_2 \\ &= \int G\,\mathrm{d}[\alpha Q_1 + (1-\alpha)Q_2] = \int G\,\mathrm{d}\mu \in \mathcal{E},\end{aligned}$$

since $\mu = \alpha Q_1 + (1-\alpha)Q_2 \in \mathcal{K}$. Further, $\mathcal{E} \neq \emptyset$, since $P \in \mathcal{K}$.

An application of Theorem 2.4.3 yields a vector $\varphi \in \mathbb{R}^d$ that separates $\mathcal{E}$ and $\{0\}$. That means,

$$\varphi' x \ge 0 \quad \text{for all } x \in \mathcal{E}$$

and

$$\varphi' x_0 > 0 \quad \text{for some } x_0 \in \mathcal{E}.$$

Since $x = E_Q(G) \in \mathcal{E}$ for any $Q \in \mathcal{K}$,

$$E_Q(\varphi' G) \ge 0, \qquad \text{for all } Q \in \mathcal{K} \tag{2.6}$$

and

$$E_{Q_0}(\varphi' G) > 0 \quad \text{for some } Q_0 \in \mathcal{K}.$$

Since $Q_0 \sim P$, $E_{Q_0}(\varphi' G) > 0$ implies that $P(\varphi' G > 0) > 0$. We will show that $\varphi' G \ge 0$ P-a.s., too. Then $(*)$ implies that $\varphi' G = 0$ P-a.s., a contradiction to $P(\varphi' G > 0) > 0$.

It remains to show $\varphi' G \ge 0$ P-a.s. To do so, we will use Lemma 2.5.5. In order to show $E_P(\varphi' G \mathbf{1}_{\{\varphi' G < 0\}}) \ge 0$ one could try to interpret $\mathbf{1}_{\{\varphi' G < 0\}}$ as a bounded P-density of a new measure Q. Then $Q \in \mathcal{K}$ and Equation (2.6) would give $E_Q(\varphi' G) \ge 0$. However, for that purpose Q has to be a probability, i.e. we have to divide by $P(\varphi' G < 0)$, which is not possible.

The trick is now to approximate the indicator $\mathbf{1}_{\{\varphi'G<0\}}$ by a sequence of functions, whose supports have nonempty intersection with the set $\{\varphi'G \geq 0\}$. Thus, define

$$f_n = (1 - 1/n)\mathbf{1}_{\{\varphi'G<0\}} + (1/n)\mathbf{1}_{\{\varphi'G\geq 0\}}, \qquad n \in \mathbb{N}.$$

The sequence $\{f_n\}$ has the following properties; the proofs being left to the reader as an exercise.

(i) $0 < f_n \leq 1$, for $n \geq 2$.

(ii) $f_n \to \mathbf{1}_{\{\varphi'G<0\}}$, $n \to \infty$, ω-pointwise.

(iii) $\int f_n \, \mathrm{d}P = (1 - 1/n)P(\varphi'G < 0) + (1/n)P(\varphi'G \geq 0) > 0$ for all $n \in \mathbb{N}$.

(iv) $\omega \mapsto \widetilde{f}_n(\omega) = f_n(\omega)/\int f_n \mathrm{d}P$, $\omega \in \Omega$, defines a bounded P-density.

Let $Q_n = \widetilde{f}_n \mathrm{d}P$ be the corresponding probability measures, i.e.

$$Q_n(A) = \int_A \widetilde{f}_n \, \mathrm{d}P, \quad A \in \mathcal{F}.$$

Now, property (i) yields $Q_n \in \mathcal{K}$, $n \geq 2$, and due to Equation (2.6) we can conclude that

$$0 \leq E_{Q_n}(\varphi'G) = E_P(\varphi'Gf_n)/\int f_n \mathrm{d}P$$

Applying the theorem of dominated convergence to the numerator of the ratio on the right-hand side leads to

$$E_P(\varphi'G\mathbf{1}_{\{\varphi'G<0\}}) = \lim_{n\to\infty} E_P(\varphi'Gf_n) \geq 0.$$

Now, $\varphi'G \geq 0$ P-a.s. by Lemma 2.5.5.

Finally, we consider the case $E_P|G| = \infty$. We have

$$0 < c = E\left(\frac{|G|}{|G|+1}\right) \leq 1.$$

Let P' be the probability measure with P-density

$$f(\omega) = \frac{\widetilde{c}}{1 + |G(\omega)|},$$

where $\widetilde{c}$ is chosen to ensure $P'(\Omega) = \int f(\omega)\,\mathrm{d}P(\omega) = 1$. Then $P' \sim P$ and

$$E_{P'}|G| = E_P\left(\frac{|G|}{|G|+1}\widetilde{c}\right) < \infty.$$

Thus, G is integrable under P', and we can conclude that there is some martingale measure P^* with $P^* \sim P'$ and bounded density. Clearly, $P^* \sim P$, too, and the P-density of P^* is also bounded. Indeed,

$$\frac{dP^*}{dP} = \frac{dP^*}{dP'} \cdot \frac{dP'}{dP} \le C$$

for some constant $C < \infty$.

Let us illustrate our main findings by two examples.

Example 2.5.7 *(Binomial model with one asset)*
Recall that for the simple binomial model the sample space can be chosen as $\Omega = \{+, -\}$, and the real probability measure P is given by $p_+ = P(\{+\}) \in (0, 1)$. The asset makes either an up movement from S_{01} to $S_{11} = uS_{01}$ occurring with probability p_+, or goes down to dS_{01} with probability $p_- = 1 - p_+$. There exists no arbitrage, if and only if $\mathcal{P} \neq \emptyset$. Clearly, each $P^* \in \mathcal{P}$ is determined by $p^* = P^*(\{+\})$ and $P^* \sim P$ is equivalent to $p^* \in (0, 1)$. In Section 1.5.2 we calculated the solution of the martingale equations,

$$E^*\left(\frac{S_{11}}{1+r}\right) = S_{01}$$

yielding

$$p^* = \frac{1+r-d}{u-d}, \quad 1 - p^* = \frac{u-(1+r)}{u-d}.$$

Now

$$0 < p^* < 1 \quad \Leftrightarrow \quad 0 < 1+r-d < u-d \quad \Leftrightarrow \quad d < 1+r < u.$$

Finally, let us calculate the Radon–Nikodym derivative of P^* w.r.t. P. We have

$$\frac{dP^*}{dP}(\{+\}) = \frac{P^*(\{+\})}{P(\{+\})} = \frac{p^*}{p} = \frac{1+r-d}{u-d} \cdot \frac{1}{p},$$

$$\frac{dP^*}{dP}(\{-\}) = \frac{1-p^*}{1-p} = \frac{u-(1+r)}{u-d} \cdot \frac{1}{1-p},$$

determining the mapping $\frac{dP^*}{dP} : \Omega \to \mathbb{R}$.

Example 2.5.8 (DISCRETE FINANCIAL MARKETS)
Let $\Omega = \{\omega_1, \ldots, \omega_n\}$ be a finite set of n scenarios for the financial market. We may assume $P(\{\omega_i\}) > 0$ for $i = 1, \ldots, n$, since otherwise we can reduce Ω. Each martingale measure is given by a probability vector $p^* = (p_1^*, \ldots, p_n^*)' \in \mathbb{R}^n$ with $p_1^*, \ldots, p_n^* \ge 0$ and $\sum_i p_i^* = 1$. It is an equivalent martingale measure, if and only if $p_i^* \in (0, 1)$ for all $i = 1, \ldots, n$. One has to solve the equations

$$S_{0i} = E^*\left(\frac{S_{1i}}{1+r}\right) = \sum_{j=1}^n p_j^* \frac{S_{1i}(\omega_j)}{1+r}, \quad i = 1, \ldots, d,$$

or, in matrix notation employing the payment matrix $\mathbf{D} = (S_{1i}(\omega_j))_{i=0,\dots,d,j=1,\dots,n}$,

$$(1+r)S_0 = \mathbf{D}p^*.$$

Thus, p^* *is a solution of the constrained linear system of equations,*

$$\mathbf{D}x = (1+r)S_0, \quad x > 0, \quad \mathbf{1}'x = 1,$$

consisting of d equations and n variables.

2.6 Arbitrage-free pricing of contingent claims

The present section collects basic facts on the arbitrage-free pricing of derivatives and, more generally, the larger class of so-called contingent claims. In particular, we show that the initial price of a portfolio that hedges a claim is unique if there exists an equivalent martingale measure.

Recall that any portfolio $\varphi \in \mathbb{R}^{d+1}$ generates a random payment $V(\varphi) = \varphi' S_1$ at time $t = 1$. Consider the set

$$\mathcal{V} = \{\varphi' S_1 : \varphi \in \mathbb{R}^{d+1}\}$$

of such random variables representing payments of portfolios. Clearly, $\mathcal{V}$ is a subset of all random variables $\Omega \to \mathbb{R}$ denoted in the following by $\mathcal{L}$. We have the following simple fact.

Lemma 2.6.1 $\mathcal{V} = \text{span}(S_{10}, \dots, S_{1d})$ *is a linear space with* $1 \in \mathcal{V}$.

By definition, for any payment $V \in \mathcal{V}$ there is some portfolio $\varphi \in \mathbb{R}^{d+1}$ generating that payment, i.e. $V = V(\varphi)$.

Definition 2.6.2 *Let* $X : \Omega \to \mathbb{R}$ *be a random variable. If there exists some* $\varphi \in \mathbb{R}^{d+1}$ *such that* $V(\varphi) = X$*, i.e.* $X \in \mathcal{V}$*, then* φ *is called a* **hedge**.

It makes sense to define the (fair) price of a payment $X \in \mathcal{V}$ as the initial capital required to initial the hedge.

Definition 2.6.3 *On an arbitrage-free market the price of a random payment* $X \in \mathcal{V}$ *is defined as*

$$\pi(X) = \varphi' S_0,$$

where $\varphi \in \mathbb{R}^{d+1}$ *is a generating portfolio (hedge), i.e.* $V(\varphi) = X$*. The associated linear mapping* $\pi : \mathcal{V} \to \mathbb{R}$ *is called the* **linear pricing rule**.

Since there may be many possible hedges, we have to check whether $\pi(X)$ is well defined. This holds true if we have an equivalent martingale measure at our disposal.

Theorem 2.6.4 *Suppose there exists an equivalent martingale measure . Then the fair prices of any two portfolios* $\varphi_1, \varphi_2 \in \mathbb{R}$ *generating a given* $X \in \mathcal{V}$ *coincide:* $\varphi_1' S_0 = \varphi_2' S_0$.

Proof. Obviously, $\delta = \varphi_1 - \varphi_2$ satisfies $V(\delta) = (\varphi_1 - \varphi_2)' S_1 = 0$. Thus, for any equivalent martingale measure $P^* \in \mathcal{P}$ we have $P^*(V(\delta) = 0) = 1$ implying $P(V(\delta) = 0) = 1$. It follows that

$$0 = E^*(V(\delta)) = E^*((\varphi_1 - \varphi_2)' S_1)$$

and thus

$$0 = E^*((\varphi_1 - \varphi_2)' S_1^*) = (\varphi_1 - \varphi_2)' E^*(S_1^*) = (\varphi_1 - \varphi_2)' S_0$$

as well.

It is now time to define rigorously what we mean by a derivative and contingent claim, respectively. Recall that a financial instrument is given by its value process. A classic portfolio given by $V_t = \varphi' S_t$ is a linear function of the assets.

Definition 2.6.5 *A random variable $X : \Omega \to \mathbb{R}$ with $0 \le X < \infty$ P-a.s, which is $\sigma(S_{11}, \ldots, S_{1d})$-measurable and thus of the form*

$$X = f(S_{11}, \ldots, S_{1d})$$

for some Borel-measurable function f, is called a **derivative** *or* **derivative asset**.

Remark 2.6.6 *For any asset X we may choose the function f minimal in the sense that $X = f(S_{1i_1}, \ldots, S_{1i_r})$ with indices $i_1, \ldots, i_r \in \{1, \ldots d\}$ such that r is minimal. Then the assets $S_{1i_1}, \ldots, S_{1i_r}$ are the underlyings.*

At this point it is important to note that the above definition of a derivative security does not cover all options, traded on a given financial market. Notice that a derivative X is $\mathcal{F}_1 = \sigma(S_{11}, \ldots, S_{1d})$ measurable but in general $\mathcal{F}_1 \subsetneq \mathcal{F}$ Then the case $\mathcal{F}_1 \ne \mathcal{F}$ gives rise to the following definition.

Definition 2.6.7 *Any (non-negative) random variable $X : (\Omega, \mathcal{F}) \to (\mathbb{R}, \mathcal{B})$, is called* **contingent claim**.

Example 2.6.8 (CLIMATE DERIVATIVES, ENERGY OPTIONS)
Since power consumption correlates with (average) temperature, there exists a market for so-called climate or weather derivatives, which are actively used by energy companies to hedge or reduce risks. Those contracts such as calls and puts, amongst others, are traded over-the-counter. Since many heatings switch off and, vice versa, air conditioning switch on at 18 °C, one considers the deviation of the average temperature from that threshold. The number of days where the average is larger than 18 °C is called heating degree days(HDD), and the number of days where on average the temperature is below 18 °C are the cooling degree days (CDD). Climate derivatives are therefore usually defined in terms of the HDD or CDD numbers.

In terms of L_p spaces, contingent claims are the elements of

$$L_0 = L_0(\Omega, \mathcal{F}, P) = \{X : (\Omega, \mathcal{F}, P) \to (\mathbb{R}, \mathcal{B}) : |X| < \infty, P - a.s.\},$$

the space of all random variables that are P-almost surely bounded. In what follows, we shall also consider the space

$$L_\infty = L_\infty(\Omega, \mathcal{F}, P) = \{X : (\Omega, \mathcal{F}, P) \to (\mathbb{R}, \mathcal{B}); \|X\|_\infty < \infty\}$$

of all uniformly bounded random variables. Here, the norm $\|X\|_\infty$ is defined by

$$\|X\|_\infty = \inf\{c > 0 : P(|X| > c) = 0\},$$

i.e. $\|X\|_\infty$ is the smallest upper bound that is not exceeded by $|X|$, almost surely. Obviously, we have the inclusion $L_\infty \subset L_0$. The following example shows that $L_0 \not\subset L_\infty$.

Example 2.6.9 *Let $\Omega = [0, 1]$ and $P = \lambda$ the Lebesgue measure. $\mathcal{F} = \mathcal{B}_{|[0,1]}$ denotes the Borel σ-field restricted to $[0, 1]$. Define $X(\omega) = 1/\omega, \quad \omega \in \Omega$, and use the convention $1/0 = \infty$. Since $\{X = \infty\} = \{0\}$ is a λ-null set, $P(|X| < \infty) = 1$. Thus, $X \in L_0([0, 1], \mathcal{B}_{|[0,1]}, \lambda)$. Clearly, for each $c > 0$ the set $\{|X| > c\} = [0, 1/c]$ has positive Lebesgue measure. Hence $\|X\|_\infty = \inf\{c > 0 : P(|X| > c)\} = \infty$, yielding $L_0 \not\subset L_\infty$.*

The basic idea to judge whether a price of a contingent claim is fair is to require that the price does not introduce arbitrage opportunities when the claim is traded as a new asset.

Definition 2.6.10 *$\pi(C)$ is called the fair price of C, if the extended market, given by the $d + 2$ price processes,*

$$\{S_{ti} : t = 0, 1\}, \quad i = 0, \ldots, d + 1,$$

with

$$S_{0,d+1} = \pi(C), \quad S_{1,d+1} = C,$$

is arbitrage-free. Let us agree to denote by $\Pi(C)$ the set of arbitrage-free prices.

Let C be a claim and $\pi(C)$ be an arbitrage-free price for C. Then the extended market is arbitrage-free, if $\pi(C)$ is the trading price. By definition, there exists some equivalent martingale measure P^* such that

$$E^*(S_{1i}^*) = S_{0i}, \quad i = 0, \ldots, d + 1.$$

Specifically, the pricing formula

$$\pi(C) = E^*(C^*)$$

follows. That means, on arbitrage-free markets we can calculate a fair (i.e. arbitrage-free) price of a claim, provided we have an equivalent martingale measure for the *extended market* at our disposal, by using the principle of risk-neutral pricing.

A simple arguments reveals that risk-neutral pricing using the equivalent martingale measures of the extended market yields all arbitrage-free prices for a claim.

Proposition 2.6.11 *Assume $\mathcal{P} \neq \emptyset$. Then $\Pi(C) = \{E^*(C^*) : P^* \in \mathcal{P}, E^*(C^*) < \infty\}$. Thus, a lower bound for the arbitrage is given by $\pi_-(C) = \inf\{E^*(C^*) : P^* \in \mathcal{P}\}$. Provided $E^*(C^*) < \infty$ for all $P^* \in \mathcal{P}$, an upper bound is $\pi_+(C) = \sup\{E^*(C^*) : P^* \in \mathcal{P}\}$.*

Proof. Let $\pi(C)$ be an arbitrage-free price on the extended market. Then there is an equivalent martingale measure for the extended market such that

$$\pi(C) = E^*(C^*) \text{ and } S_{0i} = E^*(S_{1i}^*), \ i = 0, \dots, d.$$

Thus, $\Pi(c) \subset \{E^*(C^*) : P^* \in \mathcal{P} \text{ with } E^*(C) < \infty\}$. By contrast, if $x = E^*(C^*)$ for some $P^* \in \mathcal{P}$, Clearly, $S_{0i} = E^*(S_{1i}^*)$ for $i = 0, \dots, d$. If we let $S_{0,d+1} = x$ and $S_{1,d+1} = C$, the extended market is arbitrage-free and P^* represents an equivalent martingale measure for that extended market, since $S_{0i} = E^*(S_{1i}^*)$, $i = 0, \dots, d+1$. C is replicable by the portfolio $\tilde{\varphi} = (0, \dots, 0, 1)'$ and, by virtue of Theorem 2.6.4, its fair price does not depend on $\tilde{\varphi}$, thus being given by $\tilde{\varphi}' S_0 = S_{0,d+1} = x$.

Of course, the fact that a claim C is replicable on the extended market is a triviality. If it suffices to invest into the assets of the original market, C is shown to be a derivative. Let us fix that point.

Definition 2.6.12 *A claim C is called* **attainable** *or* **replicable**, *if there exists some portfolio $\varphi = (\varphi_0, \dots, \varphi_d)' \in \mathbb{R}^{d+1}$ called a* **hedge**, *such that*

$$C = \varphi' S_1,$$

that is if $C \in \mathcal{V}$. In this case, C is a derivative.

The question arises whether arbitrage-free pricing is unique. Given what we have already learned, the answer has to be positive for attainable claims, since these are basically portfolios. It is a good exercise to repeat the above arguments to show the following proposition summarizing our findings for attainable claims.

Proposition 2.6.13 *Suppose the financial market is arbitrage-free. If C is an attainable contingent claim, the arbitrage-free price is unique and given by $\pi_C = \varphi' S_0$, where φ represents any replicating portfolio.*

Let us now study the more interesting and substantially more involved case of a claim that is not attainable. It turns out that now the arbitrage-free price is no longer unique. To show that result, we need two general results from functional analysis, whose proof goes beyond the scope of the present book. The first required result is the generalization of Theorem 2.4.1 to normed spaces.

Theorem 2.6.14 *Let $Y \subset X$ be a closed linear subspace of the normed linear space X and pick some $x_0 \in X \setminus Y$. Then there exists some linear functional $L : X \to \mathbb{R}$ with $\|L\| = 1$ satisfying*

$$L(y) = 0 \quad \textit{for all } y \in Y$$

and $L(x_0) = \text{dist}(x_0, Y)$.

The basic idea behind Theorem 2.6.14 is as follows. First, one defines the linear functional L on the subspace $Y_0 = Y \oplus \text{span}\{x_0\}$ by

$$L(y + t x_0) = t \cdot \text{dist}(x_0, Y), \ y \in Y, \ t \in \mathbb{R}.$$

Then $L(y) = 0$ for all $y \in Y$ and $L(x_0) = \text{dist}(x_0, Y)$. By virtue of the Hahn-Banach theorem, the linear mapping can be extended to X.

Let μ be a σ-finite measure. Then the $L_p = L_p(\mu)$ spaces, $1 \leq p \leq \infty$, are Banach spaces, that is normed linear spaces that are complete such that all Cauchy sequences converge. The Hölder inequality implies that for any $X \in L_p$ and any $Y \in L_q$, where $p \in [1, \infty)$ and $q \in (1, \infty]$ satisfying $\frac{1}{p} + \frac{1}{q} = 1$ with the convention that $p = 1$ if $q = \infty$, we have

$$\int XY \mathrm{d}\mu \leq \|X\|_p \|Y\|_q < \infty.$$

Therefore, if we fix some $Y \in L_q$ and consider the mapping

$$L_Y : L_p \to (\mathbb{R}, \mathcal{B})$$

defined by

$$L_Y(X) = \int XY \mathrm{d}\mu, \quad X \in L_p,$$

we obtain a linear and continuous functional L_Y on L_p. Since this construction works for all $Y \in L_q$, we are given a mapping

$$J : Y \mapsto L_Y, \quad \text{for } Y \in L_q,$$

which maps elements of L_q to its dual space. It turns out that the mapping J defined in this way is a linear isometric isomorphism. Thus, the dual space of L_p can be identified with the space L_q. It follows that all linear and continuous functionals are of the form $L_Y(X) = \int XY \mathrm{d}\mu$ for some $Y \in L_q$.

With these tools we are now in a position to prove the following theorem.

Theorem 2.6.15 *Suppose the claim C is not replicable and satisfies $E^*(C) < \infty$. Then the set of arbitrage-free prices is an open interval.*

Proof. Since the set $\mathcal{P}$ of equivalent martingale measures is convex, the set $\Pi(C) = \{E^*(C^*) : P^* \in \mathcal{P}\}$ is convex as well. Indeed, pick $P_1^*, P_2^* \in \mathcal{P}$ and $\varphi \in [0, 1]$. If we define the measure $\mu = \alpha P_1^* + (1 - \alpha) P_2^* \in \mathcal{P}$, we obtain

$$\alpha \int C^* \mathrm{d}P_1^* + (1 - \alpha) \int C^* \mathrm{d}P_2^* = \int C^* \mathrm{d}\mu \in \Pi(C).$$

Hence, $\Pi(C)$ is an intervall or empty. We show that $\Pi(C)$ is open, i.e. for any $E^*(C^*)$ there exist $\pi_- = \pi_-(C)$ and $\pi_+ = \pi_+(C)$ with $\pi_-, \pi_+ \in \Pi(C)$ and

$$\pi_- < E^*(C^*) < \pi_+.$$

First, notice that $S_{1i} \in L_1(P^*)$ for all $i = 0, \ldots, d$, since $E^*(S_{1i}) = S_{0i} \in [0, \infty)$ which implies $E^*(S_{1i}) \in [0, \infty)$. Clearly, $\mathcal{V} = \text{span}\{S_{10}, \ldots, S_{1d}\}$ is a linear subspace of $L_1(P^*)$. Since $C \in L_1(P^*) \backslash \mathcal{V}$, we may apply Theorem 2.6.14 to obtain a linear functional $L : L_1(P^*) \to \mathbb{R}$ with

$$L(V) = 0, \quad \text{for all} \quad V \in \mathcal{V},$$

and $L(C) > 0$. L is of the form

$$L(X) = \int XY \mathrm{d}P^* = E^*(XY), \quad X \in L_1(P^*),$$

for some $Y \in L_\infty(P^*)$. We will now define two equivalent martingale measures yielding the required prices π_- and π_+. Let

$$\frac{\mathrm{d}P^+}{\mathrm{d}P^*}(\omega) = 1 + Y(\omega), \quad \frac{\mathrm{d}P^-}{\mathrm{d}P^*}(\omega) = 1 - Y(\omega), \ \omega \in \Omega.$$

We may and will assume that $\|Y\|_\infty \leq \frac{1}{2}$. Then both P-densities are strictly positive on Ω, such that P^+ as well as P^- are equivalent to P. Notice that

$$\int X \mathrm{d}P^{+/-} = E^*(X(1 \pm Y)) = E^*(X) \pm L(X) = E^*(X)$$

for all $X \in \mathcal{V}$. Since $X = 1_\Omega \in \mathcal{V}$, we obtain $P^{+/-}(\Omega) = E(1_\Omega) = 1$, which verifies that P^+ and P^- are probability measures. In the same vein,

$$E^+(S_{1i}) = E^-(S_{1i}) = E^*(S_{1i}) = S_{0i}(1 + r),$$

for all i. Consequently, P^+ and P^- are equivalent martingale measures. The fair price under P^+ is

$$\pi_+ = E^+(C^*) = E^*(C^*) + \frac{L(C)}{1 + r},$$

whereas under P^- it is given by

$$\pi_- = E^-(C^*) = E^*(C^*) - \frac{L(C)}{1 + r}.$$

Since $L(C) > 0$, the result is shown.

Theorem 2.6.15 has the following corollary.

Corollary 2.6.16 *Suppose the claim C satisfies $E^*(C) < \infty$ and the set of arbitrage-free prices is not an open interval. Then C is replicable.*

2.7 Construction of martingale measures: General case

We have already learned that there may be many equivalent martingale measure. The question arises how to construct such a measure. In a finite model, one may apply linear programming, cf. Example 2.5.8. For the general case, e.g. if $\Omega = \mathbb{R}$ and $S_1(\omega) = \omega$ such that the real probability measure P on $\mathcal{F} = \mathcal{B}(\mathbb{R})$ corresponds to the distribution of the stock price S_1 at time 1 and can be identified with a distribution function F, the following construction based on the Esscher transformation can be used.

The mathematical problem is as follows. We aim at constructing explicitly a probability measure P^* such that

$$E^*(S_1^*) = S_0 \quad \Leftrightarrow \quad E^*(\Delta S) = 0,$$

where $\Delta S = S_1^* - S_0$. The basic idea behind the construction is as follows. If we could define P^* via a P-density of the form

$$f_a(\omega) = \frac{\mathrm{d}P^*}{\mathrm{d}P}(\omega) = \frac{\exp(a\omega)}{E\exp(a\Delta S)}, \quad \omega \in \Omega,$$

for some parameter a, we would obtain

$$E^*(\Delta S) = E\left(\frac{\Delta S \exp(a\Delta S)}{E[\exp(a\Delta S)]}\right) = \frac{E(\Delta S \exp(a\Delta S))}{E\exp(a\Delta S)}.$$

The right-hand side can be written as $\frac{h'(a)}{h(a)}$ if we introduce the exponential

$$h(a) = E\exp(a\Delta S) = \int \exp(a\Delta S)\,\mathrm{d}P, \quad a \in \mathbb{R}.$$

Indeed, we may differentiate under the integral sign, provided $h(a) < \infty$. Thus, for any minimizer a^* of h we obtain $E^*(\Delta S) = 0$. This means that the probability measure P^* given by the P-density $f_{a^*}(\omega) = \exp(a^*\omega)/E\exp(a^*\Delta S), \omega \in \Omega$, such that $P^*(A) = \int_A f_{a^*}(\omega)\,\mathrm{d}P(\omega)$, provides us with an equivalent martingale measure. Notice that $h(a)$ and therefore a^* can be calculated, if we know the distribution of $\Delta S = S_1^* - S_0$ or, equivalently, S_1, under the probability model P of the real world. Indeed, if $S_1 \sim F$ under P, such that $\Delta S = S_1/(1+r) - S_0 \sim H(x) = F((1+r)(x - S_0))$, we may conclude that

$$h(a) = \int \exp(ax)\,\mathrm{d}H(x),$$

The above construction requires that $E(\exp(aX))$ exists under P. Otherwise, one can perform a preliminary transformation. The following theorem and its proof provide the general result and elaborate on the technical details.

Theorem 2.7.1 *Suppose $P(S_1 > S_0) > 0$. Define the probability measure Q by*

$$\mathrm{d}Q(\omega) = c\exp(-\omega^2)\,\mathrm{d}P, \quad \textit{where} \quad c = \left(E_P(\exp(-(\Delta S)^2))\right)^{-1}.$$

Let

$$h(a) = E_Q\left(\exp(a\Delta S)\right), \quad a \in \mathbb{R},$$

denote the moment-generating function of ΔS under Q. Then $h(a) < \infty$ for all $a \in \mathbb{R}$ and there exists some a^* such that

$$h(a^*) = h^* := \inf\{h(a) : a \in \mathbb{R}\}.$$

The probability measure P^* defined by

$$\mathrm{d}P^*(\omega) = \frac{\exp(a^*\omega)}{h(a^*)}\mathrm{d}Q(\omega) \tag{2.7}$$

satisfies $P^* \sim Q \sim P$ and

$$E^*(\Delta S) = E_{P^*}(\Delta S) = 0.$$

This means, P^* is an equivalent martingale measure.

Proof. Clearly, $0 < E(\exp[-(\Delta S)^2]) \le 1$, which implies $c \in [1, \infty)$ and

$$Q(\Omega) = c \int_\Omega \exp(-x^2)\, \mathrm{d}P(x) = 1$$

by definition of c. Since Q is defined by the strictly positive density $\omega \mapsto c\exp(-\omega^2)$, we have $Q \sim P$. Notice that under the measure Q for any $a \in \mathbb{R}$

$$h(a) = E_Q(\exp(a\Delta S)) = cE_P(\exp[-(\Delta S)^2 + a\Delta S)]) \le c\exp(a^2/4) < \infty,$$

since the function $x \mapsto -x^2 + ax$ attains its maximum at $x = a/2$. This implies that $h^* = \inf\{h(a) : a \in \mathbb{R}\} < \infty$. Next, define the candidate probability measures

$$\frac{\mathrm{d}P_a^*}{\mathrm{d}Q}(\omega) = \frac{\exp(a\omega)}{h(a)}, \quad \omega \in \Omega,$$

indexed by $a \in \mathbb{R}$. Notice that $\frac{\mathrm{d}P_a^*}{\mathrm{d}P} \ge 0$ such that $P_a^* \sim Q$. For any candidate P_a^* we have $E_a^*(\Delta S) = \frac{h'(a)}{h(a)}$. Thus, each a with $h'(a) = 0$ yields an equivalent martingale measure. We claim that there indeed exists a minimizer a^*, i.e. some a^* such that

$$h(a^*) = h^* = \inf\{h(a) : a \in \mathbb{R}\}.$$

Suppose the contrary. Choose a sequence $\{a_n\}$ with $h(a_n) \downarrow h^*$, as $n \to \infty$. We claim $a_n \to \pm\infty$. Otherwise, one may extract a convergent subsequence $\{a_m\}$ with limit $\overline{a} \in (-\infty, \infty)$. But then $h(\overline{a}) = \lim h(a_m) = h^*$ by continuity of h. We have found a minimizer, namely $\overline{a}$, a contradiction. Thus, $a_n \to \pm\infty$. Define $u_n = |a_n|/a_n$ and

$$u = \lim_{n\to\infty} u_n \in \{-1, +1\}.$$

We will show that $h(a_n) = E_Q(\exp[a_n\Delta S]) \to \infty$, as $n \to \infty$, a contradiction, since $h(a_n) \downarrow h^* < \infty$ remains bounded if a_n converges. On the set $A_n = \{a_n\Delta S > \delta|a_n|\}$ we have $\exp(a_n\Delta S) \ge \exp(\delta|a_n|)$. Since $h(a_n) \ge E_Q(\exp(a_n\Delta S)1_{A_n})$, we obtain

$$h(a_n) \ge E_Q(\exp[\delta|a_n|]1_{A_n}) = \exp(\delta|a_n|)Q(A_n).$$

The right-hand side converges to ∞, provided $Q(A_n)$ is bounded away from 0. But

$$Q(A_n) = Q(\Delta S > \delta u_n) \to \varepsilon, \quad n \to \infty,$$

if $u_n \to u$ and $\delta > 0$ can be chosen such that δ/u is a continuity point of the d.f. F of ΔS. To show this, notice that $P(\Delta S > 0) > 0 \Leftrightarrow Q(\Delta S > 0) > 0$ rules out $Q(u\Delta S > \delta) = 0$ for all $\delta > 0$. Hence, for any $\varepsilon > 0$ we may find some $\delta > 0$ with $Q(u\Delta S > \delta) = \varepsilon > 0$, and δ can be chosen as a continuity point of F.

Here is a standard example, which is instructive to understand the construction.

Example 2.7.2 *Let $Y \sim N(0, 1)$ and define the probability measure P_λ by*

$$\frac{\mathrm{d}P_\lambda}{\mathrm{d}P} = \frac{\exp(\lambda Y)}{E_P\exp(\lambda Y)}.$$

Notice that the denominator is the moment-generating function of Y, which is given by $\exp(\lambda^2/2)$. *For brevity of notation, put* $E_\lambda(Z) = \int Z \, \mathrm{d}P_\lambda$ *for any random variable Z. Then for any measurable set A*

$$\begin{aligned} P_\lambda(Y \in A) &= E_\lambda(\mathbf{1}(Y \in A)) \\ &= \int \mathbf{1}(Y \in A) \frac{\mathrm{d}P_\lambda}{\mathrm{d}P} \, \mathrm{d}P \\ &= \int \mathbf{1}(Y \in A) \frac{\exp(\lambda Y)}{\exp(\lambda^2/2)} \, \mathrm{d}P \\ &= \int \mathbf{1}(y \in A) \frac{\exp(\lambda y)}{\exp(\lambda^2/2)} \, \mathrm{d}P_Y(y) \\ &= \int_A \frac{1}{\sqrt{2\pi}} \exp\left(-\frac{1}{2}(y^2 + 2\lambda y - \lambda^2) \right) \mathrm{d}y \\ &= \int_A \frac{1}{\sqrt{2\pi}} \exp\left(-\frac{1}{2}(y - \lambda)^2 \right) \mathrm{d}y. \end{aligned}$$

Hence, Y follows a $N(\lambda, 1)$*-distribution under* P_λ.

2.8 Complete financial markets

Definition 2.8.1 *A financial market is called* **complete** *if any claim* $C : \Omega \to \mathbb{R}_+$ *is attainable.*

Example 2.8.2 *Consider the binomial model for one asset and a bank account. Recall that in this case we may assume* $\Omega = \{+, -\}$. *Thus, any claim* $C : \Omega \to \mathbb{R}_+$ *is uniquely determined by* $c_+ = C(+)$ *and* $c_- = C(-)$. *C is replicable, if and only if we can solve the equations*

$$C(\omega) = \varphi' S_1(\omega) = \varphi_0(1 + r) + \varphi_1 S_{11}(\omega), \quad \omega \in \Omega,$$

where $\varphi = (\varphi_0, \varphi_1)'$ *is the replicating portfolio. Thus, we have to consider the equations*

$$c_+ = \varphi_0(1 + r) + \varphi_1 s_+, \quad c_- = \varphi_0(1 + r) + \varphi_1 s_-,$$

where $s_+ = S_{11}(+)$ *and* $s_- = S_{11}(-)$. *Obviously, the unique solution is given by*

$$\varphi_1 = \frac{c_+ - c_-}{s_+ - s_-}, \quad \varphi_0 = \frac{c_- s_+ - c_+ s_-}{(s_+ - s_-)(1 + r)},$$

provided $s_+ - s_- > 0$.

Let us now discuss the case of a finite probability space $\Omega = \{\omega_1, \ldots, \omega_n\}$, $n \in \mathbb{N}$. Recall the definition of the $((d + 1) \times n)$-dimensional payment matrix

$$\mathbf{D} = \begin{bmatrix} S_{10}(\omega_1), \ldots, S_{10}(\omega_n) \\ \vdots \qquad \vdots \\ S_{1d}(\omega_1), \ldots, S_{1d}(\omega_n) \end{bmatrix}.$$

For finite Ω we may identify a contingent claim C with the n-vector $(C(\omega_1), \ldots, C(\omega_n))'$ that we also denote by C. The claim C is attainable if and only if there is some portfolio vector $\varphi \in \mathbb{R}^{d+1}$ such that

$$\mathbf{D}'\varphi = C.$$

Obviously, for a complete financial market this system of linear equations must have a solution φ for any right side C. A standard result from linear algebra now leads us to the following result.

Proposition 2.8.3 *Assume that $\Omega = \{\omega_1, \ldots, \omega_n\}$ for some $n \in \mathbb{N}$. A financial market is complete, if and only if the payment matrix* $\mathbf{D}$ *has n independent rows, i.e. the rank of* $\mathbf{D}'$ *equals the number of states.*

As usual, the general case is more involved. To proceed recall that the set

$$\mathcal{V} = \{\varphi' S_1 : \varphi \in \mathbb{R}^{d+1}\}$$

of all attainable payments forms a linear space. Clearly, $\varphi' S_1$ is a $\sigma(S_{10}, \ldots, S_{1d})$-measurable random variable with

$$E^*|\varphi' S_1| \leq \sum_{i=0}^{d} |\varphi_i| E^*(S_{1i}) = \sum_{i=0}^{d} |\varphi_i| S_{0i}(1+r) < \infty$$

for all $P^* \in \mathcal{P}$. Thus, $\mathcal{V} \subset L^1(\Omega, \sigma(S_{11}, \ldots, S_{1d}), P^*)$. Since $P^*(0 \leq S_{1i} < \infty) = 1$, we have $S_{1i} \in L_0(\Omega, \mathcal{F}, P^*)$ for all $i = 0, \ldots, d$ and all P^* such that

$$\mathcal{V} \subset L_1(\Omega, \sigma(S_{11}, \ldots, S_{1d}), P^*) \subset L_0(\Omega, \mathcal{F}, P^*) = L_0(\Omega, \mathcal{F}, P),$$

where the equality follows from $P \sim P^*$.

Lemma 2.8.4 *For a complete market we have*

$$\mathcal{V} = L_1(\Omega, \sigma(S_{11}, \ldots, S_{1d}), P^*) = L_0(\Omega, \mathcal{F}, P^*) = L_0(\Omega, \mathcal{F}, P).$$

Proof. Let $X \in L_0 = L_0(\Omega, \mathcal{F}, P)$. We may write $X = C_1 - C_2$ for claims $C_1, C_2 \in L_0(\Omega, \mathcal{F}, P)$. By assumption, C_1 and C_2 are attainable, i.e. there exist $\varphi_1, \varphi_2 \in \mathbb{R}^{d+1}$ with $C_i = \varphi_i' S_1,\ i = 1, 2$. But then $X = (\varphi_1 - \varphi_2)' S_1 \in \mathcal{V}$.

For what follows, we need to know the dimension of the space L_p. Let us fix some integer $n \in \mathbb{N}$ and suppose there exists a partition of Ω consisting of n disjoint subsets $A_1, \ldots, A_n$ with $P(A_i) > 0, i = 1, \ldots, n$, i.e.

$$\Omega = \cup_{i=1}^{n} A_i, \quad P(A_i) > 0, \quad A_i \cap A_j = \emptyset \ (i \neq j). \tag{2.8}$$

We claim that in this case the indicators $\mathbf{1}_{A_1}, \ldots, \mathbf{1}_{A_n}$ are linearly independent elements of L_p. Indeed,

$$\alpha_1 \mathbf{1}_{A_1}(\omega) + \cdots + \alpha_n \mathbf{1}_{A_n}(\omega) = 0, \quad \forall\, \omega \in \Omega,$$

implies $\alpha_1 = \cdots = \alpha_n = 0$. Thus, $\dim(L_p) \geq n$. Now let

$$N = \sup\{n \in \mathbb{N} : \text{There are sets } A_1, \ldots, A_n \in \mathcal{F} \text{ with Equation (2.8)}\}.$$

If $N = \infty$, the dimension of L_p is infinite. Now assume $n < \infty$ and let $A_1, \ldots, A_N$ be a partion with Equation (2.8). It follows that all $A \in \{A_1, \ldots, A_N\}$ are atoms, i.e. for any $B \in \mathcal{F}$ with $B \subset A$ we have either $P(B) = 0$ or $P(B) = P(A)$. Otherwise, we may replace A by $B = A \cap B$ and $A \backslash B$ yielding a partition of $N + 1$ sets, a contradiction. Thus, $A_1, \ldots, A_N$ is a maximal partition with (2.8). Consequently, any random variable $X \in L_p$ is almost surely constant on the sets A_i. If we put $a_i = X(\omega_i), i = 1, \ldots, N$, where $\omega_i \in A_i$ is arbitrarily chosen, we may conclude that

$$X = \sum_{i=1}^{N} a_i \mathbf{1}_{A_i}, \qquad P - a.s.$$

This shows that the indicators $\mathbf{1}_{A_1}, \ldots, \mathbf{1}_{A_N}$ are a generator, thus forming a basis of the vector space L_p.

We have shown the following result

Lemma 2.8.5 *$L_p = L_p(\Omega, \mathcal{F}, P)$ is finite-dimensional, if and only if there exists a maximal partition of Ω in N atomic sets $A_1, \ldots, A_N$ such that $P(A_i) > 0$ for $i = 1, \ldots, N$.*

We are now in a postion to show the following final main result of the present section.

Theorem 2.8.6 *On an arbitrage-free financial market the following assertions hold true.*

(i) If the market is complete, there is one and only one equivalent martingale measure, i.e. $|\mathcal{P}| = 1$.

(ii) If there is one and only one equivalent martingale maesure, then $\dim \mathcal{V} = \dim L_0 \leq d + 1$.

Proof.

(i) Suppose the market is arbitrage-free and complete. Fix $A \in \mathcal{F}$. The claim $C = \mathbf{1}_A$ is replicable. By Theorem 2.6.4, the arbitrage-free price, namely $\varphi_A' S_0$, where φ_A is a replicating portfolio, is unique. Thus, the map

$$P^* \mapsto P^*(A) = E^*(C^*) = \varphi_A' S_0, \;\; P^* \in \mathcal{P},$$

does not depend on $P^* \in \mathcal{P}$. This argument holds for any $A \in \mathcal{F}$, which shows that $\mathcal{P}^*$ consists of one element.

(ii) Suppose $\mathcal{P} = \{P^*\}$. Let C be a bounded claim i.e. $C \in L_0(\Omega, \mathcal{F}, P)$. Its unique arbitrage-free price is given by $E^*(C^*) < \infty$. Due to Corollary 2.6.16, we may conclude that C is replicable. This tells us that the market is complete such that

$\dim(\mathcal{V}) = \dim(L_0)$ but $\dim(\mathcal{V}) \leq d + 1$. Thus, $(\Omega, \mathcal{F}, P)$ has at most $d + 1$ atoms $A_1, \ldots, A_{d+1}$.

The above theorem provides a strong interpretation of financial markets, which are arbitrage-free and complete: Any contingent claim has the form

$$C = \sum_{i=1}^{N} c_i \mathbf{1}_{A_i},$$

where $c_i \in (0, \infty)$, $i = 1, \ldots, N$, and $A_1, \ldots, A_N \in \mathcal{F}$ are atoms. Notice that $c_i = \infty$ can be ruled out, since otherwise $P(A_i) > 0$ for all i (which implies $P^*(A_i)$ for all i) yields $E^*(C) = \infty$. Arbitrage-free financial markets on which more general claims exist can not be complete.

We may summarize our findings as follows:

- If there exists an equivalent martingale measure, the price of any replicating portfolio is unique.
- A financial market is arbitrage-free, if and only if there exists an equivalent martingale measure.
- An arbitrage-free financial market is complete, if and only if there exists one and only one equivalent martingale measure.
- On an arbitrage-free and complete financial market any claim C with $E(C) < \infty$ can be hedged.

2.9 Notes and further reading

The selection of the material of the present chapter and its exposition are particularly influenced by Föllmer and Schied (2004), Pliska (1997) and Shiryaev (1999). There are several proofs of the fundamental theorem of asset pricing for an arbitrary probability space $(\Omega, \mathcal{F}, P)$. Theorem 2.5.6 is the Dalang–Morton–Willinger theorem for the case $T = 1$, cf. Dalang et al. (1990). The version for a finite sample space, Theorem 2.5.4, is due to Harrison and Pliska (1981). The recent book by Delbaen and Schachermayer (2006) provides a thorough discussion of such theorems and discusses the related mathematical theory. More results on the construction of martingale measures based on the Esscher transform can be found in Shiryaev (1999).

References

Dalang R.C., Morton A. and Willinger W. (1990) Equivalent martingale measures and no-arbitrage in stochastic securities market models. *Stochastics Stochastics Rep.* **29**(2), 185–201.

Delbaen F. and Schachermayer W. (2006) *The Mathematics of Arbitrage*. Springer Finance. Springer-Verlag, Berlin.

Föllmer H. and Schied A. (2004) *Stochastic Finance: An Introduction in Discrete Time*. vol. 27 of *de Gruyter Studies in Mathematics* extended edn. Walter de Gruyter & Co., Berlin.

Harrison J.M. and Pliska S.R. (1981) Martingales and stochastic integrals in the theory of continuous trading. *Stochastic Process. Appl.* **11**(3), 215–260.

Pliska S. (1997) *Introduction to Mathematical Finance*. Blackwell Publishing, Oxford.

Shiryaev A.N. (1999) *Essentials of Stochastic Finance*. vol. 3 of *Advanced Series on Statistical Science & Applied Probability*. World Scientific Publishing Co. Inc., River Edge, NJ. Facts, models, theory, Translated from the Russian manuscript by N. Kruzhilin.

3

Financial models in discrete time

Most of the data available in the financial markets are numbers or vectors describing quantities such as prices, returns and interest rates, which are observed at certain time points. Only certain objects represent functional data, for example yield curves or pricing models. The time points at which quantities are observed can be discrete or random and live on different time scales ranging from yearly or quaterly data to daily and intraday data. Consequently, even when we fix a certain financial object such as the price of a liquid stock for which price quotes are made every, say, Δth second, we have to deal with a series of numbers or vectors $X_1, X_2, \dots$, this means with a *time series* or *stochastic process in discrete time*.

Except in special cases, there is no reason to assume that the random variables forming such a time series are independent. One approach to go beyond the i.i.d. framework is to consider so-called *martingale differences* that still have the property that they are uncorrelated. They play an important role in financial statistics for dependent processes and econometrics, mainly since they are strongly related to the *martingales*, a really fundamental notion in mathematical finance, which we therefore discuss right at the beginning of this chapter. We also provide some of the most important results of the powerful theory of martingales, which we shall apply in later chapters.

However, martingale differences turn out to be uncorrelated, whereas often the value X_n at current time n is *correlated* with past observations or is even a function $m(X_{n-1}, X_{n-2}, \dots)$ of those lagged values $X_{n-1}, X_{n-2}, \dots$. If we knew that function m or could infer it from historical data, we could make more or less perfect predictions for X_n. However, this does not work on financial markets. A large number of institutional investors, speculators and individuals act and interact on the markets, which introduces randomness and noise superimposing such relationships. Large trades, unexpected economic and political news, new regulations and laws, unexpected financial statements or profit warnings, to mention just a few examples arising in finance, and reactions to (reactions to (reactions to ...)) such events affect the prices by smaller or larger shocks, which can best be modelled by random variables.

Financial Statistics and Mathematical Finance: Methods, Models and Applications, First Edition. Ansgar Steland.

Assuming a linear influence of lagged values on X_n, a realistic model is therefore to assume that

$$X_n = \alpha_1 X_{n-1} + \cdots + \alpha_p X_{n-p} + \sigma\epsilon_n,$$

for mean zero random variables ϵ_n with common variance 1 and deterministic and unknown parameters $\alpha_1, \ldots, \alpha_p \in \mathbb{R}$ and $\sigma > 0$. The ϵ_n model the random shocks that are also called innovations. This model is called an *autoregressive model of order p* and represents a basic time-series model to capture correlations in series of observations. We also discuss ARMA(p, q) models and, more generally, linear processes, which form an unifying framework to study such parametric models and represent a step towards *nonparametric* time series analysis, as the dependence structure is *coded* in an infinite-dimensional parameter.

It turns out that the above autoregressive equations indeed can be solved when the parameters satisfy certain conditions, and one can obtain a solution that is *stationary* in the weak sense, which essentially means that all moments up to the order two do not change, if we add a constant to the time index of all observations. In particular, the mean and variance of a stationary series are invariant with respect to time. Stationarity is a key assumption of many methods and theorems on discrete-time processes.

Figure 3.1 depicts four simulated time series of length 500 and illustrates that stationary series may look rather different, including series where the successive observations tend to

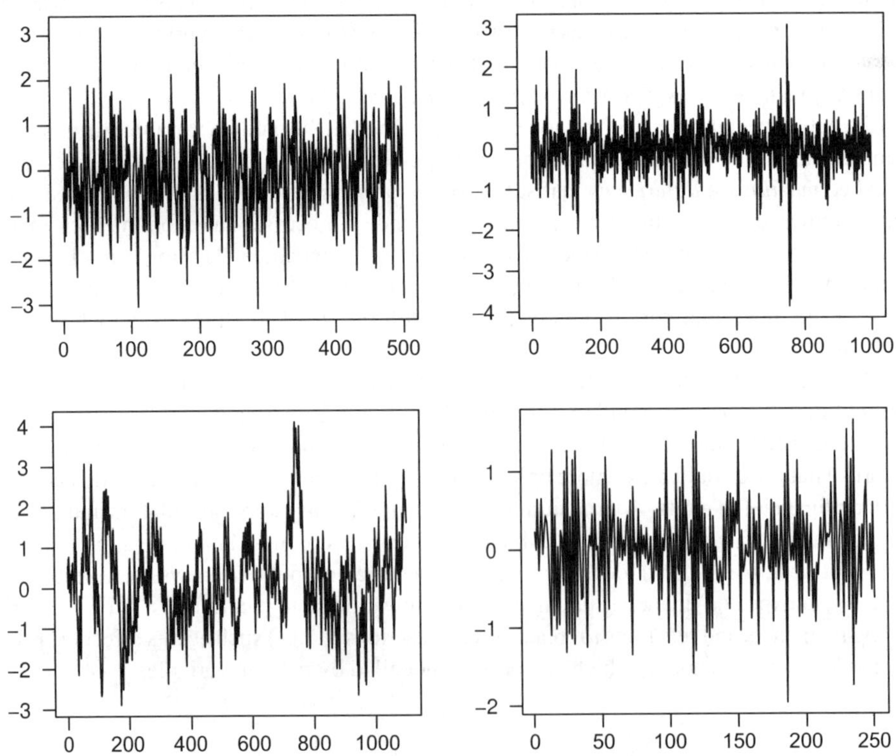

Figure 3.1 Four simulated realizations of stationary discrete-time processes (time series) showing different typical patterns.

stick together, appear to be repulsive or tend to form clusters of observations of low volatility and high volatility, respectively. GARCH models provide a convenient class of models to capture the phenomenon of volatility clusters.

If we do not believe in a parametric model as above, a nonparametric specification such as

$$X_n = m(X_{n-1}, X_{n-2}, \ldots, X_{n-p}) + \sigma(X_{n-1}, X_{n-2}, \ldots, X_{n-p})\epsilon_n \tag{3.1}$$

for possibly nonlinear functions m and σ is more appropriate. Now nonparametric methods have to be applied, which work under qualitative assumptions on m and σ such as smoothness. We discuss such methods in Chapters 8 and 9.

3.1 Adapted stochastic processes in discrete time

We have to introduce and carefully discuss some notions for time series, which are crucial for the following sections and chapters: The notion of a *filtration* as a mathematical vehicle to describe the flow of information contained in series or random variables and vectors, respectively, and the notion of an *adapted process* to capture the inherent property of time-series models for real phenomena that they do not depend on future information. But before introducing those notions, it is time to give a formal definition of our understanding of a time series.

Definition 3.1.1 *A* **stochastic process in discrete time** *is a sequence* $\{X_n : n \geq 0\}$ *of random variables*

$$X_n : (\Omega, \mathcal{F}, P) \to (\mathbb{R}, \mathcal{B}), \quad n \geq 0.$$

We need to introduce the notions of a *filtration* and a stochastic process adapted to such a filtration. Filtrations are used in probability to model the *flow of information*. Let us first recall the following elementary facts. Let $(\Omega, \mathcal{F}, P)$ be a probability space and $\mathcal{B}$ the collection of all Borel sets of $\mathbb{R}$. A function $X : \Omega \to \mathbb{R}$ is measurable (a random variable), if all sets $\{X \in A\}$, $A \in \mathcal{B}$, are elements of $\mathcal{F}$. For a discrete random variable attaining the values $x_1, x_2, \ldots$ that condition is equivalent to $\{X = x_i\} \in \mathcal{F}$, $i = 1, 2, \ldots$.

Consider a two period model and an asset, whose price increases (decreases) in each period by the factor u (d). We may use $\Omega = \{\omega_1, \ldots, \omega_4\}$ with $\omega_1 = (+, +)$, $\omega_2 = (+, -)$, $\omega_3 = (-, +)$, $\omega_4 = (-, -)$ and $\mathcal{F}_2 := \mathcal{F} = Pot(\Omega)$. Then, e.g., $A = \{\omega_1, \omega_2\}$ is the event that the asset goes up in the first period. Let

$$\mathcal{F}_0 = \{\emptyset, \Omega\}, \quad \mathcal{F}_1 = \sigma(A) = \{\emptyset, A, A^c, \Omega\}.$$

Then $\mathcal{F}_0 \subset \mathcal{F}_1 \subset \mathcal{F}_2 \subset \mathcal{F}$.

To get a better idea what $\mathcal{F}_1$-measurability means, we shall define two random variables X and Y, such that X is $\mathcal{F}_1$-measurable but Y not. Let $X : \Omega \to \mathbb{R}$ be given by $X(\omega_1) = X(\omega_2) = 100$ and $X(\omega_3) = X(\omega_4) = 50$. Notice that X is constant on the sets constituting $\mathcal{F}_1$; the outcome of X only depends on what has occured up to time $t = 1$, and that information is given by the σ-field $\mathcal{F}_1$. Formally, we have

$$\{X = 100\} = \{\omega_1, \omega_2\} = A \in \mathcal{F}_1$$

and

$$\{X = 50\} = \{\omega_3, \omega_4\} = A^c \in \mathcal{F}_1,$$

which shows that X is indeed $\mathcal{F}_1$-measurable. We even have $\mathcal{F}_1 = \sigma(X)$. $\mathcal{F}_2$ represents the more detailed information until time $t = 2$. The **flow of information** is described by $\{\emptyset, \Omega\} \subset \mathcal{F}_1 \subset \mathcal{F}_2 = \mathcal{F}$.

Consider now the random variable Y given by $Y(\omega_1) = 200$, $Y(\omega_2) = 150$, $Y(\omega_3) = 150$, $Y(\omega_4) = 100$. Y is not $\mathcal{F}_1$-measurable, since

$$\{Y = 200\} = \{\omega_1\} \notin \mathcal{F}_1,$$

but measurable w.r.t. $\mathcal{F}_2$. Moreover, $\mathcal{F}_2 = \sigma(Y)$.

Definition 3.1.2

(i) A **filtration** *on* $(\Omega, \mathcal{F})$ *is an increasing sequence*

$$\mathcal{F}_0 \subset \mathcal{F}_1 \subset \cdots \subset \mathcal{F}_n \subset \mathcal{F}, \quad n \in \mathbb{N},$$

of sub-σ-fields.

(ii) A stochastic process $\{X_n : n \geq 0\}$ *is called* **$\mathcal{F}_n$-adapted** *or* **adapted (to $\{\mathcal{F}_n\}$)**, *if X_n is $\mathcal{F}_n$-measurable for all $n \in \mathbb{N}_0$.*

For a given process $\{X_n\}$ one may always consider the **natural filtration**

$$\mathcal{F}_0 = \{\emptyset, \Omega\}, \quad \mathcal{F}_n = \sigma(X_0, \ldots, X_n), \quad n \in \mathbb{N}_0.$$

If no filtration is specified, it is common to agree that the natural one is meant.

The next notion we will use extensively in what follows is the conditional expectation of a random variable given a σ-field. For a definition and properties refer to Appendix A.3. For the reader's convenience, we shall illustrate the concept and some basic rules within the stochastic framework of a binomial tree with one stock and a bank account. To ease the generalization of what follows to the case of d stocks, let us denote by S_{t1} the time t price of the asset. Recall that, given the factors u and d, the model is determined by the one-period model

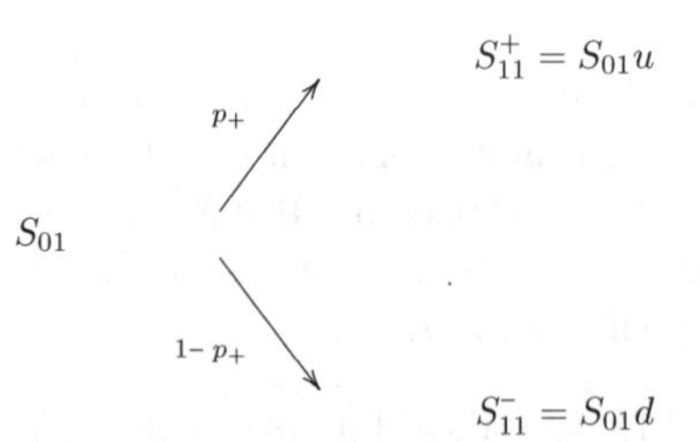

and adhering two further one-period models at both leaves yields the tree

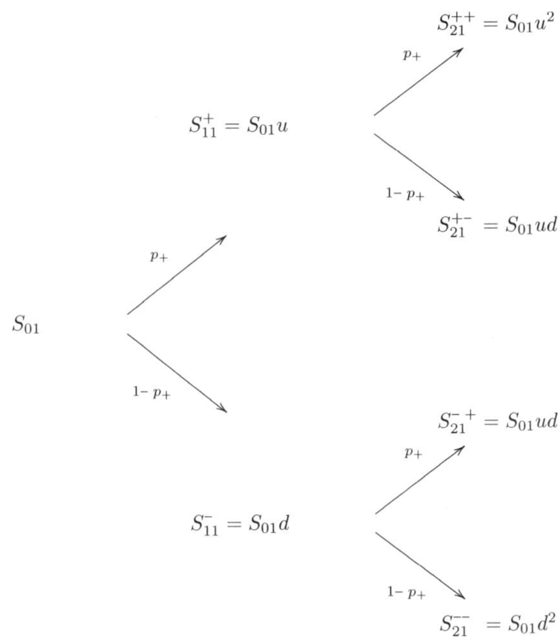

Continuing in this fashion yields the binomial model for discrete time points $t = 1, \ldots, T$.

Notice that S_{21} has discrete support $\{S_{21}^{++}, S_{21}^{+-}, S_{21}^{--}\}$ with corresponding probabilities $\{p_+^2, 2p_+(1-p_+), (1-p_+)^2\}$, describing the random experiment how the asset price evolves in the second period given it went up in the first one. Its expectation is

$$E(S_{21}) = S_{21}^{++}p_+^2 + 2S_{21}^{+-}p_+(1-p_+) + S_{21}^{--}(1-p_+)^2.$$

Put

$$\mathcal{F}_0 = \{\emptyset, \Omega\}, \mathcal{F}_1 = \sigma(S_{11}), \mathcal{F}_2 = \sigma(S_{11}, S_{22}).$$

Then $\mathcal{F}_0 \subset \mathcal{F}_1 \subset \mathcal{F}_2 \subset \mathcal{F}$. $\mathcal{F}_t$ represents the information available at time t. In general, the conditional expectation of a node is calculated as we did above. We obtained a *function* of S_{11}, i.e. a $\mathcal{F}_1$-measurable random variable. Since the events $A \in \mathcal{F}_1$ depend on $\omega \in \Omega$ only through the first coordinate, this is a function of ω_1. If $\omega_1 = +$, i.e. for $\omega = (+, -)$ and $\omega = (+, +)$, S_{21} may attain the values S_{21}^{++}, S_{21}^{+-} and we obtain $E(S_{21}|\mathcal{F}_1)(\omega) = p_+S_{21}^{++} + (1-p_+)S_{21}^{+-}$. For $\omega_1 = -$ corresponding to the outcomes $\omega \in \{(-,+), (-,-)\}$, we have $E(S_{21}|\mathcal{F}_1)(\omega) = p_+S_{21}^{-+} + (1-p_+)S_{21}^{--}$. Hereby, $E(S_{21}|\mathcal{F}_1)$ is determined as a *random variable* on Ω. By the rules of calculation for conditional expectations, as given in the appendix,

$$\begin{aligned} E(E(S_{21}|\mathcal{F}_1)) &= E(\mathbf{1}_\Omega E(S_{21}|\mathcal{F}_1)) \\ &= E(\mathbf{1}_\Omega S_{21}) \\ &= E(S_{21}). \end{aligned}$$

The above exposition shows that conditional expectations naturally arise even in the simplest models for financial markets.

In our two-period binomial model, the asset price at time t is a function of its price at time $t-1$, which simplifies calculations of conditional probabilities considerably.

Definition 3.1.3 *A process $\{X_n : n \in \mathbb{N}\}$ equipped with its natural filtration $\{\mathcal{F}_n : n \in \mathbb{N}\}$, $\mathcal{F}_n = \sigma(X_1, \dots, X_n)$, $n \in \mathbb{N}$, is called* **Markovian (Markov process)**, *if*

$$P(X_{n+1} \in A \mid \mathcal{F}_n) = P(X_{n+1} \in A \mid X_n)$$

holds true for all $A \in \mathcal{F}$.

For a Markovian process conditional probabilities of events $\{X_{n+1} \in A\}$ given its history until time n only depend on the most recent value X_n.

Example 3.1.4 *Let $\{\xi_n : n \in \mathbb{N}\}$ be i.i.d. random variables with values in $\{-1, 1\}$ and $p = P(\xi_1 = 1)$. Define $S_0 = 0$, $S_n = \sum_{i=1}^n \xi_i$, $n \in \mathbb{N}$. Then $\{S_n\}$ is a Markov process with*

$$\begin{aligned} P(S_n = j \mid S_{n-1} = i) &= P(\xi_n = j - i \mid S_{n-1} = i) \\ &= \begin{cases} p, & i = j-1, \\ 1-p, & i = j+1, \\ 0, & \textit{otherwise}, \end{cases} \end{aligned}$$

which is a function $p(i, j)$ of i and j, $i, j \in \{-n, \dots, n\}$. $p(i, j)$ is known as the transition probability that the random walk moves to the **state** *j if its previous state is i.*

Generalizing the last example, a sequence $X_0, \dots, X_T$ of random variables taking values in a set S, called a **state space**, is called a Markov chain, if

$$P(X_n = x_n \mid X_0 = x_0, \dots, X_{n-1} = x_{n-1})$$

is equal to

$$p(x_{n-1}, x_n) = P(X_n = x_n \mid X_{n-1} = x_{n-1})$$

for all $x_n \in S$ and $n = 1, \dots, T$ with $P(X_0 = x_0, \dots, X_{n-1} = x_{n-1}) > 0$. The latter restriction means that the path $x_0, \dots x_{n-1}$ is observable. The matrix $\mathbb{P} = (p(i, j))_{i,j \in S}$ is called **transition matrix**, the row vector $p_0 = (p_i)_{i \in S}$ with $p_i = P(X_0 = i)$, $i \in S$, the **initial distribution**. It is easy to see that the (marginal) distribution after one step is

$$p_j^{(1)} = P(X_1 = j) = \sum_{i \in S} P(X_1 = j \mid X_0 = i) P(X_0 = i) = (p_0 \mathbb{P})_{aj}$$

for $j \in S$, such that $p^{(1)} = (p_j^{(1)})_{j \in S}$ is given by

$$p^{(1)} = p_0 \mathbb{P}.$$

In general, after n transitions, the marginal distribution $p^{(n)} = (P(X_n = j))_{j\in S}$ is given by

$$p^{(n)} = p_0\mathbb{P}^n,$$

where $\mathbb{P}^n = (p^n(i, j))_{i,j\in S}$ is called an **n-step transition matrix**. Using $\mathbb{P}^{m+n} = \mathbb{P}^n \cdot \mathbb{P}^m$ we obatin the celebrated Chapman–Kolmogorov equations

$$p^{(m+n)}(x, y) = \sum_{z\in S} p^{(m)}(x, z)p^{(n)}(z, y).$$

3.2 Martingales and martingale differences

The following exposition serves as a motivation for martingales and provides a fundamental insight as well. Denote the time t price of an asset by S_t and let us assume that the increments $Y_t = S_t - S_{t-1}$, $t = 1, \ldots, T$, are i.i.d. with

$$P(Y_t = 1) = p \quad \text{and} \quad P(Y_t = -1) = 1 - p.$$

Put $Y_0 = 0$. Suppose an investor starts with φ_1 shares at time $t = 0$, which are held until $t = 1$. More generally, let φ_t denote the number of shares held from $t - 1$ to t. We consider **arbitrary** investment strategies where φ_t may depend on the price history, i.e.

$$\varphi_t = \varphi_t(Y_0, \ldots, Y_{t-1})$$

may be a function of $Y_1, \ldots, Y_{t-1}$ (and the constant S_0) and therefore a function of $S_0, \ldots, S_{t-1}$. If we put $\mathcal{F}_0 = \{\emptyset, \Omega\}$, $\mathcal{F}_t = \sigma(Y_1, \ldots, Y_t)$, $t \geq 1$, then φ_t is $\mathcal{F}_{t-1}$-measurable. Notice that the current value of the investment can be calculated by the recursion

$$X_t = X_{t-1} + \varphi_t(Y_0, \ldots, Y_{t-1})Y_t, \quad t = 1, \ldots, T, \quad X_0 = 0,$$

i.e.

$$X_t = \sum_{i=1}^{t} \varphi_i(Y_0, \ldots, Y_{i-1})Y_i.$$

Let us calculate the conditional expectation of X_t given the information $\mathcal{F}_{t-1} = \sigma(Y_1, \ldots, Y_{t-1})$ until time $t - 1$.

$$\begin{aligned}
E(X_t \mid \mathcal{F}_{t-1}) &= E(X_t \mid Y_0, \ldots, Y_{t-1}) \\
&= E\left(\sum_{i=1}^{t} \varphi_i(Y_0, \ldots, Y_{i-1})Y_i \mid Y_0, \ldots, Y_{t-1}\right) \\
&= \sum_{i=1}^{t} \varphi_i(Y_0, \ldots, Y_{i-1})E(Y_i \mid Y_0, \ldots, Y_{t-1}),
\end{aligned}$$

where we used the linearity of the conditional expectation and the rule that known factors can be factored out. Next, notice that by independence of $Y_0, \dots, Y_T$,

$$E(Y_i \mid Y_0, \dots, Y_{t-1}) = \begin{cases} Y_i, & i \le t-1, \\ E(Y_i) = 2p - 1, & i > t-1. \end{cases}$$

Thus, the expected time t value of the investment given the information until time $t-1$ turns out to be

$$E(X_t \mid \mathcal{F}_{t-1}) = X_{t-1} + (2p-1)\varphi_t(Y_0, \dots, Y_{t-1}).$$

All the above calculations are a.s. In the special case $p = 1/2$ we obtain $E(X_t \mid \mathcal{F}_{t-1}) = X_{t-1}$, i.e. there does not exist any investment strategy that improves in terms of the conditional expectation. However, if $p > 1/2$, we have $E(X_t \mid \mathcal{F}_{t-1}) > X_{t-1}$, provided $\varphi_t > 0$. This tells us that in this case it is preferable to be invested. Analogously, if $p < 1/2$ it is better to have a short position.

3.2.0.1 Martingales

Definition 3.2.1 *$\{X_n\}$ is called a $\mathcal{F}_n$*-**martingale** *or* **martingale (w.r.t. $\mathcal{F}_n$)**, *if*

(i) $\{X_n\}$ *is $\mathcal{F}_n$-adapted,*
(ii) $E|X_n| < \infty \; \forall n$ *and*
(iii) $E(X_{n+1} \mid \mathcal{F}_n) = X_n$, *a.s.,* $\forall\, n$.

If (i), (ii) and

(iv) $E(X_{n+1} \mid \mathcal{F}_n) \ge X_n$, *a.s.,* $\forall n$

hold true, then $\{X_n\}$ is called a $\mathcal{F}_n$-**submartingale**, *and $\{X_n\}$ is a $\mathcal{F}_n$*-**supermartingale**, *if (i), (ii) and*

(v) $E(X_{n+1} \mid \mathcal{F}_n) \le X_n$, *a.s.,* $\forall n$

are satisfied.

Remark 3.2.2 *Using the tower property of conditional expectations, it is easy to verify that for a submartingale*

$$E(X_{n+m} \mid \mathcal{F}_n) \ge X_n.$$

holds true for all n and $m \ge 0$, whereas for a supermartingale $\ge$ has to be replaced by $\le$, and a martingale satisfies

$$E(X_{n+m} \mid \mathcal{F}_n) = X_n.$$

Remark 3.2.3 *If $\{X_n\}$ is a $\mathcal{F}_n$-submartingale and $\{\mathcal{G}_n\}$ is a coarser filtration, i.e. $\mathcal{G}_n \subset \mathcal{F}_n \forall\, n$, then $\{X_n\}$ is a submartingale with respect to $\mathcal{G}_n$ as well, provided X_n is $\mathcal{G}_n$-adapted, since*

$$E(X_{n+1}|\mathcal{G}_n) = E(E(X_{n+1}|\mathcal{F}_n)|\mathcal{G}_n) \ge E(X_n|\mathcal{G}_n) = X_n.$$

Analogous assertions hold true for martingales and supermartingales.

Example 3.2.4 *Here are two important standard examples.*

(i) Let $X_1, X_2, \ldots$ be independent random variables with $E|X_i| < \infty$ and $E(X_i) = 0$ for all $i \geq 1$. Put

$$S_n = \sum_{i=1}^{n} X_i, \qquad n \in \mathbb{N}.$$

Then $\{S_n : n \in \mathbb{N}_0\}$ is a martingale with respect to the natural filtration $\mathcal{F}_n = \sigma(X_1, \ldots, X_n)$, $n \in \mathbb{N}_0$. Indeed,

$$E|S_n| \leq \sum_{i=1}^{n} |E(X_i)| \leq \sum_{i=1}^{n} E|X_i| < \infty,$$

for all $n \geq 1$. Further, for $n \geq 0$

$$\begin{aligned} E(S_{n+1}|\mathcal{F}_n) &= E(S_n + X_{n+1}|\mathcal{F}_n) \\ &= E(S_n|\mathcal{F}_n) + E(X_{n+1}|\mathcal{F}_n) \\ &= S_n, \end{aligned}$$

P-almost surely, since, first, S_n is $\mathcal{F}_n$-measurable and $E(X_{n+1}|\mathcal{F}_n) = E(X_{n+1}) = 0$, and, secondly, X_{n+1} is independent from $\mathcal{F}_n = \sigma(X_1, \ldots, X_n)$ by assumption.

(ii) Let $Z_1, \ldots, Z_n$ be i.i.d. with $E|Z_1| < \infty$ and $E(Z_1) = 1$. Put

$$T_n = \prod_{i=1}^{n} Z_i, \qquad n \geq 1.$$

We claim that T_n is a martingale with respect to the natural filtration $\mathcal{F}_n = \sigma(Z_1, \ldots, Z_n)$, $n \in \mathbb{N}_0$. First, by independence,

$$E|T_n| \leq \prod_{i=1}^{n} E|Z_i| < \infty,$$

for all n. Secondly,

$$\begin{aligned} E(T_{n+1}|\mathcal{F}_n) &= E\left(Z_{n+1} \prod_{i=1}^{n} Z_i \middle| \mathcal{F}_n \right) \\ &= E\left(Z_{n+1}|\mathcal{F}_n\right) \prod_{i=1}^{n} Z_i \\ &= E(Z_{n+1}) T_n \\ &= T_n, \end{aligned}$$

P-almost surely. Here we used the fact that $\prod_{i=1}^{n} Z_i$ is $\mathcal{F}_n$-measurable and Z_{n+1} independent from $\mathcal{F}_n$.

Part (ii) of the above example has an important application in statistics. Suppose we are given a random sample

$$X_1, \dots, X_n \overset{i.i.d.}{\sim} f_{\theta_0}(x),$$

for a member f_{θ_0} of some parametric family $\{f_\theta : \theta \in \Theta\}$ of densities, where $\Theta \subset \mathbb{R}^k$, for some $k \in \mathbb{N}$, is a k-dimensional parameter space. It is assumed that θ_0 is unknown and has to be estimated from $X_1, \dots, X_n$. The (random) function

$$L(\theta|X_1, \dots, X_n) = \prod_{i=1}^{n} f_\theta(X_i), \qquad \theta \in \Theta,$$

is called the **likelihood** and its maximizer, providing it exists, which has to be checked in an application, is the **maximum likelihood estimator** (ML estimator or m.l.e.) for θ_0. Notice that for small $\Delta x > 0$,

$$\int_{[X_i - \Delta x/2, X_i + \Delta x/2]} f_\theta(u)\,du \approx \Delta x f_\theta(X_i)$$

is the probability (in an independently repeated experiment) to get an observation close to X_i. In this sense, $L(\theta|X_1, \dots, X_n)$ can be interpreted as the probability to observe approximately the sample $X_1, \dots, X_n$, and the m.l.e. maximizes that probability given the sample.

The likelihood can also be used to handle testing problems such as

$$H_0 : \theta_0 = \theta^{(0)} \qquad \text{against} \qquad H_1 : \theta_0 = \theta^{(1)},$$

where $\theta^{(1)}$ and $\theta^{(2)}$ are known given constants. The **likelihood ratio statistic**

$$L_n = \frac{\prod_{i=1}^n f_{\theta^{(1)}}(X_i)}{\prod_{i=1}^n f_{\theta^{(0)}}(X_i)} = \prod_{i=1}^{n} \frac{f_{\theta^{(1)}}(X_i)}{f_{\theta^{(0)}}(X_i)}$$

compares the likelihood of $\theta^{(1)}$ and $\theta^{(0)}$. It is natural to reject the null hypothesis in favor of the alternative hypothesis, if L_n is large.

Part (ii) of Example 3.2.4 now tells us that L_n is a martingale under the null hypothesis. This follows from

$$\begin{aligned} E_{\theta^{(0)}}\left(\frac{f_{\theta^{(1)}}(X_i)}{f_{\theta^{(0)}}(X_i)}\right) &= \int \frac{f_{\theta^{(1)}}(x)}{f_{\theta^{(0)}}(x)} f_{\theta^{(0)}}(x)\,dx \\ &= \int f_{\theta^{(1)}}(x)\,dx = 1, \end{aligned}$$

where $E_{\theta^{(0)}}$ means that the expectation is calculated assuming that $X_i \sim f_{\theta^{(0)}}(x)$, provided we additionally assume that the parametric model under consideration ensures that the ratio variables

$$Z_i = \frac{f_{\theta^{(1)}}(X_i)}{f_{\theta^{(0)}}(X_i)}, \qquad i = 1, \dots, n,$$

are in L_1.

The next example is simple but important.

Example 3.2.5 (LÉVY MARTINGALE)
Let ξ be an arbitrary random variable with $E|\xi| < \infty$ and $\{\mathcal{F}_n : n \in \mathbb{N}_0\}$ be an arbitrary filtration. Then the sequence of conditional expectations

$$X_n = E(\xi|\mathcal{F}_n), \qquad n \in \mathbb{N}_0,$$

defines a martingale called the **Levy martingale**. *Indeed, $X_n \in L_1$, since*

$$E|X_n| = E|E(\xi|\mathcal{F}_n)| \leq E(E(|\xi| \,|\, \mathcal{F}_n) = E|\xi| < \infty,$$

for all n. The martingale property now follows from the tower property of conditional expectations, since $\mathcal{F}_n \subset \mathcal{F}_{n+1}$: For $n \in \mathbb{N}$ we have

$$\begin{aligned} E(X_{n+1}|\mathcal{F}_n) &= E(E(\xi|\mathcal{F}_{n+1})|\mathcal{F}_n)) \\ &= E(\xi|\mathcal{F}_n) \\ &= X_n. \end{aligned}$$

Theorem 3.2.6

(i) If X_n is a $\mathcal{F}_n$-submartingale, then $-X_n$ is a $\mathcal{F}_n$-supermartingale.

(ii) Let X_n be a $\mathcal{F}_n$-martingale and $\varphi : \mathbb{R} \to \mathbb{R}$ be a convex function such that $E|\varphi(X_n)| < \infty$ for all n. Then $\varphi(X_n)$ is a $\mathcal{F}_n$-submartingale.

(iii) If $\{X_n\}$ is a submartingale, then

$$E(X_n) \leq E(X_m)$$

holds for $n \leq m$.

Proof. Assertion (i) is trivial. (ii) follows from the conditional version of Jensen's inequality, cf. Theorem A.3.1 (xi).

Example 3.2.7 *Let $X_1, X_2, \ldots$ be independent with $X_i \in L_1$ and $E(X_i) = 0$ for all $i \in \mathbb{N}$. Then $S_n = \sum_{i=1}^n X_i$, $n \geq 1$, is a martingale and S_n^2 a submartingale. Further, the inequality*

$$E(S_n^2) \geq E(S_{n+1}^2)$$

holds for all n.

3.2.0.2 Martingale differences

We are now going to introduce *martingale difference sequences*, which play a prominent role in financial statistics. As we shall see later, under an additional moment condition, they form white-noise processes that allow the powerful machinery of the theory of martingales to be applied. For this reason, they are a common replacement of i.i.d. sequences in models for financial data.

Definition and Theorem 3.2.8 (Martingale Difference Sequence (MDS)) *Let $\{M_n\}$ be a $\mathcal{F}_n$-martingale. Then the sequence of random variables*

$$\epsilon_n = M_n - M_{n-1}, \quad n \geq 1,$$

is called a **martingale difference sequence** *(MDS). Any such sequence satisfies*

(i) $E|\epsilon_n| < \infty$, *for all n.*

(ii) $\{\epsilon_n\}$ *is* $\mathcal{F}_n$*-adapted.*

(iii) $E(\epsilon_n \mid \mathcal{F}_{n-1}) = 0$, *P-a.s., for all n.*

If $\{\epsilon_n\}$ *is a sequence of random variables such that (i)–(iii) are satisfied, then*

$$M_n = \sum_{i=1}^{n} \epsilon_i, \qquad n \geq 0,$$

is a $\mathcal{F}_n$*-martingale, such that* $\{\epsilon_n\}$ *is a martingale difference sequence.*

Proof. Since M_n as well as M_{n-1} are $\mathcal{F}_n$-measurable, the difference $\epsilon_n = M_n - M_{n-1}$ is $\mathcal{F}_n$-measurable, for all n. Further, $E|\epsilon_n| \leq E|M_n| + E|M_{n-1}| < \infty$ for all n. Finally,

$$\begin{aligned} E(\epsilon_n \mid \mathcal{F}_{n-1}) &= E(M_n \mid \mathcal{F}_{n-1}) - E(M_{n-1} \mid \mathcal{F}_{n-1}) \\ &= M_{n-1} - M_{n-1} = 0, \end{aligned}$$

a.s. Now let $\{\epsilon_n\}$ be a sequence such that (i)–(iii) are fulfilled. Then $M_n = \sum_{i=1}^{n} \epsilon_i$ is $\mathcal{F}_n$-measurable with $E|M_n| \leq \sum_{i=1}^{n} E|\epsilon_i| < \infty$ for all n and

$$E(M_n \mid \mathcal{F}_{n-1}) = \sum_{i=1}^{n-1} \epsilon_i + E(\epsilon_n \mid \mathcal{F}_{n-1}) = M_{n-1}, \text{ a.s.},$$

since ϵ_i is $\mathcal{F}_{n-1}$-measurable for $i \leq n-1$ such that $E(\epsilon_i \mid \mathcal{F}_{n-1}) = \epsilon_i$ a.s.

Martingale differences are centered and, provided they possess finite second moments, uncorrelated. Indeed, for any m with $m < n$ we have $\mathcal{F}_m \subset \mathcal{F}_{n-1}$ such that $E(\epsilon_n \mid \mathcal{F}_{n-1}) = 0$ a.s. implies

$$E(\epsilon_n \mid \mathcal{F}_m) = E(E(\epsilon_n \mid \mathcal{F}_{n-1}) \mid \mathcal{F}_m) = 0,$$

a.s., which in turn yields $E(\epsilon_n) = E(E(\epsilon_n \mid \mathcal{F}_m)) = 0$. As a consequence,

$$\text{Cov}(\epsilon_n, \epsilon_m) = E(\epsilon_n \epsilon_m) = E(\epsilon_m E(\epsilon_n \mid \mathcal{F}_m)) = 0,$$

if $E(\epsilon_n^2) < \infty$ and $E(\epsilon_m^2) < \infty$.

We have verified the following elementary but important result.

Lemma 3.2.9 *Let* $\{M_n\}$ *be a martingale with differences* $\{\epsilon_n\}$ *satisfying*

$$E\epsilon_n^2 < \infty$$

for all n.

(i) The ϵ_ns are centered and pairwise uncorrelated.

(ii) $\text{Var}(M_n) = \text{Var}(M_0) + \sum_{i=1}^{n} \text{Var}(\epsilon_i)$.

3.2.1 The martingale transformation

Let S_t, $t = 0, \ldots, T$, be the price process of a stock, where the initial price S_0 is assumed to be known. Denote the number of shares hold from $t-1$ to t by φ_t. We assume that φ_t is $\mathcal{F}_{t-1}$-measurable.

Definition 3.2.10 (PREDICTABLE PROCESS)
Let $\{X_n\}$ be a stochastic process in discrete time and $\{\mathcal{F}_n\}$ be a filtration. Then $\{X_n\}$ is called **predictable** *or* **previsible**, *if X_n is $\mathcal{F}_{n-1}$-measurable for all n.*

This means, the porfolios $\{\varphi_t\}$ form a predictable process in the sense of this definition. The time t value V_t of the investment strategy given by φ_t satisfies the recursion

$$V_t = V_{t-1} + \varphi_t(S_t - S_{t-1})$$

and we have the explicit formula

$$V_t = V_0 + \sum_{i=1}^{t} \varphi_i(S_i - S_{i-1})$$

for all $t = 0, \ldots, T$.

Definition 3.2.11 (DISCRETE STOCHASTIC INTEGRAL)
Let $\{X_n : n \geq 0\}$ be a $\mathcal{F}_n$-adapted process and $\{\varphi_n : n \geq 1\}$ be a predictable process. Then

$$I_n = \sum_{i=1}^{n} \varphi_i(X_i - X_{i-1}), \qquad n \geq 0,$$

is called a **discrete stochastic integral** *(with starting value I_0). It is written as*

$$I_n = \int_0^n \varphi_t \, \mathrm{d}X_t.$$

To this end, the integral notation in the above definition is formal.

Theorem 3.2.12 *Let X_n be a $\mathcal{F}_n$-martingale and φ_n be $\mathcal{F}_n$-measurable. Suppose that*

$$\varphi_n \in L_p \qquad \text{and} \qquad X_n \in L_q,$$

for all n, with $p, q \in [1, \infty]$ satisfying $\frac{1}{p} + \frac{1}{q} = 1$. Then the discrete stochastic integral is a $\mathcal{F}_n$-martingale, and the same holds true for

$$X_0 + \int_0^n \varphi_t \, \mathrm{d}X_t = X_0 + \sum_{i=1}^{n} \varphi_i(X_i - X_{i-1}),$$

for any constant X_0.

Proof. By virtue of Hölder's inequality, we have

$$E|\varphi_n X_n| \leq \|\varphi_n\|_{L_p} \|X_n\|_{L_q} < \infty,$$

for all n. Therefore

$$\begin{aligned} E|X_0 + I_n| &= E\left|X_0 + \sum_{i=1}^{n} \varphi_i(X_i - X_{i-1})\right| \\ &\leq E|X_0| + \sum_{i=1}^{n} (E|\varphi_i X_i| + E|\varphi_i X_{i-1}|) < \infty, \end{aligned}$$

which verifies that $X_0 + I_n \in L_1$ for all n. Since, first, φ_i is $\mathcal{F}_{i-1}$-measurable and therefore also $\mathcal{F}_n$-measurable, and, secondly, X_i is $\mathcal{F}_i$-measurable and therefore $\mathcal{F}_n$-measurable as well, we obtain

$$\begin{aligned} E(X_0 + I_{n+1}|\mathcal{F}_n) &= E\left(X_0 + \sum_{i=1}^{n+1} \varphi_i(X_i - X_{i-1})\middle|\mathcal{F}_n\right) \\ &= X_0 + \sum_{i=1}^{n} E(\varphi_i(X_i - X_{i-1})|\mathcal{F}_n) + E(\varphi_{n+1}(X_{n+1} - X_n)|\mathcal{F}_n). \end{aligned}$$

Notice that for $i = 1, \ldots, n$,

$$E(\varphi_i(X_i - X_{i-1})|\mathcal{F}_n) = \varphi_i(X_i - X_{i-1})$$

and

$$\begin{aligned} E(\varphi_{n+1}(X_{n+1} - X_n)|\mathcal{F}_n) &= \varphi_{n+1} E(X_{n+1} - X_n|\mathcal{F}_n) \\ &= \varphi_{n+1}[E(X_{n+1}|\mathcal{F}_n) - E(X_n|\mathcal{F}_n)] \\ &= 0. \end{aligned}$$

Therefore

$$E(X_0 + I_{n+1}|\mathcal{F}_n) = X_0 + \sum_{i=1}^{n} \varphi_i(X_i - X_{i+1}) = I_n.$$

All of the above identities hold P-a.s.

Definition 3.2.13 (Martingale Transform)
In the setting of Theorem 3.2.12, the discrete stochastic integral

$$I_n = \sum_{i=1}^{n} \varphi_i(X_i - X_{i-1}), \qquad n \geq 0,$$

is called a **martingale transform**.

3.2.2 Stopping times, optional sampling and a maximal inequality

Suppose the random series $V_1, V_2, \ldots$ represents the value of a trading strategy, say, a long position in some financial instrument. The question arises what happens if we stop trading according to some decision rule (exit strategy) N, which takes values in the natural numbers, such that we stay with the final value V_N. The corresponding value process is

$$V_1, V_2, \ldots, V_N, V_N, \ldots \tag{3.2}$$

Assuming that $\{V_n\}$ is a martingale, say, does the martingale property still apply to the stopped series (3.2)? Or is it possible to find such an exit strategy such that the series becomes a submartingale, which would imply that $E(V_N) \geq E(V_1)$, whereas without stopping we have $E(V_n) = E(V_1)$ for all n?

3.2.2.1 Stopping times

It is clear that the decision to either stop at time n or continue trading should be a function of all the information available up to and including time n, but we cannot use future information. This means, in mathematical terms, the decision must be a $\mathcal{F}_n$-measurable function. This leads to the following definition.

Definition 3.2.14 *Let $\{\mathcal{F}_n\}$ be a filtration defined on the probability space $(\Omega, \mathcal{F}, P)$. A random variable N with values in $\mathbb{N}_0$ such that*

$$\{N \leq n\} \in \mathcal{F}_n \qquad \text{for all } n \in \mathbb{N}$$

is called the **stopping time** *or* **optional time**.

Example 3.2.15 *Here are examples and counterexamples.*

(i) Let $V_1, V_2, \ldots$ be the value process of some investment strategy such as a buy-and-hold strategy or some sophisticated trading strategy. Suppose the investor stops his engagement, if the value process reaches the target wealth $\overline{v}$. Then he applies the stopping time

$$N_1 = \inf\{t \in \mathbb{N} : V_t \geq \overline{v}\}.$$

If, additionally, the investor closes all positions, if the value falls below a threshold $\underline{v}$, then he exits at time

$$N_2 = \inf\{t \in \mathbb{N} : V_t < \underline{v} \text{ or } V_t \geq \overline{v}\}.$$

It is easy to check formally, that N_1 as well as N_2 are stopping times in the sense of Definition 3.2.14.

(ii) Let $X_1, X_2, \ldots$ be independent random variables with $E|X_i| < \infty$ for all i. Put $\mathcal{F}_n = \sigma(X_1, \ldots, X_n)$, $n \in \mathbb{N}_0$, and define $S_0 = 0$ and $S_n = \sum_{i=1}^{n} X_i,\ n \in \mathbb{N}_0$. Then

$$N = \inf\{n \geq 0 : \ S_n \geq 1\}$$

is a stopping time, since

$$\{N = n\} = \{S_1 < 1, \ldots, S_{n-1} < 1, S_n \geq 1\} \in \mathcal{F}_n.$$

(iii) By contrast,

$$U = \sup\{n \geq 0 : \; S_n \geq 1\}$$

is not a stopping time. Indeed, the event $\{U = n\} = \{S_1, \ldots, S_{n-1} \in \mathbb{R}, S_n \geq 1, S_{n+1} < 1, S_{n+2} < 1, \ldots\}$ *depends on all* S_i*s.*

Definition 3.2.16 *Let* $\{X_n\}$ *be a* $\mathcal{F}_n$*-adapted process and* N *be a stopping time. Then* $X^N = \{X_n^N : \; n \geq 0\}$ *with*

$$X_n^N = X_{\min(n,N)} = \begin{cases} X_n, & n \leq N, \\ X_N, & n > N, \end{cases}$$

is called a **stopped process.**

Using the notation $a \wedge b = \min(a, b)$ we may write $X_n^N = X_{n \wedge N}$.

Lemma 3.2.17 *The stopped process can be represented as a discrete stochastic integral,*

$$X_n^N = X_0 + \int_0^n \varphi_t \, \mathrm{d}X_t = X_0 + \sum_{i=1}^n \varphi_i (X_i - X_{i-1}),$$

if we put $\varphi_n = \mathbf{1}(n \leq N), n \in \mathbb{N}$.

Proof. This is a simple but instructive exercise. We have

$$\begin{aligned} X_0 + \sum_{i=1}^n \varphi_i (X_i - X_{i-1}) &= X_0 + \sum_{i=1}^n \mathbf{1}(i \leq N)(X_i - X_{i-1}) \\ &= X_0 + \sum_{i=1}^{n \wedge N} (X_i - X_{i-1}) \\ &= \begin{cases} X_n, & n \leq N, \\ X_N, & n > N. \end{cases} \end{aligned}$$

It remains to show that φ_n is $\mathcal{F}_{n-1}$ measurable. But $\{n \leq N\} = \{N \leq n-1\}^c \in \mathcal{F}_{n-1}$.

3.2.2.2 Optional sampling theorems

Optional sampling theorems start with a submartingale or martingale X_n with respect to $\mathcal{F}_n$ and consider stopping times $N_1, N_2, \ldots$, which are almost surely ordered. This means,

$$N_1 \leq N_2 \leq \cdots$$

where each inequality holds true with probability one. The question arises whether the corresponding sequence

$$X_{N_1}, X_{N_2}, \ldots$$

is again a submartingale or a martingale, if X_n is a martingale.

Having in mind what we know about the martingale transform, we can easily establish the following first result of this type.

Theorem 3.2.18 *Let $\{X_n\}$ be a $\mathcal{F}_n$-martingale and N be a stopping time. Then the stopped process $\{X_n^N\}$ is a $\mathcal{F}_n$-martingale and*

$$E(X_n^N) = E(X_n) = E(X_1) = X_0.$$

If $\{X_n\}$ is a $\mathcal{F}_n$-submartingale (supermartingale), then the stopped process $\{X_n^N\}$ is a $\mathcal{F}_n$-submartingale (supermartingale).

Proof. $\{X_n^N\}$ is a martingale transformation associated to $\{X_n\}$, if $\varphi_n = \mathbf{1}(n \leq N)$. By boundedness of $\{\varphi_n\}$, the assertion follows from Theorem 3.2.12. The assertion on the submartingale property is shown by straightforward modifications of the proof of Theorem 3.2.12.

The optional sampling theorem has an important interpretation in finance: If an investor holds a long position in a financial instrument whose price process is a martingale such that its expected future price given the information until now coincides with its current price, then there does not exist an exit strategy that breaks that martingale property.

3.2.2.3 Optimal stopping

Suppose X_t stands for the value of some financial investment at time t, $t = 1, \ldots, T$. At time T we know X_T and can easily determine whether the investment was profitable. At time $T-1$ it makes sense to compare the current value X_{T-1} with the conditional expectation of the future value X_T given the available information $\mathcal{F}_{T-1}$. Human reasoning suggests to terminate the investment, if $X_{T-1} \geq E(X_T|\mathcal{F}_{T-1})$, and to stay invested otherwise. Is this exit strategy optimal in some sense?

Definition 3.2.19 (Snell Envelope)
If $\{X_t : t = 1, \ldots, T\}$ is adapted to a filtration $\{\mathcal{F}_t : t = 1, \ldots, T\}$ and satisfies

$$E|X_t| < \infty, \qquad t = 1, \ldots, T,$$

then the recursively defined process

$$\begin{aligned} Z_T &= X_T, \\ Z_t &= \max\{X_t, E(Z_{t+1}|\mathcal{F}_t)\}, \qquad t = 0, \ldots, T-1, \end{aligned}$$

is called the **Snell envelope of** $\{X_t\}$.

Notice that by definition the Snell envelope $\{Z_t\}$ dominates $\{X_t\}$, that is

$$Z_t \geq X_t, \qquad t = 0, \ldots, T.$$

Further,

$$Z_t \geq E(Z_{t+1}|\mathcal{F}_t), \qquad t = 0, \ldots, T,$$

which shows that $\{Z_t\}$ is a supermartingale.

Proposition 3.2.20 *The Snell envelope $\{Z_t : t = 1, \ldots, T\}$ is the smallest supermartingale that dominates $\{X_t : t = 1, \ldots, T\}$.*

Proof. Let $\{Y_t\}$ be an arbitrary supermartingale that dominates $\{X_t\}$. We show by backward induction that

$$Y_t \geq Z_t, \qquad t = 1, \ldots, T.$$

For $t = T$ we have $Y_T = X_T = Z_T$ by definition of the Snell envelope. Assume now that $Y_t \geq Z_t$. Then

$$Y_{t-1} \geq E(Y_t|\mathcal{F}_{t-1}) \geq E(Z_t|\mathcal{F}_{t-1})$$

by the supermartingale property and the fact that $Y_t \geq Z_t$. Further,

$$Y_{t-1} \geq X_{t-1},$$

since $\{Y_t\}$ dominates $\{X_t\}$. Therefore

$$Y_{t-1} \geq \max\{X_{t-1}, E(Z_t|\mathcal{F}_{t-1})\} = Z_{t-1},$$

which completes the proof.

Now define the stopping time

$$\tau^* = \min\{t \geq 0 : Z_t = X_t\},$$

which stops when $Z_t = X_t$ for the first time. Notice that

$$\{\tau^* = k\} = \{X_0 < Z_0, \ldots, X_{k-1} < Z_{k-1}, X_k = Z_k\}.$$

Further, τ^* takes values in $\{0, \ldots, T\}$, since $Z_T = X_T$.

Theorem 3.2.21 *The stopped Snell envelope*

$$Z_t^{\tau^*} = Z_{\min(t,\tau^*)}, \qquad t = 1, \ldots, T,$$

is a $\mathcal{F}_t$-martingale.

Proof. We have to show that

$$E(Z_{t+1}^{\tau^*} - Z_t^{\tau^*}|\mathcal{F}_t) = 0, \qquad t = 0, \ldots, T-1.$$

By Lemma 3.2.17, the stopped process $Z_t^{\tau^*}$ can be written as a discrete stochastic integral, such that for each t

$$\begin{aligned} Z_t^{\tau^*} &= Z_{\min(\tau^*,t)} \\ &= Z_0 + \sum_{i=1}^{t} \mathbf{1}(i \le \tau^*)(Z_{i+1} - Z_i). \end{aligned}$$

Hence,

$$Z_{t+1}^{\tau^*} - Z_t^{\tau^*} = \mathbf{1}(t+1 \le \tau^*)(Z_{t+1} - Z_t).$$

We claim that the above equation still holds true, when we replace Z_t by $E(Z_{t+1}|\mathcal{F}_t)$, that is

$$Z_{t+1}^{\tau^*} - Z_t^{\tau^*} = \mathbf{1}(t+1 \le \tau^*)(Z_{t+1} - E(Z_{t+1}|\mathcal{F}_t)) \tag{3.3}$$

for all $t = 0, \ldots, T-1$. To verify this fact, fix t. On $\{t+1 \le \tau^*\}^c = \{\tau^* \le t\}$, the right-hand side is zero and we have to check that the left-hand side vanishes as well. But

$$Z_{t+1}^{\tau^*} = Z_{\min(\tau^*,t+1)} = Z_{\tau^*}$$

and $Z_t^{\tau^*} = Z_{\tau^*}$, such that $Z_{t+1}^{\tau^*} - Z_t^{\tau^*} = 0$. On

$$\{t+1 \le \tau^*\} = \{X_1 < Z_1, \ldots, X_t < Z_t\}$$

we have

$$Z_t = \max\{X_t, E(Z_{t+1}|\mathcal{F}_t)\} = E(Z_{t+1}|\mathcal{F}_t)$$

leading to

$$\begin{aligned} Z_{t+1}^{\tau^*} - Z_t^{\tau^*} &= \mathbf{1}(t+1 \le \tau^*)(Z_{t+1} - Z_t) \\ &= \mathbf{1}(t+1 \le \tau^*)(Z_{t+1} - E(Z_{t+1}|\mathcal{F}_t)). \end{aligned}$$

Hence Equation (3.3) holds true. Since $\mathbf{1}(t+1 \le \tau^*) = \mathbf{1}_{\{\tau^* \le t\}^c}$ is $\mathcal{F}_t$-measurable, we can now conclude that

$$\begin{aligned} E(Z_{t+1}^{\tau^*} - Z_t^{\tau^*}|\mathcal{F}_t) &= \mathbf{1}(t+1 \le \tau^*)E\left((Z_{t+1} - E(Z_{t+1}|\mathcal{F}_t))|\mathcal{F}_t\right) \\ &= \mathbf{1}(t+1 \le \tau^*)\left(E(Z_{t+1}|\mathcal{F}_t) - E(Z_{t+1}|\mathcal{F}_t)\right) \\ &= 0, \end{aligned}$$

which verifies the martingale property.

Our aim is to maximize $E(X_\tau)$ among all stopping times that stop latest at time T. Let us now denote by $\mathcal{T}_{k,T}$ the set of all stopping times taking values in $\{k, \ldots, T\}$,

$$\mathcal{T}_{k,T} = \{\tau : \tau \text{ is a stopping time with } \tau(\Omega) \subset \{k, \ldots, T\}\}.$$

Notice that $\tau^* \in \mathcal{T}_{0,T}$, since $Z_T = X_T$ such that $\tau^* \le T$.

Definition 3.2.22 *A stopping time $\tau \in \mathcal{T}_{0,T}$ is called* **optimal for** $\{X_t : t = 1, \ldots, T\}$, *if it maximizes the expectation $E(X_\sigma)$ among all $\sigma \in \mathcal{T}_{0,T}$ in the sense that*

$$E(X_\tau) = \sup\{E(X_\sigma) : \sigma \in \mathcal{T}_{0,T}\}.$$

The main result on optimal stopping is as follows.

Theorem 3.2.23 *τ^* solves the optimal stopping problem for $\{X_t : t = 1, \ldots, T\}$ with*

$$Z_0 = E(X_{\tau^*}) = \sup\{E(X_\sigma) : \sigma \in \mathcal{T}_{0,T}\}.$$

Proof. Since the stopped process $\{Z_t^{\tau^*} : t = 1, \ldots, T\}$ is a martingale, its mean does not depend on $t = 0, \ldots, T$, such that

$$Z_0 = E(Z_0^{\tau^*}) = E(Z_T^{\tau^*}).$$

Further, $\tau^* \le T$ implies that

$$Z_T^{\tau^*} = Z_{\min(\tau^*, T)} = Z_{\tau^*},$$

and $Z_{\tau^*} = X_{\tau^*}$ by definition of τ^*. Hence,

$$E(X_{\tau^*}) = E(Z_{\tau^*}) = E(Z_T^{\tau^*}) = Z_0.$$

Let $\tau \in \mathcal{T}_{0,T}$ be an arbitrary stopping time. τ^* is optimal, if

$$E(X_\tau) \le E(X_{\tau^*}).$$

The stopped process $\{Z_t^\tau\}$ associated to τ is a supermartingale, since the Snell envelope $\{Z_t\}$ is a supermartingale. Therefore,

$$Z_0 = Z_0^\tau \ge E(Z_T^\tau) = E(Z_\tau),$$

where the last equality follows from $Z_{\min(\tau,T)} = Z_\tau$, since $\tau \le T$. But $\{Z_t\}$ dominates $\{X_t\}$, such that $Z_\tau \ge X_\tau$ (argue ω-wise), leading to

$$E(Z_\tau) \ge E(X_\tau).$$

This establishes the optimality of τ^* such that

$$Z_0 = \sup\{E(X_\sigma) : \sigma \in \mathcal{T}_{0,T}\}$$

follows.

Theorem 3.2.23 tells us that τ^* is indeed the optimal stopping time that maximizes the expected terminal value $E(X_\tau)$. Further, we see that the maximal value can be calculated by backward induction, since it is given by Z_0, the value of the Snell envelope at time 0.

3.2.2.4 A maximal inequality

We shall now discuss a *maximal inequality*, which bounds the probability that the maximal partial sum up to time n exceeds a given bound. Recall that the Chebychev inequality tells us that the probabiliy

$$P(|S_n| \geq \varepsilon) = P\left(\left|\sum_{i=1}^{n} X_i\right| > \varepsilon\right)$$

that a sum $S_n = \sum_{i=1}^{n} X_i$ of i.i.d. random variables $X_1, \dots, X_n$ with $E(X_1) = 0$ and $\sigma^2 = \text{Var}\,(X_1) \in (0, \infty)$ is larger in absolute value than a given $\varepsilon > 0$ can be bounded by

$$\text{Var}\,(S_n)/\varepsilon^2 = \sigma^2/(n\varepsilon^2).$$

This means that the probability that a random walk, e.g. a price process, drifts away from its mean by more than ε at time n can be effectively controlled by the second moment regardless of the underlying distribution of the increments X_i. The question arises whether such a result also holds true for the *maximal deviation* until time n. This means that we are seeking bounds for

$$P\left(\max_{1 \leq k \leq n} S_k > \varepsilon\right).$$

Inequalities dealing with such bounds are called maximal inequalities.

Lemma 3.2.24 *Let $\{X_n : n \in \mathbb{N}_0\}$ be a submartingale and N be a stopping time with $P(N \leq n) = 1$ for some $n \in \mathbb{N}$. Then $E(X_0) \leq E(X_N) \leq E(X_n)$.*

Proof. To verify the first inequality, notice $P(N \leq n) = 1$ implies that

$$E|X_N| \leq E\left(\max_{1 \leq i \leq n} |X_i|\right) \leq \sum_{i=1}^{n} E|X_i| < \infty.$$

We have

$$X_{\min(N,n)} = X_n \mathbf{1}(N > n) + X_N \mathbf{1}(N \leq n).$$

Hence,

$$\begin{aligned} X_N &= X_N \mathbf{1}(N \leq n) + X_N \mathbf{1}(N > n) \\ &= (X_{\min(N,n)} - X_n \mathbf{1}(N > n)) + X_N \mathbf{1}(N > n). \end{aligned}$$

Since $X_{\min(N,n)}, n \in \mathbb{N}$, is a submartingale, $E(X_{\min(N,n)}) \geq EX_0$, such that taking expectations in the second line of the last display yields

$$E(X_N) \geq E(X_0) - E(X_n \mathbf{1}(N > n)) + E(X_N \mathbf{1}(N > n)).$$

We may take the limit $n \to \infty$ and arrive at

$$E(X_N) \geq E(X_0) + \lim_{n \to \infty} E(X_N \mathbf{1}(N > n)) = E(X_0),$$

by an application of dominated convergence. To show the second inequality,

$$E(X_N) \leq E(X_n),$$

put $\varphi_n = \mathbf{1}(N < n) = \mathbf{1}(N \le n-1)$, $n \in \mathbb{N}$, and $\varphi_0 = 0$. φ_n is $\mathcal{F}_{n-1}$-measurable and

$$\int_0^n \varphi_t \mathrm{d}X_t = \sum_{i=1}^n \mathbf{1}(N < i)(X_i - X_{i-1}).$$

If $N \ge n$, we have $\int_0^n \varphi_t \mathrm{d}X_t = 0$, whereas for $N < n$

$$\int_0^n \varphi_t \mathrm{d}X_t = \sum_{i=N+1}^n (X_i - X_{i-1}) = X_n - X_N,$$

such that

$$\int_0^n \varphi_t \mathrm{d}X_t = X_n - X_{\min(n,N)}.$$

This identity shows that $X_n - X_{\min(n,N)}$ is a submartingale, since the discrete stochastic integral is a submartingale. Now we may conclude that

$$0 = E\left(\int_0^0 \varphi_t \mathrm{d}X_t\right) \le E\left(\int_0^n \varphi_t \mathrm{d}X_t\right) = \begin{cases} 0, & n \ge N, \\ E(X_n) - E(X_N), & N < n. \end{cases}$$

Hence, $E(X_N) \le E(X_n)$ follows.

Theorem 3.2.25 (DOOB'S MAXIMAL INEQUALITY)
Let $\{X_n\}$ be a submartingale, i.e.

$$E(X_{n+1}|\mathcal{F}_n) \ge X_n \quad P-a.s., \text{ for all } n \in \mathbb{N},$$

and $c > 0$ be a constant. Then $X_n^+ = \max(0, X_n)$ satisfies the inequalities

$$c\,P\Big(\max_{0\le k\le n} X_k^+ \ge c\Big) \le E\Big(X_n\, 1_{\{\max_{0\le k\le n} X_k^+ \ge c\}}\Big) \le EX_n^+.$$

In particular,

$$P\Big(\max_{0\le k\le n} X_k^+ \ge c\Big) \le \frac{E(X_n^+)}{c}.$$

Proof. Let $N = \inf\{1 \le k \le n : X_k \ge c\}$ where, by convention, $\inf \emptyset = 0$. Thus, N is the smallest time k such that X_k exceeds the level c, or otherwise k. On the intersection of the sets $\{N \le n\}$ and

$$A = \Big\{\max_{0\le k\le n} X_k^+ \ge c\Big\} = \Big\{\max_{0\le k\le n} X_k \ge c\Big\}$$

we have $X_N \ge c$. Therefore $E(X_N 1_A) \ge E(c 1_A) = c \cdot P(A)$. It remains to show

$$E(X_N 1_A) \le E(X_n^+) = E(\max(0, X_n)).$$

Lemma 3.2.24 tells us that $E(X_n) \ge E(X_N)$. Thus,

$$E(X_n) \ge E(X_N 1_A) + E(X_N 1_{A^c}).$$

On A^c we have $N = n$, such that $E(X_N 1_{A^c}) = E(X_n 1_{A^c})$.

Hence the above inequality is equivalent to $E(X_n 1_A) \geq E(X_N 1_A)$, since the term $E(X_n 1_{A^c})$ cancels. Finally, by definition of X_n^+, $X_n 1_A \leq X_n^+ 1_A$ yielding

$$E(X_n^+) \geq E(X_N 1_A),$$

which completes the proof.

We are now in a position to formulate the Kolmogorov inequality, which provides the generalization of Chebychev's inequality discussed at the beginning of this subsection. It appears as a special case of Doob's maximal inequality.

Corollary 3.2.26 (KOLMOGOROV INEQUALITY)
Let $\xi_1, \ldots, \xi_n$ be i.i.d. with $E(\xi_1) = 0$ and $E(\xi_1^2) < \infty$. If $S_n = \sum_{i=1}^n \xi_i$, then S_n^2 is a submartingale and for any $c = \varepsilon^2, \varepsilon > 0$,

$$P\left(\max_{1 \leq k \leq n} |S_k| \geq \varepsilon \right) = P\left(\max_{1 \leq k \leq n} |S_k^2| \geq c \right) \leq \frac{E(S_n^2)}{c} = \frac{E(S_n^2)}{\varepsilon^2}.$$

3.2.3 Extensions to $\mathbb{R}^d$

Let us extend some of the definitions and results, which are confined to processes attaining values in $\mathbb{R}$, to series $\{X_n\}$ with values in $\mathbb{R}^d$ for some integer $d > 1$. To this end, we denote by $\|\cdot\|$ an arbitrary vector norm on $\mathbb{R}^d$ and assume that we are given a sequence of random variables $X_n : (\Omega, \mathcal{F}, P) \to (\mathbb{R}^d, \mathcal{B}^d)$, where $\mathcal{B}^d$ denotes the Borel σ-field on $\mathbb{R}^d$. It is natural to say that $\{X_n\}$ is a $\mathcal{F}_n$-martingale if X_n is $\mathcal{F}_n$-adapted with $E\|X_n\| < \infty$ and $E(X_{n+1} \mid \mathcal{F}_n) = X_n$ a.s., for all n. Here and in what follows, for any random vector $X = (X_1, \ldots, X_d)'$ and a (sub-) σ-field $\mathcal{A}$ $E(X|\mathcal{A}) = (E(X_1|\mathcal{A}), \ldots, E(X_d|\mathcal{A}))'$. Clearly, notions such as predictability require no modifications.

Definition 3.2.27 *Let $\{X_n\}$ be a $\mathcal{F}_n$-adapted process with values in $\mathbb{R}^d$ and $\{\varphi_n\}$ be a predictable process also attaining values in $\mathbb{R}^d$. Then the $\mathbb{R}$-valued process $I_n : (\Omega, \mathcal{F}, P) \to (\mathbb{R}, \mathcal{B})$, defined by*

$$I_n = \sum_{i=1}^n \varphi_i'(X_i - X_{i-1}), \quad n \geq 0,$$

is called a **(multivariate) discrete stochastic integral** *and written as*

$$I_n = \int_0^n \varphi_t' \, \mathrm{d}X_t.$$

Theorem 3.2.12 carries over in the following form.

Theorem 3.2.28 *Suppose $\{X_n\}$ is a $\mathcal{F}_n$- (sub-) martingale and $\{\varphi_n\}$ $\mathcal{F}_n$-adapted. If*

$$\|\varphi_n\| \in L_p \qquad \text{and} \qquad \|X_n\| \in L_q,$$

where $p, q \in [0, \infty]$ with $\frac{1}{p} + \frac{1}{q} = 1$, then I_n is a $\mathcal{F}_n$-(sub-) martingale, and the same holds true for $X_0 + I_n$ for any constant X_0.

Proof. We may follow the proof of Theorem 3.2.12 using

$$E|X_0 + I_n| \leq E|X_0| + \sum_{i=1}^{n} E\|\varphi_i' X_i\| + \|\varphi_i' X_{i-1}\| < \infty,$$

since by virtue of Hölder's inequality

$$E\|\varphi_i' X_i\| \leq E(\|\varphi_i\| \cdot \|X_i\|) \leq \left(\int \|\varphi_i\|^p \, \mathrm{d}P \right)^{1/p} \cdot \left(\int \|X_i\|^q \, \mathrm{d}P \right)^{1/q} < \infty.$$

3.3 Stationarity

In order to be in a position to draw conclusions from a time series $X_1, X_2, \ldots$, the observations must be comparable in some appropriate sense. What are, in a certain sense, minimal assumptions to make basic formulas such as sample averages meaningful?

Suppose we observe the first T observations $X_1, \ldots, X_T$. If those observations are identically distributed, they have the same mean, $E(X_t) = E(X_1)$ for all $t = 1, \ldots, T$, and it makes sense to average them, in order to average out random fluctuations and estimate their common mean.

Next, suppose that $E(X_t) = 0$ for all $t = 1, \ldots, T$. If we want to estimate quantities that depend on more than one observation such as the covariance $E(X_t X_{t+1})$ between successive observations, that cross-moment should be the same for all $t = 1, \ldots, T-1$, in order to make averaging meaningful. This means that all pairs of successive observations

$$(X_1, X_2), (X_2, X_3), \ldots \tag{3.4}$$

should have the same mean cross-products,

$$E(X_1 X_2) = E(X_2 X_3) = \cdots.$$

A sufficient condition for that property to hold is that these pairs (3.4) have the same law. These considerations lead to the notion of weak and strict stationarity, which we discuss before introducing the most important quantities to describe serial dependence.

3.3.1 Weak and strict stationarity

Definition 3.3.1 (WEAK STATIONARITY)
Let $\{X_t\} = \{X_t : t \in \mathbb{Z}\}$ be a time series with values in $\mathbb{R}$ or $\mathbb{C}$.

(i) $\{X_t\}$ is called **strictly stationary**, *if for all $t_1, \ldots, t_n \in \mathbb{Z}$ and all $h \in \mathbb{Z}$*

$$(X_{t_1}, \ldots, X_{t_n}) \stackrel{d}{=} (X_{t_1+h}, \ldots, X_{t_n+h})$$

holds true.

(ii) A time series $\{X_t\}$ is called **L_2-time series**, *if*

$$E|X_t|^2 = E(X_t \overline{X_t}) < \infty, \qquad t \in \mathbb{Z},$$

(iii) A L_2-time series is called **(weakly) stationary**, *if*

$$E(X_t) = E(X_1), \qquad t \in \mathbb{Z},$$

and if

$$E(X_t X_{t+h}) = E(X_1 X_{1+h}),$$

for all h.

Remark 3.3.2

(i) The term stationary *refers to a weakly stationary series.*

(ii) Strict stationarity means invariance *of all finite-dimensional distributions with respect to* **time shifts** *(also called* **lags**), *$h \in \mathbb{Z}$: Consider the h-shift operator*

$$L_h(\{X_t\}) = \{X_{t+h} : t \in \mathbb{Z}\}$$

and denote by

$$\Pi_{t_1,\dots,t_n}(\{X_t\}) = (X_{t_1}, \dots, X_{t_n})$$

the projection onto the coordinates $t_1, \dots, t_n$. $\{X_t\}$ is strictly stationary, if and only if

$$\Pi_{t_1,\dots,t_n}(L_h(\{X_t\})) \stackrel{d}{=} \Pi_{t_1,\dots,t_n}(\{X_t\}),$$

for all $t_1, \dots, t_n, h \in \mathbb{Z}$.

(ii) Although many results only require weak stationarity, in practical applications one usually applies at least one procedure that makes use of a central limit theorem, and those results typically make more restrictive assumptions such as strict stationarity or that the time series is induced by an underlying sequence of i.i.d. error terms.

Definition 3.3.3 *Let $\{X_t\}$ be a weakly stationary time series. Then the function*

$$\gamma_X(h) = \text{Cov}(X_{t+h}, X_t), \qquad h \in \mathbb{Z},$$

is well defined and is called an **autocovariance function**. *Provided $\gamma_X(0) > 0$, the related function*

$$\rho_X(h) = \text{Cor}(X_1, X_{1+h}) = \frac{\gamma_X(h)}{\gamma_X(0)}, \qquad h \in \mathbb{Z},$$

is the **autocorrelation function (ACF)**.

Remark 3.3.4 *We shall denote the $T \times T$ covariance matrix of the random subvector $(X_1, \dots, X_T)$ of a L_2 series by*

$$\Gamma_X = \left(\text{Cov}(X_i, X_j)\right)_{i,j},$$

if the dependence on the dimension matters, by Γ_T. *Notice that for a weakly stationary time series*

$$\Gamma_T = \begin{pmatrix} \gamma(0) & \gamma(1) & \cdots & \gamma(T-1) \\ \gamma(1) & \gamma(0) & \cdots & \gamma(T-2). \\ \vdots & \ddots & \ddots & \vdots \\ \gamma(T-1) & \cdots & \gamma(1) & \gamma(0) \end{pmatrix}$$

It is time for some examples.

Example 3.3.5

(i) *Let* ϵ_t, $t \in \mathbb{Z}$, *be i.i.d. random variables with mean zero and common variance* $\sigma^2 \in (0, \infty)$. *It is clear that* $\{\epsilon_t\}$ *is weakly stationary. Define the process*

$$X_t = \epsilon_t - \epsilon_{t-1}, \qquad t \in \mathbb{Z},$$

of first-order differences. We have $E(X_t) = 0$ *and, since* $E(\epsilon_i \epsilon_j) = \sigma^2$ *if* $i = j$ *and* $E(\epsilon_i \epsilon_j) = 0$ *otherwise,*

$$\begin{aligned} E(X_t X_{t+h}) &= E\left[(\epsilon_t - \epsilon_{t-1})(\epsilon_{t+h} - \epsilon_{t+h-1})\right] \\ &= E(\epsilon_t \epsilon_{t+h}) - E(\epsilon_{t-1}\epsilon_{t+h}) - E(\epsilon_t \epsilon_{t+h-1}) + E(\epsilon_{t-1}\epsilon_{t+h-1}) \\ &= \begin{cases} 2\sigma^2, & h = 0, \\ -\sigma^2, & h = -1, 1, \\ 0, & |h| > 1. \end{cases} \end{aligned}$$

Hence, $\{X_t\}$ *is weakly stationary. The associated ACF is given by*

$$\rho_X(h) = \begin{cases} 1, & h = 0, \\ -1/2, & |h| = 1, \\ 0, & |h| > 1. \end{cases}$$

(ii) *Let* $\{X_t : t \in \mathbb{Z}\}$ *be a weakly stationary time series with autocovariance function*

$$\gamma(h) = E(X_0 X_h), \qquad h \in \mathbb{Z}.$$

Consider

$$Y_t := X_t + X_0, \qquad t \in \mathbb{Z}.$$

Then one easily checks that

$$\text{Cov}(Y_s, Y_t) = \gamma(0) + \gamma(s) + \gamma(t) + \gamma(s-t).$$

Hence, $\{Y_t\}$ *is not (weakly) stationary.*

(iii) *The following example looks a little odd: Let X be a random variable* $X \stackrel{d}{=} -X$ *and* $\sigma^2 = \text{Var}(X) \in (0, \infty)$. *Define the time series*

$$X_t := (-1)^t X, \qquad t \in \mathbb{N}.$$

Then $\{X_t : t \geq 1\}$ *is weakly stationary. Indeed,* $E(X) = 0$ *implies*

$$E(X_t) = 0, \qquad t \in \mathbb{N},$$

and

$$\begin{aligned} \text{Cov}(X_t,\ X_{t+k}) &= \text{Cov}\left((-1)^t X,\ (-1)^{t+k} X\right) \\ &= (-1)^{2t+k}\ \sigma^2 \\ &= \begin{cases} \sigma^2, & k \text{ even}, \\ -\sigma^2, & k \text{ odd}, \end{cases} \end{aligned}$$

Hence, $\gamma_X(t,\ t+k)$ *does not depend on t at all.*

(iv) *Nonstationary time series may possess stationary subseries. For example, consider the following periodic series*

$$X_t = \sin\left(\frac{2\pi}{r}t\right) + \epsilon_t, \qquad t \in \mathbb{Z},$$

where $\{\epsilon_t\}_{t\in\mathbb{Z}}$ *is weakly (strictly) stationary. The trend function has the period r. Define for* $0 \leq h < r$

$$Y_t = X_{t \cdot r + h}, \qquad t \in \mathbb{Z},$$

Then $\{Y_t\}_{t\in\mathbb{Z}}$ *is weakly (stricly) stationary.*

(v) *The following example shows that it is not sufficient to look at the mean and variance in order to check whether or not a time series is stationary. Let* $X_1, X_2, \ldots$ *be i.i.d. with* $E(X_1) = 0$ *and* $\sigma^2 = \text{Var}(X_1) \in (0, \infty)$. *Consider the random walk*

$$S_n = \sum_{i=1}^{n} X_i, \qquad n = 1, 2, \ldots.$$

Clearly,

$$E(S_n) = 0 \qquad \textit{and} \qquad \text{Var}(S_n) = n\sigma^2.$$

Hence, $\{S_n : n \in \mathbb{N}\}$ *is not weakly stationary, since that would imply that the variance is constant with respect to n. Let us stabilize the variance and consider*

$$S_n^* = \frac{S_n}{\sqrt{n}}, \qquad n = 1, 2, \ldots.$$

Then $E(S_n^*) = 0$ and $\text{Var}(S_n^*) = \sigma^2$. Let us check whether S_n^* is weakly stationary. We have

$$\begin{aligned}\text{Cov}(S_n^*, S_m^*) &= \frac{1}{\sqrt{nm}} E(S_n S_m) \\ &= \frac{1}{\sqrt{nm}} E\left[\left(\sum_{i=1}^{n} X_i\right)\left(\sum_{i=1}^{m} X_i\right)\right] \\ &= \frac{1}{\sqrt{nm}} E\left(\sum_{i=1}^{n}\sum_{j=1}^{m} X_i X_j\right) \\ &= \frac{1}{\sqrt{nm}} \sum_{i=1}^{n}\sum_{j=1}^{m} E(X_i X_j).\end{aligned}$$

But $E(X_i X_j) = \sigma^2$ if $i = j$ and $E(X_i X_j) = 0$ otherwise. Hence,

$$\text{Cov}(S_n^*, S_m^*) = \frac{\sigma^2}{\sqrt{nm}} \min(n, m),$$

for all $n, m \in \mathbb{N}$, which is not a function of $n - m$. This shows that $\{S_n^*\}$ is not weakly stationary.

Lemma 3.3.6 *Strictly stationary L_2 time series are weakly stationary, but there exist weakly stationary series that are not strictly stationary.*

Proof. By strict stationarity of $\{X_t\}$,

$$X_1 \stackrel{d}{=} X_t \quad \text{and} \quad (X_t, X_{t+h}) \stackrel{d}{=} (X_1, X_{1+h})$$

for all t. Hence, since $E(X_1^2) < \infty$ and $\mu = E|X_1| < \infty$, we have $E(X_t) = E(X_1)$ as well as $E(X_t X_{t+h}) = E(X_1 X_{1+h})$ for all t and h. Here is a counterexample for a weakly stationary time series, which is not strictly stationary. Take independent random variables X_t with marginals

$$X_t \sim N(0, 1/3), \qquad \text{if } t \text{ is even}$$

and

$$X_t \sim U(-1, 1), \qquad \text{if } t \text{ is odd.}$$

Obviously, $\{X_t\}$ is not strictly stationary. But $E(X_t) = 0$ and $\text{Var}(X_t) = 1/3$ for all t as well as $\gamma_X(h) = 0$ for $h > 0$, such that $\{X_t\}$ is weakly stationary.

Lemma 3.3.7 *The autocovariance function of a weakly stationary process satisfies the following properties.*

(i) *$\gamma(0)$ is real-valued and non-negative.*

(ii) *$\gamma(-h) = \overline{\gamma(h)}$ for all $h \in \mathbb{Z}$.*

(iii) $|\gamma(h)| \leq \gamma(0)$ *for all* $h \in \mathbb{Z}$.

(iv) $\gamma(h) = 0$, $|h| \geq 1$, *for an uncorrelated time series.*

Proof. Let $\{X_t\}$ be weakly stationary.

(i) Clearly, $\gamma(0) = E((X_1 - \mu)\overline{(X_1 - \mu)}) = E|X_1|^2 \in [0, \infty)$.

(ii) Let $\mu = E(X_1)$. We have

$$\begin{aligned}\gamma(-h) &= E\left((X_{t-h} - \mu)\overline{(X_t - \mu)}\right) \\ &= E(X_t - \mu)\overline{(X_{t+h} - \mu)} \\ &= E\left(\overline{\overline{(X_t - \mu)}(X_{t+h} - \mu)}\right) \\ &= \overline{E\overline{(X_t - \mu)}(X_{t+h} - \mu)} \\ &= \overline{\gamma(h)},\end{aligned}$$

for all h.

(iii) This is an application of the Cauchy–Schwarz inequality.

$$\begin{aligned}|\gamma(h)| &= |\operatorname{Cov}(X_1, X_{1+h})| \\ &\leq \sqrt{\operatorname{Var}(X_1)\operatorname{Var}(X_{1+h})} \\ &= \sqrt{\gamma(0)^2} \\ &= \gamma(0).\end{aligned}$$

(iv) Clearly, for $h \neq 0$ we have $\gamma(h) = \operatorname{Cov}(X_1, X_{1+h}) = 0$.

A common replacement for i.i.d. error terms are white-noise processes.

Definition 3.3.8 (White Noise)
Let $\{\epsilon_t\}$ *be a sequence of pairwise uncorrelated random variables,*

$$\operatorname{Cov}(\epsilon_t, \epsilon_s) = 0 \qquad \text{for } t \neq s,$$

with

$$E(\epsilon_t) = 0 \qquad \text{and} \qquad \operatorname{Var}(\epsilon_t) = \sigma^2 \in (0, \infty).$$

Then $\{\epsilon_t\}$ *is called a* **white-noise process (or series)** *or simply* **white noise**, *denoted by*

$$\epsilon_t \sim \mathrm{WN}(0, \sigma^2).$$

Remark 3.3.9

(i) A white-noise process is weakly stationary with autocovariance function

$$\gamma(h) = \begin{cases} \sigma^2, & h = 0, \\ 0, & \text{elsewhere.} \end{cases}$$

(ii) *Notice that the assumption $\epsilon_t \sim \mathrm{WN}(0, \sigma^2)$ only fixes the autocovariance function but neither specifies the marginal distributions nor the finite-dimensional distributions. Therefore, the ϵ_t are in general not independent. For these reasons, the assumption*

$$\epsilon_t \sim \mathrm{WN}(0, \sigma^2)$$

is much more general than

$$\epsilon_t \overset{i.i.d.}{\sim} N(0, \sigma^2).$$

(iii) *A $\mathcal{F}_t$-martingale difference sequence $\{\epsilon_t\}$ satisfies, by assumption,*

$$E|\epsilon_t| < \infty \qquad \text{and} \qquad E(\epsilon_t|\mathcal{F}_{t-1}) = 0$$

for all t. If, in addition, ϵ_t is a L_2 series, then $\{\epsilon_t\}$ is uncorrelated, cf. Lemma 3.2.9 and the preceding discussion. Therefore, a L_2 MDS is a white-noise process, but the MDS property provides more structure, namely that the conditional means $E(\epsilon_t|\mathcal{F}_{t-1})$ vanish, which implies $E(\epsilon_t) = 0$, whereas for a white-noise process we only have $E(\epsilon_t) = 0$, but this does not imply $E(\epsilon_t|\mathcal{F}_{t-1}) = 0$.

The autocovariance function

$$\gamma(h) = \mathrm{Cov}(X_1, X_{1+h}), \qquad h \in \mathbb{Z},$$

of a weakly stationary process $\{X_t\}$ with mean $\mu = E(X_1)$ is a meaningful approach to measure serial dependence. The weak stationarity implies that the random variables

$$(X_t - \mu)(X_{t+h} - \mu), \qquad t = 1, \ldots, T - h,$$

have the same expectation, namely $\gamma(h)$. Thus, it makes sense to average them and consider the estimator

$$\widetilde{\gamma}_T(h) = \frac{1}{T-h} \sum_{t=1}^{T-h} (X_{t+h} - \mu)(X_t - \mu),$$

provided μ is known to us. It is straightforward to check that

$$E(\widetilde{\gamma}_T(h)) = \gamma(h),$$

whatever the value of $\gamma(h)$. Hence, $\widetilde{\gamma}_T(h)$ is an unbiased estimator for $\gamma(h)$.

Since μ is unknown in practice, one replaces it by the sample mean.

Definition 3.3.10 *Let $\{X_t\}$ be a time series. Then*

$$\widehat{\gamma}_T(h) = \frac{1}{T} \sum_{t=1}^{T-h} (X_{t+h} - \overline{X}_T)(X_t - \overline{X}_T), \qquad 0 \le h \le T - 1,$$

$$\widehat{\gamma}_T(h) = 0, \qquad T - 1 < h,$$

is called an **empirical or sample autocovariance function***. If* $\widehat{\gamma}_T(0) > 0$*, then*

$$\widehat{\rho}_T(h) := \frac{\widehat{\gamma}(h)}{\widehat{\gamma}(0)}, \qquad h \geq 0,$$

is the **empirical or sample autocorrelation function (ACF).**

Notice that those estimators sum up $T - h$ terms but use the factor $1/T$ instead of $1/(T - h)$. This is common in time-series analysis, since usually T is quite large, such that the difference is negligible.

The autocovariance matrix Γ_T is then estimated by

$$\widehat{\Gamma}_T = \begin{pmatrix} \widehat{\gamma}_T(0) & \widehat{\gamma}_T(1) & \cdots & \widehat{\gamma}_T(T-1) \\ \widehat{\gamma}_T(1) & \ddots & \ddots & \vdots \\ \vdots & \ddots & \ddots & \widehat{\gamma}_T(1) \\ \widehat{\gamma}_T(T-1) & \cdots & \widehat{\gamma}_T(1) & \widehat{\gamma}_T(0). \end{pmatrix}$$

Figure 3.2 depicts the sample autocorrelation function for log returns of Credit Suisse. There seem to be no significant autocorrelations, which is a rather typical picture for daily returns of an asset. The horizontal lines correspond to confidence intervals for the sample autocorrelations. They are are based on the fact that

$$\sqrt{T}[\widehat{\gamma}_T(h) - \gamma(h)] \stackrel{d}{\to} N(0, \sigma^4)$$

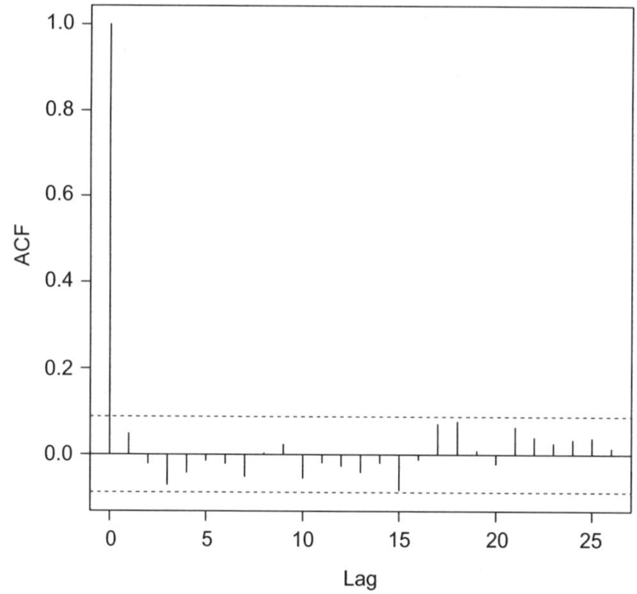

Figure 3.2 Sample autocorrelation function of the log returns for Credit Suisse from 2/10/2003 to 9/1/2005.

and

$$\sqrt{T}[\widehat{\rho}_T(h) - \rho(h)] \stackrel{d}{\to} N(0, 1),$$

as $T \to \infty$, provided the X_t are i.i.d. with common variance $\sigma^2 \in (0, \infty)$ and a finite moment of the order $4 + \delta$ for some $\delta > 0$, see Theorem 8.8.3. Therefore, an asymptotic significance test for testing the null hypothesis $H_0 : \gamma(h) = 0$ against the alternative hypothesis $H_1 : \gamma(h) \neq 0$ is given by

$$\phi = \begin{cases} 1, & \text{if } |\widehat{\rho}_T(h)| > \Phi^{-1}(1 - \alpha/2)/\sqrt{T}, \\ 0, & \text{otherwise.} \end{cases}$$

Further, a confidence interval for $\rho(h)$ is given by

$$\widehat{\rho}_T(h) \pm \frac{\Phi^{-1}(1 - \alpha/2)}{\sqrt{T}}.$$

That is, for $\alpha = 0.05$ the sample autocorrelation $\widehat{\rho}_T(h)$ should lie between $-1.96/\sqrt{T}$ and $1.96/\sqrt{T}$.

Lemma 3.3.11 *$\widehat{\Gamma}_T$ is positive semidefinite for all $T \geq 2$.*

Next, consider the dependence measure

$$P(X_t > 0, X_{t+h} > 0) = E(\mathbf{1}(X_t > 0, X_{t+h} > 0)). \tag{3.5}$$

If X_t is the return of an asset at time t, that measure considers the joint probability that two returns with time lag h have both a positive sign. We could compare that value with the corresponding probability $1/4$ we obtain when the returns are independent and symmetrically distributed around zero. Notice that weak stationarity does not ensure that Equation (3.5) is invariant under time shifts, since Equation (3.5) can not be written as a function of the mean and the autocovariance function. But we may regard Equation (3.5) as a functional τ defined on the set

$$\mathcal{P} = \{\mathbb{P} : \mathbb{P} \text{ is a probability measure on } \mathbb{R}^2\}$$

evaluated at $P_{(X_t, X_{t+h})} \in \mathcal{P}$, if we define $\tau : \mathcal{P} \to \mathbb{R}$ by

$$\tau(\mathbb{P}) = \mathbb{P}(\{(x, y) \in \mathbb{R}^2 : x > 0, y > 0\}),$$

for $\mathbb{P} \in \mathcal{P}$. Suppose now that $X_1, X_2, \ldots$ is strictly stationary. Then the random vectors

$$(X_1, X_{1+h}), (X_2, X_{2+h}), \ldots$$

have all the same distribution and therefore represent identically distributed random vectors. Thus, it makes sense to consider the estimator

$$\widetilde{\tau}_T = \frac{1}{T - h} \sum_{t=1}^{T-h} \mathbf{1}(X_t > 0, X_{t+h} > 0),$$

which is easily seen to be an unbiased estimator for $\tau(P_{(X_1, X_{1+h})})$.

The question arises whether transformed processes such as $\mathbf{1}(X_t > 0, X_{t+h} > 0)$ are again strictly stationary. Indeed, this nice property holds true.

Proposition 3.3.12 *Let $X_0, X_1, \ldots$ be a strictly stationary series and*

$$g : \mathbb{R}^{\mathbb{N}_0} \to \mathbb{R}, \qquad (x_0, x_1, \ldots) \mapsto g(x_0, x_1, \ldots)$$

be a measurable mapping. Then the series

$$Y_k = g(X_k, X_{k+1}, \ldots), \qquad k \in \mathbb{N},$$

is again strictly stationary.

Proof. We shall show that

$$P((Y_0, Y_1, \ldots) \in B) = P((Y_k, Y_{k+1}, \ldots) \in B)$$

holds true for all measurable sets $B \subset \mathbb{R}^{\mathbb{N}_0}$. This implies strict stationarity. Put $x = (x_0, x_1, \ldots) \in \mathbb{R}^{\mathbb{N}_0}$ and define the functions

$$g_k(x) = g(x_k, x_{k+1}, \ldots),$$

for $k \geq 1$. For a measurable set $B \subset \mathbb{R}^{\mathbb{N}_0}$ let

$$A = \{x : (g_0(x), g_1(x), \ldots) \in B\}.$$

Now we have, by definition of A,

$$\begin{aligned} P((Y_0, Y_1, \ldots) \in B) &= P((g_0(X_0, X_1, \ldots), g_1(X_0, X_1, \ldots), \ldots) \in B) \\ &= P((X_0, X_1, \ldots) \in A) \\ &= P((X_k, X_{k+1}, \ldots) \in A) \\ &= P((Y_k, Y_{k+1}, \ldots) \in B), \end{aligned}$$

which completes the proof.

3.4 Linear processes and ARMA models

The working horse of classic time-series analysis is the machinery developed for ARMA models. They appear as special cases of linear processes. Therefore, we start with linear processes and discuss some important basic concepts related to them such as series in the lag operator and inversion, and then specialize to ARMA processes.

3.4.1 Linear processes and the lag operator

Definition 3.4.1 *Let $\{X_n : n \in I\}$ be a set of random variables defined on a common probability space with index set $I = \mathbb{N}$ or $I = \mathbb{Z}$ and let $\{a_n : n \in I\}$ be a set of real numbers. Then the formal series*

$$Y_t = \sum_{i \in I} a_i X_{t-i}, \qquad t \in I,$$

is called a **linear process** *or* **linear filter**. *A linear process is called* **causal**, *if* $a_i = 0$ *for* $i < 0$.

Notice that in the above definition arbitrary sequences $\{X_t\}$ are allowed. However, often i.i.d. or white noise sequences are considered.

A simple example of a linear process is the time series $\{Y_t\}$ defined by

$$Y_t = a_0 X_t + a_1 X_{t-1} + \cdots + a_q X_{t-q}, \qquad t \in \mathbb{N},$$

i.e. a linear combination of $X_t, \ldots, X_{t-q}$. Such a process is called a moving average process of order q, abbreviated by MA(q).

Linear filters can be expressed via the **lag operator**, which is formally defined on the space of all sequences of real numbers and random variables, respectively, by

$$L(\{X_n\}) = \{X_{n-1}\}.$$

Here $\{X_{n-1}\} = \{X_{n-1} : n \in \mathbb{Z}\}$ denotes the shifted series, i.e.

$$L(\{X_n\}) = \{Z_n\} \qquad \text{if and only if} \qquad Z_n = X_{n-1} \text{ for all } n.$$

It is common to write the series $L(\{X_n\})$ element-wise, even if one means the whole series:

$$L(X_n) = LX_n = X_{n-1}, \qquad n \in \mathbb{Z}.$$

Obviously, the operator L is linear, this means it satisfies the following rules of calculation

(i) $L(X_n + Y_n) = L(X_n) + L(Y_n)$ for any series $\{X_n\}$ and $\{Y_n\}$.

(ii) $L(aX_n) = aL(X_n)$ for any scalar a.

We can consider the composition L^2,

$$L^2(X_n) = L(L(X_n)) = L(X_{n-1}) = X_{n-2},$$

and more generally powers of L,

$$L^k X_n = L(L^{k-1}(X_n)) = \cdots = X_{n-k}$$

for $k \geq 2$. Further, one denotes by $L^0 = 1$ the identity operator

$$1\, X_n = X_n.$$

Finally, one can also introduce negative powers by

$$L^{-k} X_n = X_{n+k}$$

for $k \geq 1$. Consequently, one can study polynomials in L such as $4L^3 - L^2 + 2L + 3$, which yields

$$(4L^3 - L^2 + 2L + 3)(X_n) = 4X_{n-3} - X_{n-2} + 2X_{n-1} + 3X_n,$$

when applied to $\{X_n\}$.

Thus, polynomials in L make sense and are called **lag polynomials**. More generally, one can also define (formal) **lag series operators**

$$\theta(L) = \sum_{i \in \mathbb{N}} \theta_i L^i$$

with coefficients $\theta_i \in \mathbb{C}$, given by

$$\theta(L)X_n = (\theta(L)(\{X_n\}))_n = \sum_i \theta_i X_{n-i}.$$

When a lag series is given by coefficients $\{\theta_i\}$, the associated lag series operator (acting on sequences) is frequently denoted by $\theta(L)$, and the notation $\theta(L)$ often means the coefficients are $\{\theta_i\}$. Although not in all instances, we shall often follow this convention.

We see that linear processes can be nicely written as $Y_n = \theta(L)X_n$ via lag operators of the form $\theta(L)$. The question arise when this formal series exists in L_2 or even a.s. and whether one gets a stationary process when $\{X_n\}$ is stationary.

Let us consider the important case of a causal linear process with white-noise innovations,

$$X_t = \sum_{k=0}^{\infty} \psi_k \epsilon_{t-k}, \qquad \epsilon_t \sim \mathrm{WN}(0, \sigma^2),$$

for some $\sigma^2 \in (0, \infty)$. We claim that X_t is stationary, provided that the condition

$$\sum_{k=0}^{\infty} \psi_k^2 < \infty$$

holds true. Indeed, under that condition the truncated series $X_t^{(n)} = \sum_{k=0}^{n} \psi_k \epsilon_{t-k}$ satisfies for $n \leq m$

$$\begin{aligned} E(X_t^{(n)} - X_t^{(m)})^2 &= \sum_{k,l=n+1}^{m} \psi_k \psi_l \operatorname{Cov}(\epsilon_{t-k}, \epsilon_{t-l}) \\ &= \sigma^2 \sum_{k=n+1}^{m} \psi_k^2, \end{aligned}$$

which tends to 0, as $n, m \to \infty$. Hence, $\{X_t^{(n)} : n \geq 1\}$ is a L_2-Cauchy sequence, such that the limit

$$X_t = \psi(L)\epsilon_t = \sum_{k=0}^{\infty} \psi_k \epsilon_{t-k},$$

exists in the L_2 sense. A similar argument ensures the validity of $E(X_t) = 0$ and the following calculation of the autocovariances:

$$\begin{aligned} \gamma_X(h) &= \operatorname{Cov}(X_t, X_{t+h}) \\ &= \sum_{k,l=0}^{\infty} \psi_k \psi_l \operatorname{Cov}(\epsilon_{t-k}, \epsilon_{t+h-l}) \\ &= \sigma^2 \sum_{k=0}^{\infty} \psi_k \psi_{k+h} \end{aligned}$$

for any lag $h \in \mathbb{Z}$. This shows that $X_t = \psi(L)\epsilon_t$ is stationary. It is worth mentioning that the convergence of the above series is straightforward for the important case that the coefficients ψ_k are nonincreasing, since then $\psi_{k+h} \leq \psi_k$.

More generally, if $\{\epsilon_t\}$ is an arbitrary stationary time series with mean μ and autocovariances $\gamma_\epsilon(h)$, X_t is stationary with mean

$$E(X_t) = \mu \sum_{k=0}^{\infty} \psi_k$$

and autocovariances given by

$$\gamma_X(h) = \sum_{k,l=0}^{\infty} \psi_k \psi_l \gamma_\epsilon(l - k + h), \qquad h \in \mathbb{Z}.$$

Definition 3.4.2 *The lag operator* $\psi(L) = \sum_i \psi_i L^i$ *is called a* L_1 **filter** *or* **absolutely summable**, *if* $\sum_i |\psi_i| < \infty$.

For L_1 filters the following results hold, cf. Theorem A.5.1.

Proposition 3.4.3 *Let* $\psi(L)$ *be an absolutely summable linear filter and* $\{X_n\}$ *be a sequence of random variables.*

(i) If $\sup_n E|X_n| < \infty$, *then* $Y_n = \psi(L)X_n$ *exists a.s. and* $EY_n = \psi(L)\mu_n$ *where* $\mu_n = E(X_n)$.

(ii) If $\sup_n EX_n^2 < \infty$, *then* $Y_n = \psi(L)X_n$ *converges in* L_2.

In particular, the above result tells us that if $\sum_i |\psi_i| < \infty$, then one can apply the filter $\psi(L) = \sum_i \psi_i L^i$ to a white-noise process $\{\epsilon_t\}$, i.e. $X_t = \psi(L)\epsilon_t$ exists (a.s.)

If we are given a lag operator

$$\psi(L) = \sum_i \theta_i L^i,$$

we may formally replace the lag operator L by a complex variable z.

Definition 3.4.4 *If* $Y_n = \sum_i \theta_i X_{n-i}$ *is a linear process with coefficients* $\{\theta_i\}$ *and associated lag operator series* $\psi(L) = \sum_i \theta_i L^i$, *then the (formal) series*

$$\psi(z) = \sum_i \theta_i z^i, \qquad z \in \mathbb{C},$$

is called a z-**transform**. *If* $\psi(L)$ *is of the form*

$$\psi(L) = 1 - \theta_1 L - \cdots - \theta_p L^p$$

for some $p \in \mathbb{N}$, *then the associated* z*-transform*

$$\psi(z) = 1 - \theta_1 z - \cdots - \theta_p z^p, \qquad z \in \mathbb{C}$$

is called a **characteristic function** *(of the filter).*

Example 3.4.5 *It is time for some examples.*

(i) *If $\psi(L) = \sum_{i=0}^{\infty} \theta_i L^i$ is a L_1-filter, i.e. $\sum_{i=0}^{\infty} |\theta_i| < \infty$, then $\psi(z)$ is a power series, which converges at least on the unit disc*

$$E = \{z \in \mathbb{C} : |z| \leq 1\}.$$

(ii) *If $p(L) = \sum_{i=-\infty}^{0} a_i L^i$, then*

$$p(z) = a_0 + \frac{a_1}{z} + \frac{a_2}{z^2} + \cdots$$

is a Laurent series. If $\{a_i\}$ is absolutely summable, then $p(z)$ exists for all $z \in \mathbb{C}$ satisfying $|\frac{1}{z}| \leq 1 \Leftrightarrow |z| \geq 1$.

Suppose we are given the following two linear processes

$$Y_n = \sum_i a_i X_{n-i},$$
$$X_n = \sum_i b_i Z_{n-i}.$$

Let

$$p(z) = \sum_i a_i z^i \quad \text{and} \quad q(z) = \sum_i b_i z^i,$$

be the corresponding z-transforms. Then we may write

$$Y_n = p(L)X_n, \qquad X_n = q(L)Z_n.$$

Formally substituting X_n by $q(L)Z_n$ in the expression for Y_n suggests the validity of the rule

$$Y_n = p(L)X_n = p(L)q(L)Z_n.$$

This means

$$Y_n = r(L)Z_n$$

with $r(L) = p(L)q(L)$ or equivalently

$$r(z) = p(z)q(z),$$

for all z where $p(z)$ and $q(z)$ are defined. This rule of calculations is indeed valid.

Proposition 3.4.6 *Let $Y_n = p(L)X_n$ and $X_n = q(L)Z_n$ be given by lag operators $p(L)$ and $q(L)$ whose z-transforms converge for $z \in \mathbb{C}$ with $|z| \leq \rho$ for some $\rho > 0$. Then $Y_n = r(L)Z_n$ where $r(L)$ is given by its z-transform*

$$r(z) = p(z)q(z) = \sum_{i,j=0}^{\infty} a_i b_j z^{i+j}, \qquad z \in \mathbb{C} \text{ with } |z| \leq \rho.$$

This means that the composition $p(L) \circ q(L)$ is given by the product

$$p(L) \cdot q(L) = \sum_{i,j=0}^{\infty} a_i b_j L^{i+j}.$$

Proof. We have

$$\begin{aligned} Y_n &= \sum_i a_i X_{n-i} \\ &= \sum_{i=0}^{\infty} a_i \sum_{j=0}^{\infty} b_j Z_{n-(i+j)} \\ &= \sum_{i=0}^{\infty} \sum_{j=0}^{\infty} a_i b_j L^{i+j} Z_n = \left(\sum_{i,j=0}^{\infty} a_i b_j L^i L^j \right) Z_n, \end{aligned}$$

which shows that $Y_n = r(L)Z_n$, where $r(L) = p(L)q(L)$. But the z-transform of $r(L)$ is, by definition, $r(z) = p(z)q(z) = \sum_{i,j=0}^{\infty} a_i b_j z^{i+j}$. It is well defined for all $z \in \mathbb{C}$ with $|z| \leq \rho$.

3.4.2 Inversion

Suppose that a time series $\{Y_n\}$ can be computed from the series $\{X_n\}$ by a L_1 filter $\theta(L) = \sum_i \theta_i L^i$, i.e.

$$Y_n = \theta(L)X_n = \sum_i \theta_i X_{n-i}, \qquad \text{for all } n.$$

The question arises whether this equation of sequences, $\{Y_n\} = \theta(L)(\{X_n\})$ can be *inverted* in the sense that there exists another (L_1?) filter $\varphi(L) = \sum_i \varphi_i L^i$ such that

$$X_n = \varphi(L)Y_n = \sum_i \varphi_i Y_{n-i}, \qquad \text{for all } n.$$

If that is the case, $\{Y_n\}$ and $\theta(L)$, respectively, is called **invertible** and $\varphi(L) = \theta^{-1}(L)$ an **inverse filter**. In this case, we can reconstruct the input $\{X_n\}$ from the output $\{Y_n\}$. Above we have learned that filters can be composed. Hence, we have

$$\begin{aligned} \{Y_n\} &= \theta(L)(\{X_n\}) \\ &= \theta(L)(\varphi(L)(\{Y_n\})) \\ &= \theta(L) \cdot \varphi(L)(\{Y_n\}), \end{aligned}$$

since $\theta(L) \circ \varphi(L) = \theta(L) \cdot \varphi(L)$. This equation can only be valid for all series, if

$$\theta(L)\varphi(L) = \text{id} = 1$$

holds. We want to derive a necessary condition for the corresponding z-transforms. Suppose that $\theta(L)$ is invertible with inverse L_1 filter $\{\varphi(L)\}$. Since $\theta(L)$ and $\varphi(L)$ are L_1 filters, the associated z-transforms exist at least for all $z \in \mathbb{C}$ with $|z| \leq 1$. Thus, their product $\theta(z)\varphi(z)$

also exists on the unit disc. This means that the product $\theta(z)\varphi(z)$ represents the constant function 1 on the unit disc, i.e.

$$\theta(z)\varphi(z) = 1, \qquad \text{for all } z \in \mathbb{C} \text{ with } |z| \leq 1. \tag{3.6}$$

It follows that

$$\varphi(z) = \frac{1}{\theta(z)}, \qquad \text{for all } |z| \leq 1.$$

Consequently, in order to obtain the inverse filter

$$\theta^{-1}(L) = \varphi(L) = \sum_i \varphi_i L^i,$$

one has to calculate the series expansion of the function $1/\theta(z)$ on the unit disc and then resubstitute $z \mapsto L$. This works, because Equation (3.6) particularly implies the following fact.

Lemma 3.4.7 *For an invertible filter the z-transform $\theta(z)$ has no roots on the unit disc $\{z \in \mathbb{C} : |z| \leq 1\}$.*

Example 3.4.8 *Let us consider the two simplest cases of linear filters and polynomial filters (lag polynomials).*

(i) Linear case: Consider

$$p(L) = 1 - aL.$$

We claim that $p(L)$ possesses an inverse filter with absolutely summable coefficients, if $|a| < 1$, i.e. if the root $1/a$ of the z-transform $p(z) = 1 - az$ lies outside the unit disc. The inverse filter is given by

$$p^{-1}(L) = \frac{1}{1 - aL} = \sum_{i=0}^{\infty} a^i L^i.$$

This fact can be shown as follows. The z-transform is $p(z) = 1 - az$ for all $z \in \mathbb{C}$. If $z \neq 1/a$, then

$$p^{-1}(z) = \frac{1}{1 - az},$$

and

$$p(z)p^{-1}(z) = 1.$$

For $|az| < 1$ if and only if $|z| < 1/|a|$, we can expand $p^{-1}(z)$ into a geometric series

$$p^{-1}(z) = \sum_{i=0}^{\infty} a^i z^i.$$

If $|a| < 1$, *the series* $p^{-1}(1) = \sum_i a^i$ *converges. Hence, the radius of convergence is at least* 1.

(ii) *Lag polynomials: The filter*

$$p(L) = 1 - a_1 L - \cdots - a_p L^p.$$

attains an inverse filter $p^{-1}(L)$ *with absolutely summable coefficients, if all roots of the characteristic polynomial*

$$p(z) = 1 - a_1 z - \cdots - a_p z^p,$$

lie outside the unit dics, i.e.

$$p(z) = 0 \quad \Rightarrow \quad |z| > 1.$$

To verify this result, recall that any polynomial $f(z) = a_0 + a_1 z + \cdots + a_n z^n$ *of degree n can be written as*

$$f(z) = a_n \prod_{i=1}^{n} (z - z_i),$$

with complex roots $z_1, \ldots, z_n$. *Thus, the characteristic polynomial can be factored as*

$$p(z) = -a_p(z - z_1) \cdots (z - z_p),$$

where $z_1, \ldots, z_p$ *denote the (complex) roots of* $p(z)$, *which all differ from* 0. *Write this factorization as*

$$p(z) = (-a_p)(-1)^p \left(1 - \frac{z}{z_1}\right) \cdots \left(1 - \frac{z}{z_p}\right) \prod_{j=1}^{p} z_j.$$

Hence, up to the constant $C = (-a_p)(-1)^p \prod_{j=1}^{p} z_j$, *the lag polynomial* $p(L)$ *is the composition of the linear filters*

$$p_i(L) = 1 - \frac{L}{z_i}, \qquad i = 1, \ldots, p.$$

We can expand (the z-transform of) each factor $\frac{1}{1 - z/z_i}$ *into a power series*

$$\frac{1}{1 - z/z_i} = \sum_{j=0}^{\infty} \left(\frac{1}{z_i}\right)^j z^j$$

with radius of convergence greater or equal to 1, *if* $|z_j| > 1$. *Since the product of series with convergence radii* ≥ 1 *is again a series with radius of convergence* ≥ 1, *we can write*

$$\frac{1}{p(z)} = \frac{1}{-a_p(-1)^p z_1 \cdots z_p} \prod_{i=1}^{p} \sum_{j=0}^{\infty} \left(\frac{1}{z_j}\right)^j z^j = \sum_{j=0}^{\infty} b_j z^j$$

for coefficients $\{b_j\}$. *These coefficients determine the inverse filter*

$$p^{-1}(L) = \sum_j b_j L^j.$$

3.4.3 AR(p) and AR(∞) processes

We are now in a position to introduce autoregressive processes

Definition 3.4.9

(i) A stochastic process $\{X_t\}$ *is called an* **autoregressive process of order** p *or* AR(p) *process, if there exists* $p \in \mathbb{N}$ *and coefficients* $\theta_1, \ldots, \theta_p \in \mathbb{R}$ *with* $\theta_p \neq 0$ *such that*

$$X_t = \theta_1 X_{t-1} + \cdots + \theta_p X_{t-p} + \epsilon_t, \qquad t \in \mathbb{Z},$$

for a white-noise process $\{\epsilon_t\}$.

(ii) If for some sequence $\{\theta_n\}$ *and a white-noise process* $\{\epsilon_n\}$

$$X_t = \sum_{i=1}^{\infty} \theta_i X_{t-i} + \epsilon_t, \qquad t \in \mathbb{Z},$$

then $\{X_t\}$ *is called* AR(∞) *process.*

(iii) $\{X_t : t \in \mathbb{N}_0\}$ *is called an* AR(p) **process (given** $X_{-1}, \ldots, X_{-p}$), *if*

$$X_t = \theta_1 X_{t-1} + \cdots + \theta_p X_{t-p} + \epsilon_t, \qquad t = 0, 1, \ldots$$

for a white-noise process $\{\epsilon_t\}$ *and coefficients* $\theta_1, \ldots, \theta_p \in \mathbb{R}$ *with* $\theta_p \neq 0$.

Notice that we do not require that $\{X_t\}$ is stationary in the above definition.

Example 3.4.10 *Consider the* AR(*1*) *equations*

$$X_t = \rho X_{t-1} + \epsilon_t, \qquad t \in \mathbb{Z},$$

with $|\rho| < 1$. *Using the definition of the lag operator, the above equation can be written as*

$$p(L) X_t = \epsilon_t,$$

where $p(L) = 1 - \rho L$. *Hence, the characteristic polynomial is given by* $p(z) = 1 - \rho z$. *All roots of* $p(z)$ *lie outside of the unit disc, such that*

$$p(z) \neq 0 \qquad \text{for all } z \in \mathbb{C} \text{ with } |z| \leq 1,$$

if and only if $|\rho| < 1$; *we have anticipated that condition above. Hence, we can invert the filter* $p(L)$ *by determining the series expansion of* $1/p(z)$, *which is, of course, given by*

$$\frac{1}{p(z)} = \sum_{i=0}^{\infty} \rho^i z^i.$$

It follows that the representation of $\{X_t\}$ as a linear process is given by

$$X_t = (1 - \rho L)^{-1}\epsilon_t = \sum_{i=0}^{\infty} \rho^i \epsilon_{t-i}. \tag{3.7}$$

This representation particularly allows us to calculate the autocovariances of X_t as

$$\begin{aligned}\gamma_X(h) &= \text{Cov}\left(\sum_{i=0}^{\infty} \rho^i \epsilon_{t+h-i}, \sum_{j=0}^{\infty} \rho^j \epsilon_{t-j}\right) \\ &= \sum_{i=0}^{\infty}\sum_{j=0}^{\infty} \rho^{i+j}\gamma_\epsilon(h - i + j).\end{aligned}$$

Observe that $\gamma_\epsilon(h - i + j) = \sigma^2$ iff. $i = h + j$ and $= 0$ otherwise, such that

$$\begin{aligned}\gamma_X(h) &= \sigma^2 \sum_{j=0}^{\infty} \rho^{2j}\rho^h \\ &= \frac{\rho^h}{1-\rho^2}\sigma^2.\end{aligned}$$

We see that the autocovariances decay at an exponential rate,

$$\alpha(h) \sim \mathrm{e}^{-h\gamma},$$

with $\gamma = -\log(\rho) > 0$.

Example 3.4.11 *Consider now the* AR(1) *process with starting value* 0, *i.e.*

$$X_0 = 0, \qquad X_t = \rho X_{t-1} + \epsilon_t, \qquad t \geq 1, \qquad \epsilon_t \sim \text{WN}(0, \sigma^2),$$

which are nonstationary but converge to a stationary process, as $t \to \infty$. Indeed, notice that $X_0 = 0$, $X_1 = \epsilon_1$, $X_2 = \rho\epsilon_1 + \epsilon_2$ and, in general,

$$X_t = \sum_{i=1}^{t} \rho^{t-i}\epsilon_i = \sum_{i=0}^{t-1} \rho^i \epsilon_{t-i}, \tag{3.8}$$

which is close to the stationary solution $\sum_{i=0}^{\infty} \rho^i \epsilon_{t-i}$ of the AR(1) *equations; the L_2 distance is*

$$\sqrt{E\left(X_t - \sum_{i=0}^{\infty} \rho^i \epsilon_{t-i}\right)^2} = \sigma\sqrt{\sum_{i \geq t} \rho^{2i}}.$$

The variance of Equation (3.8) is

$$\sigma_t^2 = \text{Var}\,(X_t) = \sigma^2 \sum_{i=0}^{t-1} \rho^{2i},$$

which depends on t. Obviously, we have

$$\sigma_t^2 \to \frac{\sigma^2}{1-\rho^2}, \qquad t \to \infty.$$

Further, for $h \geq 0$

$$\begin{aligned} \text{Cov}(X_t, X_{t+h}) &= \text{Cov}\left(\sum_{i=1}^{t} \rho^{t-i}\epsilon_i, \sum_{j=1}^{t+h} \rho^{t+h-j}\epsilon_j \right) \\ &= \sigma^2 \rho^h \sum_{i=1}^{t} \rho^{2(t-i)}. \end{aligned}$$

Thus, the autocovariances also depend on t, but they converge,

$$\text{Cov}(X_t, X_{t+h}) \to \frac{\sigma^2 \rho^h}{1-\rho^2},$$

as $t \to \infty$. For large t the differences between Equation (3.8) and the stationary solution (3.7) are small and often ignored.

The following useful result establishes conditions such that a stationary AR(p) process can be represented as a MA(∞) process, i.e. as a linear process.

Proposition 3.4.12 *An* AR(p) *process*

$$X_t = \sum_{i=1}^{p} \theta_i X_{t-i} + \epsilon_t, \qquad t \in \mathbb{Z}, \qquad \epsilon_t \sim \text{WN}(0, \sigma^2),$$

attains a representation as a stationary MA(∞) *process*

$$X_t = \sum_{i=0}^{\infty} c_i \epsilon_{t-i},$$

if all roots of the lag polynomial

$$p(z) = 1 - \theta_1 z - \cdots - \theta_p z^p, \qquad z \in \mathbb{C},$$

lie outside the unit disc.

Proof. Notice that the AR(p) equations are equivalent to

$$\epsilon_t = X_t - \sum_{i=1}^{p} \theta_i X_{t-i} = p(L) X_t, \qquad t \in \mathbb{Z}.$$

That equation is invertible, if $p(z) = 0$ implies $|z| > 1$. Then

$$X_t = p(L)^{-1}\epsilon_t = \sum_{i=0}^{\infty} c_i \epsilon_{t-i},$$

where the coefficients c_i are determined by solving

$$\sum_{i=0}^{\infty} c_i z^i \stackrel{!}{=} \frac{1}{p(z)}$$

with $\sum_i |c_i| < \infty$. This means, X_t can be obtained by applying a L_1 filter to ϵ_t, thus being stationary.

In general, the coefficients c_i of the MA(∞) representation $X_t = q(L)\epsilon_t = \sum_{i=0}^{\infty} c_i \epsilon_{t-i}$ can be calculated as follows. With $p(L) = 1 - \theta_1 L - \cdots - \theta_p L^p$ we have the identity

$$q(L)p(L) = 1,$$

which is equivalent to

$$(c_0 + c_1 L + c_2 L^2 + \cdots)(1 - \theta_1 L - \cdots - \theta_p L^p) = 1.$$

Clearly, the left-hand side equals

$$c_0 + (c_1 - \theta_1 c_0)L + (c_2 - \theta_1 c_1 - \theta_2 c_0)L^2 + \cdots + (c_i - \theta_1 c_{i-1} - \cdots - \theta_p c_{i-p})L^i + \cdots$$

Comparison of the coefficients now leads to the solution

$$\begin{aligned} c_0 &= 1 \\ c_1 &= \theta_1 c_0 = \theta_1 \\ c_2 &= \theta_1 c_1 + \theta_2 c_0 = \theta_1^2 + \theta_2. \end{aligned}$$

Put differently, for $i \geq 1$ with $c_{-p} = \cdots = c_{-1} = 0$ one has to solve the difference equations

$$c_0 = 1, \qquad c_i - \theta_1 c_{i-1} - \cdots - \theta_p c_{i-p} = 0.$$

3.4.4 ARMA processes

Definition 3.4.13 (ARMA(p, q) PROCESS)

(i) $\{X_t : t \in \mathbb{Z}\}$ *is called* ARMA(p, q) **process**, $p, q \in \mathbb{N}$, *if it is stationary and satisfies*

$$X_t - \phi_1 X_{t-1} - \cdots - \phi_p X_{t-p} = \epsilon_t + \theta_1 \epsilon_{t-1} + \theta_q \epsilon_{t-q}, \qquad t \in \mathbb{Z},$$

for constants $\phi_1, \ldots, \phi_p, \theta_1, \ldots, \theta_q \in \mathbb{R}$ *with* $\phi_p \neq 0$ *and* $\theta_q \neq 0$ *and a white-noise process* $\{\epsilon_t\}$.

(ii) $\{X_t : t \in \mathbb{Z}\}$ *is called* ARMA(p, q) **process with mean** μ, *if* $X_t - \mu \sim$ ARMA(p, q).

(iii) *A* ARMA(p, q) *process is called* **causal** *(with respect to* $Z_t \sim$ WN$(0, \sigma^2)$*), if* $X_t = \sum_{i=0}^{\infty} \psi_i Z_{t-i}$ *for some filter* $\{\psi_k : k \in \mathbb{N}_0\}$.

Define the lag polynomials

$$\phi(z) = 1 - \phi_1 z - \cdots - \phi_p z^p,$$
$$\theta(z) = 1 + \theta_1 z + \cdots + \theta_q z^q,$$

for $z \in \mathbb{C}$. ϕ is called an AR **lag polynomial**, θ is called a MA **lag polynomial** Then the ARMA(p, q) equations take the form

$$\phi(L)X_t = \theta(L)\epsilon_t.$$

In what follows, we assume that the lag polynomials have no common roots.

Theorem 3.4.14 *Let $\{X_t\}$ be a stationary* ARMA(p, q) *process with lag polynomials $\phi(z)$ and $\theta(z)$,*

$$\phi(L)X_t = \theta(L)\epsilon_t.$$

Suppose that the roots of $\phi(z)$ lie outside the unit disc. Then

$$X_t = \psi(L)\epsilon_t$$

is a stationary and causal ARMA(p, q) *process with coefficients given by $\psi(z) = \sum_{i=0}^{\infty} \psi_i z^i$, where*

$$\psi(z) = \frac{1 + \theta_1 z + \cdots + \theta_q z^q}{1 - \phi_1 z - \cdots - \phi_p z^p},$$

for $|z| \leq 1$.

Proof. For a white noise ϵ_t the series $Y_t = \theta(L)\epsilon_t$ is stationary. Provided the roots of $\phi(z)$ satisfy $|z| > 1$, we may solve $\phi(L)X_t = Y_t$.

The following standard example is instructive to understand the above result.

Example 3.4.15 *Consider a* ARMA$(1, 1)$ *process given by*

$$X_t - \phi X_{t-1} = \epsilon_t + \theta\epsilon_{t-1},$$

with coefficients $\phi \neq -\theta$ to ensure that the lag polynomials have no common root. To determine the coefficients ψ_i, $i \geq 0$, consider the equation

$$\sum_{i=0}^{\infty} \psi_i z^i = \psi(z) = \frac{\theta(z)}{\phi(z)} = \frac{1 + \theta z}{1 - \phi z}.$$

Notice that

$$\begin{aligned}\frac{1+\theta z}{1-\phi z} &= (1+\theta z)\sum_{i=0}^{\infty}(\phi z)^i \\ &= (1+\theta z)\left(1+\sum_{i=1}^{\infty}(\phi z)^i\right) \\ &= 1+\sum_{i=0}^{\infty}(\phi+\theta)\phi^{i-1}z^i.\end{aligned}$$

Hence, $\psi_0 = 1$ *and* $\psi_i = (\phi+\theta)\phi^{i-1}$*, for* $i \geq 1$.

3.5 The frequency domain

3.5.1 The spectrum

Let $\{X_t\}$ be a stationary process with

$$\begin{aligned}\mu &= E(X_t), \\ \gamma_h &= \mathrm{Cov}(X_1, X_{1+h}), \qquad h \in \mathbb{Z}.\end{aligned}$$

The function

$$g_X(z) = \sum_{k=-\infty}^{\infty} \gamma_k z^k$$

is called an **autocovariance generating function**, provided it exists. A sufficient condition for g_X to be well defined on the unit disc is the summability of the autocovariances. Notice that $\gamma_k = g_X^{(k)}(0)/k!$ for $k \in \mathbb{N}_0$.

In what follows, let us agree to define $\imath$ by $\imath^2 = -1$.

Definition 3.5.1 *Let* $\{X_t\}$ *be stationary with autocovariance function* $\{\gamma_k\}$ *satisfying* $\sum_k |\gamma_k| < \infty$*. Then*

$$f_X(\omega) = \frac{1}{2\pi} g_X(\mathrm{e}^{-\imath\omega}) = \frac{1}{2\pi}\sum_{k=-\infty}^{\infty} \gamma_k \mathrm{e}^{-\imath\omega k}, \qquad \omega \in \mathbb{R},$$

is called the **spectrum or spectral density of** $\{X_t\}$.

In time series analysis the notation f_X for a spectral density is widespread. To avoid confusion with probability densities, we shall denote them by $f_X(\omega)$. Using the formulas

$$\mathrm{e}^{-\imath\omega k} = \cos(\omega k) - \imath\sin(\omega k), \quad \mathrm{e}^{-\imath\omega k} + \mathrm{e}^{\imath\omega k} = 2\cos(\omega k),$$

which is a direct consequence of Euler's identity

$$\mathrm{e}^{\imath z} = \cos(z) + \imath\sin(z),$$

and the symmetry of the autocovariances, i.e. $\gamma_{-k} = \gamma_k$, leads to the frequently used formula

$$f_X(\omega) = \frac{1}{2\pi}\left(\gamma_0 + 2\sum_{k=1}^{\infty}\gamma_k\cos(\omega k)\right) = \frac{1}{2\pi}\sum_{k=-\infty}^{\infty}\gamma_k\cos(\omega k).$$

Since $f_X(\omega)$ is periodic with period 2π and even, i.e. $f_X(-\omega) = f_X(\omega)$, it is common to study it for $\omega \in [-\pi, \pi]$.

It is further worth mentioning that, since $\int_{-\pi}^{\pi}\cos(\omega k)\,\mathrm{d}\omega = 2\pi\mathbf{1}(k=0)$,

$$\int_{-\pi}^{\pi} f_X(\omega)\,\mathrm{d}\omega = \frac{1}{2\pi}\sum_k \gamma_k \int_{-\pi}^{\pi}\cos(\omega k)\,\mathrm{d}\omega = \gamma_0 = \mathrm{Var}\,(X_t),$$

i.e. when integrating the spectral density over $[-\pi, \pi]$, one obtains the marginal variance of the process.

For linear processes, there is a nice formula that allows the spectral density from the z-transform to be calculated.

Lemma 3.5.2 *For a linear process* $X_t = \sum_{k=0}^{\infty}\psi_k\epsilon_{t-k}$, $\epsilon_t \sim \mathrm{WN}(0, \sigma^2)$,

$$g_X(z) = \sigma^2\psi(z)\psi(z^{-1}),$$

such that the spectrum is given by

$$f_X(\omega) = \frac{\sigma^2}{2\pi}\psi(\mathrm{e}^{-\imath\omega})\psi(\mathrm{e}^{\imath\omega}),$$

where $\psi(z)$ *is the* z*-transform. More generally,* $g_X(z) = \psi(z^{-1})g_\epsilon(z)\psi(z)$ *holds.*

Proof. To calculate $g_X(z) = \sum_k \gamma_k z^k$ plug in the formula

$$\gamma_X(k) = \sum_i\sum_j \psi_i\psi_j\gamma_\epsilon(k+i-j)$$

to obtain

$$g_X(z) = \sum_k\sum_i\sum_j \psi_i\psi_j\gamma_\epsilon(k+i-j)z^k.$$

Now the substitution $h = k + i - j$, such that $z^k = z^h z^{-i} z^j$, leads to

$$g_X(z) = \left(\sum_i \psi_i z^{-i}\right)\left(\sum_h \gamma_\epsilon(h)z^h\right)\left(\sum_j \psi_j z^j\right)$$

which equals $\psi(z^{-1})g_\epsilon(z)\psi(z)$, where $g_\epsilon(z) = \sigma^2$ for a white-noise process $\{\epsilon_t\}$. Hence, the result follows.

Remark 3.5.3 *Notice that*

$$f_X(\omega) = \frac{\sigma^2}{2\pi}|\psi(\mathrm{e}^{\imath\omega})|^2,$$

since $\overline{\psi(z)} = \psi(\overline{z})$ *and* $\overline{e^{\imath\omega}} = e^{-\imath\omega}$.

Again it is instructive to illustrate the formula by means of the following standard example.

Example 3.5.4 *Consider a stationary* AR(1) *process*

$$X_t = \phi X_{t-1} + \epsilon_t, \qquad \epsilon \sim \mathrm{WN}(0, \sigma^2).$$

Then the spectrum $f_X(\omega)$ *is given by*

$$f_X(\omega) = \frac{\sigma^2}{2\pi} \frac{1}{1 - 2\phi\cos(\omega) + \phi^2}.$$

Indeed, we have

$$\begin{aligned} f_X(\omega) &= \frac{\sigma^2}{2\pi} \frac{1}{1 - \phi e^{-\imath\omega}} \frac{1}{1 - \phi e^{\imath\omega}} \\ &= \frac{\sigma^2}{2\pi} \frac{1}{|1 - \phi e^{\imath\omega}|^2} \\ &= \frac{\sigma^2}{2\pi} \frac{1}{|1 - \phi\cos(\omega) - \imath\sin(\omega)|^2} \\ &= \frac{1}{(1 - \phi\cos(\omega))^2 + \phi^2\sin^2(\omega)} \\ &= \frac{\sigma^2}{2\pi} \frac{1}{1 - 2\phi\cos(\omega) + \phi^2}. \end{aligned}$$

Let us consider an ARMA(p, q) process

$$\phi(L)X_t = \theta(L)\epsilon_t, \qquad \epsilon_t \sim \mathrm{WN}(0, \sigma^2),$$

with lag polynomials $\phi(z) = 1 - \phi_1 z - \cdots - \phi_p z^p$ and $\theta(z) = 1 + \theta_1 z + \cdots + \theta_q z^q$. Since $X_t = \frac{\theta(z)}{\phi(z)}\Big|_{z=L} \epsilon_t$, the spectrum is given by

$$f_X(\omega) = \frac{\sigma^2}{2\pi} \frac{\theta(z)}{\phi(z)} \frac{\theta(z^{-1})}{\phi(z^{-1})}\bigg|_{z=e^{-\imath\omega}}$$

which turns out to be

$$\frac{\sigma^2}{2\pi} \frac{1 + \theta_1 e^{-\imath\omega} + \cdots + \theta_q e^{-\imath q\omega}}{1 - \phi_1 e^{-\imath\omega} - \cdots - \phi_p e^{-\imath p\omega}} \frac{1 + \theta_1 e^{\imath\omega} + \cdots + \theta_q e^{\imath q\omega}}{1 - \phi_1 e^{\imath\omega} - \cdots - \phi_p e^{\imath p\omega}}$$

3.5.2 The periodogram

It is natural to estimate the spectrum by substituting the autocovariances by some empirical counterpart such as

$$\widehat{\gamma}_T(h) = \frac{1}{T} \sum_{t=1}^{T-h} (X_t - \overline{X}_T)(X_{t+h} - \overline{X}_T), \quad \widehat{\gamma}_T(-h) = \widehat{\gamma}_T(h),$$

or

$$\widetilde{\gamma}_T(h) = \frac{1}{T}\sum_{t=1}^{T-h} X_t X_{t+h}, \quad \widetilde{\gamma}_T(-h) = \widetilde{\gamma}_T(h).$$

Here $\overline{X}_T = \frac{1}{T}\sum_{t=1}^{T} X_t$.

Definition 3.5.5 *The random function*

$$I_T(\omega) = \frac{1}{2\pi}\sum_{h=-(T-1)}^{T-1} \widetilde{\gamma}_T(h)\mathrm{e}^{-\imath\omega h}$$

is called a **periodogram** *and*

$$J_T(\omega) = \frac{1}{2\pi}\sum_{h=-(T-1)}^{T-1} \widehat{\gamma}_T(h)\mathrm{e}^{-\imath\omega h}$$

is the **centered periodogram**.

Lemma 3.5.6 *A periodogram and centered periodogram satisfy*

$$I_T(\omega) = \frac{1}{2\pi T}\left|\sum_{t=1}^{T} X_t \mathrm{e}^{-\imath\omega t}\right|^2,$$

$$J_T(\omega) = \frac{1}{2\pi T}\left|\sum_{t=1}^{T} (X_t - \overline{X}_T)\mathrm{e}^{-\imath\omega t}\right|^2,$$

The following important result, a version of Herglotz' theorem, shows that the autocovariance function is determined by the spectrum.

Theorem 3.5.7 *Let $\{X_t\}$ be a stationary process with autocovariances $\{\gamma_k\}$ satisfying $\sum_k |\gamma_k| < \infty$.*

(i) $I_T(\omega)$ is asymptotically unbiased for the spectral density $f_X(\omega)$, i.e.

$$\lim_{T\to\infty} E(I_T(\omega)) = f_X(\omega).$$

(ii) The spectral density $f_X(\omega)$ is even, non-negative, continuous and determines the autocovariance function via

$$\gamma_h = \int_{-\pi}^{\pi} \cos(\omega h) f_X(\omega)\,\mathrm{d}\omega, \qquad h \in \mathbb{Z}.$$

Proof. In order to show the continuity of the function

$$f(\omega) = f_X(\omega) = \frac{1}{2\pi}\sum_t \gamma_t \cos(\omega t), \tag{3.9}$$

it suffices to show that series converges uniformly. But $|\gamma_t \cos(\omega t)| \leq |\gamma_t|$ for all t and all ω, such that $|f_X(\omega)| \leq \sum_t |\gamma_k| < \infty$. Recall that the functions

$$\varphi_0(t) = \frac{1}{\sqrt{2\pi}},$$
$$\varphi_k(t) = \frac{1}{\sqrt{\pi}} \cos(kt), \qquad k \in \mathbb{N},$$

form an orthonormal system w.r.t. the inner product

$$(f, g) = \int_{-\pi}^{\pi} f(x)g(x)\,\mathrm{d}x$$

for functions $f, g \in L_2([-\pi, \pi]; \lambda)$. Since f is continuous on $[-\pi, \pi]$ with $f(-\pi) = f(\pi)$, we have uniform convergence of the Fourier series

$$f(x) = \sum_{k=0}^{\infty} (f, \varphi_k)\varphi_k(x).$$

This means,

$$\begin{aligned} f(x) &= \frac{1}{2\pi} \int_{-\pi}^{\pi} f(t)\,\mathrm{d}t + 2\sum_{k=1}^{\infty} \frac{1}{2\pi} \int_{-\pi}^{\pi} f(t)\cos(kt)\,\mathrm{d}t \cdot \cos(kx) \\ &= \sum_{k\in\mathbb{Z}} \frac{1}{2\pi} \int_{-\pi}^{\pi} f(t)\,\mathrm{d}t \cos(kx). \end{aligned}$$

Comparing this representation with Equation (3.9), we obtain

$$\gamma_k = \int_{-\pi}^{\pi} f(t)\cos(kt)\,\mathrm{d}t,$$

for all k, since the coefficients in the Fourier series expansion are unique.

To proceed, notice that Lemma 3.5.6 immediately yields the non-negativity of $I_T(\omega)$, which implies $E(I_T(\omega)) \geq 0$ as well. Let us consider that expectation in greater detail. By linearity,

$$\begin{aligned} E(I_T(\omega)) &= \frac{1}{2\pi T} \sum_{s,t=1}^{T} E(X_s X_t)\mathrm{e}^{\imath\omega(s-t)} \\ &= \frac{1}{2\pi T} \sum_{s,t=1}^{T} \gamma_{|s-t|}\mathrm{e}^{\imath\omega(s-t)}. \end{aligned}$$

Here the T^2 elements of the symmetric matrix with entries $\gamma_{|s-t|}\mathrm{e}^{\imath\omega(s-t)}$ are summed up. Summing over all diagonals and using $\mathrm{e}^{\imath\omega(s-t)} + \mathrm{e}^{\imath\omega(t-s)} = 2\cos(\omega(s-t))$, which implies that the matrix with elements

$$a_{st} = \gamma_{|s-t|}\mathrm{e}^{\imath\omega(s-t)}$$

satisfies $a_{st} + a_{ts} = 2\gamma_{|s-t|}\cos(\omega(|s-t|))$, we obtain

$$E(I_T(\omega)) = \frac{1}{2\pi}\sum_{|t|<T}\left(1-\frac{|t|}{T}\right)\gamma_t\cos(\omega t).$$

The last expression can be written as an integral w.r.t. the counting measure $\mathrm{d}\nu(t)$ on $\mathbb{Z}$[1],

$$\int F_T(t;\omega)\,\mathrm{d}\nu(t),$$

if we put

$$F_T(t;\omega) = \begin{cases} \frac{1}{2\pi}\left(1-\frac{|t|}{T}\right)\gamma_t\cos(\omega t), & |t| \le T-1, \\ 0, & |t| > T-1. \end{cases}$$

This means that we know that

$$0 \le E(I_T(\omega)) = \int F_T(t;\omega)\,\mathrm{d}\nu(t).$$

We will now apply dominated convergence to conclude that the non-negativity also holds for the limit. To do so, notice that $|F_T(t;\omega)| \le |\gamma_t|$ for all $t \in \mathbb{Z}$ and

$$\int |\gamma_t|\,\mathrm{d}\nu(t) = \sum_{t\in\mathbb{Z}} |\gamma_t| < \infty$$

by assumption. Thus, $t \mapsto |\gamma_t|\mathbf{1}_{\mathbb{Z}}(t)$ is a $\mathrm{d}\nu$-integrable dominating function. Further, $F_T(t;\omega)$ converges pointwise, namely

$$F_T(t;\omega) = \frac{1}{2\pi}\left(1-\frac{|t|}{T}\right)\cos(\omega t) \to \frac{1}{2\pi}\gamma_t\cos(\omega t),$$

as $T \to \infty$. Therefore,

$$\begin{aligned} 0 &\le \int F_T(t;\omega)\,\mathrm{d}\nu(t) \\ &\to \int \frac{1}{2\pi}\gamma_t\cos(\omega t)\,\mathrm{d}\nu(t) \\ &= \frac{1}{2\pi}\sum_{t=-\infty}^{\infty}\gamma_t\cos(\omega t) \\ &= f_X(\omega), \end{aligned}$$

as $T \to \infty$, which completes the proof.

We have shown that the spectral density $f_X(\omega)$ is non-negative and satisfies $\int_{-\pi}^{\pi} f_X(\omega)\,\mathrm{d}\omega = \mathrm{Var}\,(X_t) < \infty$. Hence, it defines a finite measure.

---

[1] For the counting measure $\mathrm{d}\nu$ and a real-valued function we have $\int f(x)\,\mathrm{d}\nu(x) = \sum_{n\in\mathbb{Z}} f(n)$, provided the series exists.

Definition 3.5.8 *The measure on* $((-\pi, \pi], \mathcal{B}(-\pi, \pi])$ *given by*

$$\nu_X(A) = \int_A f_X(\omega)\,\mathrm{d}\omega, \qquad A \subset (-\pi, \pi] \text{ Borel-measurable},$$

is called the **spectral measure of** $\{X_t\}$ *(or* $f_X(\omega)$*). The associated (generalized) distribution function*

$$F_X(\omega) = \int_{-\pi}^{\omega} f_X(\lambda)\,\mathrm{d}\lambda, \qquad \omega \in (-\pi, \pi],$$

is called the **spectral distribution function**.

Since a spectral density can be defined as long as we are given an autocovariance function (not necessarily of a stationary process), the spectral measure and the spectral distribution function can be defined for a given autocovariance function as well.

Remark 3.5.9 *When dealing with a process* $\{X_t\}$ *taking values in* $\mathbb{C}$*, one obtains the representation*

$$\gamma_X(h) = \int_{-\pi}^{\pi} \mathrm{e}^{\imath h\omega} f_X(\omega)\,\mathrm{d}\omega,$$

or, equivalently,

$$\gamma_X(h) = \int_{-\pi}^{\pi} \mathrm{e}^{\imath h\omega}\mathrm{d}F_X(\omega) = \int_{-\pi}^{\pi} \mathrm{e}^{\imath h\omega}\mathrm{d}\nu_X(\omega).$$

One can generalize the above results and omit the assumption of a summable autocovariances, which, however, simplifies the arguments and provides more transparent proofs, since then the spectral density exists. The general result is as follows.

Theorem 3.5.10 (HERGLOTZ' LEMMA)

(i) Let $\gamma : \mathbb{Z} \to \mathbb{C}$ *be a positive semidefinite function. Then there exists a unique spectral measure* ν*, such that*

$$\gamma(h) = \int_{(-\pi,\pi]} \mathrm{e}^{\imath h\omega}\,\mathrm{d}\nu(\omega), \qquad h \in \mathbb{Z}. \tag{3.10}$$

(ii) If ν *is an arbitrary finite measure on* $(-\pi, \pi]$ *equipped with the Borel-*σ*-field, then Equation (3.10) defines a complex-valued positive semi-definite function on* $\mathbb{Z}$*, i.e. an autocovariance function of a stationary process.*

Remark 3.5.11 *The spectral distribution function* $F_X(\omega)$ *of a process* $\{X_t\}$ *with autocovariances* $\gamma_X(k)$ *satisfies*

$$F_X(\pi) = \int_{-\pi}^{\pi} f(\omega)\,\mathrm{d}\omega = \gamma_X(0) = \operatorname{Var}(X_t).$$

Since $\text{Var}(X_t) < \infty$,

$$G_X(\omega) = F_X(\omega)/F_X(\pi), \qquad \omega \in [-\pi, \pi],$$

$G_X(\omega) = 0$ *for* $\omega < -\pi$ *and* $G_X(\omega) = 1$ *for* $\omega > \pi$ *defines the distribution function of a probability measure on* $(-\pi, \pi]$. *It is related to the autocorrelations via*

$$\rho_X(h) = \frac{\gamma_X(h)}{\gamma_X(0)} = \int_{-\pi}^{\pi} e^{\imath h\omega} \, dG_X(\omega), \qquad h \in \mathbb{Z}.$$

If a process $\{X_t\}$ does not possess a spectral density, then one studies the spectral distribution function. If particularly

$$\gamma_X(h) = \sum_j \alpha_j e^{\imath \omega_j h}$$

for coefficients $\alpha_j \in \mathbb{R}$ and fixed frequencies $\omega_j \in [-\pi, \pi]$, then the formula $\gamma_X(h) = \int_{-\pi}^{\pi} e^{\imath\omega h} \, dF_X(\omega)$ shows that

$$F_X(\omega) = \sum_j \alpha_j \delta_{\omega_j}(\omega),$$

where δ_a denotes the one-point (Dirac) measure in the point a. The spectral distribution function is concentrated on the set $\{\omega_j\}$ and assigns the mass α_j to the frequency ω_j.

The following example shows that, first, the above case is not artificial, and, secondly, makes clear why we call ω_j frequencies.

Example 3.5.12 *Consider the random periodic function*

$$X_t = A\cos(\omega t) + B\sin(\omega t), \qquad t \in \mathbb{Z},$$

for a fixed frequency $\omega \in (-\pi, \pi]$ *and random and independent amplitudes A and B. Assume that* $E(A) = E(B) = 0$ *and* $\text{Var}(A) = \text{Var}(B) = \sigma^2 \in (0, \infty)$. *A straightforward calculation shows that*

$$\gamma_X(h) = \sigma^2 \cos(\omega h) = \sigma^2 \frac{e^{\imath\omega h} + e^{-\imath\omega h}}{2}.$$

Hence, the spectral distribution function is given by

$$F_X(\lambda) = \begin{cases} 0, & \lambda < -\omega, \\ \sigma^2/2, & \lambda \in [-\omega, \omega), \\ \sigma^2, & \lambda \geq \omega. \end{cases}$$

F_X *assigns the mass* $\sigma^2/2$ *to both points* $-\omega$ *and* ω.

3.6 Estimation of ARMA processes

The estimation and inference for ARMA time series is usually based on maximum likelihood by assuming that the white-noise series driving the process is an i.i.d. sequence of normally distributed random variables.

Suppose we are given a stationary ARMA(p, q) process with mean μ, such that

$$X_t - \mu - \phi_1(X_{t-1} - \mu) - \cdots - \phi_p(X_{t-p} - \mu) = \theta_1\epsilon_{t-1} + \cdots + \theta_q\epsilon_{t-q} + \epsilon_t,$$

or equivalently

$$\phi(L)(X_t - \mu) = \theta(L)\epsilon_t$$

with lag polynomials $\phi(L) = \sum_{i=0}^{p} \phi_i L^i$ and $\theta(L) = \sum_{i=0}^{q} \theta_i L^i$ with $\phi_0 = 1$ and $\theta_0 = 1$. Suppose that $\phi(z)$ has no roots on the unit disc. Then we may invert the process and obtain the representation of X_t as a linear process,

$$X_t = \mu + \psi(L)\epsilon_t,$$

where $\psi(L) = \sum_{i=0}^{\infty} \psi_i L^i$ satisfies $\psi(z) = \theta(z)/\phi(z) = \sum_{i=0}^{\infty} \psi_i z^i$, $|z| \le 1$.

Since linear processes are Gaussian processes, the finite series $X_1, \ldots, X_T$ is multivariate normal, such that

$$(X_1, \ldots, X_T) \sim N(\mu, \sigma^2\Sigma(\vartheta)),$$

where the covariance matrix $\Sigma(\vartheta)$ depends on the parameter vector

$$\vartheta = (\phi_1, \ldots, \phi_p, \theta_1, \ldots, \theta_q).$$

The exact likelihood function is then given by

$$L(\vartheta, \mu, \sigma^2|x) = \left(\frac{1}{2\pi\sigma^2}\right)^{T/2} \det(\Sigma)^{-1/2} \exp\left\{-\frac{(x - \mu\mathbf{1})'\Sigma(\vartheta)^{-1}(x - \mu\mathbf{1})}{2\sigma^2}\right\},$$

where $x = (x_1, \ldots, x_T)$ is the observed realization of $X_1, \ldots, X_T$ and $\mathbf{1} = (1, \ldots, 1)' \in \mathbb{R}^d$. There exist efficient algorithms to compute $L(\vartheta, \mu, \sigma^2|X = x)$ or the log likelihood. It is maximized over the parameter set defined by those parameter values leading to a stationary solution of the ARMA(p, q) equations.

For AR(p) models, it is also common to consider the conditional likelihood. The rule $f_{(X,Y)} = f_{Y|X} f_X$ allows us to factorize the exact likelihood for a sample $X_{-p+1}, \ldots, X_0, X_1, \ldots, X_T$, since

$$\begin{aligned} f_{(X_{-p+1},\ldots X_T)} &= f_{X_T|(X_{-p+1},\ldots,X_{T-1})} f_{(X_{-p+1},\ldots,X_{T-1})} \\ &= \ldots \\ &= \prod_{t=1}^{T} f_{X_t|(X_{t-p+1},\ldots,X_{t-1})} f_{(X_{-p+1},\ldots,X_0)}. \end{aligned}$$

Given $X_{-p+1}, \dots, X_0$, the last factor can be ignored. This means that it is sufficient to maximize the **conditional likelihood**

$$L_c(\vartheta|x_1, \dots, x_T) = \prod_{t=1}^{T} f_{X_t|(X_{t-p},\dots,X_{t-1})}(x_t|x_{t-p}, \dots, x_{t-1}; \vartheta),$$

where $\vartheta = (\theta_1, \dots, \theta_p)$, and any maximizer is called a maximum likelihood estimator. For an AR(p) model with $\mu = 0$, we have $X_t - \sum_{j=1}^{p} \theta_j X_{t-j} \sim N(0, \sigma^2)$ leading to

$$L_c(\vartheta|X_1, \dots, X_T) = \left(\frac{1}{2\pi\sigma^2}\right)^{T/2} \exp\left\{-\frac{1}{2\sigma^2}\sum_{t=1}^{T}\left(X_t - \sum_{j=1}^{p}\theta_j X_{t-j}\right)^2\right\}.$$

It follows that any minimizer of the least squares criterion

$$Q(\vartheta) = \sum_{t=1}^{T}\left(X_t - \sum_{j=1}^{p}\theta_j X_{t-j}\right)^2$$

is a conditional ML estimator and those minimizers are solutions of the system of linear equations

$$\widetilde{\Gamma}_T\vartheta + \widetilde{g}_T = 0,$$

where

$$\widetilde{\Gamma}_T = \left(\frac{1}{T}\sum_{t=1}^{T} X_{t-r}X_{t-s}\right)_{r=1,\dots,p;s=1,\dots,p}, \qquad \widetilde{g}_T = \left(\frac{1}{T}\sum_{t=1}^{T} X_t X_{t-r}\right)_{r=1}^{p}.$$

This shows that the resulting conditional least squares estimator $\widehat{\vartheta}_T$ depends on the time series through the sample autocovariances.

3.7 (G)ARCH models

The models studied so far dealt with the modeling of dependence structures and conditional first moments. We shall now study a celebrated class of models aiming at modeling conditional volatility: The seminal work of Robert Engle was awarded with the 2003 Sveriges Riksbank Prize in Economc Sciences in memory of Alfred Nobel.

It turns out that those models represent white-noise processes having the additional structure of martingale differences. To set the scene, we shall first show that any L_2 martingale can be written as a product of two factors, one having the interpretation as a conditional volatility given past information, in such a way that these models appear as natural simplifying parametric models for the conditional volatility.

Let $\{X_t : t \in \mathbb{Z}\}$ be a L_2 martingale difference sequence, that is

$$E|X_t|^2 < \infty \qquad \text{and} \qquad E(X_t|\mathcal{F}_{t-1}) = 0,$$

for all t, with respect to the natural filtration $\mathcal{F}_t = \sigma(X_s : s \leq t)$.

Definition 3.7.1 *The conditional expectation*

$$\sigma_t^2 = E(X_t^2|\mathcal{F}_{t-1}), \qquad t \in \mathbb{Z},$$

is called the **conditional variance** *and* $\sigma_t = \sqrt{\sigma_t}$ *the* **conditional volatility**, *given the past information* $\mathcal{F}_{t-1}$.

Notice that

$$\sigma_t^2 = H_t(X_{t-1}, X_{t-2}, \ldots)$$

for Borel-measurable functions H_t. In what follows, let us assume that for all t

$$h_t \geq c_t > 0, \qquad \text{almost surely,} \tag{3.11}$$

for constants c_t. Put

$$u_t = \frac{X_t}{\sigma_t}, \qquad t \in \mathbb{Z}.$$

Then, by definition

$$X_t = \sigma_t u_t, \qquad t \in \mathbb{Z}.$$

It is easy to verify the following fact.

Lemma 3.7.2 *Under Assumption (3.11)* $\{u_t\}$ *is a* $\mathcal{F}_t$*-martingale difference sequence with* $\text{Var}\,(u_t|\mathcal{F}_{t-1}) = 1$.

It follows that $\{u_t\}$ is a white-noise process. Further, we can conclude that any weakly stationary martingale difference sequence can be written in the form

$$X_t = h_t u_t$$

for some white-noise process with $E(u_1) = 0$ and $E(u_1^2) = 1$.

However, model building for the conditional volatility usually proceeds by assuming specific functional forms for h_t as a function of lagged values of X_t. It is also routinely assumed in those models that the u_t are i.i.d.(0, 1) or even $u_t \overset{i.i.d.}{\sim} N(0, 1)$. Let us start with ARCH models.

Definition 3.7.3 *Let* $\{u_t\}$ *be i.i.d.(0, 1).* $\{X_t\}$ *is called an* **autoregressive conditional heteroscedastic process** *of order p, abbreviated as* ARCH(p), $p \in \mathbb{N}$, *if*

$$X_t = \sigma_t u_t,$$
$$\sigma_t^2 = \alpha_0 + \sum_{i=1}^{p} \alpha_i X_{t-i}^2,$$

for all t, where $\alpha_0 > 0$ *and* $\alpha_1, \ldots, \alpha_p \geq 0$ *are parameters.*

Let us first consider the case $p = 1$ in some detail. Let X_0 be a random variable with $E(X_0^2) < \infty$ and define X_t, $t \geq 1$, by the recursion

$$X_t = \sqrt{\alpha_0 + \alpha_1 X_{t-1}^2}\, u_t, \qquad t = 1, 2, \ldots.$$

Notice that $X_1 = f(X_0, u_1)$, $X_2 = f(X_1, u_1) = f(f(X_0, u_1), u_1)$, and so forth, where $f(x, y) = \sqrt{\alpha_0 + \alpha_1 x y}$. This shows that X_t is a function of $X_0, u_1, \ldots, u_t$. In particular, X_t and u_r are independent, whenever $t < r$. Analogously, σ_t is a function of $X_0, u_1, \ldots, u_{t-1}$. We may conclude that

$$E(X_t | \mathcal{F}_{t-1}) = \sigma_t E(u_t) = 0,$$

which implies that $\{u_t\}$ is a martingale difference sequence, provided $E|X_t| < \infty$ (see below). Further, σ_t^2 is the conditional variance of X_t, since

$$E(X_t^2 | \mathcal{F}_{t-1}) = \sigma_t^2,$$

and the independence of u_{t+h} from X_t and σ_{t+h} implies that

$$\begin{aligned} E(X_t X_{t+h}) &= E(X_t \sigma_{t+h} u_{t+h}) \\ &= E(X_t \sigma_{t+h}) E(u_{t+h}) \\ &= 0, \end{aligned}$$

for all $h \geq 1$. Finally, notice that the sequence of second moments $E(X_t^2)$ satisfies the recursion

$$E(X_t^2) = \alpha_0 + \alpha_1 E(X_{t-1}^2),$$

which has a stationary solution given by

$$E(X_t^2) = E(X_0^2) = \frac{\alpha_0}{1 - \alpha_1}.$$

We have shown the following result.

Proposition 3.7.4 *For any random starting value X_0 with $E(X_0^2) = \frac{\alpha}{1-\alpha_1}$, the* ARCH(1) *equations have a stationary causal solution that is a white-noise process.*

In practice, ARCH(p) models with large values of p are often appropriate to fit financial return series. Observing that, by assumption of the model, σ_t^2 is given by a weighted average of the past p values $X_{t-1}^2, \ldots, X_{t-p}^2$, motivates replacement of those terms by a smaller number of past σ_ts. This idea leads to the following definition.

Definition 3.7.5 *A time series $\{X_t\}$ is called* **generalized autoregressive conditional heteroscedastic** *process of order p and q, abbreviated as* GARCH(p, q)*, $p \geq 1$ and $q \geq 0$, if*

$$\begin{aligned} X_t &= \sigma_t u_t, \\ \sigma_t^2 &= \alpha_0 + \sum_{i=1}^{p} \alpha_i X_{t-i}^2 + \sum_{j=1}^{q} \beta_j \sigma_{t-j}^2, \end{aligned}$$

where $\alpha_0, \alpha_1, \ldots, \alpha_p \geq 0$ *and* $\beta_1, \ldots, \beta_q \geq 0$ *are constants with* $\alpha_p, \beta_q > 0$ *and* $u_t \sim$ *i.i.d.*$(0, 1)$ *is independent of* $\{X_{t-k} : k \geq 1\}$.

Empirical work has shown that the intuition leading to their formulation also applies in practice: Usually GARCH(p, q) models with small p and q nicely capture persistence of high volatility (volatility clusters), whereas ARCH(p) specifications usually need high orders.

Again let us focus on the simplest model of that class given by $p = q = 1$. The GARCH$(1, 1)$ often leads to satisfactory model fits in practical applications. The following theorem provides a necessary and sufficient condition for the existence of a strictly stationary solution.

Theorem 3.7.6 *Let* $\alpha_0 > 0$ *and* $\alpha_1, \beta_1 \geq 0$. *The GARCH*$(1, 1)$ *equations admit a strictly stationary solution, if and only if*

$$E \log(\alpha_1 u_1^2 + \beta_1) < 0.$$

Then, the strictly stationary solution for σ_t^2 *is given by*

$$\sigma_t^2 = \alpha_0 \left(1 + \sum_{j=1}^{\infty} \prod_{i=1}^{j} (\alpha_1 u_{t-i}^2 + \beta_1) \right), \qquad t \in \mathbb{Z}. \tag{3.12}$$

Proof. We only show the sufficiency part and verify the representation of σ_t^2. Consider

$$\sigma_t^2 = \alpha_0 + \alpha_1 X_{t-1}^2 + \beta_1 \sigma_{t-1}^2.$$

Substituting $X_{t-1}^2 = \sigma_{t-1}^2 u_{t-1}^2$ leads to the recursion

$$\begin{aligned} \sigma_t^2 &= \alpha_0 + \alpha_1 \sigma_{t-1}^2 u_{t-1}^2 + \beta_1 \sigma_{t-1}^2 \\ &= \alpha_0 + (\alpha_1 u_{t-1}^2 + \beta_1) \sigma_{t-1}^2. \end{aligned}$$

This recursion is of the form

$$x_t^2 = \alpha_0 + \gamma_{t-1} x_{t-1}^2,$$

and we claim that it is solved by the series

$$x_t^2 = \alpha_0 \left(1 + \sum_{j=1}^{\infty} \prod_{j=1}^{i} \gamma_{t-i} \right),$$

provided the series converges. Indeed,

$$
\begin{aligned}
\alpha_0 + \gamma_{t-1} x_{t-1}^2 &= \alpha_0 + \gamma_{t-1}\alpha_0 \left(1 + \sum_{j=1}^{\infty} \prod_{i=1}^{j} \gamma_{t-1-i} \right) \\
&= \alpha_0 + \gamma_{t-1}\alpha_0 + \alpha_0 \gamma_{t-1}(\gamma_{t-2} + \gamma_{t-2}\gamma_{t-3} + \cdots) \\
&= \alpha_0 + \alpha_0(\gamma_{t-1} + \gamma_{t-1}\gamma_{t-2} + \cdots) \\
&= \alpha_0 + \alpha_0 \sum_{j=1}^{\infty} \prod_{i=1}^{j} \gamma_{t-i} \\
&= x_t^2.
\end{aligned}
$$

To verify the convergence of the random series (3.12), notice that the partial sums of the relevant series $\sum_{j=1}^{\infty} \prod_{i=1}^{j} (\alpha_1 u_{t-i}^2 + \beta_1)$ can be written as

$$
\sum_{j=1}^{n} \prod_{i=1}^{j} (\alpha_1 u_{t-i}^2 + \beta_1) = \sum_{j=1}^{n} \rho^{-j} Z_j, \qquad Z_j = \rho^j \prod_{i=1}^{j} (\alpha_1 u_{t-i}^2 + \beta_1),
$$

for any real ρ. If we choose $\rho > 1$ and show that with probability one

$$
Z_j = \rho^j \prod_{i=1}^{j} (\alpha_1 u_{t-i}^2 + \beta_1) \to 0, \qquad j \to \infty, \tag{3.13}
$$

then the above partial sum can be bounded by $\sum_{j=1}^{n} \rho^{-j} \to \frac{1}{1-\rho^{-1}}$, which establishes its a.s. convergence. To check Equation (3.13), notice that $E \log(\alpha_1 u_1^2 + \beta_1) < 0$ implies that there exists some $\rho > 1$ with

$$
\log \rho + E \log(\alpha_1 u_1^2 + \beta_1) < 0.
$$

Since u_t are i.i.d. and $\log(\alpha_1 u_1^2 + \beta_1)$ is integrable, the strong law of large numbers ensures that

$$
\log(\rho) + \frac{1}{n} \sum_{i=1}^{n} \log(\alpha_1 u_{t-i}^2 + \beta_1)
$$

converges a.s. to $\log \rho + E \log(\alpha_1 u_1^2 + \beta_1)$, as $n \to \infty$. But this implies that

$$
\log \left(\rho^n \prod_{i=1}^{n} (\alpha_1 u_{t-i}^2 + \beta_1) \right) = n \left(\log(\rho) + \frac{1}{n} \sum_{i=1}^{n} \log(\alpha_1 u_{t-i}^2 + \beta_1) \right)
$$

converges to $-\infty$, with probability 1. Hence,

$$
\rho^n \prod_{i=1}^{n} (\alpha_1 u_{t-i}^2 + \beta_1) \overset{a.s.}{\to} 0,
$$

as $n \to \infty$, which establishes Equation (3.13).

The result can be extended to GARCH(p, q) models. A strictly stationary solution $\{X_t\}$ with EX_t^2 exists, if and only if

$$\sum_{i=1}^{p} \alpha_i + \sum_{j=1}^{q} \beta_j < 1.$$

Then $\{X_t\}$ is a white noise process with marginal variance

$$E(X_t^2) = \frac{\alpha_0}{1 - \sum_{i=1}^{p} \alpha_i - \sum_{j=1}^{q} \beta_j}.$$

ARCH and GARCH models are usually estimated by relying on the likelihood approach. Assume that the innovation sequence $\{u_t\}$ satisfies

$$u_t \overset{i.i.d.}{\sim} f,$$

for some density f with mean 0 and variance 1. For a GARCH(1, 1) model, the conditional density of X_t given the past values $X_{t-1}, \ldots, X_0$ is easily seen to be

$$f_{X_t|X_{t-1},\ldots,X_0}(x_t|x_{t-1}, \ldots, x_0) = f_{X_t|X_{t-1}}(x_t|x_{t-1}) = \frac{1}{\sigma_t} f\left(\frac{x_t}{\sigma_t}\right),$$

leading to the likelihood

$$L(\alpha_0, \alpha_1, \beta_1) = \prod_{t=1}^{T} \frac{1}{\sigma_t} f\left(\frac{x_t}{\sigma_t}\right),$$

where $\sigma_t = \sqrt{\alpha_0 + \alpha_1 X_{t-1}^2 + \beta_1 \sigma_{t-1}^2}$ can be calculated recursively. Notice that we need to have a starting σ_0 that is unobservable. A common approach is to use the sample standard deviation of historical data or simply put $\sigma_0 = 0$. For a GARCH(p, q) the likelihood can be established in a similar way, but one needs starting values $X_{-p+1}, \ldots, X_0$ as well as $\sigma_{-p+1}, \ldots, \sigma_0$.

Given estimates for the unknown parameters of the selected GARCH model, one can calculate the estimators $\widehat{\sigma}_t$, for example we have $\widehat{\sigma}_t = \sqrt{\widehat{\alpha}_0 + \widehat{\alpha}_1 X_{t-1}^2 + \widehat{\sigma}_{t-1}}$ for a GARCH(1, 1). Then, one also calculates the residuals

$$\widehat{u}_t = \frac{\widehat{X}_t}{\widehat{\sigma}_t},$$

which should be approximately white noise, if the model is true. This can be checked by taking a look at the sample autocovariances of those residuals.

A plethora of modifications and extensions of the classical GARCH model has been proposed and studied in the literature. Here are only three examples. To allow for an asymmetric effect of information on volatility, it has been proposed to substitute terms like X_{t-1}^2 by $(X_{t-1} + \delta|X_{t-1}|)^2$, where the new parameter δ is constrained to lie in the interval [0, 1]. Then, the effect of X_{t-1}^2 on σ_t^2 is $(1 + \delta)^2 X_{t-1}^2$, if $X_{t-1} \geq 0$, and $(1 - \delta)^2 X_{t-1}^2$, if $X_{t-1} < 0$. The economic reasoning behind such a **leverage effect** is that when the price of a company falls, its debt-to-equity ratio increases, which should increase volatility, since volatility measures the risk associated with an investment in the stocks of the company.

The **GARCH in mean model (GARCH-M)** assumes that

$$X_t = \beta Z_t + \lambda \sigma_t + \sigma_t u_t,$$
$$\sigma_t^2 = \alpha_0 + \sum_{i=1}^{p} \alpha_i X_{t-i}^2 + \sum_{j=1}^{q} \beta_j \sigma_{t-j}^2,$$

for some regressor Z_t. The conditional mean of X_t given $\mathcal{F}_{t-1}$ is now given by $\beta Z_t + \lambda \sigma_t$ and therefore depends on the volatility.

The **exponential GARCH model** specifies the conditional variance as

$$\sigma_t^2 = \exp\left(\alpha_0 + \sum_{i=1}^{p} \alpha_i g(X_{t-i}) + \sum_{j=1}^{q} \beta_j \log(\sigma_{t-j}) \right),$$

where $\alpha_0, \ldots, \alpha_p$ and $\beta_1, \ldots, \beta_q$ are real-valued parameters. By modeling the log volatility, the parameters have not been constrained at this point. The terms $g(X_{t-i})$ are given by

$$g(X_t) = \theta X_t + \gamma(|X_t| - E|X_t|)$$

with further parameters θ and γ. One may derive a MA(∞) representation

$$\log(\sigma_t^2) = \omega_t + \sum_{k=1}^{\infty} g(X_{t-k}).$$

Provided $\sum_{k=1}^{\infty} \beta_k^2 < \infty$, σ_t^2 is strictly stationary.

3.8 Long-memory series

There is some evidence that certain financial series are affected by what is called long memory. That means that the autocorrelations decay much slower than they do for models such as AR or ARMA processes. Here we have seen that the autocovariances, say γ_k, tend to 0 at an exponential rate such as $\gamma_k \sim a^{-k}$, which rapidly decays as k gets large. Thus, one says that such models have a short memory. But if, for example, $\gamma_k \sim k^{-\beta}$ for some $\beta > 0$, the γ_k decay much slower. For $\beta > 1$ they are still summable, but even that property gets lost for $0 < \beta < 1$.

A convenient way to understand long-memory processes is to introduce first fractional differences.

3.8.1 Fractional differences

For $d \in \mathbb{N}$ the lag polynomial $(1 - L)^d$ takes differences of order d, i.e.

$$(1 - L)X_t = X_t - X_{t-1},$$
$$(1 - L)^2 X_t = X_t - 2X_{t-1} + X_{t-2},$$

and so forth. We aim at extending those differences to values $-1 < d < 1$ and will call them **fractional differences**.

For $d \in \mathbb{R}$ the (formal) series representation of the lag operator $p(L) = (1 - L)^d$ is obtained from the Taylor expansion of the function $f(z) = (1 - z)^d$ for $z \in \mathbb{C}$. It is easy to check that

$$f^{(k)}(z) = \left(\prod_{i=0}^{k-1} (d - i) \right) (-1)^k (1 - z)^{d-k}, \qquad k \in \mathbb{N}$$

leading to the **binomial series**

$$(1 - z)^d = \sum_{k=0}^{\infty} \psi_k z^k, \qquad z \in \mathbb{C}, \tag{3.14}$$

with coefficients $\psi_0 = 1$ and

$$\psi_k = \frac{f^{(k)}(0)}{k!} = \frac{(-1)^k}{k!} \prod_{i=0}^{k-1} (d - i) = (-1)^k \binom{d}{k}, \qquad k \in \mathbb{N},$$

where

$$\binom{d}{k} = \frac{d(d-1)\cdots(d-k+1)}{k!}$$

are the generalized binomial coefficients. For $d > 0$ the series converges absolutely on the unit disc, such that

$$(1 - L)^d = \sum_{k=0}^{\infty} \psi_k L^k$$

is a well-defined L_1 filter allowing us to consider linear processes of the form $(1 - L)^d \epsilon_t$, where ϵ_t is a white-noise process. More generally, processes of the form $(1 - L)^d X_t$ exist a.s., provided $\{X_t\}$ is a process with $\sup_n E|X_n| < \infty$, and also in L_2 if $\sup_n E|X_n|^2 < \infty$.

Before studying the case $-1 < d < 0$, let us collect some first properties of the coefficients ψ_k.

The coefficients ψ_k can be expressed in terms of the Gamma function Γ. For that purpose, write

$$\psi_k = \frac{1}{k!} \prod_{i=0}^{k-1} (-d + i). \tag{3.15}$$

Recall the definition of the Gamma function,

$$\Gamma(z) = \lim_{n \to \infty} \frac{n^z n!}{z(z+1)(z+2)\cdots(z+n)}, \qquad z \in \mathbb{C} \backslash \{0, -1, -2, \ldots\}.$$

For positive real z one has the integral representation

$$\Gamma(z) = \int_0^{\infty} \exp(-t) t^{z-1} \, dt.$$

The Gamma function satisfies $\Gamma(n+1) = n!$ for $n \in \mathbb{N}$ and, more generally, the recursion

$$\Gamma(z+1) = z\Gamma(z)$$

on its domain. That leads to the formula

$$\Gamma(z+n+1) = (z+n)\Gamma(z+n) = \cdots = \prod_{i=0}^{n}(z+i)\Gamma(z),$$

which allows us to express the product in Equation (3.15) by the Gamma function. Indeed, if we apply the latter with $z = -d$ and combine it with $1/k! = 1/\Gamma(k+1)$, we arrive at

$$\psi_k = \frac{\Gamma(k-d)}{\Gamma(k+1)\Gamma(-d)}, \qquad k \in \mathbb{N}.$$

The following lemma shows that the coefficients ψ_k decay as $k^{-\gamma}$, where $\gamma = d+1 > 0$. It makes use of **Sterling's formula**

$$\Gamma(x) \sim \sqrt{2\pi}\mathrm{e}^{-x+1}(x-1)^{x-1/2}, \qquad x \to \infty.$$

Lemma 3.8.1 *For $d > -1$*

$$\psi_k \sim \frac{k^{-(d+1)}}{\Gamma(-d)},$$

as $k \to \infty$.

Proof. By virtue of Sterling's formula,

$$\begin{aligned}
\psi_k &= \frac{\Gamma(k-d)}{\Gamma(k+1)\Gamma(-d)} \\
&\sim \frac{\mathrm{e}^{d-k+1}(k-d-1)^{k-d-1/2}}{\Gamma(-d)\mathrm{e}^{-k}k^{k+1/2}} \\
&= \frac{1}{\Gamma(-d)}\mathrm{e}^{d+1}\left(\frac{k-d-1}{k}\right)^k \frac{(k-d-1)^{-d-1/2}}{k^{1/2}},
\end{aligned}$$

where $\left(\frac{k-d-1}{k}\right)^k = \left(1 - \frac{d+1}{k}\right)^k \to \mathrm{e}^{-(d+1)}$, as $k \to \infty$, and

$$\begin{aligned}
\frac{(k-d-1)^{-d-1/2}}{k^{1/2}} &= (k-d-1)^{-(d+1)}\left(\frac{k-d-1}{k}\right)^{1/2} \\
&= k^{-(d+1)}\left(\frac{k-(d+1)}{k}\right)^{-(d+1)}\left(\frac{k-(d+1)}{k}\right)^{1/2} \\
&\sim k^{-(d+1)},
\end{aligned}$$

as $k \to \infty$, since d is fixed.

In order to study negative fractional powers of $1-L$ we need the following lemma.

Lemma 3.8.2 *For* $-1/2 < d < 1/2$ *we have* $\sum_{k=0}^{\infty} \psi_k^2 < \infty$, *such that* $\sum_{k=0}^{\infty} \psi_k L^k$ *is a* L_2 *filter.*

Proof. Let $\varepsilon > 0$. By Lemma 3.8.1 we can find k_0 such that $|\psi_k/(k^{-(d+1)}/\Gamma(-d))| \leq 1 + \varepsilon$ for $k \geq k_0$. We have for some constant $0 < C_1 < \infty$

$$\sum_{k=0}^{\infty} \psi_k^2 = \sum_{k=0}^{k_0-1} \psi_k^2 + \sum_{k=k_0}^{\infty} \left(\frac{1}{\Gamma(-d)} \frac{1}{k^{d+1}} \right)^2 \left(\frac{\psi_k}{k^{-(d+1)}/\Gamma(-d)} \right)^2$$
$$\leq C_1 + (1+\varepsilon)^2 \sum_{k=k_0}^{\infty} \left(\frac{1}{\Gamma(-d)} \frac{1}{k^{d+1}} \right)^2 .$$

It remains to verify that $\sum_{k=k_0}^{\infty} \left(\frac{1}{k^{d+1}} \right)^2 < \infty$. If $0 \leq d < 1/2$, then

$$\sum_{k=k_0}^{\infty} \left(\frac{1}{k^{d+1}} \right)^2 \leq \sum_{k=0}^{\infty} \frac{1}{k^{2+2d}} \leq \sum_{k=0}^{\infty} \frac{1}{k^2} < \infty.$$

If $-1/2 < d < 0$, then choose d_0 with $-1/2 < d_0 < d$. Then $2d_0 \in (-1, 0)$ and $\delta = 1 + 2d_0 > 0$. It follows that

$$\sum_{k=k_0}^{\infty} \frac{1}{k^{2+2d}} \leq \sum_{k=k_0}^{\infty} \frac{1}{k^{2+2d_0}} \leq \sum_{k=0}^{\infty} \frac{1}{k^{1+\delta}} < \infty,$$

which completes the proof.

Having the fractional difference operator $(1 - L)^d$, $d > 0$, at our disposal, which defines stationary time series when applied to white-noise processes, we want to invert that filter yielding $(1 - L)^{-d}$. For negative exponents larger than -1 the binomial series converges only for $|z| < 1$. Let us solve the crucial equation (3.6), i.e.

$$\psi(z)\theta(z) = 1,$$

for $|z| < 1$. The solution $\theta(z) = (1 - z)^{-d}$ is an analytic function for $0 < d < 1$. The associated Taylor expansions as in Equation (3.14),

$$\theta(z) = \sum_{k=0}^{\infty} \theta_k z^k, \qquad \theta_k = \theta^{(k)}(0)/k!,$$

with d replaced by $-d$ yields the coefficients

$$\theta_k = \frac{1}{k!} \prod_{i=0}^{k-1} (d+i) = \frac{\Gamma(k+d)}{\Gamma(k+1)\Gamma(d)}.$$

A further application of Sterling's approximation shows that

$$\theta_k \sim k^{d-1}/\Gamma(d), \qquad k \to \infty.$$

which also implies that $\sum_{k=0}^{\infty}\theta_k^2 < \infty$, i.e. $\theta(L)$ is a L_2 filter, if $0 < d < 1/2$. This means, for $0 < d < 1/2$

$$X_t = (1-L)^{-d}\epsilon_t = \sum_{k=0}^{\infty}\theta_k\epsilon_{t-k}, \qquad \epsilon_t \sim \mathrm{WN}(0,\sigma^2),\ \sigma^2 \in (0,\infty),$$

exists in the L_2 sense and provides a stationary solution of the equations

$$(1-L)^d X_t = \epsilon_t,$$

cf. Proposition 3.4.3.

Definition 3.8.3 (Fractionally Integrated Noise)
Let $-1/2 < d < 1/2$. A stationary solution $\{X_t\}$ of the equations

$$(1-L)^d X_t = \epsilon_t$$

for a white-noise process $\{\epsilon_t\}$ is called **fractionally integrated noise**.

We have seen in Lemma 3.5.2 that the spectral density of a linear process given by a lag operator $p(L)$ is $f(\omega) = \frac{\sigma^2}{2\pi}|p(\mathrm{e}^{\imath\omega})|^2$ from which one can also calculate the autocovariance function via the formula $\gamma(h) = \int_{-\pi}^{\pi}\cos(\omega h)f(\omega)\,\mathrm{d}\omega$, cf. Theorem 3.5.7.

For $0 < d < 1/2$ the linear filter $\psi(L) = \sum_j \psi_j L^j$ is an L_1 filter and consequently the spectral density of $X_t = (1-L)^{-d}\epsilon_t = \psi(L)\epsilon_t$ is given by

$$f_X(\omega) = \frac{\sigma^2}{2\pi}|1-\mathrm{e}^{\imath\omega}|^{-2d},$$

since $\psi(z) = (1-z)^{-d}$. Using $|1-\mathrm{e}^{\imath\omega}| = 2\sin(\omega/2)$, we obtain the formula

$$f_X(\omega) = \frac{\sigma^2}{2\pi}|1-\sin(\omega/2)|^{-2d}.$$

Let us use the last formula to calculate the autocovariances

$$\begin{aligned}\gamma_X(h) &= \int_{-\pi}^{\pi}\mathrm{e}^{\imath h\omega}f(\omega)\,\mathrm{d}\omega \\ &= \frac{\sigma^2}{\pi}\int_0^{\pi}\cos(h\omega)(2\sin(\omega/2))^{-2d}\,\mathrm{d}\omega.\end{aligned}$$

The last integral is of the form $\int_0^{\pi}\cos(hx)\sin^{z-1}(x)\,\mathrm{d}x$, which is known to equal $\frac{\pi\cos(h\pi/2)\Gamma(z+1)2^{1-z}}{z\Gamma((z+h+1)/2)\Gamma((z-h-1)/2)}$.

Lemma 3.8.4 *Let $0 < d < 1/2$ and $\epsilon_t \sim \mathrm{WN}(0,\sigma^2)$. Then the autocovariances and autocorrelations of $\{X_t\}$ are given by*

$$\gamma_X(h) = \sigma^2\frac{(-1)^h\Gamma(1-2d)}{\Gamma(h-d+1)\Gamma(1-h-d)}$$

and

$$\rho_X(h) = \frac{\Gamma(h+d)\Gamma(1-d)}{\Gamma(h-d+1)\Gamma(d)},$$

for $h \in \mathbb{Z}$. *Further,*

$$\rho_X(h) \sim h^{2d-1}\frac{\Gamma(1-d)}{\Gamma(d)}.$$

Another approach to introduce long-memory processes is as follows: A mean zero Gaussian series $\{Z_t\}$ exhibits long-range dependence, if its autocovariance function $\gamma_Z(k)$ satisfies

$$\gamma_Z(k) \sim k^{-D}L(k),$$

as $k \to \infty$, for some $0 < D < 1$ and a slowly varying function L, that is

$$\lim_{x\to\infty} \frac{L(cx)}{L(x)} = 1$$

for any $c > 0$. Then $H = 1 - D/2$ is also called the **Hurst index**. Lemma 3.8.4 asserts that a fractionally integrated series with $0 < d < 1/2$ has a Hurst index $H = 1/2 + d$.

3.8.2 Fractionally integrated processes

Let $\{X_t\}$ be a time series. If its fractional differences can be expressed as a ARMA(p, q), it is called a FARIMA(p, q) process. More precisely, one defines.

Definition 3.8.5 (FARIMA PROCESS)
A time series $\{X_t\}$ *is called a* **fractionally integrated** ARMA **of order** (p, d, q), $d \in (-1/2, 1/2)$, *if* $\{X_t\}$ *is a stationary solution of the equations*

$$\psi(L)(1-L)^d X_t = \phi(L)\epsilon_t$$

for some white noise $\{\epsilon_t\}$ *and lag polynomials* ψ, ϕ *of degrees* p, q, *respectively.*

3.9 Notes and further reading

For a thorough introduction to martingales in discrete time the textbook Williams (1991) can be recommended. A classic comprehensive reference on martingales and related limit theorems is the monograph Hall and Heyde (1980). Time-series analysis, mainly for parametric models, can be found in Brockwell and Davis (1991). For a special focus on applications to financial markets we refer to Fan and Yao (2003), Carmona (2004), Lai and Xing (2008), Tsay (2010) and Jondeau et al. (2007). The ARCH model was proposed in the seminal work Engle (1982). For the exponential GARCH model we refer to Nelson (1991). A comprehensive monograph on such models and their estimation is Straumann (2005). The basic idea of the proof of Theorem 3.7.6 seems to be due to Nelson (1990), cf. the discussion in Kreiss and Neuhaus (2006).

References

Brockwell P.J. and Davis R.A. (1991) *Time Series: Theory and Methods*. Springer Series in Statistics 2nd edn. Springer-Verlag, New York.

Carmona R.A. (2004) *Statistical Analysis of Financial Data in S-Plus*. Springer Texts in Statistics. Springer-Verlag, New York.

Engle R.F. (1982) Autoregressive conditional heteroscedasticity with estimates of the variance of United Kingdom inflation. *Econometrica* **50**(4), 987–1007.

Fan J. and Yao Q. (2003) *Nonlinear Time Series: Nonparametric and Parametric Methods*. Springer Series in Statistics. Springer-Verlag, New York.

Hall P. and Heyde C.C. (1980) *Martingale Limit Theory and its Application*. Academic Press Inc. [Harcourt Brace Jovanovich Publishers], New York. Probability and Mathematical Statistics.

Jondeau E., Poon S.H. and Rockinger M. (2007) *Financial Modeling under Non-Gaussian Distributions*. Springer Finance. Springer-Verlag London Ltd., London.

Kreiss J.P. and Neuhaus G. (2006) *Einführung in die Zeitreihenanalyse*. Springer, Berlin Heidelberg.

Lai T.L. and Xing H. (2008) *Statistical Models and Methods for Financial Markets*. Springer Texts in Statistics. Springer, New York.

Nelson D.B. (1990) Stationarity and persistence in the GARCH(1, 1) model. *Econometric Theory* **6**(3), 318–334.

Nelson D.B. (1991) Conditional heteroskedasticity in asset returns: a new approach. *Econometrica* **59**(2), 347–370.

Straumann D. (2005) *Estimation in conditionally heteroscedastic time series models* vol. 181 of *Lecture Notes in Statistics*. Springer-Verlag, Berlin.

Tsay R.S. (2010) *Analysis of Financial Time Series*. Wiley Series in Probability and Statistics 3rd edn. John Wiley & Sons Inc., Hoboken, NJ.

Williams D. (1991) *Probability with Martingales*. Cambridge Mathematical Textbooks. Cambridge University Press, Cambridge.

4

Arbitrage theory for the multiperiod model

Having the martingale theory in discrete time discussed in the previous chapter at our disposal, we are now in a position to study how to price contingent claims assuming that trading is possible at discrete fixed time points; without loss of generality these time points will be denoted by $t = 0, \dots, T$, where T denotes the number of time points and coincides with the time horizon. First, we need to extend various definitions and notions, such as *self-financing strategy*, *equivalent martingale measure* and *no-arbitrage*, from the one-period to the multiperiod setting, before we can study the question how the no-arbitrage condition relates to the existence of an equivalent martingale measure P^*. We shall see that for a financial market with a finite state space Ω we get nice explicit formulas for P^*.

After a discussion of the general theory for the multiperiod case, we turn our attention to the Cox-Ross-Rubinstein binomial model, which allows all quantities explicitly, including the (delta) hedge for a path-dependent derivative to be easily calculated. The multiperiod binomial model is also an powerful and simple vehicle to derive the celebrated Black–Scholes option pricing formula by studying the convergence in distribution of the stock price under the sequence of equivalent martingale measures appropriately constructed in the binomial model.

Finally, we study how to price American-style derivatives such as American options on a stock, where one can exercise the right to buy or sell the underlying at any time point before maturity. We shall see that the problem can be treated in a concise and transparent way by using the theory of optimal stopping discussed in the previous chapter. The results are actually constructive and lead to an algorithm for the arbitrage-free pricing of American derivatives by using a binomial tree.

Financial Statistics and Mathematical Finance: Methods, Models and Applications, First Edition. Ansgar Steland.

4.1 Definitions and preliminaries

Our model for the financial market is as follows. Let $(\Omega, \mathcal{F}, \{\mathcal{F}_t\}, P)$ be a filtered probability space, where

$$\{\emptyset, \Omega\} = \mathcal{F}_0 \subset \mathcal{F}_1 \subset \cdots \subset \mathcal{F}_T \subset \mathcal{F}$$

and P is the real probability measure. We assume that investors can invest in d assets traded at the market or deposit their money into a bank account. The prices are modeled by d $\mathcal{F}_t$-adapted processes

$$\{S_{ti} : t = 0, 1, \ldots, T\}, \qquad i = 0, \ldots, d,$$

with $0 \leq S_{ti}$ and $E(S_{ti}) < \infty$ for all t and i. As in Chapter 2, it is assumed that at each period one can borrow or deposit money at the same fixed interest rate. For simplicity of the exposition, we will confine ourselves to the case that the yield is deterministic and known to us. That means that a payment C deposited at time t growths to $C(1 + r)$ at time $t + 1$, for all $t = 0, \ldots, T - 1$, where r denotes the interest rate for each period. The bank account can be modeled equivalently by a bond given by the price process

$$S_{t0} = S_{00}(1 + r)^t, \qquad t = 0, \ldots, T,$$

where S_{00} is the nominal value of the bond.

It is worth mentioning that the theory of the present chapter remains more or less valid when considering so-called locally riskless bonds, although we confine ourselves to the above setting, in order to simplify the exposition.

Definition 4.1.1 *A locally riskless bond is given by a price process*

$$S_{t0} = \prod_{i=1}^{t}(1 + r_i), \qquad t = 0, \ldots, T,$$

for some predictable process $\{r_t\}$.

4.2 Self-financing trading strategies

Since investors can trade at times $0, \ldots, T - 1$, a trading strategy is given by the number of shares, φ_{ti}, of asset i held from time t to time $t + 1$. Clearly, φ_{Ti} are the terminal values. This means that an investor determines for each investment opportunity i his or her position at time 0. If $\varphi_{1i}S_{1i} > 0$, he has to pay $\varphi_{1i}S_{1i}$ and holds a long position, whereas he receives the amount $|\varphi_{1i}S_{1i}|$ and is short, if $\varphi_{1i}S_{1i} < 0$. In other words, $\varphi_1 = (\varphi_{10}, \ldots, \varphi_{1d})'$ is what we have called a portfolio. Then the investor proceeds and sets up a portfolio $\varphi_2 = (\varphi_{20}, \ldots, \varphi_{2d})'$ for the next period, and so forth.

Since, obviously, the φ_{ti} are determined given the information available up to and including time $t - 1$, the following definition is self-evident.

Definition 4.2.1 *A predictable process* $\{\varphi_t : t = 0, \ldots, T\}$, $\varphi_t = (\varphi_{t0}, \ldots, \varphi_{td})'$, *taking values in* $\mathbb{R}^{d+1}$ *with* $\varphi_0 = \varphi_1$ *is called a* **trading strategy**.

Notice that φ_0 is not really needed in our model. However, putting $\varphi_0 = \varphi_1$ will simplify some of the formulas we are going to derive.

Definition 4.2.2 *Let* $\{S_t : t = 0, \ldots, T\}$ *be a d-dimensional price process and* $\varphi = \{\varphi_t : t = 0, \ldots, T\}$ *be a trading strategy. Then the corresponding process* $V_t = V_t(\varphi)$ *given by*

$$V_t = \varphi_t' S_t, \qquad t = 0, \ldots T,$$

is called the **value process** *of the trading strategy* φ_t.

The starting value V_0 is the initial capital required to set up the trade. $V_1 = \varphi_1' S_1$ is the value at time 1 and, in general, $V_t = \varphi_t' S_t$ is the value of the trading strategy at time t. In the real world, a trading strategy can cover additional payments or withdrawals, e.g. income paid to the investor or additional investments from new investors. However, since such financial issues can be taken into account by introducing a new artificial asset, we will ignore them in what follows. Then the initial capital $\varphi_0' S_0$ has to be financed at time 0 and further portfolio updates have to be financed from the portfolio itself; buying more shares from one asset requires withdrawing money from the bank account or selling other assets.

At time $t \in \{1, \ldots, T-1\}$ the value of the portfolio φ_t held from $t-1$ to t, also called time t value, equals $\varphi_t' S_t$. That portfolio required the capital $\varphi_t' S_{t-1}$ at time $t-1$ when it was set up. We could realize the time t value $\varphi_t' S_t$ at time t by closing all positions. For the next period we have to determine φ_{t+1} giving rise to the costs $\varphi_{t+1}' S_t$. The net value is

$$C_t = \varphi_{t+1}' S_t - \varphi_t' S_t = (\varphi_{t+1} - \varphi_t)' S_t = \Delta \varphi_{t+1}' S_t.$$

Here and in what follows, we agree on the following definition of the difference operator

$$\Delta \varphi_{t+1} = \varphi_{t+1} - \varphi_t.$$

If $C_t > 0$, external money would be needed to finance the portfolio φ_{t+1}, whereas we could withdraw $|C_t|$ if $C_t < 0$. When $C_t = 0$, the portfolio update is self-financing.

Definition 4.2.3 *A trading strategy is called* **self-financing**, *if for all* $t = 1, \ldots, T$

$$\Delta \varphi_t' S_{t-1} = 0 \Leftrightarrow \varphi_t' S_{t-1} = \varphi_{t-1}' S_{t-1}$$

holds true.

Recall the definition of the stochastic integral in discrete time. Let φ_t and S_t be $\mathbb{R}$-valued processes, φ_t being predictable and S_t adapted. Then the discrete-time stochastic integral $\int \varphi_r \, dS_r$ denotes the $\mathbb{R}$-valued process

$$I_t = \int_0^t \varphi_r \, dS_r = \sum_{r=1}^{t} \varphi_r \Delta S_r = \sum_{r=1}^{t} \varphi_r (S_r - S_{r-1}), \qquad t = 0, \ldots, T, \tag{4.1}$$

see Definition 3.2.11. If both φ_t and S_t are k-dimensional, then $\int \varphi_r' \, dS_r$ is the $\mathbb{R}$-valued process

$$I_t = \int_0^t \varphi_r' \, dS_r = \sum_{r=1}^t \varphi_r'(S_r - S_{r-1}), \qquad t = 0, \ldots, T.$$

Theorem 4.2.4 *A trading strategy $\{\varphi_t : t = 0, \ldots, T\}$ is self-financing, if and only if for all $t = 0, \ldots, T$*

$$V_t = V_0 + \sum_{r=1}^t \varphi_r' \Delta S_r = V_0 + \int_0^t \varphi_r' \, dS_r$$

holds true, where $V_0 = \varphi_0' S_0$. In other words, a trading strategy is self-financing if and only if the associated value process is a discrete stochastic Itô integral.

Proof. For $t = 0$ the assertion is trivially satisfied. We have

$$\begin{aligned} \varphi_t \text{ self-financing} &\Leftrightarrow \Delta\varphi_t' S_{t-1} = 0, \qquad t = 1, \ldots, T \\ &\Leftrightarrow \varphi_t' S_{t-1} - \varphi_{t-1}' S_{t-1} = 0, \qquad t = 1, \ldots, T. \end{aligned}$$

Now add $\varphi_t' S_t$ and subtract $\varphi_t' S_{t-1}$ from both sides of the above equation to obtain

$$\varphi_t' S_t - \varphi_{t-1}' S_{t-1} = \varphi_t' S_t - \varphi_t' S_{t-1} = \varphi_t'(S_t - S_{t-1}).$$

The left-hand side is the $(t+1)$th summand, ΔV_t, of $V_t = V_0 + \sum_{r=1}^t \Delta V_r$. Thus, we obtain for $t = 1, \ldots, T$

$$V_t = V_0 + \sum_{r=1}^t \Delta V_r = V_0 + \int_0^t \varphi_r' \, dS_r.$$

Definition 4.2.5 *The process $S_t^* = (1, S_{t1}/S_{t0}, \ldots, S_{td}/S_{t0})'$, $t = 0, \ldots, T$, is called* **discounted price process**. *$V_t^* = \varphi_t' S_t^*$, $t = 0, \ldots, T$, is called a* **discounted value process**.

Remark 4.2.6 *If the bank account (numeraire) is a bond with face value* 1 *paying a fixed interest rate r, we have*

$$S_{ti}^* = \frac{S_{ti}}{(1+r)^t}, \qquad V_t^* = \frac{\varphi_t' S_t}{(1+r)^t}.$$

Noting that $\Delta\varphi_t' S_{t-1} = \varphi_t' S_{t-1} - \varphi_{t-1}' S_{t-1} = 0$ is equivalent to $\Delta\varphi_t' S_{t-1}^* = 0$, we immediately obtain

Proposition 4.2.7 *A trading strategy is self-financing if and only if the corresponding value process $V_t^* = V_t^*(\varphi)$ satisfies*

$$V_t^* = V_0^* + \sum_{r=1}^t \varphi_r' \Delta S_r^* = V_0^* + \int_0^t \varphi_r' \, dS_r^*$$

for $t = 0, \ldots, T$.

Notice that when discounting, the stochastic integral does not depend on φ_{t0}, $t = 1, \dots, T$, the money deposited into the bank, since $\Delta S_{t0}^* = 0$ for all t.

Theorem 4.2.8 *Let Q be a probability measure on $(\Omega, \mathcal{F})$ such that $\{S_{ti}^* : t = 0, \dots, T\}$, $i = 1, \dots, d$, are martingales under Q, and let $\{\varphi_t\}$ be a self-financing trading strategy. If the corresponding value process $V_t = \varphi_t' S_t$ is Q-integrable for all t, then V_t^* is a martingale under Q, such that*

$$E_Q(V_t^* | \mathcal{F}_s) = V_s^* \quad Q - a.s.,$$

for all $s \le t$, $E_Q(V_t^*) = V_0$ and

$$V_t^* = E_Q(V_T^* | \mathcal{F}_t) \quad Q - a.s.,$$

for all $t = 0, \dots, T$.

Proof. Since φ_t is self-financing, $V_{t-1} = \varphi_t' S_{t-1}$. Hence,

$$\varphi_{t+1}'(S_{t+1} - S_t) = V_{t+1} - \varphi_{t+1}' S_t = V_{t+1} - V_t$$

is Q-integrable, such that the martingale property of S_t^* implies

$$E_Q(\varphi_{t+1}'(S_{t+1}^* - S_t^*) | \mathcal{F}_t) = \varphi_{t+1}'(E_Q(S_{t+1}^* | \mathcal{F}_t) - S_t^*) = 0.$$

Consequently, for $t = 0, \dots, T - 1$,

$$\begin{aligned} E_Q(V_{t+1}^* | \mathcal{F}_t) &= E_Q\left(V_0^* + \sum_{i=1}^t \varphi_i' \Delta S_i^* \middle| \mathcal{F}_t \right) + E(\varphi_{t+1}'(S_{t+1}^* - S_t^*) | \mathcal{F}_t) \\ &= V_0^* + \sum_{i=1}^t \varphi_i' \Delta S_i^* \\ &= V_t^*, \end{aligned}$$

since φ_i is $\mathcal{F}_t$-predictable, i.e. φ_t is $\mathcal{F}_{t-1}$-measurable, if $i \le t$, which verifies that $\{V_t^*\}$ is a martingale under Q.

Remark 4.2.9 *Here are some sufficient conditions for $E|V_t| < \infty$, $t = 0, \dots, T$.*

(i) Ω is finite.

(ii) φ_t is bounded, i.e. $|\varphi_t| \le C$ for all $t = 0, \dots, T$.

(iii) $E|\varphi_t|^p < \infty$ and $ES_t^q < \infty$ for $p, q \in [1, \infty]$ with $p^{-1} + q^{-1} = 1$, by virtue of Hölder's inequality.

Theorem 4.2.8 provides an important insight. If we have a probability measure Q such that the price processes are martingales under Q, the discounted value process V_t^* of a self-financing trade can be calculated from the discounted final value V_T^*, namely

$$V_t^* = E_Q(V_T^* | \mathcal{F}_t).$$

Further, the time t value is given by

$$V_t = E_Q(V_T^*(1+r)^t|\mathcal{F}_t) = E_Q(V_T(1+r)^{(t-T)}|\mathcal{F}_t).$$

If a self-financing strategy is used to replicate a claim, these formulas provide the fair value of the claim. It is time to put these facts into the framework of arbitrage-freeness.

4.3 No-arbitrage and martingale measures

Definition 4.3.1 *A self-financing trading strategy $\varphi = \{\varphi_t\}$ is called an* **arbitrage opportunity** *or* **arbitrage**, *if the corresponding value process $V_t = V_t(\varphi)$ satisfies*

$$V_0 \leq 0, \quad V_T \geq 0 \ P\text{-a.s.}, \quad P(V_T > 0) > 0. \tag{4.2}$$

Remark 4.3.2

(i) Whether or not φ_t is an arbitrage depends on the probability measure P. The notion of an arbitrage is invariant on the set of all probability measures that are equivalent to P. But notice that the profit we make by an arbitrage opportunity may depend strongly on P.

(ii) Obviously, Equation (4.2) is equivalent to

$$V_0^* \leq 0, \quad V_T^* \geq 0 \ P\text{-a.s.}, \quad P(V_T^* > 0) > 0.$$

Obviously, the multiperiod model is basically a chain of one-period models. One can also establish a connection between arbitrage opportunities in the multiperiod model and the corresponding one-period models.

Theorem 4.3.3 *There exists an arbitrage opportunity in the multiperiod model if and only if there exists an arbitrage opportunity in at least one of its one-period model.*

Definition 4.3.4

(i) A probability measure Q is called a **martingale measure**, *if Q is trivial on $\mathcal{F}_0$, i.e. $Q(A) \in \{0, 1\}$ for all $A \in \mathcal{F}_0$, and when all discounted price processes $\{S_{ti}^* : t = 0, \dots, T\}$, $i = 1, \dots, d$, are $\mathcal{F}_t$-martingales under Q.*

(ii) Q is called an **equivalent martingale measure** *(EMM), if Q is a martingale measure and $Q \sim P$.*

After these preparations, we are now in a position to study the first fundamental theorem, which we prove for the case of a finite probability space. Recall that for finite $\Omega = \{\omega_1, \dots, \omega_n\}$, $n \in \mathbb{N}$, we may identify random variables X with vectors $(X(\omega_1), \dots, X(\omega_n))$, which will allow us to argue in a similar manner as in the one-period model.

Theorem 4.3.5 *A financial market is arbitrage-free, if and only if there exists some equivalent martingale measure.*

Proof. When $\Omega = \{\omega_1, \dots, \omega_n\}$ for some $n \in \mathbb{N}$ and ω_i with $p_i := P(\{\omega_i\}) > 0$ for $i = 1, \dots, n$, we may identify random variables with n-dimensional vectors. We can extend the idea of the proof of Theorem 2.5.4 to the multiperiod setting. Consider

$$U = \left\{(-V_0^*(\varphi), V_T^*(\varphi)) : \varphi = \{\varphi_t\} \text{ is predictable}\right\} \subset \mathbb{R}^{n+1}.$$

where the random variable $V_T^*(\varphi)$ is identified with the vector $(V_T^*(\varphi)(\omega_1), \dots, V_T^*(\varphi)(\omega_n))$, whereas $V_0^*(\varphi)$ is a constant. Notice that U is a linear subspace of $\mathbb{R}^{n+1}$. Let

$$M = \left\{(y_0, \dots, y_n)' \in \mathbb{R}^{n+1} : y_i \geq 0, i = 0, \dots, n, y_j > 0 \text{ for some } j\right\}.$$

We have $0 \notin M$ and the market is arbitrage-free, if and only if $U \cap M = \emptyset$. Further,

$$K = \{y \in M : y'\mathbf{1} = 1\}, \quad \mathbf{1} = (1, \dots, 1)' \in \mathbb{R}^{n+1},$$

is a nonempty, compact and convex subset of M satisfying $U \cap K = \emptyset$. Theorem 2.4.5 allows us to separate K and U: there is some vector $\lambda = (\lambda_0, \dots, \lambda_n)' \in \mathbb{R}^{n+1}$, such that

$$\begin{aligned} \lambda' x &= 0, \quad \forall x \in U, \\ \lambda' x &> 0, \quad \forall x \in K. \end{aligned}$$

Since the unit vectors e_j are in K, $\lambda_j = \lambda' e_j > 0$, $j = 0, \dots, n$, follows. Thus, the probability measure

$$P^*\left(\{\omega_j\}\right) := \frac{\lambda_j}{\sum_{k=1}^n \lambda_k}, \quad j = 1, \dots, n,$$

is equivalent to P. We claim that P^* is the equivalent martingale measure we are looking for. We have

$$\begin{aligned} E^*|S_{ti}| &= \sum_{j=1}^n \lambda_j |S_{ti}(\omega_j)| \Big/ \sum_{k=1}^n \lambda_k \\ &= \sum_{j=1}^n p_j |S_{ti}(\omega_j)| \frac{\lambda_j}{p_j \sum_{k=1}^n \lambda_k} \\ &\leq C \cdot E|S_{ti}| < \infty, \end{aligned}$$

where $C = \max_{1 \leq j \leq n} \frac{\lambda_j}{p_j} \left(\sum_{k=1}^n \lambda_k \right)^{-1} < \infty$. Noting that $S_{t0} = S_{00}(1 + r)^t$, it follows that $E^*(S_{ti}^*) < \infty$ as well. It remains to show that

$$E^*(S_{ti}^* - S_{si}^* | \mathcal{F}_s) = 0, \quad P^*\text{-a.s.},$$

for $s < t$ and $i = 1, \dots, d$. By definition, 0 is a version of the conditional expectation of $S_{ti}^* - S_{si}^*$ given $\mathcal{F}_s$, if for all $A \in \mathcal{F}_s$

$$0 = \int_A 0 \, \mathrm{d}P^* = \int_A (S_{ti}^* - S_{si}^*) \, \mathrm{d}P^* = E^*[\mathbf{1}_A (S_{ti}^* - S_{si}^*)].$$

In order to verify that equation, let $A \in \mathcal{F}_s$ and define the trading strategy $\widetilde{\varphi}_u = (\widetilde{\varphi}_{u0}, \dots, \widetilde{\varphi}_{ud})'$, $0 \leq u \leq T$, with

$$\widetilde{\varphi}_{uj}(\omega) = \mathbf{1}_A(\omega)\mathbf{1}_{\{s+1 \leq u \leq t\}}\mathbf{1}_{\{j=i\}} \quad \omega \in \Omega, \quad j = 0, \dots, d,$$

and consider the corresponding stochastic integral in discrete time

$$\begin{aligned} Z_u &= \int_0^u \widetilde{\varphi}_r' \, \mathrm{d}S_r^* = \sum_{r=1}^{u} \widetilde{\varphi}_r'(S_r^* - S_{r-1}^*) \\ &= \sum_{r=s+1}^{\min(u,t)} \mathbf{1}_A(S_{ri}^* - S_{r-1,i}^*), \quad u = 0, \dots, T. \end{aligned}$$

Then (by telescoping) $Z_T = \mathbf{1}_A(S_{ti}^* - S_{si}^*)$. Clearly, $(-Z_0, Z_T) \in U$, such that

$$\sum_{\omega_j \in \Omega} \lambda_j Z_T(\omega_j) = 0.$$

Thus, using the definition of P^*

$$E^*(\mathbf{1}_A(S_{ti}^* - S_{si}^*)) = E^*(Z_T) = \sum_{j=1}^{n} \lambda_j Z_T(\omega_j) \Big/ \sum_{k=1}^{n} \lambda_k = 0.$$

This shows that $\{S_t^*\}$ is a P^*-martingale.

4.4 European claims on arbitrage-free markets

Let us confine our study to European claims C paying a random payment C at maturity T, which depends on the underlying(s). It is time to distinguish formally contingent claims and derivatives.

Definition 4.4.1 *A non-negative random variable C, i.e. a $\mathcal{F} - \mathcal{B}$-measurable mapping $\Omega \to [0, \infty)$, is called a* **claim** *or* **contingent claim**. *A contingent claim C is called a* **derivative** *of the underlyings $i_1, \dots, i_j$, $1 \leq j \leq d$, if $C = g(S_{Ti_1}, \dots, S_{Ti_j})$ for some measurable full rank function $g : \mathbb{R}^j \to \mathbb{R}$. Recall at this point that a vector function $g = (g_1, \dots, g_j)$ is of full rank, if there is no Borel-measurable function h such that $g_i = h(g_1, \dots, g_{i-1}, g_{i+1}, \dots, g_j)$ for any $i = 1, \dots, j$.*

Definition 4.4.2 *An European claim is called* **replicable** *or* **attainable**, *if there is some trading strategy $\varphi = \{\varphi_t\}$, whose value at maturity T coincides with C, a.s., i.e.*

$$C = \varphi_T' S_T, \ P\text{-a.s.}$$

$\varphi = \{\varphi_t\}$ is called a **replicating (trading) strategy**.

If $\varphi = \{\varphi_t\}$ is a replicating strategy and $V_t = V_t(\varphi) = \varphi_t' S_t$, $1 \le t \le T$, denotes the corresponding value process, then

$$C = V_T = V_0 + \sum_{t=1}^{T} \varphi_t'(S_t - S_{t-1}),$$

or, equivalently, in terms of the discounted quantities,

$$C^* = V_T^* = V_0^* + \sum_{t=1}^{T} \varphi_t'(S_t^* - S_{t-1}^*).$$

Vice versa, such a representation implies that C is attainable and φ is a replicating trading strategy.

The following basic result asserts that the discounted value process of a replicable claim is, a.s., the Lévy martingale of the discounted claim. This important result implies that the discounted value can be calculated without knowing a replicating trading strategy.

Theorem 4.4.3 *Any attainable contingent claim C with $E(C) < \infty$ is P^*-integrable for each $P^* \in \mathcal{P}$. Further, the discounted value process $V_t^* = V_t^*(\varphi)$ of an arbitrary replicating trading strategy satisfies*

$$V_t^* = E^*(C^* | \mathcal{F}_t), \; P^*\text{-a.s.}, \; t = 1, \ldots, T.$$

In particular, V_t^* is a non-negative $\mathcal{F}_t$-martingale under P^*.

Proof. Notice that Theorem 4.2.8 remains true, if the generalized conditional expectations of V_t exists, for which $V_t \ge 0$ P-a.s. is a sufficient condition we want to show now. Noting that $V_T = C \ge 0$, we apply backward induction and therefore assume that $V_t \ge 0$, P-a.s. Let $\varphi_t(c) = \varphi_t \mathbf{1}_{\{|\varphi_t| \le c\}}$, $c > 0$, and notice that

$$\varphi_t = \lim_{c \to \infty} \varphi_t(c) \quad \text{and} \quad X_{t-1} = \lim_{c \to \infty} X_{t-1} \mathbf{1}_{\{|\varphi_t| \le c\}},$$

ω-wise. First, notice that

$$V_{t-1}^* = V_t^* - \varphi_t'(S_t^* - S_{t-1}^*) \ge -\varphi_t'(S_t^* - S_{t-1}^*)$$

and

$$\begin{aligned} E^*(V_{t-1}^* \mathbf{1}_{\{|\varphi_{t-1}| \le c\}}) &= E^*(\mathbf{1}_{\{|\varphi_{t-1}| \le c\}} \varphi_{t-1}' S_{t-1}^*) \\ &= E^*(\varphi_{t-1}(c)' S_{t-1}^*) \\ &\le c \cdot \sum_{i=0}^{d} E^* |S_{ti}^*| < \infty. \end{aligned}$$

Thus, we obtain

$$
\begin{aligned}
V_{t-1}^* \mathbf{1}_{\{|\varphi_t| \le c\}} &= E^*(V_{t-1}^* \mathbf{1}_{\{|\varphi_t| \le c\}} | \mathcal{F}_{t-1}) \\
&\ge E^*(-\mathbf{1}_{\{|\varphi_t| \le c\}} \varphi_t'(S_t^* - S_{t-1}^*) | \mathcal{F}_{t-1}) \\
&= -\varphi_t(c)' E^*(S_t^* - S_{t-1}^* | \mathcal{F}_{t-1}) \\
&= 0.
\end{aligned}
$$

Taking the limit $c \to \infty$ now yields $V_{t-1}^* \ge 0$, P^*-a.s., which implies $V_t \ge 0$, P-a.s., for all t. Hence, $E(V_t^* | \mathcal{F}_{t-1})$ is well defined.

On the set $A_c = \{|\varphi_t| \le c\}$ we have $\varphi_t = \varphi_t(c)$ such that for any $c > 0$ and almost all $\omega \in A_c$

$$
\begin{aligned}
E^*(V_t^* | \mathcal{F}_{t-1})(\omega) - V_{t-1}^*(\omega) &= E^*(V_t^* - V_{t-1}^* | \mathcal{F}_{t-1})(\omega) \\
&= E^*(\varphi_t(c)'(S_t^* - S_{t-1}^*) | \mathcal{F}_{t-1})(\omega) \\
&= \varphi_t(c)' E^*(S_t^* - S_{t-1}^* | \mathcal{F}_{t-1})(\omega) \\
&= 0.
\end{aligned}
$$

Thus, $\mathbf{1}_{\Omega \setminus A_c} \cdot 0 + \mathbf{1}_{A_c} \cdot V_{t-1}^*$ is a version of $E(V_t^* | \mathcal{F}_{t-1})$, cf. A.3.1 (x). Now, by dominated convergence, $c \to \infty$ yields the generalized martingale property of $\{V_t^*\}$. It remains to show that $V_t^* = E^*(C^* | \mathcal{F}_t)$. We have for $t < T$

$$
\begin{aligned}
V_t^* &= E^*(V_{t+1}^* | \mathcal{F}_t) \\
&= E^*(\cdots E^*(V_T^* | \mathcal{F}_{T-1}) \cdots | \mathcal{F}_t) \\
&= E^*(\cdots E^*(C^* | \mathcal{F}_{T-1}) \cdots | \mathcal{F}_t) \\
&= E^*(C^* | \mathcal{F}_t),
\end{aligned}
$$

which completes the proof.

Definition 4.4.4 *Suppose we are given an arbitrage-free market with d risky assets and a claim C. $\pi(C) \ge 0$ is called an* **arbitrage-free price of** C, *if there exists some adapted stochastic process $\{S_{t,d+1}^* : t = 0, \dots, T\}$, such that*

(i) $S_{0,d+1}^* = \pi(C)$, *P-a.s.;*

(ii) $S_{t,d+1}^* \ge 0$, *P-a.s., for $t = 0, \dots, T$;*

(iii) $S_{T,d+1}^* = C^*$, *P-a.s.; and*

(iv) the extended market consisting of the $d + 2$ price processes $\{S_{t0}\}, \dots, \{S_{t,d+1}\}$ is arbitrage-free.

If an attainable claim is traded at a price p that differs from the initial capital $V_0 = \varphi_0' S_0 = E^*(C^*)$ required to initiate the replicating hedge, arbitrageurs can earn riskless profits. For example, if $p > V_0$, the claim is sold at time 0 and V_0 is invested to acquire the hedge, whose value is always non-negative. The difference $p - V_0 > 0$ earns the riskless profit. At maturity, the claim is settled using the payoff of the hedge.

Theorem 4.4.5 *The set of arbitrage-free prices of a claim C is given by $\Pi(C) = \{E^*(C^*) : P^* \in \mathcal{P},\ E^*(C^*) < \infty\}$*

Proof. '$\subset$': Let p be an arbitrage-free price. Then the extended market is arbitrage-free. By Theorem 4.3.5 this holds true, if and only if $\mathcal{P} \neq \emptyset$. Therefore, there is some $P^* \in \mathcal{P}$, such that the $d+1$ processes $\{S_{ti}^* : t = 0, \ldots, T\}$, $i = 1, \ldots, d+1$, are P^*-martingales. We have to show that $p = E^*(C^*)$. By the martingale property,

$$p = S_{0,d+1}^* = E^*(S_{T,d+1}^* | \mathcal{F}_0) = E^*(C^* | \mathcal{F}_0) = E^*(C^*),$$

since $S_{T,d+1} = C$.

'$\supset$': Let $p \in \{E^*(C^*) : P^* \in \mathcal{P},\ E^*(C^*) < \infty\}$. Then $p = E^*(C^*)$ for some $P^* \in \mathcal{P}$. Fix that P^* and define $\{S_{t,d+1}^* : t = 0, \ldots, T\}$ by the Lévy martingale

$$S_{t,d+1}^* = E^*(C^* | \mathcal{F}_t), \qquad t = 0, \ldots, T.$$

By definition, $S_{t,d+1}^*$ is $\mathcal{F}_t$-adapted with expectation $p = E^*(C^*)$. From Theorem 4.4.3 we know that $S_{t,d+1}^*$ is non-negative and coincides with the discounted value process, a.s. Since the $d+1$ processes $\{S_{ti}^* : t = 0, \ldots, T\}$, $i = 1, \ldots, d+1$, are $\mathcal{F}_t$-martingales under P^*, P^* is an equivalent martingale measure for the extended market. By Theorem 4.3.5, this holds true if and only if the extended market is arbitrage-free. Hence, all conditions of Definition 4.4.4 are checked.

Definition 4.4.6 *An arbitrage-free financial market is called* **complete**, *if all contingent claims are attainable.*

In a complete market all claims can be hedged by self-financing trading strategies. We have learned in Chapter 2 that in a one-period model the linear space of portfolios can be identified with the linear space $L_\infty(\Omega, \mathcal{F}, P)$ of bounded random variables. Its dimension is bounded by $d+1$. In the multiperiod model, a similar result holds true.

Theorem 4.4.7 *On an arbitrage-free market where all bounded claims are attainable, the following assertions hold true:*

(i) The market is complete.

(ii) The dimension of $L_\infty(\Omega, \mathcal{F}, P)$ is bounded by $(d+1)^T$.

As a consequence of Theorem 4.4.7, if trading is restricted to discrete time instants, any arbitrage-free and complete market can be represented by a tree: At each time point t the price processes attain values in a finite set of at most $d+1$ values. In other words, if an arbitrage-free market can not be presented as a tree, there may be contingent claims that can not be hedged. Theorem 4.4.3 told us that the value process of a claim, which can be hedged by a self-financing trading strategy, satisfies

$$V_t^* = E^*(C^* | \mathcal{F}_t) \quad \text{for all } P^* \in \mathcal{P}.$$

What happens if the claim can not be hedged? Is the right-hand side still constant in P^*?

Theorem 4.4.8 *Let C be an European contingent claim. Then the following assertions are equivalent:*

(i) C is attainable by a self-financing trading strategy.

(ii) For each equivalent martingale measure the conditional expectation $E^*(C^*|\mathcal{F}_t)$ has the same value, P-a.s., for all $t = 0, \dots, T$.

Proof. $(i) \Rightarrow (ii)$: This is Theorem 4.4.3.

$(ii) \Rightarrow (i)$: Suppose that C is not attainable. Then $C^* \notin \mathcal{V} = \text{span}\{\varphi' S_T^* : \varphi \in \mathbb{R}^{d+1}\}$. Fix some arbitrary $P^* \in \mathcal{P}$ and let

$$(X, Y) = E^*(XY), \quad X, Y \in L_2(P^*).$$

We have the decomposition $C^* = C_{\mathcal{V}}^* + C_{\mathcal{V}^\perp}^*$, where $C_{\mathcal{V}}^*$ ($C_{\mathcal{V}^\perp}^*$) is the orthogonal projection on $\mathcal{V}$ ($\mathcal{V}^\perp$). Define

$$\widetilde{P}(\{\omega\}) = \left(1 + \frac{C_{\mathcal{V}^\perp}^*(\omega)}{\sup |C_{\mathcal{V}^\perp}^*|^2}\right) P^*(\{\omega\}).$$

Then

$$\widetilde{P}(\Omega) = P^*(\Omega) + E^*\left(\frac{C_{\mathcal{V}^\perp}^*}{\sup |C_{\mathcal{V}^\perp}^*|^2} \mathbf{1}_\Omega\right) = P^*(\Omega) = 1,$$

since $\mathbf{1}_\Omega \perp C_{\mathcal{V}^\perp}^*$ (recall that $\mathbf{1}_\Omega$ is attainable such that $\mathbf{1}_\Omega \in \mathcal{V}$). Further, $P^* \sim \widetilde{P}$, such that $\widetilde{P}$ is an EMM. For any $Z \in \mathcal{V}$

$$E_{\widetilde{P}}(Z) = E^*(Z) + \frac{1}{\sup |C_{\mathcal{V}^\perp}^*|^2} E^*(C_{\mathcal{V}^\perp}^* Z), = E^*(Z),$$

yielding $E_{\widetilde{P}}(C_{\mathcal{V}}^*) = E^*(C_{\mathcal{V}}^*)$. But

$$E_{\widetilde{P}}(C_{\mathcal{V}^\perp}^*) = E^*(C_{\mathcal{V}^\perp}^*) + \frac{1}{\sup |C_{\mathcal{V}^\perp}^*|^2} E^*(C_{\mathcal{V}^\perp}^{*2}) > E^*(C_{\mathcal{V}^\perp}^*).$$

It follows that $E_{\widetilde{P}}(C^*|\mathcal{F}_0) = E_{\widetilde{P}}(C^*) > E^*(C^*) = E^*(C^*|\mathcal{F}_0)$, a contradiction.

By definition, a complete financial market is arbitrage-free, such that we have at least one EMM at our disposal. We claim that there is exactly one EMM, if the market is complete. The proof is as follows: If $A \in \mathcal{F}$ then $C = \mathbf{1}_A$ is a bounded claim, which is attainable. Its fair (i.e. arbitrage-free) price $E^*(C^*)$ is unique by the above theorem. In other words, the mapping

$$\Phi : \mathcal{P} \to \mathbb{R}, \ P^* \mapsto E^*(\mathbf{1}_A) = P^*(A), \ P^* \in \mathcal{P}$$

is constant. Hence, $\mathcal{P} = \{P^*\}$. By contrast, if $\mathcal{P} = \{P^*\}$, then the set $\{E^*(C^*) : P^* \in \mathcal{P}, E^*(C^*) < \infty\}$ of arbitrage-free prices consists of one element. If C were not attainable, we had more than one arbitrage-free price, a contradiction. Thus, we have shown the following fundamental theorem.

Theorem 4.4.9 *An arbitrage-free market is complete if and only if there is exactly one equivalent martingale measure.*

4.5 The martingale representation theorem in discrete time

The present subsection is short but important.

Theorem 4.5.1 *(MARTINGALE REPRESENTATION THEOREM)*
Equivalent are

(i) There exists one and only one equivalent martingale measure P^*.

(ii) Any P^*-martingale $\{M_t : t = 0, \ldots, T\}$ can be represented as a stochastic integral in discrete time w.r.t $\{S_t^* : t = 0, \ldots, T\}$, i.e. there is some predictable process $\{\varphi_t : t = 0, \ldots, T\}$ and an adapted pocess such that

$$M_t = M_0 + \sum_{i=1}^{t} \varphi_i'(S_i^* - S_{i-1}^*) = M_0 + \int_0^t \varphi_r' \, \mathrm{d}S_r^*.$$

Proof. (*i*) ⇒ (*ii*) Let $\{M_t\}$ be a P^*-martingale. Consider the decomposition $M_t = M_t^+ - M_t^-$. $M_T^+, M_T^- \geq 0$ are discounted claims with $EM_T^+ < \infty$ and $EM_T^- < \infty$. Since $\mathcal{P} = \{P^*\}$, the market is complete, such that M_T^+ and M_T^- are attainable by self-financing trading strategies. Hence, there are predictable processes $\{\varphi_t^+\}$ und $\{\varphi_t^-\}$, such that $(\varphi_T^+)' S_T = M_T^+$ and $(\varphi_T^-)' S_T = M_T^-$. The discounted versions of the processes $V_t^+ = (\varphi_t^+)' S_t$ and $V_t^- = (\varphi_t^-)' S_t$ are martingales under P^* by Theorem 4.4.3. But then

$$\left(V_t^+\right)^* = E^*\left((V_T^+)^* | \mathcal{F}_t\right) = E^*(M_T^+ | \mathcal{F}_t)$$
$$\left(V_t^-\right)^* = E^*\left((V_T^-)^* | \mathcal{F}_t\right) = E^*(M_T^- | \mathcal{F}_t)$$

for all $t = 0, \ldots, T$. Now, we may conclude that

$$\begin{aligned} M_t &= E^*(M_T^+ - M_T^- | \mathcal{F}_t) \\ &= (V_t^+)^* - (V_t^-)^* \\ &= V_0^+ - V_0^- + \sum_{i=1}^{t} (\varphi_i^+ - \varphi_i^-)'(S_i^* - S_{i-1}^*), \end{aligned}$$

since we have learned that any value process for a trading strategy $\{\varphi_t\}$ can be written as $V_t = V_0 + \sum_{i=1}^{t} \varphi_i'(S_i - S_{i-1})$.

(*ii*) ⇒ (*i*) Let C be a contingent claim with $E(C) < \infty$ and consider the value process $E(C|S_t^*)$, which is a P^*-martingale, cf. Theorem 4.4.3. By assumption, that martingale can be written as $C_0 + \int_0^t \varphi_r' \, \mathrm{d}S_r^*$ for some predictable process $\{\varphi_t\}$, which yields the replicating trading strategy for the claim C.

It is remarkable that such a general probabilistic result, namely that any P^*-martingale in discrete time can be represented as a discrete stochastic integral with respect to the process $\{S_t^*\}$, is related to mathematical finance in such a natural way. Further, every ingredient and any step of the proof has a clear and intuitive economic interpretation.

4.6 The Cox–Ross–Rubinstein binomial model

The well known binomial model of Cox, Ross and Rubinstein, abbreviated as the CRR model, considers a financial market with one share and a bank account. In each period, the price of the share is described by a one-period binomial model. Formally, let

$$\Omega = \{+, -\}^T = \{\omega = (\omega_1, \ldots, \omega_T)' : \omega_t \in \{+, -\}, t = 1, \ldots, T\}.$$

Then, the price process $\{S_t : t = 0, \ldots, T\}$ is given by

$$S_t(\omega) = \begin{cases} S_{t-1}(\omega)u, & \omega_t = +, \\ S_{t-1}(\omega)d, & \omega_t = -, \end{cases}$$

for $\omega = (\omega_1, \ldots, \omega_T)' \in \Omega$ and $t = 1, \ldots, T$. S_0 is the asset price at time 0, d the down factor and u the up factor. The corresponding returns

$$R_t = \frac{S_t - S_{t-1}}{S_{t-1}}, \qquad t = 1, \ldots, T,$$

take values in the set $\{r_-, r_+\}$, where $r_+ = u - 1$ and $r_- = d - 1$. Notice that

$$\{R_t = r_+\} = \{\omega \in \Omega : \omega_t = +\} \qquad \text{and} \qquad \{R_t = r_-\} = \{\omega \in \Omega : \omega_t = -\}.$$

We consider the natural filtration

$$\mathcal{F}_0 = \{\emptyset, \Omega\}, \quad \mathcal{F}_t = \sigma(S_0, \ldots, S_t), \quad t = 1, \ldots T,$$

and put $\mathcal{F} = \mathcal{F}_T = \text{Pot}\,(\Omega)$. For what follows, it is convenient to list some useful formulas. We have

$$S_t = S_0 \prod_{i=1}^{t} (1 + R_i), \qquad S_t = S_{t-1}(1 + R_t),$$

$$S_t^* = S_0^* \prod_{i=1}^{t} \frac{(1 + R_i)}{1 + r}, \qquad S_t^* = S_{t-1}^* \frac{1 + R_t}{1 + r}.$$

Clearly, r is the fixed interest rate of the bank account in each period.

The following theorem provides the no-arbitrage conditions of the CRR model and explicit formulas for the equivalent martingale measure.

Theorem 4.6.1 *Suppose* $P(\{\omega\}) > 0$ *for all* $\omega \in \Omega$.

(i) The CRR model is arbitrage-free if and only if

$$r_- < r < r+ \Leftrightarrow d < 1 + r < u.$$

In this case completeness follows as well.

(ii) If the no-arbitrage condition $d < 1 + r < u$ holds true, the unique equivalent martingale measure P^* is given by

$$P^*(\{(\omega\}) = (p^*)^k(1 - p^*)^{T-k}, \qquad k = 0, \ldots, T,$$

for $\omega = (\omega_1, \ldots, \omega_T)' \in \Omega$, where $k = k(\omega) = \sum_{t=1}^T \mathbf{1}_{\{\omega_t = +\}}$ and

$$p^* = \frac{r - r_-}{r_+ - r_-} = \frac{1 + r - d}{u - d} \in (0, 1).$$

Further, $\frac{\mathrm{d}P^*}{\mathrm{d}P}(\omega) = \frac{(p^*)^k(1-p^*)^{T-k}}{P(\{\omega\})}$, $\omega \in \Omega$. Finally, the returns $R_1, \ldots, R_T$ are i.i.d under P^* with $P^*(R_1 = r_+) = p^*$.

Proof. Suppose the model is arbitrage-free. Then there is some $P^* \in \mathcal{P}$ such that S_t^* is a P^*-martingale. The martingale condition

$$S_t^* = E^*(S_{t+1}^* | \mathcal{F}_t)$$

is equivalent to

$$1 + r = E^*(1 + R_{t+1} | \mathcal{F}_t),$$

since $S_t^* = S_{t-1}^* \frac{R_{t+1}}{1+r}$. Define $p_t^* = P^*(1 + R_{t+1} = 1 + r_+ | \mathcal{F}_t)$. Then the above equation can be written as

$$\begin{aligned} 1 + r &= (1 + r_+)p_t^* + (1 + r_-)(1 - p_t^*) \\ &= p_t^*(r_+ - r_-) + (1 + r_-). \end{aligned}$$

Since $r_- < r_+$, that equation has the unique and time independent solution

$$p^* = p_t^* = \frac{r - r_-}{r_+ - r_-} = \frac{1 + r - d}{u - d}$$

for any $t = 1, \ldots, T$. Also, notice that $p^* \in [0, 1]$ if and only if $r_- \le r \le r_+$ and $p^* \in (0, 1)$ if and only if $r_- < r < r_+$. Let us now verify the representation of P^*. Notice that $\mathcal{F}_t = \sigma(S_0, \ldots, S_t) = \sigma(R_1, \ldots, R_t)$. Hence,

$$P^*(R_{t+1} = r_+ | R_1, \ldots, R_t) = P^*(R_{t+1} = r_+ | \mathcal{F}_t) = p^*$$

shows that the conditional distribution of R_{t+1} given $R_1, \ldots, R_t$ coincides with $P^*(R_{t+1} = r_+)$. Consequently, R_{t+1} is independent of $R_1, \ldots, R_t$, which implies that $R_1, \ldots, R_T$ are i.i.d. under P^*. If we define $f(+) = r_+$ and $f(-) = r_-$, we have

$$\begin{aligned} P^*(\{\omega\}) &= P^*(R_1 = f(\omega_1), \ldots, R_T = f(\omega_T)) \\ &= (p^*)^k(1 - p^*)^{T-k} \end{aligned}$$

for $\omega = (\omega_1, \ldots, \omega_T)' \in \Omega$. Thus, P^* is uniquely determined such that $\mathcal{P} = \{P^*\}$.

Notice that under the equivalent martingale measure the returns are i.i.d., although that must not be true under the real probability measure. Indeed, the evidence from statistical analyses that financial returns are dependent is overwhelming. Further, a direct calculation shows that $E^*(R_1) = r_+ p^* + r_-(1 - p^*) = r$, which is in agreement with the fact that P^* can be used to price random future payments by risk-neutral evaluation.

The fact that $R_1, \dots, R_T$ are i.i.d. under P^* immediately yields the following result on the distribution of the asset price.

Lemma 4.6.2 *Under the equivalent martingale measure, we have*

$$P^*\left(S_t = S_0 u^k d^{t-k}\right) = \binom{t}{k} \left(p^*\right)^k (1 - p^*)^{t-k},$$

for $k = 0, \dots, t$ *and* $t = 0, \dots, T$.

Now we can easily calculate the fair price of an European claim depending on S_T.

Lemma 4.6.3 *Let* $C = f(S_T)$ *be a derivative, where* $f : \mathbb{R} \to [0, \infty)$ *is a Borel function. Then the arbitrage-free price of* C *is given by*

$$\pi(C) = \sum_{k=0}^{T} \frac{f(S_0 u^k d^{T-k})}{(1 + r)^T} \binom{T}{k} (p^*)^k (1 - p^*)^{T-k}.$$

Proof. Notice that $f(S_T)$ takes the value $f(S_0 u^k d^{T-k})$ with probability $\binom{T}{k}(p^*)^k(1 - p^*)^{T-k}$, $k = 0, \dots, T$.

The simple structure of the CRR model allows us to calculate explicitly the whole value process V_t, which turns out to be a function of S_t, provided we know the model parameters u, d and p^*.

Theorem 4.6.4 *Let* $C = f(S_T)$ *as above. Then, the value process of any replicating trading strategy satisfies*

$$V_t = \sum_{k=0}^{T-t} \frac{f(S_t u^k d^{T-t-k})}{(1 + r)^{T-t}} \binom{T - t}{k} (p^*)^k (1 - p^*)^{T-t-k}.$$

Notice that V_t *is a function of* S_t.

Proof. We use the representation

$$S_T = S_t \cdot \prod_{i=t+1}^{T} (1 + R_i).$$

Clearly, S_t is $\mathcal{F}_t$-measurable. Since $R_{t+1}, \dots, R_T$ are independent of $\mathcal{F}_t$, we obtain

$$P^*\left(\prod_{i=t+1}^{T} (1 + R_i) = u^k d^{T-t-k}\right) = \binom{T - t}{k} (p^*)^k (1 - p^*)^{T-t-k}.$$

As a consequence,

$$\begin{aligned}
V_t^* &= E^*(C^* \mid \mathcal{F}_t) \\
&= E^*\left(\frac{f(S_T)}{(1+r)^T} \,\middle|\, \mathcal{F}_t \right) \\
&= E^*_{(R_{t+1},\dots,R_T)}\left(\frac{f\left(S_t \prod_{i=t+1}^{T}(1+R_i)\right)}{(1+r)^T} \,\middle|\, \mathcal{F}_t \right) \\
&= \sum_{k=0}^{T-t} \frac{f(S_t u^k d^{T-t-k})}{(1+r)^T} \binom{T-t}{k} (p^*)^k (1-p^*)^{T-t-k}.
\end{aligned}$$

Noting that $V_t = (1+r)^t V_t^*$, the assertion follows.

We may even calculate the value process of a path-dependent derivative, whose time t value is a function $v_t(S_0, \dots, S_t)$ of the stock price up to time t.

Lemma 4.6.5 *Suppose $C = f(S_0, \dots, S_T)$ for some Borel function $f : \mathbb{R}^{T+1} \to [0, \infty)$. Define for $x_1, \dots, x_t \ge 0$*

$$v_t(x_0, \dots, x_t) = E^* f\left(x_0, \dots, x_t, x_t(1+R_{t+1}), \dots, x_t \prod_{i=t+1}^{T}(1+R_i) \right),$$

which equals

$$\int \cdots \int f\left(x_0, \dots, x_t, x_t(1+r_{t+1}), \dots, x_t \prod_{i=t+1}^{T}(1+r_i) \right) \mathrm{d}F^*(r_1) \cdots \mathrm{d}F^*(r_T),$$

where F^* denotes the d.f. of R_1 under P^*. Then the discounted value process is given by

$$V_t^* = E^*(C^* \mid \mathcal{F}_t) = \frac{v_t(S_0, \dots, S_t)}{(1+r)^T}.$$

Proof. Again we use the formula $S_{t+j} = S_t \prod_{i=t+1}^{t+j}(1+R_i)$. Under P^* the factor $\prod_{i=t+1}^{T}(1+R_i)$ is independent of $\mathcal{F}_t = \sigma(S_0, \dots, S_t)$. Thus,

$$\begin{aligned}
(1+r)^T V_t^* &= E^*(C \mid \mathcal{F}_t) \\
&= E^*(f(S_0, \dots, S_T \mid \mathcal{F}_t) \\
&= E^*(C(S_0, \dots, S_T) \mid \mathcal{F}_t) \\
&= E^*\left(C\left(S_0, \dots, S_t, S_t(1+R_{t+1}), \dots, S_t \prod_{i=t+1}^{T}(1+R_i) \right) \right) \\
&= v_t(S_0, \dots, S_t)
\end{aligned}$$

by definition of V_t. Now the assertion is obvious.

The above formulas are simple. However, in practice binomial models are often used with a large number T of periods in order to discretize a given time interval. This is also done since sometimes positions are checked and updated when required on a fine intraday time scale, e.g. every five minutes. Then the required computing time can be substantial, if a large number of derivatives has to be priced.

Being in position to calculate the value process yields a way to determine a hedge. Consider the sequence

$$S_0 = V_0^*, V_1^*, \ldots, V_T^* = C^*,$$

where the V_t^* are calculated according to our above results. We have to find a predictable process $\{\varphi_t\}$, such that the corresponding value process replicates the above values. This means that we have to find a self-financing trading strategy φ_t such that

$$V_t^* = V_{t-1}^* + \varphi_t(S_t - S_{t-1})$$

for all $t = 0, \ldots, T-1$, i.e.

$$V_t^*(\omega) = V_{t-1}^*(\omega) + \varphi_t(\omega)(S_t(\omega) - S_{t-1}(\omega)), \quad \forall \omega \in \Omega.$$

At time $t-1$, the $V_{t-1}^*(\omega)$ and $S_{t-1}^*(\omega)$ are known and $\varphi_t(\omega)$ depends on $\omega = (\omega_1, \ldots, \omega_T)$ only through $(\omega_1, \ldots, \omega_{t-1})$ and is therefore fixed as well. $S_t(\omega)$ can take the values $S_{t,+} = S_t(+)$ and $S_{t,-} = S_t(-)$, and $V_t(\omega)$ the values $V_{t,+} = V_t(+)$ and $V_{t,-} = V_t(-)$, respectively. We arrive at the following two equations

$$\begin{aligned} V_{t,+}^* - V_{t-1}^* &= \varphi_t(S_{t,+}^* - S_{t-1}^*) \\ V_{t,-}^* - V_{t-1}^* &= \varphi_t(S_{t,-}^* - S_{t-1}^*), \end{aligned}$$

which are easy to solve provided $S_{t,+}^* \neq S_{t,-}^*$.

Definition and Theorem 4.6.6 *Let C be a path-dependent derivative of the European type. Then we obtain a self-financing hedge, called a* **delta hedge**, *when investing*

$$\varphi_t = \frac{V_{t,+}^* - V_{t,-}^*}{S_{t,+}^* - S_{t,-}^*} = \frac{V_{t,+} - V_{t,-}}{S_{t,+} - S_{t,-}},$$

or $\frac{\Delta V_t}{\Delta S_t}$ in symbolic notation, into the underlying and deposit the rest, $V_{t-1} - \varphi_t S_{t-1}$, at the riskless interest rate at the bank account or borrow it from the bank.

Proof. Subtract the second equation from the first one to eliminate V_{t-1}^* and S_{t-1}^*. This easily leads to the equation

$$V_{t,+}^* - V_{t,-}^* = \varphi_t(S_{t,+}^* - S_{t,-}^*),$$

which is solved by the delta hedge. At time $t-1$, we may sell the shares we have yielding the revenue $V_{t-1} = \varphi_{t-1} S_{t-1}$. The new amount of shares costs $\varphi_t S_{t-1}$. Consequently, the difference $V_{t-1} - \varphi_t S_{t-1}$ is either financed by a loan or deposited into the bank account.

The delta hedge can be regarded as the workhorse for hedging derivatives. Notice that we need to be in a position to calculate the current value or, to be more precise, the change of the current value. For this reason we provided such formulas for the most important cases.

4.7 The Black–Scholes formula

The present section is devoted to a careful derivation of the famous Black–Scholes formula of the arbitrage-free price of an European call option with maturity T. That formula can be found as the limit price when considering a sequence of binomial models for the stock price with N periods covering the fixed time span $[0, T]$, as N approaches ∞. Posing this question is interesting in its own right for the following reasons. First, as already mentioned above, if N is large and many options have to be priced, then the computational costs can be substantial. Secondly, such a result implies that by increasing N we approach a well-defined limit. Indeed, it turns out that an appropriate and rather natural choice of the model parameters leads to a log normal distribution for the stock price resulting in the Black–Scholes option price formula.

Recall that for $t \in \mathbb{N}$

$$S_t = S_0 \prod_{i=1}^{t} (1 + R_i) \quad \text{with} \quad R_i = \frac{S_i - S_{i-1}}{S_{i-1}},$$

such that

$$\log(S_t) = \log(S_0) + \sum_{i=1}^{t} \log(1 + R_i).$$

If t is large and the returns are i.i.d., the central limit theorem suggests that $\log(S_t)$ is approximately normal. Let us try to anticipate the limit distribution of the stock price by assuming that the returns satisfy

$$\log(1 + R_i) \overset{i.i.d.}{\sim} N(\mu, \sigma^2).$$

Then,

$$\log(S_t) \sim N(\log(S_0) + \mu t, \sigma^2 t), \quad t \in \mathbb{N},$$

and that distribution also makes sense in continuous time.

For simplicity of our exposition, let us assume that T is an integer. We divide the time span $[0, T]$ into $N = nT$ equidistant subintervals given by the time points

$$t_k = t_{nk} = \frac{k}{n}, \quad k = 0, \ldots, nT,$$

at which trading is possible. This means, for each $n \in \mathbb{N}$ we consider a binomial model with nT periods, the nth binomial model, which spans $[0, T]$. Let S_{nk} denote the stock price within the nth binomial model after k periods, i.e. at calendar time k/n, $k = 0, \ldots, nT$. To any $t \in [0, T]$ we may associate the nearest time point $t^* \le t$ at which trading is possible, namely $t^* = \frac{\lfloor nt \rfloor}{n}$, since $\lfloor nt \rfloor / n \le t \le (\lfloor nt \rfloor + 1)/n$. Thus, we define

$$S_n(t) = S_{n, \lfloor nt \rfloor}, \quad t \in [0, T].$$

In what follows, we fix $t \in [0, T]$ and study the log of $S_n(t)$, i.e.

$$\log S_n(t) = \log(S_0) + \sum_{i=1}^{\lfloor nt \rfloor} \log(1 + R_{ni}).$$

Denote by P_n^* the equivalent probability measure of the nth binomial model. We want to study whether the distribution of $\log S_n(t)$ converges in distribution, presumably to a log normal law, under the sequence $\{P_n^*\}$ of equivalent probability measures. If such a central limit theorem holds true, the limit

$$F^*(x) = \lim_{n \to \infty} P_n^* \left(\frac{1}{\sqrt{n}} (\log S_n(t) - \log S_0) \le x \right)$$

exists for all $x \in \mathbb{R}$.

Let us first determine how to select the parameters r_n, d_n and u_n of the nth binomial model. Let r denote the continuous interest rate in the real world, which corresponds to one unit of time (usually one year), such that a unit investment grows to e^{rT}. Clearly, we have to select r_n, the fixed interest rate for each of the n periods, in such a way that the binomial model leads to the same accumulated value. This can be ensured by the choice

$$r_n = e^{r/n} - 1$$

implying

$$(1 + r_n)^{nT} = \left(e^{r/n}\right)^{nT} = e^{rT}.$$

The model parameters u_n and d_n have to satisfy the no-arbitrage condition

$$0 < d_n < 1 + r_n < u_n, \quad n \in \mathbb{N}.$$

In addition, we confine ourselves to recombinant binomial models where

$$d_n u_n = 1, \quad n \in \mathbb{N},$$

holds true. But then we may choose them as

$$u_n = e^{\alpha_n}, \quad d_n = e^{-\alpha_n}, \quad n \in \mathbb{N},$$

for some sequence $\{\alpha_n\}$ of positive numbers. Notice that the summands $\log(1 + R_{ni})$ of $\log S_n(t)$ are i.i.d. under P_n^* with

$$p_n^* = P_n^* ((\log(1 + R_{ni}) = \log(u_n)) = P_n^* (\log(1 + R_{ni}) = \alpha_n).$$

Hence under P_n^*

$$\log S_n(t) \stackrel{d}{=} \log(S_0) + \sum_{i=0}^{\lfloor nt \rfloor} Z_{ni},$$

if $Z_{n1}, \ldots, Z_{nn}$ are $\{-\alpha_n, \alpha_n\}$-valued random variables with $P_n^*(Z_{n1} = \alpha_n) = p_n^*$. Recall that the standard form of the central limit theorem makes an assertion on $\frac{1}{\sqrt{n}} \sum_{j=1}^{n} \xi_j$ with i.i.d.

random variables $\xi_1, \xi_2, \ldots$ with mean zero and finite positive variance. Thus, we introduce random variables $Y_{n1}, \ldots, Y_{nn}$ that are i.i.d. under P_n^* with

$$P_n^*(Y_{n1} = \sigma/\sqrt{n}) = 1 - P_n^*(Y_{n1} = -\sigma/\sqrt{n}) = p_n^*,$$

for some constant $\sigma > 0$. Then,

$$\log S_n(t) \stackrel{d}{=} \log(S_0) + \frac{1}{\sqrt{n}} \sum_{i=1}^{\lfloor nt \rfloor} Y_{ni}.$$

To summarize, we select the parameters as follows:

$$\begin{aligned} r_n &= \mathrm{e}^{r/n} - 1 \\ d_n &= \mathrm{e}^{-\sigma/\sqrt{n}}, \quad \text{such that} \quad r_{n-} = \mathrm{e}^{-\sigma/\sqrt{n}} - 1 \\ u_n &= \mathrm{e}^{\sigma/\sqrt{n}}, \quad \text{such that} \quad r_{n+} = \mathrm{e}^{\sigma/\sqrt{n}} - 1 \end{aligned} \tag{4.3}$$

Consequently, the equivalent martingale measure of the nth binomial model is given by

$$p_n^* = \frac{r_n - r_{n-}}{r_{n+} - r_{n-}} = \frac{\mathrm{e}^{r/n} - \mathrm{e}^{-\sigma/\sqrt{n}}}{\mathrm{e}^{\sigma/\sqrt{n}} - \mathrm{e}^{-\sigma/\sqrt{n}}}. \tag{4.4}$$

Does p_n^* converge to some nice value?

Lemma 4.7.1 *We have*

(i) $\lim_{n\to\infty} p_n^* = \frac{1}{2}$ *and*

(ii) $\lim_{n\to\infty} \sqrt{n}(2p_n^* - 1) = \frac{r}{\sigma} - \frac{\sigma}{2}$.

Proof. It suffices to show (ii). Notice that

$$\sqrt{n}(2p_n^* - 1) = \sqrt{n}\,\frac{2(\mathrm{e}^{r/n} - 1) - (\mathrm{e}^{\sigma/\sqrt{n}} + \mathrm{e}^{-\sigma/\sqrt{n}} - 2)}{\mathrm{e}^{\sigma/\sqrt{n}} - \mathrm{e}^{-\sigma/\sqrt{n}}}$$

where, by virtue of the Taylor expansion $\mathrm{e}^x = 1 + \sum_{k=1}^{\infty} \frac{x^k}{k!}$,

$$\begin{aligned} \mathrm{e}^{r/n} - 1 &= \frac{r}{n} + O\left(\frac{1}{n^2}\right), \\ \mathrm{e}^{\sigma/\sqrt{n}} - 1 &= \frac{\sigma}{\sqrt{n}} + \frac{1}{2}\frac{\sigma^2}{n} + \frac{1}{6}\frac{\sigma^3}{n^{3/2}} + O\left(\frac{1}{n^2}\right), \\ \mathrm{e}^{-\sigma/\sqrt{n}} - 1 &= -\frac{\sigma}{\sqrt{n}} + \frac{1}{2}\frac{\sigma^2}{n} - \frac{1}{6}\frac{\sigma^3}{n^{3/2}} + O\left(\frac{1}{n^2}\right), \end{aligned}$$

implying

$$\begin{aligned} e^{\sigma/\sqrt{n}} + e^{-\sigma/\sqrt{n}} - 2 &= \frac{\sigma^2}{n} + O\left(\frac{1}{n^2}\right), \\ e^{\sigma/\sqrt{n}} - e^{-\sigma/\sqrt{n}} &= \frac{2\sigma}{\sqrt{n}} + \frac{1}{3}\frac{\sigma^3}{n^{3/2}} + O\left(\frac{1}{n^2}\right). \end{aligned}$$

We arrive at

$$\begin{aligned} \sqrt{n}(2p_n^* - 1) &= \frac{n}{\sqrt{n}} \frac{\frac{2r}{n} + O\left(\frac{1}{n^2}\right) - \left(\frac{\sigma^2}{n} + O\left(\frac{1}{n^2}\right)\right)}{\frac{2\sigma}{\sqrt{n}} + \frac{1}{3}\frac{\sigma^3}{n^{3/2}} + O\left(\frac{1}{n^2}\right)} \\ &\overset{n\to\infty}{\to} \frac{2r - \sigma^2}{2\sigma} = \frac{r}{\sigma} - \frac{\sigma}{2}. \end{aligned}$$

Noticing that in a recombinant binomial model the trajectories of the price process are determined by the number of up movements, which follows a binomial distribution, we shall apply the following version of the central limit theorem.

Theorem 4.7.2 *Let $Y_n, n \in \mathbb{N}$, be a sequence of binomially distributed random variables with parameters n and $p_n \in (0, 1)$ satisfying*

$$\lim_{n\to\infty} p_n = p \in (0, 1).$$

Then,

$$\frac{Y_n - np_n}{\sqrt{np_n(1 - p_n)}} \underset{n\to\infty}{\overset{d}{\to}} N(0, 1),$$

as $n \to \infty$.

Proof. We may assume that

$$Y_n = \sum_{i=1}^{n} X_{ni}$$

with $X_{n1}, \ldots, X_{nn} \overset{i.i.d}{\sim} \text{Bin}(p_n)$ for each $n \in \mathbb{N}$. Clearly, the random variables $X_{ni} - p_n, i = 1, \ldots, n$, are bounded by 2 and have mean 0. Further, $s_n^2 = \text{Var}(Y_n) = np_n(1 - p_n) \to \infty$, as $n \to \infty$. Thus, Liapounov's condition with $\delta = 1$ is satisfied, since

$$\begin{aligned} \frac{1}{s_n^3} \sum_{i=1}^{n} E|X_{ni} - p_n|^3 &\leq \frac{2}{s_n^3} \sum_{i=1}^{n} E(X_{ni} - p_n)^2 \\ &\leq \frac{2}{s_n} \to 0, \end{aligned}$$

as $n \to \infty$.

Theorem 4.7.3 (Convergence in Distribution of the Binomial Model) *Under the sequence* $\{P_n^*\}$ *of equivalent martingale models given by Equation (4.4), we have for each t*

$$S_n(t) \xrightarrow[n\to\infty]{d} S_t = S_0 \exp\left(\sigma B_t + \left(r - \frac{\sigma^2}{2}\right) t\right),$$

as $n \to \infty$, *provided the model parameters are chosen according to Equation (4.3). Here,* $B_t \sim N(0, t)$, *under the limiting probability measure* P^*, *such that* $S_n(t)$ *is asymptotically log normal, i.e.*

$$\log S_n(t) \xrightarrow[n\to\infty]{d} \log(S_0) + \left(r - \frac{\sigma^2}{2}\right) t + \sigma B_t, \tag{4.5}$$

as $n \to \infty$.

Proof. We have the representation

$$S_n(t) = S_0 u_n^{K_{\lfloor nt \rfloor}} d_n^{\lfloor nt \rfloor - K_{\lfloor nt \rfloor}}$$

where

$$K_{\lfloor nt \rfloor} \sim \text{Bin}(\lfloor nt \rfloor, p_n^*)$$

is the number of up movements until time $\lfloor nt \rfloor$. Since $f(z) = e^z$, $z \in \mathbb{R}$, is a continuous function, it suffices to show Equation (4.5). By definition we have

$$\log u_n = -\log d_n = \frac{\sigma}{\sqrt{n}},$$

such that

$$\log S_n(t) = \log(S_0) + \frac{2\sigma}{\sqrt{n}} K_{\lfloor nt \rfloor} - \frac{\sigma \lfloor nt \rfloor}{\sqrt{n}}.$$

Standardizing $K_{\lfloor nt \rfloor}$ in the above expression yields

$$\begin{aligned}\log S_n(t) = \log(S_0) &+ \frac{2\sigma\sqrt{\lfloor nt \rfloor p_n^*(1-p_n^*)}}{\sqrt{n}} \cdot \frac{K_{\lfloor nt \rfloor} - \lfloor nt \rfloor p_n^*}{\sqrt{\lfloor nt \rfloor p_n^*(1-p_n^*)}} \\ &+ \frac{2\sigma \lfloor nt \rfloor p_n^*}{\sqrt{n}} - \frac{\sigma \lfloor nt \rfloor}{\sqrt{n}}.\end{aligned}$$

The third and fourth terms can be simplified to

$$\sigma \frac{\lfloor nt \rfloor}{n} \sqrt{n}(2p_n^* - 1),$$

which converges to $t(r - \sigma^2/2)$, by virtue of Lemma 4.7.1. Since $p_n^* \to 1/2, n \to \infty$, Theorem 4.7.2 yields

$$\frac{K_{\lfloor nt \rfloor} - \lfloor nt \rfloor p_n^*}{\sqrt{\lfloor nt \rfloor p_n^*(1-p_n^*)}} \xrightarrow[n\to\infty]{d} N(0, 1),$$

as $n \to \infty$. Finally, we have

$$\lim_{n\to\infty} \frac{2\sigma\sqrt{\lfloor nt \rfloor p_n^*(1-p_n^*)}}{\sqrt{n}} = \sigma\sqrt{t}.$$

Now, the Lemma of Slutzky leads to

$$\log S_n(t) \xrightarrow{d} \log(S_0) + \sigma\sqrt{t}\cdot U + \left(r - \frac{\sigma^2}{2}\right)t,$$

as $n \to \infty$, if $U \sim N(0, 1)$. Putting $B_t = \sqrt{t}U \sim N(0, t)$ yields the assertion.

This representation of the distributional limit is reminiscent of the geometric Brownian motion. Theorem 4.7.3 shows that the distributional limit is given by a geometric Brownian motion at each fixed time point t. Indeed, one can establish the stronger result that distributional convergence to that process also holds true in the sense of weak convergence of random (càdlàg) functions, see Appendix B.2.

These findings suggest that in a continuous-time model the stock price should be modeled as

$$S_t = S_0 \exp\left(\sigma B_t + \left(r - \frac{\sigma^2}{2}\right)t\right), \quad t \in [0, T].$$

We are naturally led to the famous Black–Scholes model. The discounted stock price is then given by

$$S_t^* = S_t e^{-rt} = S_0 \exp\left(\sigma B_t - \frac{\sigma^2}{2}t\right), \quad t \in [0, T],$$

and follows a log normal law. Notice that the payoff function $f(x) = \max(K - x, 0)$ of the put option is a continuous function that is bounded by K. Hence, the distributional convergence obtained in Theorem 4.7.3 implies that

$$E^*\left(\max(K - S_n(T), 0)e^{-rT}\right) \to_{n\to\infty} E^*\left(\max(K - S_T, 0)e^{-rT}\right).$$

Recall that the put-call parity relates the put price and the price of an European call given by the payoff function $C(S_T) = \max(S_T - K, 0)$,

$$E^*\left(\max(K - S_n(T), 0)e^{-rT}\right) = E^*\left(C(S_T)e^{-rT}\right) - S_0 + Ke^{-rT},$$

we also obtain the convergence of the arbitrage-free price of a European call

$$E^*\left(C(S_T)e^{-rT}\right) \to_{n\to\infty} E^*\left(\max(S_T - K, 0)e^{-rT}\right),$$

but we calculated the expectation at the right-hand side in Section 1.5.5 leading to

$$E^*\left(C(S_T)e^{-rT}\right) = S_0\Phi(d_1) - Ke^{-rT}\Phi(d_2),$$

with d_1 and d_2 as defined in Section 1.5.5.

4.8 American options and contingent claims

It remains to discuss how to price American-style contingent claims that can be exercised at an arbitrary time point before they expire. We shall first provide the relevant theoretical results that rely on the theory of optimal stopping discussed in Chapter 3 but are, however, constructive in that we will see how one can construct a hedge and calculate the arbitrage-free price.

In practice, the pricing of American claims is often done within the binomial model framework. Therefore, we provide the details and the resulting algorithm.

4.8.1 Arbitrage-free pricing and the optimal exercise strategy

Recall that the holder of an American contingent claim with maturity T can exercise the option at any time point $t = 1, \dots, T$. When exercising at time t, he receives the payment F_t. In the case of an American call on a stock S_t, the **intrinsic value** or **no-exercise value** is

$$S_t - K$$

and the resulting payment is

$$F_t = \max\{S_t - K, 0\},$$

since the option is worthless, if $S_t \leq K$. Most of the results discussed here apply to arbitrary American-style derivatives,

$$F_t = f(S_t), \qquad t = 0, \dots, T,$$

f a measurable function, or even arbitrary claims where $\{F_t\}$ is a $\mathcal{F}_t$-adapted process. We have to answer the following questions:

1. What is the fair price of an American-style claim?
2. When should we exercise in order to maximize the mean terminal value?

In what follows, we assume that the financial market is arbitrage-free and complete. Suppose that $\{\varphi_t\}$ is a trading strategy that hedges the associated value process $\{V_t\}$ of the American claim, such that

$$V_t = V_t(\varphi), \qquad t = 0, \dots, T.$$

Then the arbitrage-free price x we are seeking is given by the initial capital required to set up the hedge. We have

$$V_0 = x$$
$$V_t \geq F_t, \qquad t = 1, \dots, T.$$

For a perfect hedge, the above inequalities become equalities. Denote the equivalent martingale measure by P^*. The discounted value process

$$M_t = V_t^* = \mathrm{e}^{-rt} V_t, \qquad t = 0, \dots, T,$$

is a $\mathcal{F}_t$-martingale under P^*. Hence, for any stopping time τ,

$$E^*(V_\tau^*) = M_0 = V_0 \tag{4.6}$$

holds true, as shown in Theorem 3.2.18. Since the hedge $\{\varphi_t\}$ ensures that

$$V_t \geq F_t, \qquad \text{for all } t,$$

we may conclude that

$$V_t^* = \mathrm{e}^{-rt} V_t \geq \mathrm{e}^{-rt} F_t = F_t^*$$

for all t, and therefore

$$V_\tau^* \geq F_\tau^*$$

as well for any stopping time $\tau : \Omega \to \mathbb{N}_0$. Combining this inequality with Equation (4.6), we obtain

$$V_0 = E^*(\mathrm{e}^{-r\tau} V_\tau) \geq E^*(\mathrm{e}^{-r\tau} F_\tau).$$

In particular it follows that the arbitrage-free price satisfies

$$x \geq E^*(\mathrm{e}^{-r\tau} F_\tau)$$

for any stopping time $\tau \in \mathcal{T}_{0,T}$, where

$$\mathcal{T}_{0,T} = \{\tau : \tau \text{ is a stopping time with } \tau(\Omega) \subset \{0, \ldots, T\}\}$$

is the set of all stopping times stopping latest at time T. In other words,

$$x \geq \sup_{\tau \in \mathcal{T}_{0,T}} E^*(F_\tau^*)$$

with equality if the hedge is perfect. That is, we have to determine the optimal value of the optimal stopping problem where one seeks $\tau^* \in \mathcal{T}_{0,T}$ such that $E^*(F_{\tau^*}^*)$ is maximal. We have solved that problem in Section 3.2.2.3. There we have seen that we need to determine the Snell envelope of $\{F_t^* : t = 0, \ldots, T\}$ defined recursively by

$$\begin{aligned} Z_T &= F_T^*, \\ Z_t &= \max\{F_t^*, E^*(F_{t+1}^* | \mathcal{F}_t)\}, \qquad t = 0, \ldots, T-1. \end{aligned}$$

Then, according to Theorem 3.2.23 the optimal stopping time is given by the first time point where the discounted payment F_t^* is at least as large as $E^*(F_{t+1}^* | \mathcal{F}_t)$, for any P-almost all states $\omega \in \Omega$, such that $F_t^* = Z_t$. In other words,

$$\tau^* = \min\{t \geq 0 : Z_t = F_t^*\}$$

satisfies

$$Z_0 = E^*(Z_{\tau^*}) = \sup_{\tau \in \mathcal{T}_{0,T}} E^*(Z_\tau).$$

Further, the optimal value and thus the arbitrage-free price is given by

$$x = Z_0.$$

The fact that the hedge has to ensure $V_t^* \geq F_t^*$ with equality for a perfect hedge suggests that V_t^* is already the Snell envelope. Indeed, this holds true as shown in the following proposition, whose proof is constructive in that we see how one can construct the hedge in practice.

Proposition 4.8.1 *The Snell envelope of the discounted payment F_t^*, $t = 0, \ldots, T$, associated to an American-style contingent claim is given by the discounted value process V_t^*, $t = 0, \ldots, T$, for a perfect hedge.*

Proof. We have $F_T = V_T$, by definition. At time $T - 1$ we can either exercise the claim and receive the payment F_{T-1} or hold it until time T. We can apply Theorem 4.4.3 with $C^* = F_T^*$ to see that the discounted time $T - 1$ value, V_{T-1}^*, of the contingent claim is given by the conditional expectation $E^*(F_T^*|\mathcal{F}_{T-1})$. Hence, the no-exercise value at time $T - 1$ is given by

$$N_{t-1} = \mathrm{e}^{r(T-1)} E^*(F_T^*|\mathcal{F}_{T-1}).$$

Notice that N_{T-1} is also the arbitrage-free price of a one-period call option, i.e. with maturity T, providing the payment F_T. One can use such a portfolio replicating such an option to construct the hedge. Further, when knowing N_{t-1}, we can easily decide in favor of early exercise, if $F_{T-1} \geq N_{T-1}$; if $F_{T-1} \leq N_{T-1}$, we keep the position in the claim. Therefore, it follows that the hedge has to guarantee

$$V_{T-1} = \max\{F_{T-1}, \mathrm{e}^{r(T-1)} E^*(F_T^*|\mathcal{F}_{T-1})\},$$

or when discounting those figures

$$V_{T-1}^* = \max\{F_{T-1}^*, E^*(F_T^*|\mathcal{F}_{T-1})\}.$$

Using exactly the same arguments, backward induction shows that

$$V_t = \max\{F_t, \mathrm{e}^{-r} E^*(F_{t+1}|\mathcal{F}_t)\}$$

and

$$V_t^* = \max\{F_t^*, E^*(F_{t+1}^*|\mathcal{F}_t)\}$$

hold true for $t = 0, \ldots, T - 1$, where we used the fact that $F_{t+1}^* = \mathrm{e}^{-r(t+1)} F_{t+1}$ such that

$$\mathrm{e}^{rt} E^*(F_{t+1}^*|\mathcal{F}_t) = \mathrm{e}^{-r} E^*(F_{t+1}|\mathcal{F}_t)$$

Hence, $\{V_t^*\}$ is the Snell envelope of $\{F_t^*\}$. Notice also that the above arguments show that one can construct the hedge by backward recursion.

4.8.2 Pricing American options using binomial trees

Recall that in an N-period binomial model the stock price can only go down or up in each period,

$$S_t(\omega) = \begin{cases} S_{t-1}u, & \omega_t = +, \\ S_{t-1}d, & \omega_t = -, \end{cases}$$

where $\omega = (\omega_1, \ldots, \omega_N) \in \Omega = \{+, -\}^N$, u is the up factor and d the down factor. If the time step is Δ, the equivalent martingale measure is given by

$$p^* = \frac{e^{r\Delta} - e^{-\sigma\Delta}}{e^{\sigma\Delta} - e^{-\sigma\Delta}},$$

for some $\sigma > 0$. In Section 4.7 we put $\Delta = 1/\sqrt{n}$, where n denotes the number of periods corresponding to one unit of time (usually one year). Notice that in the nth period we have n nodes, say $(n, 1), \ldots, (n, n)$, which are uniquely determined by the number of up movements of the stock price. Let us introduce the following notation

$$\begin{aligned} S_{ni} &: \text{stock price in node } (n, i); \\ F_{ni}, F_{ni}^* &: \text{(discounted) payment of the option in node } (n, i) \text{ when exercising;} \\ N_{ni} &: \text{the no-exercise value in node } (n, i); \\ V_{ni}, V_{ni}^* &: \text{(discounted) value in node } (n, i); \\ E_{ni} &: 1 \text{ if the options is exercised in node } (n, i), 0 \text{ otherwise.} \end{aligned}$$

Let us arrange them according to this natural order, such that node (n, i) corresponds to i up movements. The following figure illustrates this arrangement.

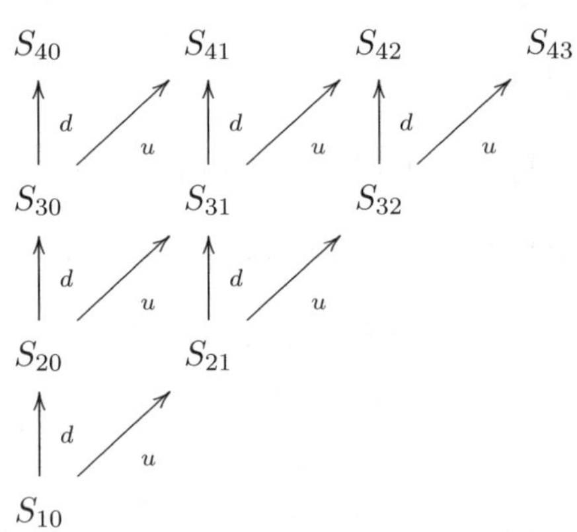

According to the model, we have for $i = 0, \ldots, n-1,\ n = 1, \ldots, N$

$$S_{ni} = S_0 d^i u^{n-i}$$

and

$$F_{ni} = f(S_0 d^i u^{n-i}),$$

where the latter equals

$$F_{ni} = \max\{S_0 d^i u^{n-i} - K, 0\}$$

for an American call option. The no-exercise value in node (n, i) is

$$N_{ni} = \mathrm{e}^{-r\Delta} \left(p^* F_{n+1,j+1} + (1 - p^*) F_{n+1,j} \right)$$

such that

$$V_{ni} = \max\{F_{ni}, \mathrm{e}^{-rt} \left(p^* F_{n+1,j+1} + (1 - p^*) F_{n+1,j} \right)\}.$$

Finally, we put $E_{ni} = 1$, if $F_{ni} \geq \mathrm{e}^{-rt} \left(p^* F_{n+1,j+1} + (1 - p^*) F_{n+1,j} \right)$, that is when the option is exercised in node (n, i).

The binomial tree is now calculated by starting at the leaf nodes and going back all the way until the root node is reached. The price of the option is then V_0.

4.9 Notes and further reading

A standard reference for arbitrage theory in discrete time is Föllmer and Schied (2004), on which the present chapter draws. In particular, Theorem 4.4.7 is proved there and Theorem 4.3.3 is a simplified version of (Föllmer and Schied, 2004, Proposition 5.11). For comprehensive treatments we also refer to Shiryaev (1999) and Bingham and Kiesel (2004) and the more elementary presentations of Pliska (1997) and Shreve (2004). The multiperiod binomial model dates back to Cox et al. (1976) and still provides the framework for option pricing in practice, when closed-form formulas are not available. The construction of martingale measures based on the Esscher transform can be extended to the multiperiod case, and we refer to Shiryaev (1999) for a detailed account of the related theory.

References

Bingham N.H. and Kiesel R. (2004) *Risk-Neutral Valuation: Pricing and Hedging of Financial Derivatives*. Springer Finance 2nd edn. Springer-Verlag London Ltd., London.

Cox J.C., Ross S.A. and Rubinstein M. (1976) *Option pricing: A simplified approach*. Journal of Financial Economics.

Föllmer H. and Schied A. (2004) *Stochastic Finance: An Introduction in Discrete Time*. vol. 27 of *de Gruyter Studies in Mathematics* extended edn. Walter de Gruyter & Co., Berlin.

Pliska S. (1997) *Introduction to Mathematical Finance*. Blackwell Publishing, Oxford.

Shiryaev A.N. (1999) *Essentials of Stochastic Finance: Facts, Models, Theory*. vol. 3 of *Advanced Series on Statistical Science & Applied Probability*. World Scientific Publishing Co. Inc., River Edge, NJ. Translated from the Russian manuscript by N. Kruzhilin.

Shreve S.E. (2004) *Stochastic Calculus for Finance. I: The Binomial Asset Pricing Model*. Springer Finance. Springer-Verlag, New York.

Notes and further reading

References

5

Brownian motion and related processes in continuous time

This chapter is devoted to a basic introduction to stochastic processes in continuous time, particularly to Brownian motion, its properties and some related processes. The study of Brownian motion, discovered empirically by the botanist Robert Brown in 1827, is a highlight of modern science. He studied microscopic particles dispensed in a fluid and observed that they move in an irregular fashion. Albert Einstein developed in 1905 a physical theory and the related mathematical solution and calculus. He explained the irregular movement by the impact of the much smaller fluid molecules. Similar ideas were published by Marian Smoluchowski in 1906. However, in 1900 Louis Bachelier used the one-dimensional version $t \mapsto B_t, t \geq 0$, of the Brownian motion to model stock prices. The economic reasoning is quite similar. The stream of orders to buy or sell the stock result in small changes (ups and downs) of the stock prices. The rigorous mathematical foundation is due to Norbert Wiener, who established the existence of Brownian motion that is therefore also called Wiener process. He constructed the Wiener measure that describes the distribution of Brownian motion as a function of t.

Brownian motion has a couple of specific properties that appear puzzling at first glance. The trajectories are almost surely continuous, but nowhere differentiable. Further, the length of the curve $t \mapsto B_t, t \in [a, b], a < b$, is infinite for any interval. Here, we confine ourselves to a discussion of some useful general notions for stochastic processes in continuous time, including the extension of the definition of (semi-/super-) martingales to the continuous-time framework, a formal definition of Brownian motion and its most important basic properties and rules of calculation. In addition, we introduce some related processes such as the Brownian bridge, geometric Brownian motion and Lévy processes.

5.1 Preliminaries

Let $I \subset [0, \infty)$ be an interval, e.g. $I = [0, T]$.

Financial Statistics and Mathematical Finance: Methods, Models and Applications, First Edition. Ansgar Steland.

Definition 5.1.1 *Let $d \in \mathbb{N}$. A family $\{X_t : t \in I\}$ of random variables (if $d = 1$) and random vectors (if $d > 1$), respectively,*

$$X_t : (\Omega, \mathcal{F}, P) \to (\mathbb{R}^d, \mathcal{B}^d), \quad t \in I,$$

is called a **stochastic process in continuous time**.

Throughout the book we shall use both notations $X_t(\omega)$ and $X(t, \omega)$. Again, we shall simply write $\{X_t\}$ if the index set I can be inferred from the context or by simple reasoning.

The behavior of a stochastic process as a function of the time variable t plays a crucial role, both in applications as well as in theory.

Definition 5.1.2

(i) For a stochastic process $\{X_t : t \in I\}$ the function $t \mapsto X(t, \omega)$ for a fixed $\omega \in \Omega$ is called a **(sample) path**, **trajectory** *or* **realization** *of X.*

(ii) $\{X_t\}$ is called a **left/right continuous** *or* **continuous** *process, if the function $t \mapsto X(t, \omega)$ is left/right continuous or continuous for each $\omega \in \Omega$. If these properties hold for all $\omega \in \Omega \backslash N$, N a measurable null set, we say that $\{X_t\}$ is left/right continuous a.s. or continuous a.s.*

(iii) A right continuous process with existing left-hand limits (a.s.) is called a **càdlàg** *process.*

The term càdlàg is a French acronym for *continu à droite, limite à gauche* that simply means right-continuous with left limits. The following notions directly carry over from the discrete case to case that the time index is continuous.

Definition 5.1.3

(i) A family $\{\mathcal{F}_t : t \in I\}$ of sub-σ-fields is called **filtration**, *if*

$$\mathcal{F}_s \subset \mathcal{F}_t \subset \mathcal{F}, \quad \text{for all } s, t \in I \quad \text{with} \quad s < t.$$

(ii) A stochastic process $\{X_t : t \in I\}$ is called **$\mathcal{F}_t$-adapted**, *if X_t is $\mathcal{F}_t$-measurable for any $t \in I$.*

(iii) The filtration given by $\mathcal{F}_t = \sigma(X_s : s \leq t) = \sigma(\cup_{s \leq t} X_s^{-1}(\mathcal{B}^d))$ is called **natural filtration**.

(iv) A probability space $(\Omega, \mathcal{F}, P)$ endowed with a filtration $\{\mathcal{F}_t\}$ is called a **filtered probability space** *and is denoted by $(\Omega, \mathcal{F}, \{\mathcal{F}_t\}, P)$.*

Definition 5.1.4 *Let $\{M_t : t \in I\}$ be a $\mathcal{F}_t$-adapted stochastic process with*

(i)

$$E|M_t| < \infty, \forall\, t \in I.$$

Then,

(ii) $\{M_t\}$ *is called a* **martingale**, *if* $M_s = E(M_t \mid \mathcal{F}_s)$ *(P-a.s.),* $\forall s, t \in$ *with* $s \leq t$.

(iii) $\{M_t\}$ *is called a* **submartingale**, *if* $M_s \leq E(M_t \mid \mathcal{F}_s)$ *(P-a.s.),* $\forall s, t \in I$ *with* $s \leq t$.

(iv) $\{M_t\}$ *is called a* **supermartingale**, *if* $M_s \geq E(M_t \mid \mathcal{F}_s)$ *(P-a.s.),* $\forall s, t \in I$ *with* $s \leq t$.

It is definitely time for some examples.

Example 5.1.5

(i) *For a random variable A defined on* $(\Omega, \mathcal{F})$ *let*

$$X_t(\omega) = X(t, \omega) = A(\omega) \cdot \sin(t), \quad t \in [0, 2\pi], \ \omega \in \Omega.$$

Then $\{X_t : t \in [0, 2\pi]\}$ *is a process with continuous paths, i.e. a continuous process.*

(ii) *Let* $X_1, \ldots, X_n$ *be a random sample defined on* $(\Omega, \mathcal{F}, P)$ *and consider the empirical distribution function*

$$F_n(t) = \frac{1}{n} \sum_{i=1}^{n} \mathbf{1}(X_i \leq t), \qquad t \in \mathbb{R}.$$

$\{F_n(t) : t \in \mathbb{R}\}$ *is a càdlàg process, since all trajectories* $t \mapsto F_n(t, \omega)$, ω *fixed, are right continuous step functions.*

(iii) *For a process* $\{X_t : t \in [0, T]\}$ *consider the natural filtration* $\mathcal{F}_t = \sigma(X_s : s \leq t)$, $t \in [0, T]$. *Then the process*

$$Y_t := X_t^2 - t, \quad t \in I,$$

is $\mathcal{F}_t$*-adapted.*

(iv) *Suppose* $\{X_t : t \in I\}$, $I = [0, T]$, *is a continuous process and let* $\mathcal{F}_t = \sigma(X_s : s \leq t)$. *Since* $s \mapsto X_s(\omega)$ *is continuous for each fixed* $\omega \in \Omega$, *we may define*

$$Z_t(\omega) = \int_0^t X_s(\omega)\, \mathrm{d}s, \quad t \in I,$$

for any $\omega \in \Omega$, *where the integral is understood in the Riemann sense. By construction, it is* $\mathcal{F}_t$*-adapted. We shall study such stochastic integrals in the next chapter in greater detail.*

We shall not stress questions of measurability, but notice that the above definition does not require that the functions $t \mapsto X_t(\omega)$, $t \in I$, and $X(\cdot, \cdot) : I \times \Omega \to \mathbb{R}$ are measurable. The stochastic process X is called **measurable**, if the latter mapping is measurable when $I \times \Omega$ is equipped with the σ-field $\mathcal{B}(I) \times \mathcal{F}$ and $\mathbb{R}^d$ with the Borel-σ-field $\mathcal{B}^d$. Having in mind that in economic applications adapted processes matter it is natural to require that at each fixed time instant t the mapping $X(s, \omega)$ for $s \in [0, t]$ and all ω is measurable with respect to events that are determined up to and including time t. This gives rise to the following definition. A process is called **progressively measurable** or **progressive**, if for any fixed t the mapping

$$X : [0, t] \times \Omega \to \mathbb{R}^d, \qquad (s, \omega) \mapsto X(s, \omega), \qquad s \in [0, t], \omega \in \Omega,$$

is $\mathcal{B}([0, t]) \times \mathcal{F}_t$ - measurable, i.e. if $[0, t] \times \Omega$ is equipped with the σ-field $\mathcal{B}([0, t]) \times \mathcal{F}_t$. Since $\mathcal{B}([0, t]) \times \mathcal{F}_t \subset \mathcal{B}(I) \times \mathcal{F}$, progressive processes are measurable.

For applications left and right continuous processes are of particular importance. The following result ensures that for these classes measurability problems do not occur.

Lemma 5.1.6 *Any left or right continuous adapted process is progressively measurable.*

Proof. For a right continuous adapted process $X = \{X_t\}$ define

$$X_{nt}(s, \omega) = \sum_{i=1}^{n} X(it/n, \omega)\mathbf{1}_{((i-1)t/n, it/n]}(s).$$

Fix t. Notice that for any Borel set A

$$E_t = \{(s, \omega) : 0 \le s \le t, \omega \in \Omega, X_{nt}(s, \omega) \in A\}$$

is the disjoint union of $\{0\} \times \{\omega : X(0, \omega) \in A\}$ and $\cup_{i=1}^{n}((i-1)t/n, it/n] \times \{\omega : X(it/n, \omega) \in A\}$. Since X is adapted, $E_t \in \mathcal{B}([0, t]) \times \mathcal{F}_t$. This shows that X_{nt} is progressive. By right continuity, $X_{nt}(s, \omega) \to X(s, \omega)$, as $n \to \infty$, pointwise for all $s \in [0, t]$ and $\omega \in \Omega$. Since limits of measurable functions are measurable, we have verified that X is progressive.

If (X, Y) is a bivariate random vector with $EX^2 < \infty$ and $EX^2 < \infty$, the quantities

$$\mu_X = E(X), \mu_Y = E(Y), \sigma_X^2 = \text{Var}(X), \sigma_Y^2 = \text{Var}(Y)$$

and

$$\rho_{XY} = \text{Cov}(X, Y)$$

exist. They provide a lot of information on the behavior of (X, Y). For example, Chebechev's inequality implies that $|X - \mu_X| \le k\sigma_X$ occurs with probability larger than $1 - 1/k^2$, for any constant k, and X and Y are located on a straight line a.s., if $|\rho_{XY}| = 1$.

The corresponding extensions of the notions *mean, variance, covariance* and *((strict) stationarity*) to a stochastic process are as follows.

Definition 5.1.7 *Let $\{X_t : t \in I\}$ be a stochastic process.*

(i) A process $\{X_t\}$ is called a **second-order process** *or L_2* **process**, *if $EX_t^2 < \infty$ for all t.*

(ii) If $E|X_t| < \infty$ for all t, $\mu(t) = EX(t)$ is called a **mean function**.

(iii) $\gamma_X(s, t) = \text{Cov}(X_s, X_t)$, $t, s \in I$, is called the **(auto) covariance function of** X.

(iv) $r_X(s, t) = E(X_s X_t)$, $s, t \in I$, is called a **correlation function**.

(v) $\{X_t\}$ *is* **(weakly) stationary**, *if* $\gamma_X(s,t) = \widetilde{\gamma}_X(|s-t|)$ *for some function* $\widetilde{\gamma}$.

(vi) $\{X_t\}$ *is* **strictly stationary**, *if*

$$(X_{t_1}, \ldots, X_{t_n}) \stackrel{d}{=} (X_{t_1+h}, \ldots, X_{t_n+h})$$

for all $t_1 < \cdots < t_n$, $n \geq 1$, *and* $h \in \mathbb{Z}$.

5.2 Brownian motion

5.2.1 Definition and basic properties

Definition 5.2.1 (Brownian Motion, Wiener Process)
A stochastic process $\{B(t) = B_t,\ t \in [0,\infty)\}$ *is called* **Brownian motion** *(BM) or a* **Wiener process**, *if for some constant* $\sigma > 0$ *the following properties are satisfied.*

(i) *For any time points* $0 \leq t_0 < t_1 < \cdots < t_n$, $n \in \mathbb{N}$, *the increments*

$$B(t_0),\ B(t_1) - B(t_0), \ldots, B(t_n) - B(t_{n-1})$$

are stochastically independent.

(ii) *For all* $t, h \geq 0$ *the increments are normally distributed,*

$$B(t+h) - B(t) \sim N(0, \sigma^2 h).$$

Equivalently, for all $0 \leq s \leq t$ *we have* $B(t) - B(s) \sim N(0,\ \sigma^2(t-s))$.

(iii) *The paths* $t \mapsto B(t,\omega)$, $\omega \in \Omega$, *are continuous.*

(iv) $B(0) = 0$ *(BM starts in* 0*).*

If $\sigma = 1$, *then* $\{B_t\}$ *is called a* **standard Brownian motion**.

In the literature, a Brownian motion also often refers to a standard Brownian motion.

Suppose we are given a Brownian motion $\{B_t\}$. We may then consider the natural filtration $\mathcal{F}_t = \sigma(B_s : 0 \leq s \leq t)$, $t \geq 0$, making B_t adapted. Often, one wants to consider larger filtrations containing more information than the path of the Brownian motion of interest, e.g. $\mathcal{F}_t = \sigma(B_s, X_s : 0 \leq s \leq t)$ for some other process $\{X_t\}$ defined on the same probability space. This gives rise to the following more general definition:

Definition 5.2.2 $\{B_t\}$ *is called* **Brownian motion** *or a* **Wiener process w.r.t. a filtration** $\mathcal{F}_t$, *if the following properties hold true.*

(i) B_t *is* $\mathcal{F}_t$*-adapted.*

(ii) *For all* $s \leq t$ *the increment* $B_t - B_s$ *is independent from the* σ*-field* $\mathcal{F}_s$.

(iii) *For all* $s \leq t$ *we have* $B_t - B_s \stackrel{d}{=} B_{t-s} - B_0 \sim N(0, \sigma(t-s))$ *for some* $\sigma > 0$.

If $\sigma = 1$, *then* $\{B_t\}$ *is a* **standard Brownian motion w.r.t.** $\mathcal{F}_t$.

By definition, the value B_t of a Brownian motion as well as its increments $B_t - B_s$, $s \leq t$, are normally distributed. As we shall see below, the independence of the increments automatically implies that any random vector $(B_{t_1}, \ldots, B_{t_k})$, where $t_1, \ldots, t_k, k \in \mathbb{N}$, are fixed time points, corresponding to Brownian motion sampled at k time instants, is multivariate normal.

Definition 5.2.3 (Gaussian Process)
A stochastic process $\{X_t : t \in I\}$ is called a **Gaussian process**, *if for all $t_1, \ldots, t_k$, $k \in \mathbb{N}$, the random vector $(X_{t_1}, \ldots, X_{t_k})$ obtained by sampling the process at these time points satisfies*

$$(X_{t_1}, \ldots, X_{t_k}) \sim N\left(\mu_{t_1,\ldots,t_k}, \Sigma_{t_1,\ldots,t_k}\right)$$

with

$$\mu = \mu_{t_1,\ldots,t_k} = E(X_{t_1}, \ldots, X_{t_n})',$$
$$\Sigma = \Sigma_{t_1,\ldots,t_k} = \left(\mathrm{Cov}(X_{t_i}, X_{t_j})\right)_{1 \leq i,j \leq k}.$$

If $\Sigma_{t_1,\ldots,t_k}$ is invertible, then $(X_{t_1}, \ldots, X_{t_k})$ attains the density

$$f(x_1, \ldots, x_k) = \frac{1}{(2\pi)^{n/2}\sqrt{\det(\Sigma)}} \exp\left\{-\frac{1}{2}(x-\mu)'\Sigma^{-1}(x-\mu)\right\},$$

for $x = (x_1, \ldots, x_k) \in \mathbb{R}^k$.

Example 5.2.4 *Brownian motion is a Gaussian process with mean function*

$$m(t) = EB(t) = 0, \qquad t \geq 0.$$

For $0 \leq s \leq t$ the autocovariance function is given by

$$\begin{aligned}
\gamma(s,t) &= \mathrm{Cov}\big(B(s), B(t)\big) \\
&= \mathrm{Cov}\big(B(s), B(s) + [B(t) - B(s)]\big) \\
&= \mathrm{Cov}\big(B(s), B(s)\big) \\
&= \sigma^2 s,
\end{aligned}$$

by independence of $B(s)$ and $B(t) - B(s)$. Therefore

$$\gamma(s,t) = \sigma^2(s \wedge t) = \sigma^2 \min(s,t).$$

Let us consider the random vector

$$(B_{t_1}, \ldots, B_{t_n}) \tag{5.1}$$

obtained by sampling a standard Brownian motion at n fixed time points $t_1 < \cdots < t_n$. It is clear that this vector is normal, since it is a linear transformation of the random vector

$$(B_{t_1}, B_{t_2} - B_{t_1}, \ldots, B_{t_n} - B_{t_{n-1}})$$

whose coordinates are independent and normally distributed by the defining properties of Brownian motion.

We are interested in calculating its n-dimensional density. Let us first consider the case $n = 2$ and calculate the conditional distribution of B_{s+t} given $B_s = x$. To introduce the notion of a transition density, let $\{X_t\}$ be some process taking values in a set $\mathcal{X} \subset \mathbb{R}$. If there is some measurable function $\varphi(x, y)$ such that

$$P(X_{s+t} \leq z | X_s = x) = \int_{-\infty}^{z} \varphi(x, y)\, \mathrm{d}y, \qquad z \in \mathbb{R}, x \in \mathcal{X},$$

then φ is called a **transition density** of $\{X_t\}$. More generally, let us fix time instants $t_1 < \cdots < t_n$ and consider

$$\begin{aligned} &P(B_{t_n} \leq x_n \mid B_{t_i} = x_i, i = 0, \ldots, n-1) \\ &\quad = P(B_{t_n} - B_{t_{n-1}} \leq x_n - B_{t_{n-1}} \mid B_{t_i} = x_i, i = 0, \ldots, n-1) \\ &\quad = P(B_{t_n} - B_{t_{n-1}} \leq x_n - x_{n-1}) \\ &\quad = \int_{-\infty}^{x_n - x_{n-1}} \frac{1}{\sqrt{2\pi(t_n - t_{n-1})}} \exp\left(-\frac{u^2}{2(t_n - t_{n-1})}\right) du. \end{aligned}$$

Put $x_n = z$, $t_n = s + t$, $t_{n-1} = s$ and $n = 2$ to see that

$$\begin{aligned} P(B_{s+t} \leq z \mid B_s = x) &= \int_{-\infty}^{z-x} \frac{1}{\sqrt{2\pi t}} \exp\left(-\frac{u^2}{2t}\right) du \\ &\underset{y=x+u}{\overset{\uparrow}{=}} \int_{-\infty}^{z} \frac{1}{\sqrt{2\pi t}} \exp\left(-\frac{(x-y)^2}{2t}\right) \mathrm{d}y. \end{aligned}$$

Hence, for each $t > 0$ and $x \in \mathbb{R}$ the function

$$y \mapsto p(t, x, y) := \varphi_{(0,t)}(x - y) = \frac{1}{\sqrt{2\pi t}} \exp\left(-\frac{(x-y)^2}{2t}\right), \qquad y \in \mathbb{R},$$

is a density of B_{s+t} given $B_s = x$, where $\varphi_{(\mu,\sigma^2)}$ denotes the density of the normal distribution with mean $\mu \in \mathbb{R}$ and variance $\sigma^2 > 0$. Recalling that $B_s \sim \varphi_{(0,s)}$, we find that

$$(x, y) \mapsto \varphi_{(0,s)}(x)\varphi_{(0,t)}(x - y) \text{ is a density of } (B_s, B_{s+t}),$$

the random vector obtained by considering two **projections** of the form $\pi_t(B) = B_t$. This product form of the joint density extends to the case of a finite number of such projections, yielding the **finite-dimensional distributions (fidis)** of Brownian motion. Before discussing this point, let us briefly provide a simple argument that the random vector (5.1), denoted by X in what follows, is normally distributed. First, notice that

$$\Sigma = \mathrm{Cov}(X) = \mathrm{diag}(t_1, t_2 - t_1, \ldots, t_n - t_{n-1}).$$

Let

$$A = \begin{pmatrix} 1 & 0 & 0 & \cdots & 0 \\ 1 & 1 & 0 & \cdots & 0 \\ \vdots & & & & \vdots \\ 1 & 1 & 1 & \cdots & 1 \end{pmatrix}.$$

Then

$$(B_{t_1}, \cdots, B_{t_n})' = AX \sim N(0, \Sigma^*),$$

where

$$\Sigma^* = A\Sigma A' = \begin{pmatrix} t_1 & t_1 & \cdots & \cdots & \cdots & t_1 \\ t_1 & t_2 - t_1 & t_2 & \cdots & \cdots & t_2 \\ t_1 & t_2 & t_3 - t_2 & t_3 & \cdots & t_3 \\ \vdots & \vdots & & \ddots & & \vdots \\ \vdots & \vdots & & & \ddots & \vdots \\ t_1 & t_2 & \cdots & \cdots & \cdots & t_n - t_{n-1} \end{pmatrix}.$$

Lemma 5.2.5 (FINITE-DIMENSIONAL DISTRIBUTIONS)
Fix $0 = t_0 < t_1 < \cdots < t_n$, $n \in \mathbb{N}$. Let $\{B_t\}$ be a Brownian motion (with start in 0). Then $(B_{t_1}, \ldots, B_{t_n})$ attains the density

$$(x_1, \ldots, x_n) \mapsto \varphi_n(x_1, \ldots, x_n) = \prod_{i=1}^{n} \varphi_{(0,\sigma(t_i - t_{i-1}))}(x_i - x_{i-1}),$$

for $(x_1, \ldots, x_n) \in \mathbb{R}^n$, where $x_0 = 0$.

Proof. We establish the formula by induction, the case $n = 1$ being trivial, since $B_{t_1} \sim N(0, t_1 - t_0)$, where here and in the rest of the proof we assume $\sigma = 1$. Thus, let us show the induction step $n \mapsto n + 1$. By conditioning on $(B_{t_1}, \ldots, B_{t_n}) = (x_1, \ldots, x_n)$, we see that for any Borel sets $A_1, \ldots, A_{n+1}$ the probability $P(B_{t_1} \in A_1, \ldots, B_{t_{n+1}} \in A_{n+1})$ can be calculated as

$$\int_{A_1} \cdots \int_{A_{n+1}} P(B_{t_{n+1}} \in A_{n+1} | (B_{t_1}, \ldots, B_{t_n}) = (x_1, \ldots, x_n))\, \mathrm{d}P_{(B_{t_1},\ldots,B_{t_n})}(x_1, \ldots, x_n).$$

Recall the following facts. If $(B_{t_1}, \ldots, B_{t_n})$ admits a density $\varphi_n(x_1, \ldots, x_n)$, then

$$P(B_{t_1} \in A_1, \ldots, B_{t_n} \in A_n) = \int_{A_1} \cdots \int_{A_n} \varphi_n(x_1, \ldots, x_n)\, \mathrm{d}x_1 \ldots \mathrm{d}x_n$$

and for any $\mathrm{d}P_{(B_{t_1},\ldots,B_{t_n})}$-measurable function $h(x_1, \ldots, x_n)$, the expectation

$$Eh(B_{t_1}, \ldots, B_{t_n}) = \int \cdots \int h(x_1, \ldots, x_n)\, \mathrm{d}P_{(B_{t_1},\ldots,B_{t_n})}(x_1, \ldots, x_n)$$

equals

$$\int \cdots \int h(x_1, \ldots, x_n)\varphi_n(x_1, \ldots, x_n)\mathrm{d}x_1 \ldots \mathrm{d}x_n.$$

We now use these facts with $A_i = (-\infty, x_i]$, $i = 1, \ldots, n+1$, to obtain

$$\begin{aligned} P(B_{t_{n+1}} \le x_{n+1} | (B_{t_1}, \ldots, B_{t_n}) = (x_1, \ldots, x_n)) \\ = P(B_{t_{n+1}} - B_{t_n} \le x_{n+1} - x_n | B_{t_i} = x_i, i = 1, \ldots, n) \\ = \int_{-\infty}^{x_{n+1}-x_n} \varphi_{(0,t_{n+1}-t_n)}(z)\,\mathrm{d}z = \int_{-\infty}^{x_{n+1}} \varphi_{(0,t_{n+1}-t_n)}(u - x_n)\,\mathrm{d}u. \end{aligned}$$

Putting things together and making use of the assumption we obtain

$$\begin{aligned} &P(B_{t_1} \le x_1, \ldots, B_{t_n} \le x_n, B_{t_{n+1}} \le x_{n+1}) \\ &= \int_{-\infty}^{x_1} \cdots \int_{-\infty}^{x_n} \left\{ \int_{-\infty}^{x_{n+1}} \varphi_{(0,t_{n+1}-t_n)}(u - x_n)\,\mathrm{d}u \right\} \mathrm{d}P_{(B_{t_1},\ldots,B_{t_n})}(x_1, \ldots, x_n) \\ &= \int_{-\infty}^{x_1} \cdots \int_{-\infty}^{x_n} \int_{-\infty}^{x_{n+1}} \varphi_{(0,t_{n+1}-t_n)}(u - x_n)\,\mathrm{d}u \prod_{i=1}^{n} \varphi_{(0,t_i-t_{i-1})}(x_i - x_{i-1})\mathrm{d}x_1 \ldots \mathrm{d}x_n \\ &= \int_{-\infty}^{x_1} \cdots \int_{-\infty}^{x_{n+1}} \prod_{i=1}^{n+1} \varphi_{(0,t_i-t_{i-1})}(x_i - x_{i-1})\mathrm{d}x_1 \ldots \mathrm{d}x_{n+1}, \end{aligned}$$

which completes the proof.

Lemma 5.2.5 allows us to calculate the probability of random events, as long as they depend on Brownian motion only through its values at a finite number of time points. The question arises whether this still holds true for events depending on, say, the whole path. One can show that this indeed holds true, since the finite-dimensional distributions determine the distribution of a process.

Definition 5.2.6 (Brownian Bridge)
Let $\{B_t : t \in [0, 1]\}$ be a standard Brownian motion defined. Then the process

$$B^0(t) = B_t - tB(1), \qquad t \in [0, 1]$$

is called a **Brownian bridge**.

It is clear that $E(B_t^0) = 0$ for all t, and a direct calculation shows that

$$\mathrm{Cov}(B_t, B_s) = s(1 - t)$$

for $0 \le s < t \le 1$. Further, B^0 is a Gaussian process. Next, fix some $s \in (0, 1)$ and notice that

$$E(B_s^0 B_1) = E(B_s B_1 - sB_1^2) = \min(s, 1) - s = 0,$$

which shows that B_s^0 and B_1 are independent. Similarly, for any $0 < s_1 < \cdots < s_n < 1$ the random vector $(B_{s_1}^0, \ldots, B_{s_n}^0)$ obtained by projecting B^0 onto those time points is independent of $B(1)$, which shows that $\{B_s^0 : s \in (0, 1)\}$ is independent of B_1.

Brownian motion itself and many derived stochastic processes are martingales. The following proposition lists the most important three martingales related to Brownian motion, with respect to the natural filtration $\mathcal{F}_t = \sigma(B_s : s \le t)$.

Proposition 5.2.7 *Let $\{B_t : t \ge 0\}$ be a standard Brownian motion (w.r.t $\{\mathcal{F}_t\}$). Then the following processes are $\mathcal{F}_t$-martingales.*

(i) $\{B_t : t \ge 0\}$.

(ii) $\{B_t^2 - t : t \ge 0\}$.

(ii) $\left\{\exp\left(\sigma B_t - \frac{\sigma^2}{2} t\right) : t \ge 0\right\}$, $\sigma > 0$.

Proof. Let us first check (i). Let $s \le t$ and notice that $E(B_t - B_s) = 0$. Since B_s is $\mathcal{F}_s$-measurable and the increment $B_t - B_s$ is independent of $\mathcal{F}_s$, we have

$$E(B_t \mid \mathcal{F}_s) = B_s + E(B_t - B_s \mid \mathcal{F}_s) = B_s + E(B_t - B_s) = B_s.$$

To verify (ii) notice that

$$\begin{aligned} E(B_t^2 \mid \mathcal{F}_s) &= E\big((B_s + B_t - B_s)^2 \mid \mathcal{F}_s\big) \\ &= B_s^2 + 2B_s \underbrace{E\,(B_t - B_s \mid \mathcal{F}_s)}_{=0} + E\big((B_t - B_s)^2 \mid \mathcal{F}_s\big) \\ &= B_s^2 + \underbrace{E\big((B_t - B_s)^2 \mid \mathcal{F}_s\big)}_{=t-s} \\ &= B_s^2 + (t - s), \end{aligned}$$

yielding $E(B_t^2 - t \mid \mathcal{F}_s) = B_s^2 - s$. Finally, (iii) follows from the fact that for $s \le t$

$$\begin{aligned} E(\mathrm{e}^{\sigma B_t} \mid \mathcal{F}_s) &= E(\mathrm{e}^{\sigma(B_s + B_t - B_s)} \mid \mathcal{F}_s) \\ &= \mathrm{e}^{\sigma Bs} \cdot E\big(\mathrm{e}^{\sigma(B_t - B_s)} \mid \mathcal{F}_s\big) \\ &= \mathrm{e}^{\sigma B_s} E(\mathrm{e}^{\sigma(B_t - B_s)}). \end{aligned}$$

We have to calculate $E(\mathrm{e}^{\sigma(B_t - B_s)}) = E(\mathrm{e}^{\widetilde{Z}})$ where $\widetilde{Z} \stackrel{d}{=} \sigma(B_t - B_s) \sim N\big(0, \sigma^2(t - s)\big)$. First consider $E(\mathrm{e}^{bZ})$ for $Z \sim N(0, 1)$ and $b \in \mathbb{R}$:

$$\begin{aligned} E(\mathrm{e}^{bZ}) &= \int \mathrm{e}^{bx} \frac{1}{\sqrt{2\pi}} \mathrm{e}^{-x^2/2}\, \mathrm{d}x \\ &= \int \frac{1}{\sqrt{2\pi}} \exp\left(\frac{-(x - b)^2 + b^2}{2}\right) \mathrm{d}x \\ &= \exp\left(\frac{b^2}{2}\right) \underbrace{\int \frac{1}{\sqrt{2\pi}} \exp\left(\frac{-(x - b)^2}{2}\right) \mathrm{d}x}_{=1} \\ &= \exp\left(\frac{b^2}{2}\right). \end{aligned}$$

Now put $b = \sigma\sqrt{t-s}$ to conclude that

$$E(e^{\sigma B_t} \mid \mathcal{F}_s) = e^{\sigma B_s} e^{\frac{\sigma^2(t-s)}{2}}.$$

Thus, we arrive at

$$E\left(e^{\sigma B_t - \frac{\sigma^2}{2}t} \middle| \mathcal{F}_s\right) = e^{\sigma B_s - \frac{\sigma^2}{2}s}.$$

The process appearing in Proposition 5.2.7 is of particular importance.

Definition 5.2.8 (GEOMETRIC BROWNIAN MOTION)
The stochastic process

$$X_t(\omega) = X_0 \exp(\mu t + \sigma B_t(\omega)), \quad t > 0, \omega \in \Omega,$$

is called **geometric Brownian motion** *with drift* $\mu \in \mathbb{R}$ *and volatility* $\sigma > 0$*. Here,* B_t *is a standard Brownian motion.*

Corollary 5.2.9 *A geometric Brownian motion with drift* $\mu = -\frac{\sigma^2}{2}$ *and volatility* $\sigma > 0$ *is a* $\mathcal{F}_t$*-martingale.*

Suppose we are given a Brownian motion $\{B_t : t \in [0, 1]\}$. Our above results on the finite-dimensional distributions describe its probabilistic behavior. What happens if we zoom in the process corresponding to $t \in [0, \lambda]$, i.e. if we consider $B_{\lambda t}$ for $t \in [0, 1]$?

Lemma 5.2.10 (SCALING PROPERTY)
Let $B = \{B_t\}$ *be a standard Brownian motion (with start in* 0*) and fix* $\lambda > 0$*. Then,*

$$\{B_{\lambda t} : t \geq 0\} \stackrel{d}{=} \{\lambda^{1/2} B_t : t \geq 0\}.$$

More generally, for $t_1 < \cdots < t_n$*,* $n \in \mathbb{N}$*, the random vectors* $(B_{\lambda t_1}, \ldots, B_{\lambda t_n})$ *and* $\lambda^{1/2}(B_{t_1}, \ldots, B_{t_n})$ *are equal in distribution. In particular,* $\{\lambda^{-1/2} B_{\lambda t} : t \geq 0\}$ *is again a standard Brownian motion.*

Definition 5.2.11 *A* $\mathcal{F}_t$*-adapted process* $\{X_t\}$ *taking values in a set* $\mathcal{X} \subset \mathbb{R}$ *is called a* **Markov process w.r.t.** $\mathcal{F}_t$ *if for any measurable function* $f : \mathcal{X} \to \mathbb{R}$ *there exists a measurable function* $g : \mathcal{X} \to \mathbb{R}$*, such that*

$$E(f(X_t) \mid \mathcal{F}_s) = g(X_s)$$

holds true for all $0 \leq s \leq t \leq T$*.*

Theorem 5.2.12 *A Brownian motion* $\{B(t) : t \geq 0\}$ *with respect to* $\mathcal{F}_t = \sigma(B_s : s \leq t)$*,* $t \geq 0$*, is a Markov process.*

Proof. Notice that given $\mathcal{F}_s$ the value of B_s is fixed, say $B_s = x$, such that

$$E(f(B_t) \mid \mathcal{F}_s)(\omega) = E(f(x + (B_t - B_s)) \mid \mathcal{F}_s)(\omega)$$

for all $\omega \in \Omega$. Since $B_t - B_s$ is independent of $\mathcal{F}_s$, the last expression is given by

$$E(f(x + B_t - B_s)) = \int f(x+y)\varphi_{(0,\sigma(t-s))}(y)\,\mathrm{d}y = g(x),$$

which is a measurable function of x.

5.2.2 Brownian motion and the central limit theorem

Let $\{B_t\}$ be a Brownian motion defined on a probability space $(\Omega, \mathcal{F}, P)$. Define the random variables $\xi_{ni} : (\Omega, \mathcal{F}, P) \to (\mathbb{R}, \mathcal{B})$ by

$$\xi_{ni}(\omega) = \sqrt{n} \cdot \left(B\left(\tfrac{i}{n}, \omega\right) - B\left(\tfrac{i-1}{n}, \omega\right)\right), \quad i = 1, \dots, n.$$

The ξ_{ni}s have the following properties:

(i) $\xi_{n1}, \dots, \xi_{nn}$ are independent for each $n \in \mathbb{N}$.

(ii) $\xi_{ni} \sim N(0, \sigma^2)$, $1 \le i \le n$, $n \in \mathbb{N}$.

Further, by definition

$$\frac{1}{\sqrt{n}} \sum_{i=1}^{k} \xi_{ni} = B_{k/n}\,,$$

such that

$$\frac{1}{\sqrt{n}} \sum_{i=1}^{\lfloor nt \rfloor} \xi_{ni} = B_{\lfloor nt \rfloor/n} \underset{n\to\infty}{\longrightarrow} B_t \sim N(0, \sigma^2 t).$$

If we put aside the concrete definition of the ξ_{ni}s and just notice that properties (i) and (ii) imply that ξ_{ni}, $i = 1, \dots, n$, $n \ge 1$, forms a triangular array of row-wise i.i.d. random variables, we may apply the central limit theorem and obtain that, for any fixed t, $\lfloor nt \rfloor^{-1/2} \sum_{i=1}^{\lfloor nt \rfloor} \xi_{ni}$ converges in distribution to $N(0, \sigma^2)$, which implies that

$$\frac{1}{\sqrt{n}} \sum_{i=1}^{\lfloor nt \rfloor} \xi_{ni} \stackrel{d}{\to} B_t \sim N(0, \sigma^2 t), \tag{5.2}$$

as $n \to \infty$. This observation indicates that Brownian motion should provide the correct asymptotic limit for such scaled partial sums in the sense of distributional convergence (of the fidis or even in a more general sense) under quite general assumptions. The above observations give rise to the following definition.

Definition 5.2.13 *Let $\{X_{ni} : i = 1, \dots, n, n \ge 1\}$ be an array of random variables. Then the process*

$$S_n(t) = \frac{1}{\sqrt{n}} \sum_{i=1}^{\lfloor nt \rfloor} X_{ni}, \qquad t \in [0, 1],$$

is called a **partial sum process.**

Before relating our above findings to the results on the convergence of the stock price process in binomial models which led us to the famous Black–Scholes formula, it is worth discussing the indicated extension of the CLT in the sense that the asymptotic distributional behavior of the partial sum process is given by Brownian motion. An appropriate extension of the notion of convergence in distribution is the notion of weak convergence. A sequence $\{X, X_n\}$ of stochastic processes taking values in some separable and complete metric space (S, d) with metric d converges weakly, denoted by

$$X_n \Rightarrow X,$$

as $n \to \infty$, if $\int f(X_n)\,\mathrm{d}P \to \int f(X)\,\mathrm{d}P$, as $n \to \infty$, holds true for all bounded functions $f : S \to \mathbb{R}$ that are continuous with respect to the metric d. If $S = \mathbb{R}$ and $d(x) = |x|$, $x \in \mathbb{R}$, one obtains the convergence in distribution of random variables. The famous invariance principle of Donsker now asserts that

$$S_n \Rightarrow B,$$

as $n \to \infty$, provide the increments are i.i.d. with mean zero and variance one. Then it follows that $P(S_n \in A)$ can be approximated by $P(B \in A)$ for a sufficiently rich class of sets A consisting of càdlàg functions. It also follows that one may conclude from this fact that $P(\varphi(S_n) \in A)$ converges to $P(\varphi(B) \in A)$ for a continuous mapping f. At this point, we shall not go into the rather involved details and refer to Appendix B.1.

Taking those results for granted, we are in a position to get a deeper understanding of the central limit theorem for the stock price process that formed the basis of our derivation of the Black–Scholes price formula.

Revisiting Theorem 4.7.3: Let us first recall the assertion of that theorem. Under the sequence $\{P_n\}$, where P_n denotes the equivalent martingale measure of the nth binomial model, the stock price process $S_n(t)$ converges in distribution to the random variable

$$S_t = S_0 \exp\left(\left(r - \frac{\sigma^2}{2}\right) t + \sigma B_t\right), \qquad t > 0.$$

Here, $\{B_t\}$ is a standard Brownian motion defined on $(\Omega, \mathcal{F}, P^*)$, i.e. under the probability measure P^* the Brownian motion B has drift 0 and volatility 1, and P^* is the limit of P_n determined by the limit $\lim_{n\to\infty} p_n = 1/2$. Consequently, the discounted stock price

$$S_n^*(t) = \mathrm{e}^{-rt} S_n(t)$$

converges (point wise) in distribution to a geometric Brownian motion,

$$S_n^*(t) \xrightarrow{d} S_t^* = S_0 \exp(-t\sigma^2/2 + \sigma B),$$

as $n \to \infty$. By Corollary 5.2.9, the limit process is a martingale under P^*. Using the more involved calculus of weak convergence on Skorohod spaces outlined above, one can show that the convergence in distribution of Theorem 4.7.3 extends to weak convergence of the (discounted) partial sum price process $\{S_n^*(t) : t \in [0, T]\}$ to geometric Brownian motion $\{S_t^* : t \in [0, T]\}$, as the number n of binomial models converges to ∞, thus approximating continuous-time trading.

These results suggest that the geometric Brownian motion is an appropriate idealized mathematical continuous-time model for the stock price. Further, there exists a probability measure such that the discounted stock price is a martingale.

5.2.3 Path properties

Definition 5.2.14 *Let $\{X_t\}$ and $\{Y_t\}$ be two stochastic processes defined on a probability space $(\Omega, \mathcal{F}, P)$. $\{Y_t\}$ is called a* **modification** *or a* **version** *of $\{X_t\}$, if*

$$P(X_t \neq Y_t) = 0 \quad \forall\, t \in I.$$

In this case, $\{X_t\}$ and $\{Y_t\}$ are also called **equivalent processes.**

Lemma 5.2.15 *If $\{Y_t\}$ is a modification of $\{X_t\}$, then they are equal in distribution, i.e. their finite-dimensional (marginal) distributions (fidis) coincide.*

Proof. We show the stronger assertion that

$$P(X_{t_1} = Y_{t_1}, \ldots, X_{t_n} = Y_{t_n}) = 1 \;\; \forall\, 0 \leq t_1, \ldots, t_n \leq T,\; n \in \mathbb{N}. \qquad (*)$$

Then the fidis coincide. Suppose $(*)$ does not hold. Then $P(X_{t_1} = Y_{t_1}, \ldots, X_{t_n} = Y_{t_n}) < 1$. It follows that there exists a set $\Omega_0 \in \mathcal{F}$ with $P(\Omega_0) > 0$, such that

$$\forall\, \omega \in \Omega_0 : \quad \exists\, j(\omega) \in \{1, \ldots, n\} : \quad X_{t_{j(\omega)}}(\omega) \neq Y_{t_{j(\omega)}}(\omega).$$

Clearly, $\Omega_0 = \bigcup_{i=1}^{n} A_i$, if $A_i = \{X_{t_i} \neq Y_{t_i}\}$, with $P(A_i) > 0$ for at least one i.

Example 5.2.16 *This example shows that one may change a process, e.g. a Brownian motion, at a random time point without affecting its distribution. So, let $\{B_t\}$ be a Brownian motion and $U \sim \exp(1)$. Consider the process*

$$V_t = \begin{cases} B_t, & U \neq t, \\ B_t + 1, & U = t, \end{cases}$$

for $t \geq 0$. Notice that the paths of V_t are no longer continuous. By Lemma 5.2.15, in order to show that $\{V_t\} \stackrel{d}{=} \{B_t\}$, it suffices to check that $\{V_t\}$ is a modification of B_t. We have

$$P(V_t \neq B_t) = P(U = t) = 0,$$

for all $t \geq 0$. Hence, the fidis coincide, although $\{\omega : V_t(\omega) = B_t(\omega)$ for all $t \in [0, \infty)\} = \emptyset$, such that

$$P\big(V_t = B_t : \forall\, t \in [0, \infty)\big) = 0,$$

there is not a single realization of those processes with equal paths.

We see that equivalent processes do not necessarily have the same paths.

Definition 5.2.17 *$\{X_t\}$ and $\{Y_t\}$ are called* **indistinguishable**, *if*

$$P(X_t = Y_t, \ \forall t \in I) = 1.$$

5.2.4 Brownian motion in higher dimensions

Let $B(t)$ be a (standard) Brownian motion and let $B_1(t), \ldots, B_n(t)$ be independent copies of $B(t)$. It is natural to call the process

$$t \mapsto (B_1(t), \ldots, B_n(t))', \qquad t \geq 0,$$

with trajectories in the Euclidean space $\mathbb{R}^n$ n-dimensional Brownian motion. Clearly, this process has independent coordinates by construction. By considering linear combinations

$$X_i(t) = a_{i1}B_1(t) + \cdots + a_{in}B_n(t), \qquad i = 1, \ldots, m,$$

for constants $a_{ij} \in \mathbb{R}$, we obtain correlated Gaussian processes. Notice that

$$X(t) = (X_1(t), \ldots, X_n(t))' = \mathbf{A}B(t),$$

if we put

$$\mathbf{A} = \begin{bmatrix} \mathbf{a}_1' \\ \vdots \\ \mathbf{a}_m' \end{bmatrix},$$

where $\mathbf{a}_i' = (a_{i1}, \ldots, a_{in})$, $i = 1, \ldots, m$.

In particular,

$$\begin{aligned} \mathrm{Cov}(X_i(s), X_j(t)) &= \mathrm{Cov}(a_{i1}B_1(s) + \cdots + a_{in}B_n(s), a_{j1}B_1(t) + \cdots a_{jn}B_n(t)) \\ &= a_{i1}a_{j1}\,\mathrm{Cov}(B_1(s), B_1(t)) + \cdots + a_{in}a_{jn}\,\mathrm{Cov}(B_n(s), B_n(t)) \\ &= (a_{i1}a_{j1} + \cdots + a_{in}a_{jn})\min(s, t) \\ &= \mathbf{a}_i'\mathbf{a}_j \min(s, t), \end{aligned}$$

by independence of $B_1, \ldots, B_n$. If we denote the $m \times m$ matrix of these covariances by $\mathrm{Cov}(X(s), X(t))$, we obtain the formula

$$\mathrm{Cov}(X(s), X(t)) = \min(s, t)\mathbf{A}\mathbf{A}'.$$

The general definition of n-dimensional Brownian motion is as follows.

Definition 5.2.18 *An n-dimensional stochastic process $t \mapsto \mathbf{B}(t) = (B_1(t), \ldots, B_n(t))'$, where $\{B_i(t) : t \geq 0\}$, $i = 1, \ldots, n$, are univariate stochastic processes, is called* **n-dimensional Brownian motion with covariance matrix Σ**, *if*

(i) $E(\mathbf{B}(t)) = 0$ for all $t \geq 0$.

(ii) The finite-dimensional distributions are Gaussian.

(iii) The covariances are given by

$$\mathrm{Cov}(B_i(s), B_j(t)) = E(B_i(s)B_j(t)) = \sigma_{ij} \min(s, t)$$

for all $1 \leq i, j \leq n$ *and all* $s, t \geq 0$, *where* σ_{ij} *are the elements of* Σ.

5.3 Continuity and differentiability

The following famous theorem provides a simple moment criterion in order to determine the smoothness of the paths of a stochastic process.

Theorem 5.3.1 (Kolmogorov–Chentsov)
Let $\{X_t : t > 0\}$ *be a* $\mathbb{R}$*-valued process. If for each* $T > 0$ *there are constants* $\alpha, \beta, C > 0$ *with*

$$E|X_s - X_t|^\alpha \leq C \cdot |t - s|^{1+\beta}, \quad \forall\, s, t \in [0, T],$$

then there exists a modification of $\{X_t\}$ *with locally Hölder-continuous paths of order* γ *for any* $\gamma \in (0, \beta/\alpha)$.

Let us apply Chentsov's criterion to a standard Brownian motion B_t. The fact that

$$B_t - B_s \sim N(0, |t - s|) \stackrel{d}{=} \sqrt{|t - s|} \cdot U, \quad U \sim N(0, 1),$$

implies that

$$E|B_t - B_s|^{2n} = |t - s|^n E|U|^n.$$

For $n = 2$ we obtain $E|B_t - B_s|^4 = |t - s|^2$. Hence, we may conclude that there exists a Brownian motion with Hölder-continuous paths. If we put $\alpha = 2n$ and $\beta = n - 1$, then

$$\frac{\beta}{\alpha} = \frac{n-1}{2n} \uparrow \frac{1}{2}.$$

Therefore, for all $0 < \gamma < 1/2$ Brownian motion is Hölder-continuous of order γ.

We have seen that Brownian motion has continuous sample paths. This means that $\lim_{s\to t} B(s) = B(t)$ holds true (for a modification).

Definition 5.3.2

(i) A stochastic process $\{X_t\}$ *taking values in* $\mathbb{R}$ *is called* **continuous in probability**, *if*

$$P\left(\lim_{s\to t} X_s = X_t\right) = 1.$$

(ii) A stochastic L_2*-process* $\{X_t\}$ *taking values in* $\mathbb{R}$ *is* **continuous in the mean square** *or* **m.s. continuous**, *if*

$$\lim_{s\to t} X_s = X_t \qquad \text{in } L_2 \text{ for all } t,$$

i.e.

$$\lim_{s \to t} E|X_s - X_t|^2 = 0.$$

Obviously, the definition can be easily extended to processes with values in $\mathbb{R}^d$ by replacing $|\cdot|$ by a vector norm.

Whether or not a process is m.s. continuous can be easily analyzed by investigating the correlation function $r_X(s, t) = E(X_s X_t)$.

Lemma 5.3.3 *A process X_t is m.s. continuous, if and only if*

$$r_X(u, v) \to r_X(t, t), \qquad if (u, v) \to (t, t). \tag{5.3}$$

In particular, a stationary process is m.s. continuous, if $r_X(t) = r_X(0, t)$ is continuous at 0.

Proof. The sufficiency follows from the fact that

$$E(X_s - X_t)^2 = r_X(s, s) - 2r_X(s, t) + r_X(t, t),$$

since $s \to t$ implies that $(s, t) \to (t, t)$ which in turn yields $r_X(s, t) \to r_X(t, t)$ by Equation (5.3). The necessity is shown as follows. Notice that

$$|r_X(u, v) - r_X(t, t)| = |E(X_u X_v) - E(X_t X_t)|$$

is not larger than

$$|E(X_u X_v) - E(X_t X_u)| + |E(X_t X_u) - E(X_t X_t)|.$$

Applying the Cauchy–Schwarz inequality to the first term yields

$$|E(X_u X_v) - E(X_t X_u)| = |E(X_u[X_v - X_t])| \le \sqrt{E(X_u^2)E(X_v - X_t)^2},$$

which converges to 0 if $(u, v) \to (t, t)$. The second term is estimated analogously. The sufficiency of Equation (5.3) follows.

5.4 Self-similarity and fractional Brownian motion

Let B_t, $t \ge 0$, be a standard Brownian motion. Recall the scaling property, Lemma 5.2.10, which asserts that for any $\lambda > 0$

$$\{B_{\lambda t} : t \ge 0\} \stackrel{d}{=} \{\lambda^{1/2} B_t : t \ge 0\}. \tag{5.4}$$

This means that a change of the time scale $t \mapsto \lambda t$ has the same effect as a change of state space $x \mapsto \lambda^{1/2} x$. A process for which such a change of the time scale is equivalent (in the sense of equality in distribution) to some change of the state space is called *self-similar*. Precisely:

Definition 5.4.1 (SELF-SIMILARITY)
A process $\{X_t : t \ge 0\}$ is a called **self-similar** *if for each $\lambda > 0$ there exists some $\nu = \nu(\lambda) > 0$ such that*

$$\{X_{\lambda t} : t \ge 0\} \stackrel{d}{=} \{\nu X_t : t \ge 0\}.$$

When $\nu = \lambda^H$, H *is called the* **Hurst exponent** *and* $\{X_t : t \geq 0\}$ *is called* **self-similar with Hurst exponent** H. $D = 1/H$ *is called the* **(statistical) fractal dimension** *of* $\{X_t\}$.

The self-similarity of Brownian motion actually results from its covariance function, since

$$E(B_{\lambda s}, B_{\lambda t}) = \min(\lambda s, \lambda t) = \lambda \min(s, t) = \lambda E(B_s, B_t) = E(\lambda^{1/2} B_s)(\lambda^{1/2} B_t),$$

which carries over to the finite-dimensional distributions leading to Equation (5.4).

The question arises whether there is there some covariance function $c(s, t)$ such that

$$c(\lambda s, \lambda t) = \lambda^H c(s, t),$$

for some $H \neq 1/2$? The answer is positive: Consider

$$c_H(s, t) = \frac{1}{2}\left(|s|^{2H} + |t|^{2H} - |s - t|^{2H}\right), \qquad s, t \in \mathbb{R}. \tag{5.5}$$

For $0 < H \leq 1$ the function $c_H(s, t)$ is non-negative definite and obviously satisfies

$$c_H(\lambda s, \lambda t) = \lambda^{2H} c_H(s, t), \qquad s, t \in \mathbb{R}.$$

For any non-negative function one may define a mean zero Gaussian process possessing that function as its covariance function. So suppose that $\{X_t : t \geq 0\}$ is a process with covariance function $c_H(s, t)$. Then, repeating the arguments for Brownian motion,

$$E(X_{\lambda s} X_{\lambda t}) = c_H(\lambda s, \lambda t) = \lambda^{2H} E(X_s X_t) = E(\lambda^H X_s)(\lambda^H X_t),$$

leading to

$$\{X_{\lambda t} : t \geq 0\} \stackrel{d}{=} \{\lambda^H X_t : t \geq 0\}.$$

That is, the process $\{X_t\}$ is self-similar with Hurst exponent H. Does there exist a continuous version? Let us check the Kolmogorov–Chentsov criterion. We have for all $0 \leq s, t$

$$\begin{aligned} E|X_t - X_s|^2 &= E(X_t^2) - 2E(X_t X_s) + E(X_s^2) \\ &= c_H(t, t) + c_H(s, s) - 2c_H(s, t) = |t - s|^{2H}. \end{aligned}$$

It follows that for $H > 1/2$ the Kolmogorov–Chentsov criterion is satisfied with $\alpha = 2$ and $\beta = 2H - 1$. Before treating the case $H < 1/2$, notice that $\{X_t\}$ has stationary increments with zero mean and variance

$$E\left(B_t^H - B_s^H\right)^2 = |t - s|^{2H},$$

that is

$$B_t^H - B_s^H \sim N(0, |t - s|^{2H}).$$

This implies that the moments of the increments are given by

$$E\left(B_t^H - B_s^H\right)^{2k} = \frac{2k!}{k!2^k}|t - s|^{2Hk}.$$

For given $0 < H \leq 1/2$ select k such that $2Hk > 1$ to verify the Kolomogorov–Chentsov criterion.

Definition 5.4.2 (Fractional Brownian Motion)
A continuous Gaussian process with mean zero and covariance function (5.5) is called **standard fractional Brownian motion**.

Notice that $H = 1/2$ yields a standard Brownian motion.

A fractional Brownian motion B_t^H, $t \geq 0$, with parameter $0 < H \leq 1$ therefore has the scaling property

$$\{B_{\lambda t}^H : t \geq 0\} \stackrel{d}{=} \{\lambda^H B_t^H : t \geq 0\}.$$

Let us summarize: A standard fractional Brownian motion starts in 0, has stationary normally distributed increments and is self-similar with Hurst exponent H.

The fact that the differences are stationary but dependent gives rise to the following definition.

Definition 5.4.3 (Fractional Gaussian Noise)
Let $\{B_t^H : t \geq 0\}$ *be a fractional Brownian motion with Hurst exponent* $0 < H < 1$. *Then the sequence*

$$\xi_t = B_{t+1}^H - B_t^H, \qquad t = 0, 1, \ldots$$

is called **standard fractional Gaussian noise**.

5.5 Counting processes

5.5.1 The Poisson process

Suppose $\tau_1, \tau_2, \ldots$ are random times. For example, the τ_i may indicate the time instants when a certain event occurs such as price shifts due to unexpected news. There are different ways to represent those random times by a single function. For example, by an indicator process

$$I_t = \sum_{i=1}^{\infty} \mathbf{1}_{\{\tau_i\}}(t), \qquad t \geq 0.$$

Then, $I_t = 1$ if and only if $t = \tau_i$ for some $i \in \mathbb{N}$. But usually one also wants to count how many events already occurred and therefore considers the **counting process** or **point process** associated to the random times,

$$N_t = \sum_{i=1}^{\infty} \mathbf{1}(\tau_i \leq t), \qquad t \geq 0.$$

If $\sup_n \tau_n = \infty$, then N_t is called a **counting process without explosion**. Obviously, N_t is the number of events that occurred up to time t. Notice that N_t is a càdlàg function, more precisely a right-continuous piece-wise constant function with jumps of size 1 at the random times τ_i.

It is easy to check that N_t is adapted, if and only if all τ_i are stopping times.

Notice that if $\tau_1 \leq \tau_2 \leq \cdots$, we have

$$\tau_i \leq t \Leftrightarrow N_t \geq i, \qquad \tau_i \leq t < \tau_{i+1} \Leftrightarrow N_t = u.$$

Definition 5.5.1 *A counting process* $\{N_t : t \geq 0\}$ *is called a* **Poisson process** *if* N_t *is adapted, is without explosion and has stationary and independent increments that follow a Poisson distribution such that*

$$N_t - N_0 \sim Poi(\lambda t), \qquad t \geq 0,$$

for some $\lambda \in (0, \infty)$. λ *is called the* **intensity parameter**. *The corresponding centered process*

$$N_t^c = N_t - \lambda t, \qquad t \geq 0,$$

is the **compensated Poisson process**.

The probability to observe k jumps in $[0, t]$ is given by

$$P(N_t = k) = \frac{(\lambda t)^k}{k!} \mathrm{e}^{-\lambda t}, \qquad k = 0, 1, \ldots,$$

and the mean, variance and covariance function of N_t are given by

$$E(N_t) = \lambda t, \qquad \mathrm{Var}\,(N_t) = \lambda t,$$

and

$$\mathrm{Cov}(N_s, N_t) = \min(s, t)\lambda,$$

for $s, t \geq 0$. Notice that $E(N_t)/t = \lambda$ is the mean number of jumps (per unit of time). It is also called the **mean arrival rate**.

It is easy to verify the following result.

Lemma 5.5.2 N_t^c *is a martingale with mean* 0.

Since the Poi(λ)-distribution has the characteristic function $u \mapsto \exp(\lambda[\mathrm{e}^{iu} - 1])$, the characteristic function $\varphi_{N_t}(u) = E(\mathrm{e}^{uiN_t})$, $u \in \mathbb{R}$, of a Poisson process is given by

$$\varphi_{N_t}(u) = \exp(\lambda t[\mathrm{e}^{iu} - 1]), \qquad u \in \mathbb{R},$$

and for a compensated Poisson process we obtain

$$\varphi_{N_t^c}(u) = E\left(\mathrm{e}^{iu(N_t - \lambda t)}\right) = \exp(\lambda t[\mathrm{e}^{iu} - 1 - iu]), \qquad u \in \mathbb{R}.$$

5.5.2 The compound Poisson process

Recall our introducing example of random time points at which the market processes unexpected information that really affects the prices. Suppose that the ith event occurring at time τ_i leads to a price movement Y_i. Then the additive effect of all such unexpected events on the stock price is given by

$$X_t = \sum_{i=1}^{N_t} Y_i,$$

since at time t exactly N_t events have already occurred with jump sizes $Y_1, \ldots, Y_t$.

Definition 5.5.3 *If* $\{N_t : t \geq 0\}$ *is a Poisson process and* $\{Y_n : n \in \mathbb{N}\}$ *is a sequence of i.i.d. random variables such that* $\{N_t\}$ *and* $\{Y_n\}$ *are independent, then* $X_t = \sum_{i=1}^{N_t} Y_i$, $t \geq 0$, *is called a* **compound Poisson process**. *The corresponding centered process*

$$X_t^c = X_t - \lambda t E(Y_1), \qquad t \geq 0,$$

is called a **compensated Poisson process**.

The mean, variance and covariance function of a compound Poisson process with $Y_1 \in L_2$ are given by

$$E(X_t) = \lambda t E(Y_1), \qquad \operatorname{Var}(X_t) = \lambda t E\left(Y_1^2\right)$$

and

$$\operatorname{Cov}(X_s, X_u) = \lambda \min(s, u) E\left(Y_1^2\right).$$

Clearly, the sample paths of a compound Poisson process are not continuous. However, the following result shows that the probability of observing an increment $X_t - X_s$ larger than $\varepsilon > 0$ is small for $s \approx t$ and tends to 0, if $s \to t$.

Lemma 5.5.4 *A compound Poisson process is continuous in probability.*

Proof. Fix $\varepsilon > 0$. Observe that $|X_t - X_s| = |\sum_{i=N_s+1}^{N_t} Y_i| = 0$ if there is no jump, i.e.

$$\{\omega : N_t(\omega) = N_s(\omega)\} \subset \{\omega : |X_t(\omega) - X_s(\omega)| = 0\},$$

such that $\{|X_t - X_s| > 0\} \subset \{N_t > N_s\}$. We obtain

$$P(|X_t - X_s| \geq \varepsilon) \leq P(N_t - N_s > 0) = 1 - \exp(-\lambda(t - s)) \to 0,$$

if $s \to t$.

Lemma 5.5.5 *A compound Poisson process* $X_t = \sum_{i=1}^{N_t} Y_i$, $t \geq 0$, *has stationary increments that are independent of each other and of the past.*

Proof. Notice that for $s < t$

$$X_t - X_s = \sum_{i=N_s+1}^{N_t} Y_i = \sum_{i=1}^{N_t - N_s} Y_{N_s+i} \stackrel{d}{=} \sum_{i=1}^{N_t - N_s} Y_i.$$

Using $N_t - N_s \stackrel{d}{=} N_{t-s}$, we may further deduce that

$$X_t - X_s \stackrel{d}{=} \sum_{i=1}^{N_{t-s}} Y_i = X_{t-s},$$

which shows the stationarity of the increments. Now let us check that $X_s - X_r$ and $X_t - X_s$ are independent, if $0 \leq r < s < t$. A similar argument as above yields for all Borel sets A and B

$$P(X_s - X_r \in A, X_t - X_s \in B) = P\left(\sum_{i=1}^{N_s - N_r} Y_i \in A, \sum_{i=1}^{N_t - N_s} Y_i' \in B\right),$$

where $Y_1', Y_2', \ldots$ are i.i.d. copies of Y_1, which are independent of $Y_1, Y_2, \ldots$ and $\{N_t\}$. Since $N_s - N_r$ and $N_t - N_s$ are independent of each other as well as independent of $\{Y_i, Y_i'\}$, we arrive at

$$P\left(\sum_{i=1}^{N_s - N_r} Y_i \in A, \sum_{i=1}^{N_t - N_s} Y_i' \in B\right) = P\left(\sum_{i=1}^{N_s - N_r} Y_i \in A\right) P\left(\sum_{i=1}^{N_t - N_s} Y_i' \in B\right).$$

Hence, the increments $X_s - X_r$ and $X_t - X_s$ are independent.

Let us calculate the characteristic function $\varphi_{X_t}(u) = E(\mathrm{e}^{iuX_t})$ of a compound Poisson process. Let φ_Y be the characteristic function of Y_1. By conditioning on $N_t = k$, we obtain

$$\begin{aligned}
\varphi_{X_t}(u) &= E\left(E\left(\mathrm{e}^{iu\sum_{j=1}^{N_t} Y_j} \mid N_t\right)\right) \\
&= \sum_{k=0}^{\infty} E\left(\mathrm{e}^{iu\sum_{j=1}^{k} Y_j}\right) P(N_t = k) \\
&= \mathrm{e}^{-\lambda t} \sum_{k=0}^{\infty} \frac{(\varphi_Y(u)\lambda t)^k}{k!} \\
&= \exp(\lambda t[\varphi_Y(u) - 1]),
\end{aligned}$$

for $u \in \mathbb{R}$. From this formula we immediately get the characteristic function of a compensated compound Poisson process,

$$\varphi_{X_t^c}(u) = E\left(\mathrm{e}^{iu(X_t - \lambda t E(Y_1))}\right) = \exp\left(\lambda t[\varphi_Y(u) - 1 - iuE(Y_1)]\right).$$

As a preparation for our treatment of Lévy processes, we express this formula in terms of the distribution function F of the jump sizes. We simply have to plug in the formulas $\varphi_Y(u) = \int_{-\infty}^{\infty} \mathrm{e}^{ius}\,\mathrm{d}F(s)$, $1 = \int_{-\infty}^{\infty} \mathrm{d}F(s)$ and $E(Y_1) = \int_{-\infty}^{\infty} s\,\mathrm{d}F(s)$ to obtain

$$\varphi_{X_t^c}(u) = \exp\left(\lambda t \int_{-\infty}^{\infty} (\mathrm{e}^{ius} - 1 - ius)\,\mathrm{d}F(s)\right), \qquad u \in \mathbb{R}.$$

When the jumps Y_j are normally distributed with mean $\mu \in \mathbb{R}$ and variance $\sigma^2 > 0$, then one may calculate the density f_{X_t} of the compound Poisson process explicitly. Indeed, we have

$$P\left(\sum_{j=1}^{k} Y_j \leq x \,\middle|\, N_t = k\right) = P\left(\sum_{j=1}^{k} Y_j \leq x\right) = \Phi_{(k\mu, k\sigma^2)}(x),$$

such that

$$P(X_t \le x) = \sum_{k=0}^{\infty} P(X_t \le x \mid N_t = k)P(N_t = k)$$

$$= \sum_{k=0}^{\infty} \Phi_{(k\mu, k\sigma^2)}(x) \frac{(\lambda t)^k}{k!} \mathrm{e}^{-\lambda t},$$

leading to the series representation

$$f_{X_t}(x) = \sum_{k=0}^{\infty} \frac{1}{\sqrt{2\pi k\sigma^2}} \exp\left(-\frac{1}{2}\frac{(x-k\mu)^2}{k\sigma^2}\right) \frac{(\lambda t)^k}{k!} \mathrm{e}^{-\lambda t}, \qquad x \in \mathbb{R}.$$

5.6 Lévy processes

We have learned that a Brownian motion with drift, which can be represented as $\mu t + \sigma B_t$, $t \ge 0$, for constants $\mu \in \mathbb{R}$ and $\sigma > 0$ and a standard Brownian motion $\{B_t\}$, is a member of the class of processes with (i) continuous trajectories and (ii) stationary increments that are independent of the past. In order to allow for processes with jumps it is natural to relax requirement (i) to continuity in probability.

Definition 5.6.1 *An adapted stochastic process $\{X_t\}$ with $X_0 = 0$ a.s. is a* **Lévy process**, *if the following conditions hold.*

(i) $\{X_t\}$ has stationary (homogeneous) increments that are independent of the past, that is $X_{t+s} - X_t$ is independent of $\mathcal{F}_t$ and $X_{t+s} - X_t \stackrel{d}{=} X_s$, for $s < t$.

(ii) $\{X_t\}$ is continuous in probability.

Less generally, a process $\{X_t\}$ with $X_0 = 0$ a.s. having stationary and independent increments which is continuous in probability, is called an **intrinsic Lévy process**.

One can show that for a given Lévy process there exists a modification with càdlàg paths. Thus, the latter property is often incorporated into the definition. The extension to the d-dimensional case is straightforward; then there exists a modification where all coordinate processes have càdlàg paths.

It is easy to verify that the sum of Lévy processes is again a Lévy process.

Lemma 5.6.2 *If X_t, $t \ge 0$, and Y_t, $t \ge 0$, are Lévy processes, then $X_t + Y_t$, $t \ge 0$, is also a Lévy process.*

It follows that a Brownian motion with drift plus a compound Poisson process,

$$X_t = \mu t + \sigma B_t + \sum_{i=1}^{N_t} Y_i, \qquad t \ge 0,$$

where $\{N_t \ge 0\}$ as well as $\{Y_n : n \in \mathbb{N}\}$ are independent of $\{B_t : t \ge 0\}$, is a Lévy process. Clearly, X_t is a càdlàg process.

A Lévy process $\{X_t\}$ has stationary and independent increments. Therefore, the marginal distributions P_{X_t} are infinitely divisible. As a consequence, the characteristic function $\varphi(u) = E(e^{\imath u X_t})$, $u \in \mathbb{R}$, satisfies

$$\varphi_{t+s}(u) = \varphi_t(u)\varphi_s(u),$$

which implies that

$$\varphi_t(u) = \exp(t\psi(u))$$

for some function $\psi(\theta)$, the cumulant generating function. By the Lévy–Khintchine formule (1.6),

$$\varphi_t(u) = \exp\left\{\imath u b_t - \frac{1}{2}uC_t\theta + \int_{\mathbb{R}^d} \left(e^{\imath ux} - 1 - \imath ux\mathbf{1}(|x| \leq 1)\right) d\nu(x)\right\},$$

with triplet (b, C, ν). Combining the last two equations, it follows that

$$B_t = tb, \qquad C_t = tC, \qquad \nu_t(x) = t\nu(x)$$

and

$$\psi(u) = \imath\theta' b - \frac{1}{2}uC\theta + \int_{\mathbb{R}^d} \left(e^{\imath ux} - 1 - \imath ux\mathbf{1}(|x| \leq 1)\right) \nu(dx).$$

Recall the formula

$$\varphi_{X_t^c}(u) = \exp\left(\lambda t \int_{-\infty}^{\infty} \left(e^{ius} - 1 - ius\right) dF(s)\right), \qquad u \in \mathbb{R}.$$

of the characteristic function of a compensated compound Poisson process. If we put $d\nu(s) = \frac{dF(s)}{\lambda}$, we see that X_t^c is a Lévy process.

Finally, let us consider the following standard construction of a Lévy with infinite Lévy measure. Let $N_t^{(k)}$, $k \geq 1$, be a sequence of independent Poisson processes with parameters $\lambda_k > 0$, $k \in \mathbb{N}$, and let μ_k, $k \geq 1$, be a sequence of real numbers. Then, for each $N \in \mathbb{N}$,

$$X_t^{(N)} = \sum_{k=1}^{N} \mu_k \left(N_t^{(k)} - \lambda_k t\right)$$

is a Lévy process with Lévy measure (check)

$$\nu^{(N)} = \sum_{k=1}^{N} \lambda_k \delta_{\{\mu_k\}}$$

and characteristic function

$$\varphi_t^{(N)}(u) = \exp\left\{t \int_{-\infty}^{\infty} \left(e^{\imath ux} - 1 - \imath ux\right) \nu^{(N)}(dx)\right\}.$$

The L_2-limit

$$X_t = \sum_{k=1}^{\infty} \mu_k \left(N_t^{(k)} - \lambda_k t\right)$$

exists, if we assume that

$$\sum_{k=1}^{\infty} \lambda_k \mu_k^2 < \infty,$$

and is also a Lévy process and has the Lévy measure

$$\nu(x) = \sum_{k=1}^{\infty} \lambda_k \delta_{\{\mu_k\}}(x),$$

which satisfies $\nu(\mathbb{R}) = \sum_k \lambda_k$, which may be infinite.

5.7 Notes and further reading

The seminal papers on Brownian motion are Bachelier (1900), Einstein (1905) and Wiener (1923). Elementary introductions are Wiersema (2008) and Mikosch (1998). Comprehensive expositions of notions and properties can be found in Grigoriu (2002) and Shiryaev (1999), where the latter discusses extensively the relation to finance. Thorough and more advanced probabilistic treatments are Durrett (1996), Karatzas and Shreve (1991) and Revuz and Yor (1999). A sound introduction from an econometric perspective is Davidson (1994).

References

Bachelier L. (1900) Théorie de la spéculation. *Annales Scientifiques de l' É.N.S.* 3e **17**, 21–86.

Davidson J. (1994) *Stochastic Limit Theory: An Introduction for Econometricians*. Advanced Texts in Econometrics. The Clarendon Press Oxford University Press, New York.

Durrett R. (1996) *Probability: Theory and Examples*. 2nd edn. Duxbury Press, Belmont, CA.

Einstein A. (1905) On the movement of small particles suspended in a stationary liquid demanded by the molecular-kinetic theory of heat. *Ann. Phys.* **17**, 549–560.

Grigoriu M. (2002) *Stochastic Calculus: Applications in Science and Engineering*. Birkhäuser Boston Inc., Boston, MA.

Karatzas I. and Shreve S.E. (1991) *Brownian Motion and Stochastic Calculus*. vol. 113 of *Graduate Texts in Mathematics* 2nd edn. Springer-Verlag, New York.

Mikosch T. (1998) *Elementary Stochastic Calculus—with Finance in View*. vol. 6 of *Advanced Series on Statistical Science & Applied Probability*. World Scientific Publishing Co. Inc., River Edge, NJ.

Revuz D. and Yor M. (1999) *Continuous Martingales and Brownian Motion*. vol. 293 of *Grundlehren der Mathematischen Wissenschaften [Fundamental Principles of Mathematical Sciences]* 3rd edn. Springer-Verlag, Berlin.

Shiryaev A.N. (1999) *Essentials of Stochastic Finance: Facts, Models, Theory*. vol. 3 of *Advanced Series on Statistical Science & Applied Probability*. World Scientific Publishing Co. Inc., River Edge, NJ. Translated from the Russian manuscript by N. Kruzhilin.

Wiener N. (1923) Differential spaces. *J. Math. Phys.* **2**, 131–174.

Wiersema U. (2008) *Brownian Motion Calculus*. John Wiley & Sons Inc., New York.

6

Itô calculus

This chapter provides a basic introduction to stochastic integration and Itô's calculus. The first major goal is to generalize the discrete martingale transform

$$I_t = \int \varphi_r \, \mathrm{d}S_r = \sum_{r=1}^{t} \varphi_r (S_r - S_{r-1}), \tag{6.1}$$

which can be interpreted as the value process of a self-financing trading strategy, see formula (4.1) in Section 4.2, to continuous time in an appropriate way when the stock price S_t is a Brownian motion. This can not be done ω-pathwise by relying on the Riemann–Stieltjes approach. Generalizing the work of Norbert Wiener, Kiyoshi Itô had the brilliant idea to define the integral as the mean square limit of appropriately defined Riemann–Stieltjes sums of the form (6.1). After a review of the stochastic Stieltjes integral, we go through the necessary steps to define the Itô integral for a sufficiently large class of integrands. We confine the exposition to Brownian motion as an integrator, but briefly discuss how one can integrate with respect to more general processes.

Then we introduce the class of Itô processes. Roughly speaking, such a process is simply the sum of a Riemann integral $\int_0^t \mu_s \, \mathrm{d}s$, where μ_s is random but integrable, and an Itô integral. The famous Itô formula asserts that a smooth function of Brownian motion can be represented as an Itô process, and, more generally, a smooth function of an Itô process is again an Itô process. We apply those results to identify the stochastic differential equation solved by certain basic stochastic processes such as (generalized) geometric Brownian motion and the Ornstein–Uhlenbeck process, which provide building blocks for financial models.

Lastly, we briefly introduce ergodic diffusion processes, an important subclass covering various models used in finance, and discuss the Euler approximation scheme that allows us to determine numerical approximations to the solutions in those cases, where explicit formulas are not available. The Euler scheme also provides the basis for the statistical estimation of discretely observed diffusion processes.

Financial Statistics and Mathematical Finance: Methods, Models and Applications, First Edition. Ansgar Steland.

6.1 Total and quadratic variation

The (total) variation aims at measuring the local oscillation of a function. Suppose $f : [0, T] \to \mathbb{R}$ is a continuously differentiable function that is strictly increasing on the interval $[0, t_1]$, strictly decreasing on $[t_1, t_2]$ and again strictly increasing throughout $[t_2, T]$, where $0 < t_1 < t_2 < T$. If f attains its maximum at t_1 and its minimum at t_2, it makes sense to measure the oscillation by the number

$$\int |\mathrm{d}f| = V(f) = f(t_1) - f(0) + (f(t_1) - f(t_2)) + (f(T) - f(t_2)).$$

Notice that $\int |\mathrm{d}f|$ can be calculated as

$$\int |\mathrm{d}f| = \int_0^{t_1} f'(r)\,\mathrm{d}r + \int_{t_1}^{t_2} -f'(r)\,\mathrm{d}r + \int_{t_2}^{T} f'(r)\,\mathrm{d}r = \int_0^T |f'(r)|\,\mathrm{d}r.$$

More generally, let us define the total variation as follows.

Definition 6.1.1 *The* **(total) variation** *of a function $f : [0, T] \to \mathbb{R}$ is defined as*

$$\int |\mathrm{d}f| = V(f) = \sup \sum_{i=0}^{n-1} |f(t_{i+1}) - f(t_i)|,$$

where we allow for the value $+\infty$. The sup *is taken over all partitions $\Pi_n : 0 = t_0 < \cdots < t_n = T$, $n \in \mathbb{N}$. f is of* **bounded variation,** *if $\int |\mathrm{d}f| < \infty$.*

It is easy to see that the set of all functions of bounded variation forms a vector space denoted by $BV = BV([0, T])$.

Remark 6.1.2

(i) It is straightforward to generalize Definition 6.1.1 to a function defined on $\mathbb{R}$ or $[0, \infty)$.

(ii) A function $f : [0, \infty) \to \mathbb{R}$ has **locally bounded variation,** *if its restriction $f\,|_{[0,t]}: [0, t] \to \mathbb{R}$ is of bounded variation for all $0 < t < \infty$.*

(iii) If f has a continuous derivative, it is easy to see that $\int |\mathrm{d}f| = \int_0^T |f'(r)|\,\mathrm{d}r$.

(iv) **Hahn–Jordan decomposition:** *Any function f of bounded variation can be written as $f = f^+ - f^-$, where f^+, f^- are non decreasing functions.*

Definition 6.1.3 *Let $f : [0, T] \to \mathbb{R}$ be a function. The* **quadratic variation** *of f is defined as*

$$[f, f](t) = [f, f]_t = \lim_{||\Pi_n|| \to 0} \sum_{i=0}^{n-1} [f(t_{i+1}) - f(t_i)]^2,$$

provided that limit exists (in $[0, \infty]$*). Here* $||\Pi_n||$ *denotes the* **size** *or* **mesh**

$$||\Pi_n|| = \max_{0 \le i \le n-1} t_{i+1} - t_i|$$

of a partition $\Pi_n : 0 = t_0 < \cdots < t_n = t$.

Notice that in Definition 6.1.3 the points of the partition Π_n may depend on n. For the sake of simplicity of presentation, we omit the n in our notation.

Lemma 6.1.4 *If* f *has a continuous derivative* f' *with* $\int_0^T |f'(r)|\,\mathrm{d}r < \infty$*, then* $[f, f]_t = 0$ *for all* $t \in [0, T]$.

Proof. Fix some partition $\Pi_n : 0 = t_0 < \cdots < t_n = t$. By applying the mean value theorem, we can find points $t_i^* \in [t_i, t_{i+1}]$, such that

$$\begin{aligned}\sum_{i=0}^{n-1}(f(t_{i+1}) - f(t_i))^2 &= \sum_{i=0}^{n-1}|f'(t_i^*)|^2(t_{i+1} - t_i)^2 \\ &\le ||\Pi_n|| \sum_{i=0}^{n-1}|f'(t_i^*)|^2(t_{i+1} - t_i).\end{aligned}$$

Here we used the simple fact that $(t_{i+1} - t_i)^2 \le ||\Pi_n||(t_{i+1} - t_i)$. It follows that

$$\begin{aligned}[f, f]_t &= \lim_{\|\Pi_n\| \to 0} \sum_{i=0}^{n-1}(f(t_{i+1}) - f(t_i))^2 \\ &\le \lim_{\|\Pi_n\| \to 0} ||\Pi_n|| \cdot \lim_{\|\Pi_n\| \to 0} \sum_{i=0}^{n-1}|f'(t_i^*)|^2(t_{i+1} - t_i) = 0.\end{aligned}$$

Definition 6.1.5 *Let* $\{X_t : t \in [0, T]\}$ *be some stochastic process in continuous time and* $0 < t_0 < t_1 < \cdots < t_n < T$*. Then*

$$[X, X]_n(t) = [X, X]_{nt} = \sum_{i=0}^{n-1}(X_{t_{i+1}} - X_{t_i})^2$$

is called the **sampled quadratic variation** *(w.r.t the partition* $\Pi_n : \{t_i : i = 0, \ldots, n\}$*).*

Theorem 6.1.6 *The sampled quadratic variation*

$$Q_{nt} = Q_n(t) = \sum_{i=0}^{n-1}(B_{t_{i+1}} - B_{t_i})^2$$

of a Brownian motion $\{B_s : s \in [0, t]\}$ *converges in* L_2 *and almost surely to* t,

$$Q_n(t) \overset{L_2, a.s}{\to} t,$$

as $n \to \infty$*, along any sequence of partitions* $\{t_i\}$ *of* $[0, t]$.

Proof. We provide a proof for the L_2 convergence. Notice that $E(Q_{nt} - t)^2 = E(Q_{nt} - EQ_{nt})^2 + (EQ_{nt} - t)^2$. Hence, it suffices to show $EQ_{nt} \to t$ and $\operatorname{Var}(Q_{nt}) \to 0$, $n \to \infty$. Notice that the summands of Q_n are independent random variables with $E(B_{t_{i+1}} - B_{t_i})^2 = \operatorname{Var}(B_{t_{i+1}} - B_{t_i}) = t_{i+1} - t_i$ yielding $E(Q_{nt}) = \sum_{i=0}^{n-1}(t_{i+1} - t_i) = t$.
To calculate the variance, recall that $EU^4 = 3\sigma^4$ if $U \sim N(0, \sigma^2)$. Thus,

$$\begin{aligned}\operatorname{Var}\left((B_{t_{i+1}} - B_{t_i})^2\right) &= E\left((B_{t_{i+1}} - B_{t_i})^2 - (t_{i+1} - t_i)\right)^2 \\ &= E(B_{t_{i+1}} - B_{t_i})^4 - 2(t_{i+1} - t_i)E(B_{t_{i+1}} - B_{t_i})^2 + (t_{i+1} - t_i)^2 \\ &= 3(t_{i+1} - t_i)^2 - 2(t_{i+1} - t_i)^2 + (t_{i+1} - t_i)^2 \\ &= 2(t_{i+1} - t_i)^2.\end{aligned}$$

We obtain

$$\operatorname{Var}(Q_{nt}) = \sum_{i=0}^{n-1} \operatorname{Var}\left((B_{t_{i+1}} - B_{t_i})^2\right) \le 2||\Pi_n|| \sum_{i=0}^{n-1}(t_{i+1} - t_i),$$

yielding $\operatorname{Var}(Q_{nt}) \le 2||\Pi_n|| \cdot t \to 0$, as $n \to \infty$.

Recall that a sufficient criterion for the existence of the Riemann integral $\int_a^b f(g(t))\,dt$ is that g is integrable and f uniformly continuous on $[a, b]$. Since Brownian motion is almost surely continuous and thus integrable, the random variable

$$Z(\omega) = \int_a^b f(B(t, \omega))\,dt, \qquad \omega \in \Omega,$$

is almost surely well defined, such that

$$\sum_{i=0}^{n-1} f(B_{t_i})(t_{i+1} - t_i) \longrightarrow \int_a^b f(B_t)\,dt,$$

as $n \to \infty$, a.s, for a partition $\Pi_n : \{t_i\}$ with $||\Pi_n|| \to 0$. The next result shows that in the above display one may replace $(t_{i+1} - t_i)$ by the independent random variables $(B_{t_{i+1}} - B_{t_i})^2$, which have expectations $(t_{i+1} - t_i)$, and still obtains convergence in L_2 under a fairly general condition on f.

Theorem 6.1.7 *Let f be uniformly continuous with $\int_0^t Ef^2(B_s)\,ds < \infty$. Then*

$$\sum_{i=0}^{n-1} f(B_{t_i})(B_{t_{i+1}} - B_{t_i})^2 \xrightarrow{L_2} \int_0^t f(B_s)\,ds,$$

as $n \to \infty$, along any sequence of partitions Π_n of $[0, t]$ such that $||\Pi_n|| \to 0$, $n \to \infty$.

Proof. Let

$$A_n = E\left(\sum_{i=0}^{n-1} f(B_{t_i})(B_{t_{i+1}} - B_{t_i})^2 - \int_0^t f(B_s)\,\mathrm{d}s\right)$$

$$B_n = \mathrm{Var}\left(\sum_{i=0}^{n-1} f(B_{t_i})(B_{t_{i+1}} - B_{t_i})^2\right).$$

We show $A_n \to 0$ and $B_n \to 0$, as $n \to \infty$. Consider first A_n. We have $A_n = A_{n1} + A_{n2}$, if

$$A_{n1} = E\left(\sum_{i=0}^{n-1} \left[f(B_{t_i})(B_{t_{i+1}} - B_{t_i})^2 - f(B_{t_i})(t_{i+1} - t_i)\right]\right)$$

$$A_{n2} = E\left(\sum_{i=0}^{n-1} f(B_{t_i})(t_{i+1} - t_i) - \int_0^t f(B_s)\,\mathrm{d}s\right).$$

Our assumptions on f guarantee that $A_{n_2} \to 0$, $n \to \infty$. Since $E\left((B_{t_{i+1}} - B_{t_i})^2 \,|\, \mathcal{F}_{t_i}\right) = E(B_{t_{i+1}} - B_{t_i})^2 = (t_{i+1} - t_i)$,

$$A_{n1} = \sum_{i=0}^{n-1} E\left(E\left(f(B_{t_i})(B_{t_{i+1}} - B_{t_i})^2 - f(B_{t_i})(t_{i+1} - t_i) \,|\, \mathcal{F}_{t_i}\right)\right) = 0.$$

It remains to discuss B_n. Introduce

$$\begin{aligned} D_i &= B_{t_{i+1}} - B_{t_i}, \\ V_i &= D_i^2 - (t_{i+1} - t_i), \\ U_{ij} &= f(B_{t_i}) f(B_{t_j}) \left(D_i^2 - (t_{i+1} - t_i)\right). \end{aligned}$$

Then

$$\begin{aligned} B_n &= \sum_{i,j=0}^{n-1} E\left[f(B_{t_i}) f(B_{t_j}) \left(D_i^2 - (t_{i+1} - t_i)\right)\left(D_j^2 - (t_{j+1} - t_j)\right)\right] \\ &= \sum_{i,j=0}^{n-1} E\left(U_{ij} V_j\right). \end{aligned}$$

If $i < j$, U_{ij} and V_j are independent, since U_{ij} is $\mathcal{F}_i$-measurable and D_j^2 is independent of $\mathcal{F}_i$. Using $EV_j = 0$, it follows that all those terms vanish. A similar argument applies to the cases with $i > j$. Thus, it remains to calculate

$$B_n = \sum_{i=0}^{n-1} E\left[f^2(B_{t_i})(D_i^2 - (t_{i+1} - t_i))^2\right].$$

By $\mathcal{F}_{t_i}$-measurability of $f(B_{t_i})$ and independence of D_i^2 from $\mathcal{F}_{t_i}$, we obtain

$$E\left(f^2(B_{t_i})(D_i^2 - (t_{i+1} - t_i))^2\right) = E\left(f^2(B_{t_i})E[(D_i^2 - (t_{i+1} - t_i))^2 \mid \mathcal{F}_{t_i}]\right)$$
$$= E\left(f^2(B_{t_i})E\left[D_i^2 - (t_{i+1} - t_i)\right]^2\right),$$

where $E\left(D_i^2 - (t_{i+1} - t_i)\right)^2 = 2(t_{i+1} - t_i)^2$, as shown in the proof of Theorem 6.1.6. It follows that

$$B_n = 2\sum_{i=0}^{n-1} E\left(f^2(B_{t_i})\right)(t_{i+1} - t_i)^2 \leq 2||\Pi_n|| \sum_{i=0}^{n-1} g(t_i)(t_{i+1} - t_i)$$

with $g(t) = Ef^2(B_t)$, $t \in [0, T]$. Notice that

$$g(t) = \int Ef^2(B_t)\,\mathrm{d}t = \int_{-\infty}^{+\infty} f^2(x)\varphi_{(0,t)}(x)\,\mathrm{d}x,$$

where $\varphi_{(0,t)}(x)$ denotes the density of the $N(0, t)$ distribution. By assumption, $\int_0^t g(t)\,\mathrm{d}t < \infty$, such that the Riemann sum $\sum_{i=0}^{n-1} g(t_i)(t_{i+1} - t_i)$ converges to $\int_0^t f^2(B_s)\,\mathrm{d}s$, as $n \to \infty$. Now $B_n \to 0$, as $n \to \infty$, follows.

6.2 Stochastic Stieltjes integration

In a Riemann sum $\sum_{i=0}^n f(t_i^*)(t_{i+1} - t_i)$ the function values are weighted by the mass $(t_{i+1} - t_i)$ assigned to the intervals $(t_i, t_{i+1}]$ by the Lebesgue measure λ and the (generalized) distribution function $\mathrm{id}(x) = x$, $x \in \mathbb{R}$, respectively. Taking this viewpoint, it is a natural idea to replace the distribution function id by another (generalized) distribution function of a finite (signed) measure. Having in mind Remark 6.1.2 (iv), the following generalization is evident. If $f : [0, T] \to \mathbb{R}$ is continuous and $H : [0, T] \to \mathbb{R}$ of bounded variation, then we may define an integral

$$\int_0^T f(t)\,\mathrm{d}H(t)$$

as the limit of the so-called Riemann-Stieltjes (RS) sums

$$\sum_{i=0}^{n-1} f(t_i^*)(H(t_{i+1}) - H(t_i)), \qquad t_i^* \in [t_i, t_{i+1}]. \tag{6.2}$$

Definition 6.2.1 *If the RS sums (6.2) converges for any sequence of partitions Π_n with $||\Pi_n|| \to 0$, $n \to \infty$, f is called RS integrable and*

$$\int_0^T f\,\mathrm{d}H = \int_0^T f(t)\,\mathrm{d}H(t) = \lim_{n\to\infty} \sum_{i=0}^{n-1} f(t_i^*)(H(t_{i+1}) - H(t_i))$$

is called the **Riemann–Stieltjes (RS) integral of** f **w.r.t.** H. H *is called an* **integrator**.

Clearly, RS integrals such as $\int_a^b f(t)\,\mathrm{d}H(t)$ and $\int_a^\infty f(t)\,\mathrm{d}H(t)$ can be defined analogously, $0 \le a \le b < \infty$. The RS integral satisfies the following properties

Proposition 6.2.2 *Let f be RS integrable w.r.t. H and $0 \le a \le b < \infty$.*

(i) $f \mapsto \int_a^b f\,\mathrm{d}H$, *$f$ RS integrable, is a linear mapping.*

(ii) $\left|\int_a^b f\,\mathrm{d}H\right| \le ||f||_\infty \cdot \int |\mathrm{d}H|$.

(iii) $\int_a^b f\,\mathrm{d}(H_1 + H_2) = \int_a^b f\,\mathrm{d}H_1 + \int_a^b f\,\mathrm{d}H_2$ *for all functions H_1, H_2 of bounded variation.*

(iv) $\int_a^b f\,\mathrm{d}(\alpha H) = \alpha \int_a^b f\,\mathrm{d}H$ *for all $\alpha \in \mathbb{R}$.*

(v) f is RS integrable w.r.t. H, if and only if H is RS integrable w.r.t. f. Then the following integration by parts formula is valid.

$$\int_a^b f(x)\,\mathrm{d}H(x) = f(x)H(x)\Big|_a^b - \int_a^b H(x-)\,\mathrm{d}f(x).$$

(vi) If f is continuous and H is differentiable with derivative $h = H'$, then the RS integral $\int_a^b f\,\mathrm{d}H$ can be calculated as a Riemann integral, namely

$$\int_a^b f\,\mathrm{d}H = \int_a^b f(x)h(x)\,\mathrm{d}x.$$

(vii) If H is the (generalized) distribution function of a finite measure μ_H with discrete support $\{x_i : i \in I\} \subset [a, b]$ for some discrete set I[1]*, then*

$$\int_a^b f\,\mathrm{d}H = \sum_{i \in I} f(x_i)\mu_H(\{x_i\}).$$

Notice that the rules (iii) and (iv) assert that $H \mapsto \int f\,\mathrm{d}H$, H of bounded variation, is a linear mapping. Property (vii) allows us to represent many quantities arising in statistics as RS integrals. For example, the arithmetic mean $\overline{x} = n^{-1}\sum_{i=1}^n x_i$ of n points $x_1, \ldots, x_n \in \mathbb{R}$ is a RS integral w.r.t. the empirical distribution function

$$F_n(x) = \frac{1}{n}\sum_{i=1}^n \mathbf{1}(x_i \le x), \qquad x \in \mathbb{R}.$$

[1] I is a discrete set, if all $x \in I$ are isolated points, i.e. for all $x \in I$ there exists a neighborhood U_x such that $U_x \cap I = \{x\}$.

Indeed, noting that the associated measure $\mathrm{d}F_n$ satisfies $\mathrm{d}F_n(\{x_i\}) = \frac{1}{n}$, we obtain

$$\overline{x} = \sum_{i=1}^{n} x_i \mathrm{d}F_n(\{x_i\}) = \int x\,\mathrm{d}F_n(x).$$

For this reason, the RS integral (w.r.t. the empirical distribution function) plays an important role in statistics.

What happens if we integrate Brownian motion B w.r.t. a function H of bounded variation? In the definition of the RS integral we may replace f by some random process, as long as the convergence of the RS sum holds true in some stochastic sense, e.g. pathwise or in the mean square L_2 sense. The case of a Brownian motion is of particular importance.

Theorem 6.2.3 *Let $\{X(t) : t \in [a, b]\}$ be a Gaussian process with a.s. continuous trajectories. Denote*

$$K(s, t) = \mathrm{Cov}(X(s), X(t)), \quad a \leq s, t \leq b.$$

Let H be a function of bounded variation on each subinterval $[c, d] \subset (a, b)$. If

$$\sigma^2 = \int_a^b \int_a^b K(s, t)\,\mathrm{d}H(s)\,\mathrm{d}H(t) \in (0, \infty),$$

then the random variable

$$\int_a^b X\,\mathrm{d}H = \int_a^b X(t)\,\mathrm{d}H(t)$$

exisits a.s. and satisfies

$$\int_a^b X\,\mathrm{d}H \sim N(0, \sigma^2).$$

Proof. W.l.o.g. we assume $a = 0$ and $b = 1$. Let $\delta \in (0, 1)$. Then $\int_\delta^{1-\delta} X\,\mathrm{d}H \to \int_0^1 X\,\mathrm{d}H$, as $\delta \to 0$. On the other hand

$$\int_\delta^{1-\delta} X\,\mathrm{d}H = \lim_{n\to\infty} \sum_{i=\lfloor n\delta \rfloor}^{\lfloor n(1-\delta)\rfloor} X\left(\frac{i}{n}\right)\left[H\left(\frac{i}{n}\right) - H\left(\frac{i-1}{n}\right)\right].$$

The sum on the right-hand side follows a $N(0, \sigma_n^2)$ distribution, where

$$\sigma_n^2 = \sum_{i,j=\lfloor n\delta \rfloor}^{\lfloor n(1-\delta)\rfloor} K\left(\frac{i}{n}, \frac{j}{n}\right)\left[H\left(\frac{i}{n}\right) - H\left(\frac{i-1}{n}\right)\right]\left[H\left(\frac{j}{n}\right) - H\left(\frac{j-1}{n}\right)\right]$$

$$\longrightarrow \sigma^2(\delta) = \int_{\delta}^{1-\delta}\int_{\delta}^{1-\delta} K(s,t)\,\mathrm{d}H(s)\,\mathrm{d}H(t),$$

as $n \to \infty$. Since

$$N(0, \sigma_n^2) \underset{n\to\infty}{\overset{\mathrm{d}}{\to}} N\left(0, \sigma^2(\delta)\right) \underset{\delta\to 0}{\overset{\mathrm{d}}{\to}} N\left(0, \sigma^2\right),$$

the assertion follows.

Example 6.2.4

(i) *For a Brownian motion $B(t), t \in [0, 1]$, we have*

$$K(s,t) = \mathrm{Cov}(B(s), B(t)) = \min(s,t),\ s,t \in [0,1],$$

such that

$$\sigma^2 = \int_0^1\int_0^1 K(s,t)\,\mathrm{d}s\,\mathrm{d}t = \frac{1}{3}.$$

Thus, the random variable $Z(\omega) = \int_0^1 B(t,\omega)\,\mathrm{d}t$, $\omega \in \Omega$, is normally distributed with mean 0 and variance $1/3$.

(ii) *For each fixed t we have*

$$\int_0^t X(s)\,\mathrm{d}H(s) \sim \sigma B_t$$

where B_t denotes a standard Brownian motion, if

$$\sigma^2 = \frac{1}{t}\int_0^t\int_0^t K(s,t)\,\mathrm{d}H(s)\,\mathrm{d}H(t).$$

More generally, the integral $\int_a^b X(t)\,\mathrm{d}H(t)$ exists in the sense of the mean square (m.s.) convergence of the corresponding RS sum, if X is m.s. continuous, i.e. $E(X(t) - X(s))^2 \to 0$, if $|t - s| \to 0$, and H is of bounded variation. Let us list some further rules and properties holding true under those conditions:

(i) $\left\|\int_a^b X(t)\,\mathrm{d}H(t)\right\|_{L_2} \le \sup_{t\in[a,b]} \|X(t)\|_{L_2} \int |\mathrm{d}H|$.

(ii) $t \mapsto \int_a^t X(s)\,\mathrm{d}s$ exists and is m.s. differentiable on $[a, b]$ with

$$\frac{d}{\mathrm{d}t}\int_a^t X(s)\,\mathrm{d}s = X(t), \qquad t \in [a, b].$$

(iii) If X' is m.s. continuous on $[a, b]$, then $\int_a^b X'(t)\,\mathrm{d}t$ exists and equals $X(b) - X(a)$.

(iv) Let $\mu(t) = EX(t)$, $t \in [a, b]$. If $\int_a^b X(t)\,\mathrm{d}H(t)$ or $\int_a^b H(t)\,\mathrm{d}X(t)$ exists in the mean square sense, then the integrals $\int_a^b H(t)\,\mathrm{d}\mu(t)$ and $\int_a^b \mu(t)\,\mathrm{d}H(t)$ exist and

$$E\left(\int_a^b H(t)\,\mathrm{d}X(t)\right) = \int_a^b H(t)\,\mathrm{d}\mu(t), \quad E\left(\int_a^b X(t)\,\mathrm{d}H(t)\right) = \int_a^b \mu(t)\,\mathrm{d}H(t).$$

6.3 The Itô integral

So far, we are in a position to define an integral as long as the integrator has bounded variation. The question arises whether one may use Brownian motion as an integrator. If we want to integrate a bounded variation function f with respect to Brownian motion, we can use the integration by parts formula, Proposition 6.2.2 (iv), to define $\int_a^b f(t)\,\mathrm{d}B(t)$ by

$$\int_a^b f(t)\,\mathrm{d}B(t) = f(t)B(t)|_a^b - \int_a^b B(t)\,\mathrm{d}f(t).$$

For more general deterministic functions $f \in L_2([a, b])$ Norbert Wiener studied the following approach. The step functions $f_n(t) = f_0\mathbf{1}_{\{a\}}(t) + \sum_{i=1}^{n} f_i\mathbf{1}_{(t_i,t_{i+1}]}(t)$, $t \in \mathbb{R}$, form a dense subset of $L_2([a, b])$ w.r.t. the metric

$$\mathrm{d}(f, g) = \int_a^b f(t)g(t)\,\mathrm{d}t, \qquad f, g \in L_2([a, b]).$$

For such elementary functions one puts

$$\int_a^b f_n(t)\,\mathrm{d}B(t) = \sum_{i=1}^{n} f_i[B(t_{i+1}) - B(t_i)].$$

The right-hand side is a normally distributed random variable with mean 0 and variance $\sum_{i=1}^{n} f_i^2(t_{i+1} - t_i)$ converging to $\int_a^b f(t)\,\mathrm{d}t$, if $f_i = f(t_i)$ and $\int_a^b f^2(t)\,\mathrm{d}t < \infty$, as the size of the partition tends to 0, as $n \to \infty$. Thus, one can define $\int_a^b f(t)\,\mathrm{d}B(t)$ as the normally distributed random variable to which $\int_a^b f_n(t)\mathrm{d}B(t)$ converges in mean square, which is known to exist.

It was Kiyoshi Itô who observed that this general idea can be extended to a much larger class of integrands, which particularly covers those random processes naturally appearing in mathematical finance when tracing the (discounted) value of a self-financing trading strategy. Here, integrands of the form

$$I_t = I_0 + \sum_{i=1}^{t} \varphi_i (S_i^* - S_{i-1}^*), \qquad t = 0, \ldots, T,$$

arise, where $\{\varphi_t : t = 0, \ldots, T\}$ is predictable and $\{S_t : t = 0, \ldots, T\}$ adapted. Recall that φ_t is predictable, if φ_t is $\mathcal{F}_{t-1}$-measurable for all t. To be in agreement with common notation used in the literature on Itô integration, let us define $\{\widetilde{\varphi}_t\}$ by

$$\widetilde{\varphi}_t = \varphi_{t+1}, \qquad t = 0, \ldots, T-1.$$

Then $I_t = I_0 + \sum_{i=1}^{t} \widetilde{\varphi}_{i-1}(S_i^* - S_{i-1}^*)$. For a trading strategy, $\widetilde{\varphi}_{i-1} (= \varphi_i)$ is the number of shares constantly held during $(i-1, i]$. $\widetilde{\varphi}_{i-1}$ is $\mathcal{F}_{i-1}$-measurable and, although formally being adapted, it still makes sense to call it predictable, having in mind that the function $t \mapsto \widetilde{\varphi}_{i-1} \mathbf{1}_{(i-1,i]}(t)$ attains the value $\widetilde{\varphi}_{i-1}$ if $t \in (i-1, i]$ and vanishes otherwise. The random left-continuous step function

$$t \mapsto \widetilde{\varphi}_0(\omega) \mathbf{1}_{\{0\}}(t) + \sum_{i=1}^{t} \widetilde{\varphi}_{i-1}(\omega) \mathbf{1}_{(i-1,i]}(t), \qquad t \in [0, T],$$

gives at each time point $t \in [0, T]$ the number of shares held.

Definition 6.3.1 *A stochastic process of the form*

$$H_t(\omega) = H_0 \mathbf{1}_{\{0\}}(t) + \sum_{i=0}^{n-1} H_i(\omega) \mathbf{1}_{(t_i, t_{i+1}]}(t),$$

where $0 = t_0 < \cdots < t_n = T$ is a partition of $[0, T]$ and H_i are $\mathcal{F}_i$-measurable random variables, is called a **simple predictable process**. *The class of such processes is denoted by $\mathcal{H}$.*

Notice that *any* such simple predictable process can be interpreted as a trading strategy. For such simple predictable integrands, the stochastic integral is defined as follows.

Definition 6.3.2 *Let $\{H_t : t \in [0, T]\}$ be a simple predictable process. Then the process $\{I_t : t \in [0, T]\}$ defined by*

$$I_t = \sum_{i=0}^{k-1} H_i \left[B(t_{i+1}) - B(t_i) \right] + H_k \left[B(t) - B(t_k) \right], \quad t_k \le t < t_{k+1},$$

is called the **Itô integral of** *H and is denoted by*

$$I(H) = \int H \, dB = \int H(s) \, dB(s) = \left\{ \int_0^t H_s \, dB_s : t \in [0, T] \right\}.$$

Remark 6.3.3

(i) Notice that $I(H)$ denotes a family of stochastic integrals.

(ii) Using $s \wedge t = \min(s, t)$ we may write $I(H)$ in the compact form

$$\int_0^t H_s \, \mathrm{d}B_s = \sum_{i=0}^{n-1} H_i[B(t \wedge t_{i+1}) - B(t \wedge t_i)], \qquad t \in [0, T].$$

(iii) By definition, for $t \in \{t_0, \dots, t_n\}$ the stochastic integral $\int_0^t H_s \, \mathrm{d}B_s$ reproduces the sums $\sum_{i=0}^{k-1} H_i[B(t_{i+1}) - B(t_i)]$, $k = 1, \dots, n$.

The next important theorem shows that the Itô integral is a martingale. By virtue of Remark 6.3.3 (iii), the proof is more or less a repetition of calculations used to verify that the discrete stochastic integral is a martingale.

Theorem 6.3.4 (Martingale property and Itô isometry) *Let $\{B_t : t \in [0, T]\}$ be a standard Brownian motion, $\mathcal{F}_t = \sigma(B_s : s \leq t)$ and let $\{H_t\}$ be a simple predictable process. Then the Itô integral $\int_0^t H_s \, \mathrm{d}B_s$, $t \in [0, T]$, is a martingale. Further, for all $t \in [0, T]$ Itô's isometry property holds true*

$$E\left(\int_0^t H_r \, \mathrm{d}B_r\right)^2 = E\int_0^t H_r^2 \, \mathrm{d}r = \int_0^t E H_r^2 \, \mathrm{d}r.$$

Proof. Suppose $t_k < s < t \leq t_{k+1}$, i.e. both s and t lie in one interval $(t_k, t_{k+1}]$ of the partition $\{t_i\}$ underlying the simple predictable function H. Then

$$\int_0^t H_r \, \mathrm{d}B_r = \sum_{i=0}^{k-1} H_i(B_{t_{i+1}} - B_{t_i}) + H_k(B_t - B_{t_k}) \tag{6.3}$$

and $\int_0^s H_r \, \mathrm{d}B_r$ is obtained by simply replacing B_t by B_s. Clearly, the sum is $\mathcal{F}_s$-measurable and the same applies to H_k and B_{t_k}. Thus,

$$E\left(\int_0^t H_r \, \mathrm{d}B_r \middle| \mathcal{F}_s\right) = \sum_{i=0}^{k-1} H_i(B_{t_{i+1}} - B_{t_i}) + H_k(E(B_t|\mathcal{F}_s) - B_{t_k}).$$

But $E(B_t|\mathcal{F}_s) = B_s$, a.s. If $t_{k-1} < s \leq t_k$, i.e. s and t lie in adjacent intervals, the first $k-1$ terms of the sum in Equation (6.3) are fixed when conditioning on $\mathcal{F}_s$. We have to calculate the conditional expectation of the remaining terms in Equation (6.3), i.e.

$$E(H_{k-1}(B_{t_k} - B_{t_{k-1}})|\mathcal{F}_s) = H_{k-1}(E(B_{t_k}|\mathcal{F}_s) - B_{t_{k-1}}) = H_{k-1}(B_s - B_{t_{k-1}})$$

and, since H_k is $\mathcal{F}_{t_{k-1}} \subset \mathcal{F}_s$ measurable and $E(B_{t_k}|\mathcal{F}_s) = B_s$ a.s.,

$$E(H_k(B_t - B_{t_k})|\mathcal{F}_s) = H_k(B_s - E(B_{t_k}|\mathcal{F}_s)) = 0.$$

Collecting terms we see that the result also holds in this case. Using the same arguments, it is now easy to check the result for the general situation where s and t lie in arbitrary intervals.

We aim at extending the Itô integral to integrands being elements of the class

$$\mathcal{L} = \{Y : [0, T] \times \Omega \to \mathbb{R} \mid Y\ \mathcal{F}_t\text{-adapted and left continuous with } \|Y\|_{\mathcal{L}} < \infty\},$$

where the norm is defined by

$$||Y||_{\mathcal{L}} = \sqrt{\int_0^T E(Y_t^2)\,\mathrm{d}t}, \qquad Y \in \mathcal{L}.$$

Notice that $||Y||_{\mathcal{L}}$ is the L_2 norm of Y w.r.t. the probability measure $\mathrm{d}P \otimes \mathrm{d}\lambda$, where λ denotes the Lebesgue measure, since

$$\|Y\|_{\mathcal{L}}^2 = \int_0^T \int_\Omega Y(t, \omega)^2\,\mathrm{d}P(\omega)\,\mathrm{d}t.$$

As an example, let us verify that a standard Brownian motion $B = \{B_t : t \in [0, T]\}$ belongs to $\mathcal{L}$. We have

$$E\left(\int_0^T B_t^2\,\mathrm{d}t\right) = \int_0^T E\left(B_t^2\right)\mathrm{d}t = \int_0^T t\,\mathrm{d}t = \frac{T^2}{2} < \infty,$$

such that $B \in \mathcal{L}$.

If $H \in \mathcal{H}$, then the Itô isometry shows that

$$||H||_{\mathcal{L}}^2 = \int_0^T E\left(H_t^2\right)\mathrm{d}t = E\left(\int_0^T H_t\,\mathrm{d}B_t\right)^2 = ||I(H)||_{L_2}^2,$$

where $||\cdot||_{L_2}$ denotes the L_2 norm on $(\Omega, \mathcal{A}, P)$. In other words, the mapping

$$I : \mathcal{H} \to L_2(\Omega, \mathcal{A}, P), \qquad H \mapsto I(H),$$

is an isometry, which justifies the name of Theorem 6.3.4. The next step is to show that one may approximate any process in $\mathcal{L}$ by simple predictable processes.

Lemma 6.3.5 *$\mathcal{H}$ is a dense subset of $\mathcal{L}$, i.e. for any $Y \in \mathcal{L}$ there is some sequence $\{H_n\} \subset \mathcal{H}$, such that*

$$\int_0^T E\left(H_n(s) - Y(s)\right)^2\,\mathrm{d}s \underset{n\to\infty}{\longrightarrow} 0.$$

Proof. Suppose Y is continuous and bounded by some constant $C > 0$. Then the step function given by

$$H_n(t) = Y(0)\mathbf{1}_{\{0\}}(t) + \sum_{i=0}^{n-1} Y\left(\frac{iT}{2^n}\right)\mathbf{1}_{\left(\frac{iT}{2^n}, \frac{(i+1)T}{2^n}\right]}(t)$$

satisfies $H_n(t, \omega) \to Y(t, \omega)$, as $n \to \infty$, for all $t \in [0, T]$ and $\omega \in \Omega$. By dominated convergence, we may conclude that

$$\int_0^T \int_\Omega (H_n(t, \omega) - Y(t, \omega))^2 \, \mathrm{d}P(\omega) \, \mathrm{d}t \underset{n \to \infty}{\longrightarrow} 0.$$

The details for the general case are a bit technical and can be found in many textbooks.

Before proceeding, let us calculate an important special Itô integral explicitly.

Example 6.3.6 (The integral $\int B \, \mathrm{d}B$) *Suppose $F(t)$ attains a derivative $f(t) = F'(t)$. Then,*

$$\int_0^T F(t) \, \mathrm{d}F(t) = \int_0^T F(t) f(t) \, \mathrm{d}t = \int_0^T z \, \mathrm{d}z = \frac{T}{2}$$

by a change of variable ($z = F(t)$, $\mathrm{d}z/\mathrm{d}t = f(t)$). What happens, if we replace $F(t)$ by a Brownian motion $B(t)$? In order to calculate the associated Itô integral, we apply the above lemma. More specifically, we shall verify directly its validity for Brownian motion. So, consider the sequence of partitions

$$\Pi_n : 0 = t_0 < \ldots < t_n = T, \qquad t_i = \frac{iT}{n}, \; i = 0, \ldots, T,$$

which satisfy with $||\Pi_n|| \to 0$. For brevity of notation put

$$W_i = B_{t_i}, \quad i = 0, \ldots, n,$$

and notice that $E(W_i) = 0$ and $E(W_i^2) = t_i$. We claim that the simple, left continuous and predictable step function

$$H_n(t) = W_0 \mathbf{1}_{\{0\}}(t) + \sum_{i=0}^{n-1} W_i \mathbf{1}_{(t_i, t_{i+1}]}(t), \quad t \in [0, T],$$

satisfies $||H_n - B||_{\mathcal{L}} \to 0$, as $n \to \infty$. Noticing that $H_n(t) = B(t_j)$ for $t \in [t_j, t_{j+1}]$, by the independence of the random variables

$$\int_{t_j}^{t_{j+1}} (H_n(t) - B(t))^2 \, \mathrm{d}t, \qquad j = 0, \ldots, n-1,$$

we obtain

$$\begin{aligned}\Delta_n &= E\left(\int_0^T (H_n(t) - B(t))^2 \, \mathrm{d}t\right)\\ &= E\left(\sum_{j=0}^{n-1} \int_{t_j}^{t_{j+1}} (H_n(t) - B(t))^2 \, \mathrm{d}t\right)\\ &= \sum_{j=0}^{n-1} \int_{t_j}^{t_{j+1}} E(B_{t_j} - B_t)^2 \, \mathrm{d}t,\end{aligned}$$

where, of course, $E(B_{t_j} - B_t)^2 = (t - t_j)$. *This leads to*

$$\Delta_n = \sum_{j=0}^{n-1} \left(\frac{t^2}{2} - t_j t\right)\Bigg|_{t_j}^{t_{j+1}} = \frac{1}{2}\frac{T^2}{n} \to 0,$$

as $n \to \infty$, *which verifies that* $||H_n - B||_{\mathcal{L}} \to 0$, *as* $n \to \infty$.

To calculate $\int_0^T H_n(s)\,\mathrm{d}B_s$, *we go back to Definition 6.3.2*

$$\int_0^T H_n(s)\,\mathrm{d}B_s = \sum_{i=0}^{n-1} W_i(W_{i+1} - W_i) = \sum_{i=0}^{n-1} W_i W_{i+1} - \sum_{i=0}^{n-1} W_i^2.$$

Using

$$W_i W_{i+1} = -\frac{(W_{i+1} - W_i)^2 - W_{i+1}^2 - W_i^2}{2}$$

we obtain

$$\int_0^T H_n(s)\,\mathrm{d}B_s = -\frac{1}{2}\sum_{i=0}^{n-1}(W_{i+1} - W_i)^2 + \frac{1}{2}\sum_{i=0}^{n-1}\left(W_{i+1}^2 - W_i^2\right).$$

The second sum collapses to $W_n^2/2 = B_T^2/2$, *since* $W_{t_n} = B_T$ *and* $W_0 = B_0 = 0$. *By the properties of Brownian motion,* $W_{i+1} - W_i \sim \mathcal{N}(0, t_{i+1} - t_i)$ *such that*

$$E\left(\sum_{i=0}^{n-1}(W_{i+1} - W_i)^2\right) = \sum_{i=0}^{n-1}(t_{i+1} - t_i) = T. \tag{6.4}$$

Further,

$$(W_{i+1} - W_i)^2 \overset{d}{=} \left(\sqrt{t_{i+1} - t_i}\,U\right)^2 \overset{d}{=} (t_{i+1} - t_i)\,V,$$

if $U \sim \mathcal{N}(0,1)$ and $V \sim \chi_1^2$. Recalling that Var $(V) = 2$, *we obtain*

$$\operatorname{Var}\left(\sum_{i=0}^{n-1}(W_{i+1}-W_i)^2\right) = 2\sum_{i=0}^{n-1}(t_{i+1}-t_i)^2 \le 2\cdot ||\Pi_n||\cdot T \to 0, \tag{6.5}$$

by independence of the increments $W_{i+1}-W_i$, $i=0,\ldots,n-1$. Equations (6.4) and (6.5) imply that

$$S_n = \sum_{i=0}^{n-1}(W_{i+1}-W_i)^2 \xrightarrow{L_2} T,$$

as $n \to \infty$. Consequently, putting things together we arrive at

$$\int_0^T H_n(s)\,\mathrm{d}B_s = \frac{1}{2}\left(B_T^2 - S_n\right) \xrightarrow{L_2} \frac{B_T^2 - T}{2},$$

as $n \to \infty$. That means that the sequence of Itô integrals $\int_0^T H_n(s)\,\mathrm{d}B_s$ converges in L_2 to a well-defined random varibale, namely $\frac{B_T^2 - T}{2}$. Hence, it makes sense to define that L_2 limit as the integral $\int_0^T B_s\,\mathrm{d}B_s$, that is to put

$$\int_0^T B_s\,\mathrm{d}B_s := \frac{B_T^2}{2} - \frac{T}{2}.$$

Then we obtain the puzzling result that elementary rules such as $\int_0^T F\,\mathrm{d}F = \frac{T^2}{2}$ no longer hold, if one integrates with respect to Brownian motion.

The results from the above example provide the guide as to how we can extend the definition of the Itô integral to integrands of the class $\mathcal{L}$: For a given $Y \in \mathcal{L}$ choose a sequence $\{H_n\} \subset \mathcal{H}$ with $||Y - H_n||_{\mathcal{L}} \to 0$, as $n \to \infty$. Clearly, $\{H_n\}$ is a Cauchy sequence, i.e.

$$||H_n - H_m||_{\mathcal{L}}^2 = \int_0^T E\left(H_n(t) - H_m(t)\right)^2\,\mathrm{d}t \to 0,$$

as $n, m \to \infty$. The linearity and isometry of I now yields the following chain of equalities

$$\begin{aligned}\left\| \int_0^T H_n \, \mathrm{d}B - \int_0^T H_m \, \mathrm{d}B \right\|_{L_2}^2 &= E\,(I(H_n) - I(H_m))^2 \\ &= ||I\,(H_n) - I\,(H_m)\,||_{L_2}^2 \\ &= ||I\,(H_n - H_m)\,||_{L_2}^2 \\ &= ||H_n - H_m||_{\mathcal{L}}^2 \to 0,\end{aligned}$$

as $n, m \to \infty$. Therefore, the sequence $\int_0^T H_n \, \mathrm{d}B$, $n \in \mathbb{N}$, of Itô integrals converges in L_2. Since L_2 spaces are complete, there exists a random variable $I^* \in L_2(\Omega, \mathcal{A}, P)$ such that $\int_0^T H_n \, \mathrm{d}B \to I^*$, as $n \to \infty$, in L_2. That L_2 limit, which is almost surely unique, is the object we were looking for, namely the stochastic integral $\int_0^T Y \, \mathrm{d}B$. Let us summarize our findings.

Definition and Theorem 6.3.7 *Let $\{Y_t\}$ be a left continuous $\mathcal{F}_t$-adapted process with*

$$\int_0^T E\left(Y_t^2\right) \mathrm{d}t < \infty.$$

Then, there is some sequence $\{H_n\}$ of simple predictable processes approximating $\{Y_t\}$ in the L_2 sense with respect to the measure $\mathrm{d}P \otimes \mathrm{d}t$, such that the sequence $\int_0^T H_n \, \mathrm{d}B$, $n \geq 1$, converges in L_2. The corresponding limit is denoted by

$$I(Y) = \int_0^T Y_t \, \mathrm{d}B_t$$

and is called the **stochastic Itô integral** *of Y w.r.t. Brownian motion.*

Clearly, the above construction works in the same vein when replacing the interval $[0, T]$ by $[a, b]$, $a, b \in \mathbb{R}$. We shall now establish some properties of the Itô integral and, along the way, study the Itô integral $\int_0^t Y_s \, \mathrm{d}B_s$ as a process in continuous time t. For the sake of brevity, we will not prove all results and refer to the literature.

Lemma 6.3.8 *Let $\{Y_t : t \in [a, b]\}$ be a left continuous process with $\int_a^b E\left(Y_t^2\right) \mathrm{d}t < \infty$.*

(i) If Z is bounded and $\mathcal{F}_a$-measurable, then $ZY \in \mathcal{L}$ and

$$\int_a^b ZY \, \mathrm{d}B = Z \int_a^b Y \, \mathrm{d}B.$$

(ii) $E\left(\int_a^b Y \, \mathrm{d}B \,\middle|\, \mathcal{F}_a \right) = 0.$

(iii) $E\left(\left|\int_a^b Y\,\mathrm{d}B\right|^2 \,\middle|\, \mathcal{F}_a\right) = E\left(\int_a^b Y_t^2\,\mathrm{d}t \,\middle|\, \mathcal{F}_a\right) = \int_a^b E\left(Y_t^2 \mid \mathcal{F}_a\right)\,\mathrm{d}t.$

Proof. See Friedman (1975), Theorem 2.8 and Lemma 2.9.

For what follows, we need a version of Doob's maximal inequality 3.2.25, which refers to processes in continuous time.

Theorem 6.3.9 *Let $\{X_t : t \in [0, T]\}$ be a left or right continuous submartingale. Then for $p \geq 1$,*

$$P\left(\sup_{0\leq s\leq t} |X_s| > \lambda\right) \leq \frac{E|X_t|^p}{\lambda^p}$$

for all $\lambda > 0$, and for $p > 1$,

$$\left\|\sup_{0\leq s\leq t} |X_s|\right\|_p \leq \frac{p}{p-1} \sup_{0\leq s\leq t} \|X_s\|_p.$$

Proof. Let us verify the first assertion. Basically, the result carries over to the continuous case, since for left or right continuous submartingales $\sup_{0\leq s\leq t} |X_s| = \sup_{s\in A} |X_s|$, where $A \subset [0, t]$ is a dense countable subset. Choose an increasing sequence of finite sets A_n such that $A = \cup_n A_n$. Then for each n, since the sup is a max,

$$P\left(\sup_{s\in A_n} |X_s| > \lambda\right) \leq \frac{\sup_{s\in A_n} E|X_s|^p}{\lambda} \leq \frac{\sup_{0\leq s\leq t} E|X_s|^p}{\lambda}.$$

Noticing that $\left\{\sup_{s\in A_n} |X_s| > \lambda\right\} \uparrow \left\{\sup_{s\in A} |X_s| > \lambda\right\}$, as $n \to \infty$, the result follows from

$$P\left(\sup_{s\in A} |X_s| > \lambda\right) = \lim_{n\to\infty} P\left(\sup_{s\in A_n} |X_s| > \lambda\right).$$

Theorem 6.3.10 *Let $Y \in \mathcal{L}$. Then there exists a continuous modification of the Itô integral.*

Proof. Let $\{H_n\} \subset \mathcal{H}$ be such that $||H_n - Y||_{\mathcal{L}} \to 0$, as $n \to \infty$, and put

$$I_n(t) = \int_0^t H_n(s)\,\mathrm{d}B_s,\ n \geq 1.$$

We claim that there is some subsequence I_{n_k}, $k \geq 1$, which converges uniformly in $t \in [0, T]$. As a result, the (unique) limit is a continuous function. Notice that $\{I_n(t) : t \in [0, T]\}$ is a martingale for each n. Doob's maximal inequality 3.2.25 yields for $n, m \in \mathbb{N}$ and $\varepsilon > 0$

$$P\left(\sup_{t\in[0,T]} |I_n(t) - I_m(t)| > \varepsilon\right) \leq \frac{E\,|I_n(T) - I_m(T)|^2}{\varepsilon^2},$$

where $E\,|I_n(T) - I_m(T)|^2 = ||I_n(T) - I_m(T)||_{L_2}^2 \to 0$, as $n, m \to \infty$. Hence, we may find some subsequence $n_k \to \infty$ such that

$$P\left(\sup_{t\in[0,T]} \left|I_{n_{k+1}} - I_{n_k}\right| > \frac{1}{2^k}\right) < \frac{1}{2^k}.$$

Now we can apply the Borel–Cantelli Lemma to obtain that

$$N = \left\{\sup_{t\in[0,T]} \left|I_{n_{k+1}} - I_{n_k}\right| > 2^{-k} \text{ i.o.}\right\}$$

is a null set, such that $\sup_{t\in[0,T]} \left|I_{n_{k+1}} - I_{n_k}\right| < 2^{-k}$ holds true for k large enough. This shows the uniform convergence along a subsequence.

Remark 6.3.11 *When considering an Itô integral $\int_0^t Y_s \,\mathrm{d}B_s$, $t \in [0, T]$, one always means its continuous version.*

Let us now summarize the most important properties and rules of the Itô integral.

Theorem 6.3.12 *Let $\{X_t\}$ and $\{Y_t\}$ be left continuous processes of the class $\mathcal{L}$ and $0 \le a \le b \le c \le T$.*

(i) $\int_a^c X_t \,\mathrm{d}B_t = \int_a^b X_t \,\mathrm{d}B_t + \int_b^c X_t \,\mathrm{d}B_t.$

(ii) $\int_a^b (\lambda X_t + \mu Y_t)\,\mathrm{d}B_t = \lambda \int_a^b X_t \,\mathrm{d}B_t + \mu \int_a^b Y_t \,\mathrm{d}B_t$ *for any* $\lambda, \mu \in \mathbb{R}$.

(iii) $M_t = \int_0^t X_s \,\mathrm{d}B_s$, $t \in [0, T]$, *is a martingale.*

(iv) The Itô integral is centered, i.e. $E\left(\int_a^b X_s \,\mathrm{d}B_s\right) = 0$ *and is isometric, i.e.* $E\left(\int_a^b X_s \,\mathrm{d}B_s\right)^2 = \int_a^b E\,X_s^2 \,\mathrm{d}s.$

(v) $\left[\int_a^b X_s \,\mathrm{d}B_s, \int_a^b X_s \,\mathrm{d}B_s\right]_t = \int_a^t X_s^2 \,\mathrm{d}s$, $t \in [a, b]$.

(vi) If $\{X_t\}$ is continuous, then

$$\sum_{k=0}^{N_n-1} X_{t_{nk}}[B(t_{n,k+1}) - B(t_{nk})] \xrightarrow{P} \int_a^b X_s \,\mathrm{d}B_s,$$

as $n \to \infty$, for any sequence Π_n of partitions

$$\Pi_n : a = t_{n0} < \ldots < t_{nN_n} = b$$

of $[a, b]$ with $||\Pi_n|| \to 0$, $n \to \infty$.

(vii) For any $\varepsilon > 0$ and $c > 0$

$$P\left(\left|\int_a^b X_t \,\mathrm{d}B_t\right| > \varepsilon\right) \leq P\left(\int_a^b X_t^2 \,\mathrm{d}t > c\right) + \frac{c}{\varepsilon^2}.$$

(viii) Let $\{X, X_n\} \subset \mathcal{L}$ be a sequence with

$$\int_a^b |X_n(t) - X(t)|^2 \,\mathrm{d}t \xrightarrow{P} 0,$$

as $n \to \infty$. Then

$$\sup_{a \leq t \leq b} \left|\int_a^t X_n(s) \,\mathrm{d}B(s) - \int_a^t X(s) \,\mathrm{d}B(s)\right| \xrightarrow{P} 0,$$

as $n \to \infty$.

(ix) For any $\lambda > 0$

$$P\left(\sup_{a \leq t \leq T} \left|\int_a^t X_s \,\mathrm{d}B_s\right| > \lambda\right) \leq \frac{1}{\lambda^2} E \int_a^b X_s^2 \,\mathrm{d}s.$$

Proof. (i)–(iv) are left to the reader. We shall verify the formula (v) for the quadratic variation for simple predictable integrands. W.l.o.g. let $[a, b] = [0, t]$ and put $I(t) = \int_0^t X_s \,\mathrm{d}B_s$. Suppose X_t has the representation

$$X_s = X_0 + \sum_{i=0}^{n-1} \widetilde{X}_i \mathbf{1}_{(t_i, t_{i+1}]}(s)$$

for $\mathcal{F}_{t_i}$-measurable random variables $\widetilde{X}_i$. Then,

$$[I, I]_t = \sum_{i=0}^{n-1} \left[\int_{t_i}^{\cdot} X_s \,\mathrm{d}B_s, \int_{t_i}^{\cdot} X_s \,\mathrm{d}B_s\right]_{t_i}^{t_{i+1}},$$

where $[\cdot,\cdot]_a^b$ denotes quadratic variation over $[a,b]$. For a partition $\Pi_i : t_i = s_0^{(i)} < \cdots < s_{m_i}^{(i)} = t_{i+1}$ of the subinterval $[t_i, t_{i+1}]$ we have

$$\sum_{j=0}^{m_i-1} \left[I\left(s_{j+1}^{(i)}\right) - I\left(s_j^{(i)}\right) \right]^2 = \widetilde{X}_i \sum_{j=0}^{m_i-1} \left[B\left(s_{j+1}^{(i)}\right) - B\left(s_j^{(i)}\right) \right]^2,$$

which converges a.s. to $\widetilde{X}_i^2(t_{i+1} - t_i)$, as $||\Pi_i|| \to 0$. Thus, we arrive at

$$[I, I]_t = \sum_{i=0}^{n-1} \widetilde{X}_i(t_{i+1} - t_i) = \int_0^t X_s \, \mathrm{d}s.$$

For proofs of the remaining facts we refer to, e.g., Friedman (1975).

Due to their importance, it is worth discussing the following two issues.

Martingale property:

Theorem 6.3.12 (iii) shows that an Itô integral such as $\int_0^t X_s \, \mathrm{d}B_s$ is a mean zero martingale with variance $\sigma_t^2 = \int_0^t E\, X_s^2 \, \mathrm{d}s$. In particular,

$$E\left(\int_0^t X_r \, \mathrm{d}B_r \middle| \mathcal{F}_s \right) = \int_0^s X_r \, \mathrm{d}B_r$$

holds true for $s \leq t$. That property holds true for *any* integrand $\{X_t\}$.

Variance and quadratic variation:

It is important to distinguish the variance and the quadratic variation. According to property (v), the latter is given by

$$S_t^2(\omega) = \int_0^t X^2(s,\omega) \, \mathrm{d}s, \qquad \omega \in \Omega.$$

The quadratic variation measures the risk along a path (for any fixed ω) and is usually a non constant random variable. If for some path we take large positions X_s, the risk is high. The variance $\sigma_t^2 = \operatorname{Var}\left(\int_0^t X_s \, \mathrm{d}B_s\right) = E\left(\int_0^t X_s \, \mathrm{d}B_s\right)^2$ satisfies

$$\sigma_t^2 = E S_t^2 = \int S(t,\omega)^2 \, \mathrm{d}P(\omega),$$

(apply Itô's isometry rule), i.e. it averages the quadratic variation over all paths using the probability measure P.

Let us close this subsection by discussing how one can generalize the Itô integral. So far, we have defined the stochastic integral for Brownian motion acting as an integrator and for processes whose sample paths are of bounded variation. Suppose we want to integrate w.r.t. a process X that can be decomposed in the form $X_t = X_0 + B_t + A_T$, where B_t is a Brownian motion and A_t a bounded variation process. Such a process belongs to the class of **semimartingales**, which are those processes allowing for a decomposition $X_t = X_0 +$

$M_t + A_t$, where M_t is a martingale and A_t has bounded variation. Then, it is natural to set

$$\int H \, \mathrm{d}X = \int H \, \mathrm{d}B + \int H \, \mathrm{d}A,$$

provided both integrals at the right-hand side are well defined.

We adopted the classic L_2 approach to introduce the stochastic Itô integral, which requires that the predictable integrands f satisfy $\int_0^T Ef^2(t) \, \mathrm{d}t < \infty$. This is not always the case. In the present section we outline a more general approach that allows us to integrate functions f which are only almost surely finite but have the nice property of being càdlàg, i.e. right-continuous with existing left-hand limits.

Denote the underlying probability space by $(\Omega, \mathcal{F}, P)$ and let L_0 denote the class of random variables X that are a.s. finite, i.e. $P(X < \infty) = 1$. Again, the starting point are simple predictable functions,

$$H_n(t) = H_0 \mathbf{1}_{\{0\}} + \sum_{i=0}^{n-1} H_i \mathbf{1}_{(T_i, T_{i+1}]}, \qquad t \in [0, T],$$

where now the $\mathcal{F}_i$-measurable random variables H_i are elements of the space L_0. Further, the points $0 = T_0 < T_1 < \cdots < T_n = T$ partitioning the interval $[0, T]$ may be random, as indicated by our notation, but they are assumed to be stopping times, i.e. T_i is $\mathcal{F}_i$-measurable. Clearly, deterministic partitions appear as special cases.

Given an adapted càdlàg process X, we define the stochastic integral $\int H \, \mathrm{d}X$ as usual, i.e.

$$I(H) = \int_0^T H_s \, \mathrm{d}X_s = \sum_{i=0}^{n-1} H_i (X_{T_{i+1}} - X_{T_i})$$

and, more generally,

$$I(H)_t = \int_0^t H_s \, \mathrm{d}X_s = H_0 X_0 + \sum_{i=0}^{n-1} H_i (X_{t \wedge T_{i+1}} - X_{t \wedge T_i}),$$

for $t \in [0, T]$. Denote the class of such elementary functions by $\mathcal{E}$. Suppose we are given a càdlàg function H and know that we can find a sequence $\{H_n\} \subset \mathcal{E}$, such that H_n converges in probability to H uniformly on each interval $[0, t]$ and uniformly in $\omega \in \Omega$, i.e.

$$\|H_n - H\|_\infty = \sup_{s \in [0,t]} \sup_{\omega \in \Omega} |H_n(s, \omega) - H(s, \omega)| \to 0, \qquad n \to \infty. \tag{6.6}$$

An integrator is called nice if this already implies that $I(H_n) = \int H_n \, \mathrm{d}X$ converges in probability to a random variable we shall denote by $I(H) = \int H \, \mathrm{d}X$, as $n \to \infty$. If X is a semimartingale, i.e. $X = M + A$ for some martingale M and a process A being of bounded variation, the following important result holds true.

Theorem 6.3.13 *If X is a semimartingale, then $I : \mathcal{E} \to L_0$ is a continuous linear mapping, when we equip $\mathcal{E}$ with the supnorm as in Equation (6.6) and L_0 with any metric inducing convergence in probability.*[2]

In other words, for a semimartingale X we may extend the notion of the stochastic integral to integrands H appearing as limits of sequences $\{H_n\}$ of the space $\mathcal{E}$ of simple predictable functions, since then there exists a random variable $I^* \in L_0$ that we shall denote by $\int H \, \mathrm{d}X$ such that for any $\varepsilon > 0$

$$\lim_{n\to\infty} \|H_n - H\|_\infty \qquad \Rightarrow \qquad \lim_{n\to\infty} P\left(\left|\int H_n \, \mathrm{d}X - \int H \, \mathrm{d}X\right| > \varepsilon\right) = 0.$$

This procedure yields the stochastic integral as a random variable corresponding to integration from 0 to T. In order to obtain the stochastic integral as a process, one considers for a simple predictable function H_n as above the mapping

$$I(H_n)(t, \omega) = \sum_{i=0}^{n-1} H_i(\omega)\big(X_{\min(t,T_{i+1})}(\omega) - X_{\min(t,T_i)}(\omega)\big), \quad t \in [0, T], \omega \in \Omega,$$

i.e. for given $0 \le t \le T$ one takes only those summands with $T_i \le t$. This gives a mapping I defined on the space $\mathcal{E}$ and attaining values in the space of stochastic processes with trajectories being elements of the Skorohod space $D[0, T]$, the latter consisting of all functions defined on $[0, T]$ that are right-continuous with existing left-hand limits. We equip that space with the uniform convergence.

6.4 Quadratic covariation

We have seen in Theorem 6.1.6 that the L_2 limit of the observed quadratic variation of Brownian motion converges to the identity, i.e.

$$[B, B]_t = t.$$

Combining this fact with the formula obtained in Example 6.3.6 gives

$$[B, B]_t = B_t^2 - 2\int_0^t B_s \, \mathrm{d}B_s.$$

That result suggests the following definition.

Definition 6.4.1 *Let $\{X_t\}$ and $\{Y_t\}$ be two processes such that the Itô integrals $\int X \, \mathrm{d}Y$ and $\int Y \, \mathrm{d}X$ exist. Then*

$$[X, Y]_t = X_t Y_t - \int_0^t X \, \mathrm{d}Y - \int_0^t Y \, \mathrm{d}X, \quad x \in [0, T],$$

is called the **quadratic covariation** *or* **bracket process**.

---

[2] Convergence in probability can be metricized, e.g. by virtue of the Prohorov metric.

We immediately obtain an integration by parts rule.

Corollary 6.4.2 (Integration by parts)

$$\int_0^t X \,\mathrm{d}Y = XY\big|_0^t - [X, Y]_t - \int_0^t Y \,\mathrm{d}X.$$

The quadratic covariation can be calculated as the L_2 limit of the corresponding observed quadratic covariation.

Definition and Proposition 6.4.3 *For any partition* $\Pi_n : 0 = t_0 < \ldots < t_n = t$ *with* $||\Pi_n|| \to 0$ *the* **observed quadratic covariation**

$$[X, Y]_n(t) = [X, Y]_{nt} = \sum_{i=0}^{n-1} \left(X_{t_{i+1}} - X_{t_i}\right) \left(Y_{t_{i+1}} - Y_{t_i}\right)$$

converges in L_2 *to* $[X, Y]_t$*. The convergence is uniform in* $t \in [0, T]$.

Proof. Using $(b - a)(d - c) = bd - ac - a(d - c) - (b - a)c$ leads to

$$\begin{aligned}[X, Y]_n(t) &= \sum_{i=0}^{n-1} X_{t_{i+1}} Y_{t_{i+1}} - \sum_{i=0}^{n-1} X_{t_i} Y_{t_i} \\ &\quad - \sum_{i=0}^{n-1} Y_{t_i} \left(X_{t_{i+1}} - X_{t_i}\right) - \sum_{i=0}^{n-1} X_{t_i} \left(Y_{t_{i+1}} - Y_{t_i}\right) \\ &\to X_t Y_t - X_0 Y_0 - \int_0^t Y \,\mathrm{d}X - \int_0^t X \,\mathrm{d}Y,\end{aligned}$$

as $n \to \infty$, since the first two sums collapse to $X_{t_n} Y_{t_n} - X_{t_0} Y_{t_0} = X_t Y_t - X_0 Y_0$, whereas the third and fourth sum are the Itô integrals w.r.t. the approximating step function

$$X_n(t) = \mathbf{1}_{\{0\}} X_0 + \sum_{i=0}^{n-1} X_{t_i} \mathbf{1}_{(t_i, t_{i+1}]}(t),$$

$$Y_n(t) = \mathbf{1}_{\{0\}} Y_0 + \sum_{i=0}^{n-1} Y_{t_i} \mathbf{1}_{(t_i, t_{i+1}]}(t).$$

By virtue of Theorem 6.3.12, the convergence is uniform in $t \in [0, T]$.

6.5 Itô's formula

Let f be a smooth function and B_t be a Brownian motion. Our aim is to derive a formula for the stochastic process $f(B_t)$.

Fix an arbitrary partition $0 = t_0 < t_1 < \ldots < t_n = t$ of $[0, t]$. Clearly, we may write

$$f(B_t) - f(B_0) = \sum_{i=0}^{n-1} [f(B_{t_{i+1}}) - f(B_{t_i})].$$

Consider the ith summand and plug in a Taylor expansion of f taking into account the linear as well as the quadratic term to obtain

$$f(B_{t_{i+1}}) - f(B_{t_i}) = f'(B_{t_i})(B_{t_{i+1}} - B_{t_i}) + \frac{1}{2} f''(B_{t_i})(B_{t_{i+1}} - B_{t_i})^2 + R.$$

To this end, let us ignore the remainder term R. Taking sums at both sides yields

$$f(B_t) - f(B_0) \approx \sum_{i=0}^{n-1} f'(B_{t_i})(B_{t_{i+1}} - B_{t_i}) + \frac{1}{2} \sum_{i=0}^{n-1} f''(B_{t_i})(B_{t_{i+1}} - B_{t_i})^2.$$

The first sum converges (in L_2) to the Itô integral $\int_0^t f'(B_s)\,\mathrm{d}B_s$. By Theorem 6.1.7, the second sum due to the quadratic term of the Taylor expansion converges to $\int_0^t f''(B_s)\,\mathrm{d}s$. We shall see that a rigorous treatment of the remainder term confirms this conjecture. To do so, let us recall the following version of Taylor's theorem:

If $f \in C^2$, the space of twice differentiable functions, we have

$$f(y) = f(x) + f'(x)(y - x) + \frac{1}{2} f''(\xi)(y - x)^2$$

for some point ξ between x and y. We may write this in the form

$$f(y) = f(x) + f'(x)(y - x) + \frac{1}{2} f''(x)(y - x)^2 + R_x$$

with a remainder term

$$R_x = \frac{1}{2}(f''(\xi) - f''(x))(y - x)^2.$$

Theorem 6.5.1 (Itô's Formula)
Suppose that $f : \mathbb{R} \to \mathbb{R}$ is twice continuously differentiable. Then for each t,

$$f(B_t) - f(B_0) = \int_0^t f'(B_s)\,\mathrm{d}B_s + \frac{1}{2} \int_0^t f''(B_s)\,\mathrm{d}s.$$

Proof. Let $\Pi_n : 0 = t_0 < \ldots < t_n = t$ be a sequence of partitions of $[0, t]$ with $||\Pi_n|| \to 0$, as $n \to \infty$. A Taylor expansion of f applied to each summand of the representation

$$f(B_t) - f(B_0) = \sum_{i=0}^{n-1} f(B_{t_{i+1}}) - f(B_{t_i})$$

yields

$$f(B_{t_{i+1}}) - f(B_{t_i}) = f'(B_{t_i})(B_{t_{i+1}} - B_{t_i}) + \frac{1}{2} f''(B_{t_i})(B_{t_{i+1}} - B_{t_i})^2 + R_{t_i}$$

with

$$R_{t_i} = \frac{1}{2}(f''(\xi_i) - f''(B_{t_i}))(B_{t_{i+1}} - B_{t_i})^2,$$

ξ_i is a (random) point between B_{t_i} and $B_{t_{i+1}}$. By continuity of B_t, the set

$$K_\omega = \{B_t(\omega) : t \in [0, T]\}$$

is compact for each $\omega \in \Omega$. Hence, f'' is uniformly continuous on K_ω. Let $\varepsilon > 0$ and $\delta = \delta(\varepsilon) > 0$ be such that

$$|B_s - B_r| < \delta \Longrightarrow |f''(B_s) - f''(B_r)| < \varepsilon, \quad 0 \leq r, s \leq t.$$

Since $||\Pi_n|| \to 0$, as $n \to \infty$, we can find some $n_0 = n_0(\delta)$ such that $||\Pi_n|| < \delta$ for $n \geq n_0$ as well as $\sup_{n \geq n_0} ||\Pi_n|| \leq \delta$. Therefore, we may estimate the ith remainder by

$$|R_{t_i}| = \frac{1}{2}|f''(\xi_i) - f''(B_{t_i})|(B_{t_{i+1}} - B_{t_i})^2 \leq \frac{\varepsilon}{2}(B_{t_{i+1}} - B_{t_i})^2.$$

Consequently,

$$\sum_{i=0}^{n-1} |R_{t_i}| \leq \frac{\varepsilon}{2} \sum_{i=0}^{n-1} (B_{t_{i+1}} - B_{t_i})^2 \longrightarrow \frac{\varepsilon}{2} \cdot t,$$

in L_2 and a.s. by Theorem 6.1.6.

Example 6.5.2 *Consider* $Y_t = \frac{B_t^2}{2}$, *i.e.* $f(B_t)$ *with* $f(x) = \frac{x^2}{2}$, $x \in \mathbb{R}$. *Itô's formula yields*

$$\begin{aligned} f(B_t) &= f(B_0) + \int_0^t f'(B_s)\,\mathrm{d}B_s + \frac{1}{2}\int_0^t f''(B_s)\,\mathrm{d}s \\ &= f(0) + \int_0^t B_s\,\mathrm{d}B_s + \frac{1}{2}\int_0^t 1\,\mathrm{d}s \\ &= \int_0^t B_s\,\mathrm{d}B_s + \frac{t}{2}. \end{aligned}$$

Again, we are led to the formula

$$\int_0^t B_s\,\mathrm{d}B_s = \frac{1}{2}\left(B_t^2 - t\right).$$

Example 6.5.3 *Let B_t be a Brownian motion with respect to a filtration $\mathcal{F}_t$, and let μ_t and σ_t be $\mathcal{F}_t$-adapted processes. Concerning σ_t we also assume that $\sigma_t \geq 0$ and*

$$E \int_0^T \sigma_t^2 \, \mathrm{d}t < \infty.$$

Then, the Itô integral

$$X_t = \int_0^t \left(\mu_s - \frac{\sigma_s^2}{2} \right) \mathrm{d}s + \int_0^t \sigma_s \, \mathrm{d}B_s$$

is well defined. Consider the process

$$S_t = S_0 \exp \left\{ \int_0^t \left(\mu_s - \frac{\sigma_s^2}{2} \right) \mathrm{d}s + \int_0^t \sigma_s \, \mathrm{d}B_s \right\}, \qquad t \in [0, T],$$

is well defined. Notice that

$$S_t = f(X_t)$$

with $f(x) = S_0 \mathrm{e}^x = f'(x) = f''(x)$. An application of Itô's formula shows that

$$S_t = \int_0^t \mu_u S_u \, \mathrm{d}u + \int_0^t \sigma_u S_u \, \mathrm{d}B_u, \qquad t \in [0, T].$$

6.6 Itô processes

We shall now introduce an important class of stochastic processes.

Definition 6.6.1 *Let $B_t, t \geq 0$, be a Brownian motion with respect to the filtration $\{\mathcal{F}_t\}$. $\{\mu_t\}$ and $\{\sigma_t\}$ are assumed to be $\mathcal{F}_t$-adapted and $\mathcal{D}_T \otimes \mathcal{F}_T$-measurable processes with*

$$\int_0^T |\mu_s| \, \mathrm{d}s < \infty \quad \text{and} \quad \int_0^T |\sigma_s|^2 \, \mathrm{d}s < \infty,$$

almost surely. Then the Itô integral $\int \sigma_s \, \mathrm{d}B_s$ is well defined and a process $\{X_t\}$ satisfying

$$X_t = X_0 + \int_0^t \mu_s \, \mathrm{d}s + \int_0^t \sigma_s \, \mathrm{d}B_s$$

is called an **Itô process** *with random starting value X_0. μ_t is called the* **drift** *and σ_t is called the* **volatility**.

The integral definition of X_t is often written in differential form

$$\mathrm{d}X_t = \mu_t \, \mathrm{d}t + \sigma_t \, \mathrm{d}B_t.$$

We are now in a position to establish the important fact that the logarithm of a geometric Brownian motion belongs to the class of Itô processes. Indeed, it corresponds to those Itô processes having a constant drift as well as a constant volatility.

Example 6.6.2 (Geometric Brownian Motion)
Recall that $\{S_t\}$ is a geometric Brownian motion if

$$S_t = S_0 \exp(\mu t + \sigma B_t)$$

for fixed constants $\mu \in \mathbb{R}$ and $\sigma > 0$. Put $X_t = \ln S_t$. Then,

$$X_t = X_0 + \mu t + \sigma B_t$$

is an Itô process with drift $\mu_t = \mu$ and volatility $\sigma_t = \sigma$ for all t. In differential notation

$$\mathrm{d}X_t = \mu\,\mathrm{d}t + \sigma\,\mathrm{d}B_t.$$

Lemma 6.6.3 *The quadratic variation of an Itô process*

$$X_t = \int_0^t \mu_s\,\mathrm{d}s + \int_0^t \sigma_s\,\mathrm{d}B_s$$

is given by

$$[X, X]_t = \int_0^t \sigma_s^2\,\mathrm{d}s.$$

Proof. For brevity of notation put

$$m_t = \int_0^t \mu_s\,\mathrm{d}s \quad \text{and} \quad s_t = \int_0^t \sigma_s\,\mathrm{d}B_s.$$

Let $\Pi_n : 0 = t_0 < \ldots < t_n = t$ be a partition of $[0, t]$. Then,

$$\begin{aligned}\sum_{i=0}^{n-1}(X_{t_{i+1}} - X_{t_i})^2 &= \sum_{i=0}^{n-1}(s_{t_{i+1}} - s_{t_i})^2 + \sum_{i=0}^{n-1}(m_{t_{i+1}} - m_{t_i})^2 \\ &\quad + 2\sum_{i=0}^{n-1}(s_{t_{i+1}} - s_{t_i})(m_{t_{i+1}} - m_{t_i}).\end{aligned}$$

The first term converges a.s. to $\int_0^t \sigma_s^2\,\mathrm{d}s$, by Theorem 6.3.12. The second term can be bounded by

$$\max_{0\le i\le n-1} |m_{t_{i+1}} - m_{t_i}| \cdot \sum_{i=0}^{n-1} |m_{t_{i+1}} - m_{t_i}|$$

where

$$\sum_{i=0}^{n-1} |m_{t_{i+1}} - m_{t_i}| = \sum_{i=0}^{n-1} \left| \int_{t_i}^{t_{i+1}} \mu_s\,\mathrm{d}s \right| \le \int_0^t |\mu_s|\,\mathrm{d}s \tag{6.7}$$

and

$$\max_{0 \le i \le n-1} |m_{t_{i+1}} - m_{t_i}| \to 0, \text{ if } ||\Pi_n|| \to 0,$$

by continuity of $M : u \mapsto \int_0^u \mu_s \, \mathrm{d}s$, which implies that M is uniformly continuous on $[0, t]$. Finally, the last term is not larger than

$$2 \max_{0 \le i \le n-1} |s_{t_{i+1}} - s_{t_i}| \cdot \sum_{i=0}^{n-1} |m_{t_{i+1}} - m_{t_i}|.$$

Noting that s_t is continuous and using Equation (6.7) completes the proof.

Corollary 6.6.4 *If $\{X_t\}$ is an Itô process such that $F_t = [X, X]_t$ is differentiable with derivative $F_t' = \sigma_t^2$, then integrals w.r.t. $[X, X]_t$ can be calculated by*

$$\int_0^t Z_s \, \mathrm{d}[X, X]_s = \int_0^t Z_s \sigma_s^2 \, \mathrm{d}s,$$

for any process $\{Z_t\}$ such that the integrals exist.

We also need the following result that tells us how to integrate a proper adapted process w.r.t. an Itô process.

Theorem 6.6.5 *Let*

$$X_t = X_0 + \int_0^t \mu_s \, \mathrm{d}s + \int_0^t \sigma_s \, \mathrm{d}B_s$$

be an Itô process and Z_t be an adapted process. If

$$E \int_0^t Z_s^2 \sigma_s^2 \, \mathrm{d}s < \infty \quad \textit{and} \quad \int_0^t |Z_s \mu_s| \, \mathrm{d}s < \infty$$

hold true for any $t \in [0, T]$, then

$$\int_0^t Z_s \, \mathrm{d}X_s = \int_0^t Z_s \mu_s \, \mathrm{d}s + \int_0^t Z_s \sigma_s \, \mathrm{d}B_s, \quad 0 \le t \le T.$$

Proof. We show the result for the case that $\{Z_t\}$, $\{\mu_t\}$ and $\{\sigma_t\}$ are simple predictable w.r.t. the partition $0 = t_0 < \ldots < t_n = T$. Then $\mu_t = \mu_{t_i}$ and $\sigma_t = \sigma_{t_i}$ for $t \in (t_i, t_{i+1}]$. Hence,

$$\begin{aligned} X_{t_{i+1}} - X_{t_i} &= \int_{t_i}^{t_{i+1}} \mu_s \, \mathrm{d}s + \int_{t_i}^{t_{i+1}} \sigma_s \, \mathrm{d}B_s \\ &= \mu_{t_i}(t_{i+1} - t_i) + \sigma_{t_i}(B_{t_{i+1}} - B_{t_i}), \end{aligned}$$

yielding

$$\begin{aligned}\int_0^t Z_s \,\mathrm{d}X_s &= \sum_{i=0}^{n-1} Z_{t_i}(X_{t_{i+1}} - X_{t_i}) \\ &= \sum_{i=0}^{n-1} Z_{t_i}\mu_{t_i}(t_{i+1} - t_i) + \sum_{i=0}^{n-1} Z_{t_i}\sigma_{t_i}(B_{t_{i+1}} - B_{t_i}) \\ &= \int_0^t Z_s\mu_s \,\mathrm{d}s + \int_0^t Z_s\sigma_s \,\mathrm{d}B_s.\end{aligned}$$

We are now in a position to formulate Itô's formula for Itô processes.

Theorem 6.6.6 (Itô's Formula for Itô Processes)
Let $\{X_t\}$ be an Itô process and $f(t, x)$ be a function with continuous partial derivatives $f_t(t, x)$, $f_x(t, x)$ and $f_{xx}(t, x)$. Then,

$$\begin{aligned}f(t, X_t) &= f(0, X_0) + \int_0^t f_t(s, X_s)\,\mathrm{d}s + \int_0^t f_x(s, X_s)\,\mathrm{d}X_s + \frac{1}{2}\int_0^t f_{xx}(s, X_s)\,\mathrm{d}[X, X]_s \\ &= f(0, X_0) + \int_0^t \left[f_t(s, X_s) + f_x(s, X_s)\mu_s + \frac{1}{2} f_{xx}(s, X_s)\sigma_s^2\right] \mathrm{d}s + \int_0^t f_x(s, X_s)\sigma_s \,\mathrm{d}B_s,\end{aligned}$$

for $t \in [0, T]$. In other words, $f(t, X_t)$ is again an Itô process with drift

$$\widetilde{\mu}_s = f_t(s, X_s) + f_x(s, X_s)\mu_s + \frac{1}{2} f_{xx}(s, X_s)\sigma_s^2$$

and volatility

$$\widetilde{\sigma}_s = f_x(s, X_s)\sigma_s.$$

This theorem tells us that any smooth function of an Itô process is again an Itô process and the way how we map the value X_t to the new process may depend on the time coordinate t.

Example 6.6.7 *Consider a geometric Brownian motion*

$$S_t = S_0 \exp(\mu t + \sigma B_t),$$

where $\mu \in \mathbb{R}$ and $\sigma > 0$ are constants. We already know from Example 6.6.2 that $X_t = \log S_t$ is an Itô process with drift $\mu_s = \mu$ and volatility $\sigma_s = \sigma$. Let us apply Itô's formula to $S_t = f(X_t)$

with $f(x) = \mathrm{e}^x$. *Then* $S_t = f(X_t)$ *is also an Itô process with drift*

$$\begin{aligned}\widetilde{\mu}_s &= f_x(X_s)\mu_s + \frac{1}{2} f_{xx}(X_s)\sigma_s^2 \\ &= S_s\mu_s + \frac{1}{2} S_s\sigma_s^2 \\ &= \left(\mu + \frac{\sigma^2}{2}\right) S_s\end{aligned}$$

and volatility $\widetilde{\sigma}_s = f_x(X_s)\sigma_s = S_s\sigma$, *such that*

$$S_t = S_0 + \left(\mu + \frac{\sigma^2}{2}\right) \int_0^t S_r \, \mathrm{d}r + \sigma \int_0^t S_r \, \mathrm{d}B_r.$$

As a differential

$$\mathrm{d}S_t = \left(\mu + \frac{\sigma^2}{2}\right) S_t \, \mathrm{d}t + \sigma S_t \, \mathrm{d}B_t.$$

We have found that a geometric Brownian motion allows the representation

$$S_t = S_0 + \left(\mu + \frac{\sigma^2}{2}\right) \int_0^t S_r \, \mathrm{d}r + \sigma \int_0^t S_r \, \mathrm{d}B_r.$$

If we put $\mu = -\frac{\sigma^2}{2}$, the drift term vanishes and we obtain $S_t = S_0 + \sigma \int_0^t S_r \, \mathrm{d}B_r$, a martingale.

Corollary 6.6.8 *A geometric Brownian motion*

$$S_t = S_0 \exp(\mu t + \sigma B_t)$$

with

$$\mu = -\frac{\sigma^2}{2}$$

is a martingale.

Example 6.6.9 (Generalized Brownian Motion)
Let μ_t *and* σ_t *be adapted processes with*

$$\int_0^T |\mu_t| \, \mathrm{d}t < \infty \qquad \textit{and} \qquad \int_0^T \sigma_t^2 \, \mathrm{d}t < \infty,$$

almost surely. Consider the process

$$S_t = S_0 \exp\left\{ \int_0^t \left(\mu_r - \frac{\sigma_r^2}{2}\right) \mathrm{d}r + \int_0^t \sigma_r \, \mathrm{d}B_r \right\}, \qquad t \in [0, T].$$

Then

$$X_t = \log S_t = \log S_0 + \int_0^t \left(\mu_r - \frac{\sigma_r^2}{2} \right) \mathrm{d}r + \int_0^t \sigma_r \, \mathrm{d}B_r, \qquad t \in [0, T],$$

is an Itô process with drift $\overline{\mu}_t$ *and vola* $\overline{\sigma}_t$ *given by*

$$\overline{\mu}_t = \mu_t - \frac{\sigma_t^2}{2}, \qquad \overline{\sigma}_t = \sigma_t.$$

Since $S_t = \mathrm{e}^{X_t}$*, Itô's formula shows that* S_t *is again an Itô process with drift*

$$\begin{aligned}\widetilde{\mu}_t &= f_x(X_t)\overline{\mu}_t + \frac{1}{2} f_{xx}(X_t)\sigma_t^2 \\ &= \mathrm{e}^{X_t} \left(\mu_t - \frac{\sigma_t^2}{2} \right) + \frac{\sigma_t^2}{2} S_t \\ &= S_t \mu_t,\end{aligned}$$

and vola

$$\widetilde{\sigma}_t = f_x(X_t)\sigma_t = S_t \sigma_t.$$

Thus, we obtain the representation

$$S_t = S_0 + \int_0^t \mu_r S_r \, \mathrm{d}r + \int_0^t \sigma_r S_r \, \mathrm{d}B_r, \qquad t \in [0, T].$$

In other words, $\{S_t\}$ *solves the stochastic differential equation*

$$\mathrm{d}S_t = \mu_t S_t \, \mathrm{d}t + \sigma_t S_t \, \mathrm{d}B_t.$$

Example 6.6.10 (Vasicek's Model for Interest Rates)
The process

$$X_t = \mathrm{e}^{-bt} X_0 + \frac{a}{b}(1 - \mathrm{e}^{-bt}) + \sigma \mathrm{e}^{-bt} \int_0^t \mathrm{e}^{bs} \, \mathrm{d}B_s, \quad t \in [0, T], \tag{6.8}$$

where X_0 *is a fixed starting value and* $a \geq 0$ *as well as* $b, \sigma > 0$ *are parameters, is often used to model interest rates. It depends on Brownian motion* B_t *via the process*

$$Z_t = \int_0^t \mathrm{e}^{bs} \, \mathrm{d}B_s, \quad t \in [0, T].$$

Let $f : [0, T] \times \mathbb{R} \to \mathbb{R}$ *be given by*

$$f(t, x) = \mathrm{e}^{-bt} X_0 + \frac{a}{b}(1 - \mathrm{e}^{-bt}) + \sigma \mathrm{e}^{-bt} x.$$

Then the partial derivatives are given by

$$\begin{aligned} f_t(s, x) &= -be^{-bs}X_0 + ae^{-bs} - \sigma be^{-bs}x, \\ f_x(s, x) &= \sigma e^{-bs}, \\ f_{xx}(s, x) &= 0. \end{aligned}$$

$X_t = f(t, Z_t)$ *is an Itô process with drift*

$$\widetilde{\mu}_s = f_t(s, Z_s) + \frac{1}{2} f_{xx}(s, Z_s)e^{2bs} = a - bf(s, Z_s) = a - bX_s$$

and volatility

$$\widetilde{\sigma}_s = f_x(s, Z_s)e^{bs} = \sigma.$$

Hence,

$$X_t = X_0 + \int_0^t (a - bX_s)\,ds + \sigma B_t,$$

i.e. $dX_t = (a - bX_t)\,dt + \sigma\,dB_t$. *That equation is a stochastic differential equation where* X_t *appears at both sides of the equation. We may understand Equation (6.8) as an explicit solution of the above differential equation. The latter allows for an intuitive interpretation of the model. The drift term depends on the level. Compared to the geometric Brownian motion, where the drift is constant, the correction term* $-bX_t$ *is present if* $b > 0$. *High values of* X_t *lead to a reduced local trend. The drift is positive for* $X_t < a/b$ *and negative for* $X_t > a/b$.

For $a = 0$ we obtain the **Langevin stochastic differential equation**

$$dX_t = -bX_t\,dt + \sigma dB_t,$$

which is solved by the **Ornstein–Uhlenbeck process**

$$X_t = e^{-bt}X_0 + \sigma e^{-bt} \int_0^t e^{bs}\,dB_s, \qquad t \geq 0.$$

Clearly, $\{X_t\}$ is a Gaussian process with mean $E(X_t) = e^{-bt}X_0 \to 0$, as $t \to \infty$, if $b > 0$. To calculate the autocovariance function,

$$\gamma_X(s, t) = \text{Cov}(X_s, X_t), \qquad s \leq t,$$

we may assume $X_0 = 0$. Let us approximate X_s and X_t by sums

$$\begin{aligned} X_s^{(n)} &= \sigma e^{-bs} \sum_{i=1}^{n} e^{bs_{i-1}}[B_{s_i} - B_{s_{i-1}}], \\ X_t^{(n)} &= \sigma e^{-bt} \sum_{i=1}^{n+m} e^{bs_{i-1}}[B_{s_i} - B_{s_{i-1}}], \end{aligned}$$

where $s_0 < \cdots < s_{n+m}$ is a partition of $[0, t]$ and $s_0 < \cdots s_n$ a partition of $[0, s]$. Clearly, those approximations converge to X_s and X_t, respectively, in L_2, as $n \to \infty$. It holds that

$$\gamma_X(s,t) = \lim_{n\to\infty} \operatorname{Cov}\left(X_s^{(n)}, X_t^{(n)}\right),$$

by continuity of the inner product. But

$$\begin{aligned}\operatorname{Cov}\left(X_s^{(n)}, X_t^{(n)}\right) &= \sigma^2 \mathrm{e}^{-b(s+t)} \sum_{i=1}^{n} \mathrm{e}^{2bs_{i-1}}(s_i - s_{i-1}) \\ &\to_{n\to\infty} \sigma^2 \mathrm{e}^{-b(s+t)} \int_0^t \mathrm{e}^{2bs}\,\mathrm{d}s \\ &= \frac{\sigma^2}{2b}\left(\mathrm{e}^{-b(s-t)} - \mathrm{e}^{-b(s+t)}\right).\end{aligned}$$

In particular,

$$\operatorname{Var}(X_t) = \frac{\sigma^2}{2b}(1 - \mathrm{e}^{-2bt}) \to \frac{\sigma^2}{2b},$$

as $t \to \infty$. Since $\{X_t\}$ is Gaussian, the candidate for a stationary solution, which is then also strictly stationary, corresponds to the random initial condition

$$X_0 \sim N\left(0, \frac{\sigma^2}{2b}\right), \tag{6.9}$$

independent from $\{B_t : t \geq 0\}$. Then $\operatorname{Cov}(\mathrm{e}^{-bs}X_0, \mathrm{e}^{-bt}X_0) = \frac{\sigma^2}{2b}\mathrm{e}^{-b(s+t)}$, leading to

$$\operatorname{Cov}(X_s, X_t) = \frac{\sigma^2}{2b}\mathrm{e}^{-b|t-s|}, \qquad s, t \geq 0, \tag{6.10}$$

which is a function of $|t-s|$ thus establishing the existence of a stationary solution.

Proposition 6.6.11 *The Langevin stochastic differential equation attains a strictly stationary solution given by the initial value (6.9). That solution has mean zero and autocovariance function (6.10).*

6.7 Diffusion processes and ergodicity

A solution $\{X_t\}$ of a stochastic differential equation of the form

$$\mathrm{d}X_t = \mu(X_t)\,\mathrm{d}t + \sigma(X_t)\,\mathrm{d}B_t, \quad , X_0 = x_0, \qquad t \geq 0,$$

is called **Itô diffusion**. Up to now, we have calculated the solutions of such equations by using Itô's formula, and the solutions were (automatically) functions of the underlying Brownian motion and of the coefficient functions $\mu(x)$ and $\sigma(x)$. In general, a solution with these properties is called a **strong solution**. A strong solution exists, if the functions $\mu(x)$ and $\sigma(x)$ satisfy a global Lipschitz condition

$$|\mu(x) - \mu(y)| + |\sigma(x) - \sigma(y)| \leq L|x - y|,$$

for some constant L. Such a solution may be stationary, and the Ornstein–Uhlenbeck provides an example.

A diffusion is called **ergodic** if there exists some measure π, called an **invariant (probability) measure**, such that

$$\frac{1}{t}\int_0^t h(X_s)\,\mathrm{d}s = \int h(x)\,\mathrm{d}\pi(x)$$

with probability one for all functions h with $\int |h(x)|\,\mathrm{d}\pi(x) < \infty$. Sufficient conditions for a strong solution to be ergodic are as follows: Define the scale density by

$$s(y) = \exp\left(-2\int_0^y \frac{\mu(x)}{\sigma^2(x)}\,\mathrm{d}x\right), \qquad -\infty < y < \infty,$$

and assume that the scale function

$$S(x) = \int_0^x s(y)\,\mathrm{d}y,$$

satisfies

$$S(x) \to \pm\infty,$$

as $x \to \pm\infty$. Introduce the norming constant

$$K = \int \frac{1}{\sigma^2(x)s(x)}\,\mathrm{d}x < \infty.$$

Then, the unique stationary measure is given by the **invariant density**

$$\mathrm{d}\mu(x) = \frac{1}{K\sigma^2(x)}\exp\left(2\int_0^x \frac{\mu(t)}{\sigma^2(t)}\,\mathrm{d}t\right)\,\mathrm{d}x.$$

Further, if the initial condition is chosen as

$$X_0 \sim \mu,$$

then the corresponding strong solution is strictly stationary. In this situation, it is common to denote that strictly stationary and ergodic solution by $\{X_t\}$.

Example 6.7.1 *For the Ornstein–Uhlenbeck process we have identified the unique stationary measure as a normal distribution with mean zero and variance* $\frac{\sigma^2}{2b}$*, that is* μ *is given by*

$$\mu((a, b]) = \int_a^b \varphi_{(0, \frac{\sigma^2}{2b})}(x)\,\mathrm{d}x, \qquad a \leq b.$$

The geometric Brownian motion, the Ornstein–Uhlenbeck process as well as the CIR model appear as special cases of the family of stochastic differential equations

$$\mathrm{d}X_t = (\alpha + \beta X_t)\,\mathrm{d}t + \sigma X_t^{\gamma}\,\mathrm{d}B_t,$$

parametrized by α, β, σ and γ, known as the Chan–Karolyi–Longstaff–Sanders (CKLS) model in finance.

Here is a list of special cases and how they are referred to.

(i) Merton: $\mathrm{d}X_t = \alpha \mathrm{d}t + \sigma \mathrm{d}B_t$;

(ii) Vasicek: $\mathrm{d}X_t = (\alpha + \beta X_t)\mathrm{d}t + \sigma \mathrm{d}B_t$;

(iii) CIR square root: $\mathrm{d}X_t = (\alpha + \beta X_t)\mathrm{d}t + \sigma\sqrt{X_t}\mathrm{d}B_t$;

(iv) Brennon-Schwartz: $\mathrm{d}X_t = (\alpha + \beta X_t)\mathrm{d}t + \sigma X_t \mathrm{d}B_t$;

(v) CIR variable rate: $\mathrm{d}X_t = \sigma X_t^{3/2}\mathrm{d}B_t$;

(vi) CEV (constant elasticity of variance): $\mathrm{d}X_t = \beta X_t \mathrm{d}t + \sigma X_t^{\gamma}\mathrm{d}B_t$.

6.8 Numerical approximations and statistical estimation

Consider a time-inhomogeneous Itô diffusion

$$X_t = \int_0^t \mu(s, X_s)\,\mathrm{d}s + \int_0^t \sigma(s, X_s)\,\mathrm{d}B_s, \qquad t \geq 0,$$

where $\mu(t, x)$ and $\sigma(t, x)$ are deterministic functions and B_t is a Brownian motion. In differential notation, we have

$$\mathrm{d}X_t = \mu(t, X_t)\mathrm{d}t + \sigma(t, X_t)\mathrm{d}B_t. \tag{6.11}$$

In general, explicit solutions are not available and one has to rely on numerical approximations of a solution $\{X_t\}$, that is a process in discrete time $t \in \{t_1, t_2, \dots\}$, which approximates the solution in an appropriate sense. Also, notice that in reality we cannot observe full trajectories and a continuous-time model is an idealized mathematical description. We can only observe a snapshot taken at discrete time points $t_1, t_2, \dots$. Let us assume that these time points are equidistant with time step Δ_n,

$$t_i = t_{ni} = t_0 + i\Delta_n, \qquad i = 1, 2, \dots,$$

and denote the corresponding sample (snapshot) by

$$X_i^{(n)} = X_{t_i}, \qquad i = 1, \dots, T,$$

of the process $\{X_t : t \geq 0\}$. Assuming that Δ_n is small justifies the approximations

$$\int_{t_{i-1}}^{t_i} \mu(s, X_s)\,\mathrm{d}s \approx \mu(t_{i-1}, X_{t_{i-1}})(t_i - t_{i-1})$$

and

$$\int_{t_{i-1}}^{t_i} \sigma(s, X_s)\,\mathrm{d}B_s \approx \sigma(t_{i-1}, X_{t_{i-1}})(B_{t_{i-1}} - B_{t_i}),$$

where $t_i - t_{i-1} = \Delta_n$ for all i and $B_{t_{i-1}} - B_{t_i} \sim \sqrt{\Delta}N(0, 1)$.

These approximations and the SDE (6.11) lead to the **Euler approximation scheme**

$$\Delta X_i^{(n)} = \mu(t_0 + (i-1)\Delta_n, X_{i-1}^{(n)})\Delta_n + \sigma(t_0 + (i-1)\Delta_n, X_{i-1}^{(n)})\sqrt{\Delta_n}\epsilon_i, \tag{6.12}$$

where

$$\Delta X_i^{(n)} = X_i^{(n)} - X_{i-1}^{(n)}$$

and ϵ_i are i.i.d. standard normal random variables.

For the special case of a time-homogeneous diffusion,

$$\mathrm{d}X_t = \mu(X_t)\mathrm{d}t + \sigma(X_t)\mathrm{d}B_t,$$

the Euler approximation scheme (6.12) is given by

$$\Delta X_i^{(n)} = \mu\left(X_{i-1}^{(n)}\right)\Delta_n + \sigma\left(X_{i-1}^{(n)}\right)\sqrt{\Delta_n}\epsilon_i, \tag{6.13}$$

where $X_{t_i}^{(n)} = X_i^{(n)}$ is the approximation at time $t_i = t_0 + i\Delta_n$. Between the grid points, one constructs $X_t^{(n)}$ by piecewise constant interpolation or linear approximation.

Observe that Equation (6.13) fits the framework of the general nonparametric regression model (3.1), which we will further discuss in Chapter 9.

The following results shows that the above approximation scheme has a strong convergence rate $\Delta_n^{1/2}$, under a Lipschitz and linear growth condition.

Theorem 6.8.1 *Suppose that $E|X_0|^2 < \infty$, $\|X_0 - X_0^{(n)}\|_2 = O(\delta^{1/2})$. If the Lipschitz condition*

$$|\mu(t,x) - \mu(t,y)| + |\sigma(t,x) - \sigma(t,y)| = K_1|x-y|$$

the linear growth condition

$$|a(t,x)| + |b(t,x)| \leq K_2(1+|x|),$$

as well as

$$|a(s,x) - a(t,x)| + |b(s,x) - b(t,x)| \leq K_3(1+|x|)|s-t|^{1/2}$$

hold true for all $s, t \in [0, T]$ and $x, y \in \mathbb{R}^d$, where the constants K_1, K_2, K_3 do not depend on Δ_n, then the Euler approximation $X_t^{(n)}$ satisfies the uniform estimate

$$\sup_{t\in[0,T]} E\left(\left|X_t - X_t^{(n)}\right|\right) = O(\Delta_n^{1/2}).$$

6.9 Notes and further reading

Various monographs and textbooks were valuable sources for this chapter. For non random integration with respect to functions of bounded variation we refer to Lang (1993). A classic reference to stochastic integration is Friedman (1975). A concise treatment of the L_2 theory of Itô integration can be found in Øksendal (2003). Klebaner (2005) provides a careful introduction to the issues raised in this chapter. The proof of Itô's formula is adopted from

Klenke (2008). The nice and intuitive treatment of Shreve (2004) inspired the present exposition as well. For a reader not being interested in a rigorous mathematical treatment where all results are shown, the comprehensive as well as demonstrative book of Grigoriu (2002) can be recommended for additional reading as well as the concise and didactic treatment of Mikosch (1998). An advanced text towards stochastic analysis is Protter (2005), from which various elegant arguments, e.g. the derivation of the integration by parts formula, are taken. Our discussion on how to generalize the stochastic integral to càdlàg processes also follows Protter's approach, see also Kurtz and Protter (1996) and Jacod and Shiryaev (2003). The CKLS model is due to Chan et al. (1992). Theorem 6.8.1 is (Kloeden and Platen, 1992, Theorem 10.2.2).

References

Chan K.C., Longstaff F.A. and Sanders A.B. (1992) An empirical comparison of alternative models of the short-term interest rate. *Journal of Finance* **47**(3), 1209–1227.

Friedman A. (1975) *Stochastic Differential Equations and Applications. Vol. 1.* Academic Press [Harcourt Brace Jovanovich Publishers], New York. Probability and Mathematical Statistics, Vol. 28.

Grigoriu M. (2002) *Stochastic Calculus: Applications in Science and Engineering*. Birkhäuser Boston Inc., Boston, MA.

Jacod J. and Shiryaev A.N. (2003) *Limit Theorems for Stochastic Processes*. vol. 288 of *Grundlehren der Mathematischen Wissenschaften [Fundamental Principles of Mathematical Sciences]* 2nd edn. Springer-Verlag, Berlin.

Klebaner F.C. (2005) *Introduction to Stochastic Calculus with Applications*. 2nd edn. Imperial College Press, London.

Klenke A. (2008) *Probability Theory*. Universitext. Springer-Verlag London Ltd., London. A comprehensive course, Translated from the 2006 German original.

Kloeden P.E. and Platen E. (1992) *Numerical Solution of Stochastic Differential Equations*. vol. 23 of *Applications of Mathematics (New York)*. Springer-Verlag, Berlin.

Kurtz T.G. and Protter P.E. (1996) *Weak Convergence of Stochastic Integrals and Differential Equations*. vol. 1627 of *Lecture Notes in Math.*. Springer, Berlin.

Lang S. (1993) *Real and Functional Analysis*. vol. 142 of *Graduate Texts in Mathematics*. 3rd edn. Springer-Verlag, New York.

Mikosch T. (1998) *Elementary Stochastic Calculus—with Finance in View*. vol. 6 of *Advanced Series on Statistical Science & Applied Probability*. World Scientific Publishing Co. Inc., River Edge, NJ.

Øksendal B. (2003) *Stochastic Differential Equations: An Introduction with Applications*. Universitext 6th edn. Springer-Verlag, Berlin.

Protter P.E. (2005) *Stochastic Integration and Differential Equations*. vol. 21 of *Stochastic Modelling and Applied Probability*. Springer-Verlag, Berlin. 2nd edition. Version 2.1, Corrected third printing.

Shreve S.E. (2004) *Stochastic Calculus for Finance. II: Continuous-time Models*. Springer Finance. Springer-Verlag, New York.

7

The Black–Scholes model

The Black–Scholes model assumes that one can either invest money in a stock whose future prices are random or deposit it at a bank account. Compared to the discrete-time models we have studied so far, the striking difference is that we now model the financial market by stochastic processes in continuous time and allow for continuous trading. Consequently, all relevant quantities such as the stock price or the trading strategy are now stochastic processes in continuous time as well.

We first discuss the classic model formulation, where the bank account pays a fixed and known interest rate. It turns out that the model is arbitrage-free and complete and the unique equivalent martingale measure constructed by virtue of Girsanov's theorem leads to the famous Black–Scholes formula for the arbitrage-free price of a European call option already discussed in the first chapter. The model can be extended quite easily to the case that the volatility of the stock price process is still deterministic but depends on time. Lastly, we briefly discuss the generalized Black–Scholes model, where the stock price volatility as well as the interest rate may be random processes.

7.1 The model and first properties

To simplify the exposition in the continuous-time world, we shall change our standard notation. Up to now, when using double indices, the first one was the time index and the second referred to the financial instrument. This means, S_{ti} denoted the time t price of the ith financial instrument. From now on, we use the notation S_{it}. The reason is that we have to work extensively with integrands with respect to time for a fixed instrument. In such a situation, it is more natural to write $\int_0^t S_{1u}\,du$ instead of $\int_0^t S_{u1}\,du$.

The model due to Black and Scholes is now as follows. We are given a bank account $\{S_{0t} : t \in [0, T]\}$ for depositing and lending money at a fixed interest rate r and a risky asset, say an exchange-traded stock, $\{S_{1t} : t \in [0, T]\}$ defined on a filtered probability space $(\Omega, \mathcal{F}, \{\mathcal{F}_t\}, P)$. As in previous chapters, P stands for the real probability measure. It is

Financial Statistics and Mathematical Finance: Methods, Models and Applications, First Edition. Ansgar Steland.

assumed that the processes $\{S_{0t}\}$ and $\{S_{1t}\}$ are given by

$$S_{0t} = \mathrm{e}^{rt}, \quad t \in [0, T],$$
$$S_{1t} = S_{10} \exp\left(\left(\mu - \frac{\sigma^2}{2}\right) t + \sigma B_t\right), \quad t \in [0, T],$$

for constants $\mu \in \mathbb{R}$, $\sigma > 0$ and S_{10}. Here, $\{B_t : t \in [0, T]\}$ is a standard Brownian motion under P. Notice that $\{S_{1t}\}$ is a geometric Brownian motion with drift parameter $\mu - \sigma^2/2$ and volatility σ. Clearly, the bank account satisfies the integral equation

$$S_{0t} = S_{00} + r \int_0^t S_{0u}\,\mathrm{d}u, \quad t \in [0, T],$$

or equivalently

$$\mathrm{d}S_{0t} = r S_{0t}\,\mathrm{d}t,$$

and has the explicit representation

$$S_{0t} = \mathrm{e}^{rt}, \qquad t \in [0, T].$$

According to Example 6.6.7, $\{S_{1t}\}$ attains the representation

$$S_{1t} = S_{10} + \mu \int_0^t S_{1u}\,\mathrm{d}u + \sigma \int_0^t S_{1u}\,\mathrm{d}B_u, \quad t \in [0, T],$$

i.e.

$$\mathrm{d}S_{1t} = \mu S_{1t}\,\mathrm{d}t + \sigma S_{1t}\,\mathrm{d}B_t.$$

We interpret the latter model equation as follows: During a small time interval $[t, t + \mathrm{d}t]$, the stock price change is approximately $\mu S_{1t}\mathrm{d}t$, a deterministic quantity since S_{1t} is known at time t, which is disturbed by a random noise term $\sigma S_{1t}\sqrt{\mathrm{d}t} Z$, where

$$Z = \frac{B_{t+\mathrm{d}t} - B_t}{\sqrt{\mathrm{d}t}} \sim N(0, \mathrm{d}t)$$

is independent of S_{1t}. Put differently, the relative price change, that is the return, is modeled by a linear trend disturbed by a normally distributed error term.

$$\frac{\mathrm{d}S_{1t}}{S_{1t}} \approx \mu \mathrm{d}t + \sigma\sqrt{\mathrm{d}t} Z.$$

It is common practice to measure time in years. Then μ is the mean rate of growth of the stock per year and σ the volatility per year. Let us briefly discuss how to estimate those parameters in practice. Notice that the log price of the stock,

$$\log S_{1t} = \log S_{10} + \left(\mu - \frac{\sigma^2}{2}\right) t + \sigma B_t$$

follows a normal distribution with mean $\log S_{10} + \left(\mu - \frac{\sigma^2}{2}\right)$ and variance $\sigma^2 t$, and the log returns corresponding to time lag Δ,

$$R_t = R_{t\Delta} = \log S_{1,t+\Delta} - \log S_{1t} = \left(\mu - \frac{\sigma^2}{2}\right)\Delta + \sigma(B_{t+\Delta} - B_t)$$

are normal with mean $\left(\mu - \frac{\sigma^2}{2}\right)\Delta$ and variance $\Delta\sigma^2$. When sampling the process at an equidistant grid $t_i = i\Delta, i = 1, \dots, n$, the corresponding log returns $R_1, \dots R_n$, are independent and identically normal with those parameters. Hence, the volatility parameter σ can be estimated by

$$\widehat{\sigma}_n = \frac{\mathrm{sd}_R}{\sqrt{\Delta}}$$

where

$$\mathrm{sd}_R^2 = \frac{1}{n-1}\sum_{t=1}^{n}(R_t - \overline{R}_n)^2, \qquad \overline{R}_n = \frac{1}{n}\sum_{t=1}^{n} R_t,$$

and

$$\widehat{\mu}_n = \frac{1}{\Delta}\overline{R}_n + \frac{\mathrm{sd}_R^2}{2} \approx \frac{1}{\Delta}\overline{R}_n.$$

Figure 7.1 depicts simulated trajectories of the Black–Scholes model with parameters μ and σ estimated from a real price series, which is added to the plot for comparison.

Let us now consider the discounted stock price process

$$\begin{aligned} S_{1t}^* &= S_{0t}^{-1} S_{1t} \\ &= \mathrm{e}^{-rt} S_{1t} \\ &= S_{10}\exp\left(\left(\mu - r - \frac{\sigma^2}{2}\right)t + \sigma B_t\right). \end{aligned}$$

The question arises, whether $\{S_{1t}^*\}$ is an Itô process. Noting that S_{1t}^* is again a geometric Brownian motion, we immediately obtain the following result of Example 6.6.7.

Lemma 7.1.1 *The discounted stock price process $\{S_{1t}^*\}$ is an Itô process,*

$$S_{1t}^* = S_{10} + (\mu - r)\int_0^t S_{1u}^*\,\mathrm{d}u + \sigma\int_0^t S_{1u}^*\,\mathrm{d}B_u,$$

or equivalently,

$$\mathrm{d}S_{1t}^* = (\mu - r)S_{1t}^*\,\mathrm{d}t + \sigma S_{1t}^*\,\mathrm{d}B_t.$$

We shall allow that an investor can trade continuously until the time horizon T. However, we have to impose some regularity conditions on the continuous-time process describing such a trade. First, as in the discrete-time case, we will require that the number of shares held at

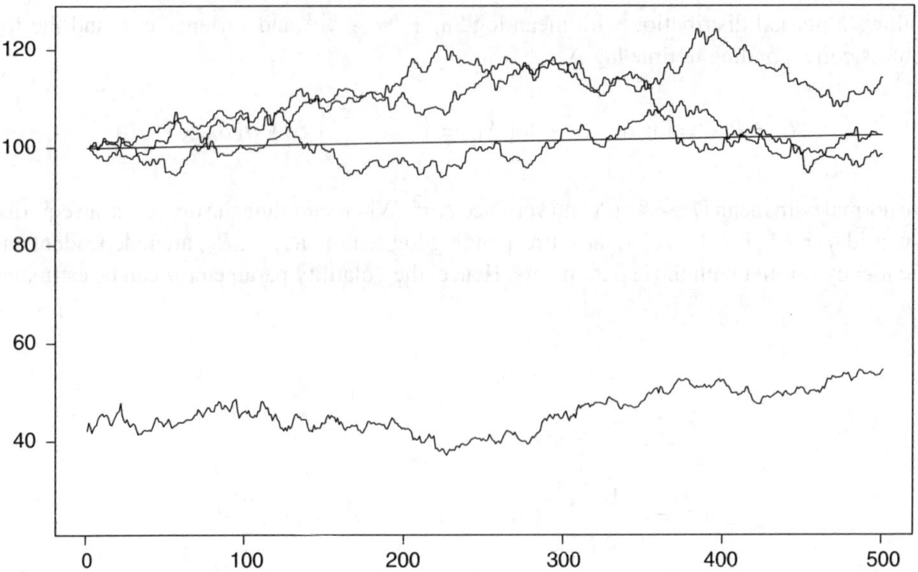

Figure 7.1 Simulated trajectories of the Black–Scholes model. The straight line is the bank account, the trajectories starting at 100 are three independent replications of the stock price. The parameters are estimated from daily log returns of Credit Suisse, whose price process has been added for comparison, data from R's free EU Stocks Data Set.

time t does not use future information. Secondly, we impose a condition ensuring that the associated (stochastic) integrals are well defined.

Definition 7.1.2 *A* **trading strategy (portfolio)** *is a bivariate stochastic process*

$$\varphi_t = (\alpha_t, \beta_t), \quad t \in [0, T],$$

with the following properties.

(i) $\varphi : [0, T] \times \Omega \to \mathbb{R}^2$ *is* $(\mathcal{B}([0, T]) \otimes \mathcal{F}_t)$*-measurable.*

(ii) φ *is* $\mathcal{F}_t$*-adapted.*

(iii) $\int_0^T \alpha_t^2 \, \mathrm{d}t < \infty$ *and* $\int_0^T |\beta_t| \, \mathrm{d}t < \infty$.

Given a trading strategy $\{(\alpha_t, \beta_t) : t \in [0, T]\}$, α_t is interpreted as the number of shares of the risky asset shared at time t, and β_t is the amount of money deposited at the bank.

Notice that condition (iii) guarantees that the stochastic integrals

$$\int_0^t \beta_u \, \mathrm{d}S_{0u} = r \int_0^t \beta_u S_{0u} \, \mathrm{d}u \quad \text{since } \left(S_{0u}' = r\mathrm{e}^{ru}\right)$$

and

$$\int_0^t \alpha_u \, \mathrm{d}S_{1u} = \mu \int_0^t \alpha_u S_{1u} \, \mathrm{d}u + \sigma \int_0^t \alpha_u S_{1u} \, \mathrm{d}B_u$$

are defined.

For clarity of the following definition, let us recall the interpretation of an Itô integral such as $\int_0^t \alpha_u \, \mathrm{d}S_{1u}$ in mathematical finance, which is crucial to understand the economic meaning of the calculus. It represents the value process of a time-dependent position α_u, the integrand, in the tradable financial instrument S_{1u}, the integrator. The position at time u is α_u. During the infinitesimal period of time $[u, u + \Delta]$ the position's change in value is given by $\alpha_u \, \mathrm{d}S_{1u}$ ($\approx \alpha_u(S_{1,u+\Delta} - S_{1u})$). The value process associated to a trading strategy $\varphi = (\alpha, \beta)$ is given by

$$V_t(\varphi) = \alpha_t S_{1t} + \beta_t S_{0t}, \quad t \in [0, T].$$

$V_0(\varphi)$ are the initial costs required to set up the trade $\{\varphi_t\}$.

Definition 7.1.3 *A trading strategy $\{\varphi_t\}$ is called* **self-financing**, *if the value process satisfies*

$$V_t(\varphi) = V_0(\varphi) + \int_0^t \alpha_u \, \mathrm{d}S_{1u} + \int_0^t \beta_u \, \mathrm{d}S_{0u}, \quad t \in [0, T],$$

or, equivalently, $\mathrm{d}V_t(\varphi) = \alpha_t \, \mathrm{d}S_{1t} + \beta_t \, \mathrm{d}S_{0t}$.

A self-financing strategy is characterized by the fact that the change in value of the portfolio is the result of the corresponding changes in value of the positions held in the stock and the bank account.

Lemma 7.1.4 *Let $\varphi = \{\varphi_t\}$ be a self-financing trading strategy. Then, the value process $\{V_t(\varphi)\}$ is an Itô process,*

$$V_t(\varphi) = V_0(\varphi) + \int_0^t \mu_V(u) \, \mathrm{d}u + \int_0^t \sigma_V(u) \, \mathrm{d}B_u, \quad t \in [0, T],$$

with drift

$$\mu_V(t) = \mu \alpha_t S_{1t} + r \beta_t S_{0t}$$

and volatility

$$\sigma_V(t) = \sigma \alpha_t S_{1t}.$$

Proof. Since φ_t is self-financing, $V_t = V_t(\varphi)$ satisfies

$$V_t = V_0 + \int_0^t \alpha_u \, \mathrm{d}S_{1u} + \int_0^t \beta_u \, \mathrm{d}S_{0u}, \tag{7.1}$$

where

$$S_{0t} = \mathrm{e}^{rt} = S_{00} + \int_0^t rS_{0u}\,\mathrm{d}u$$

and

$$S_{1t} = S_{10} + \int_0^t \mu S_{1u}\,\mathrm{d}u + \int_0^t \sigma S_{1u}\,\mathrm{d}B_u$$

are Itô processes. Let us calculate the stochastic integrals appearing in Equation (7.1). Noting that in both cases the integrator is an Itô integral, we may apply Theorem 6.6.5, yielding

$$\int_0^t \alpha_u\,\mathrm{d}S_{1u} = \int_0^t \alpha_u \mu S_{1u}\,\mathrm{d}u + \int_0^t \alpha_u \sigma S_{1u}\,\mathrm{d}B_u$$

and

$$\int_0^t \beta_u\,\mathrm{d}S_{0u} = r\int_0^t \beta_u \mathrm{e}^{ru}\,\mathrm{d}u.$$

The following result shows that the discounted value process of a self-financing strategy has a specific structure.

Proposition 7.1.5 *A trading strategy is self-financing, if and only if*

$$V_t^*(\varphi) = V_0(\varphi) + \int_0^t \alpha_u\,\mathrm{d}S_{1u}^*, \quad t \in [0, T],$$

where

$$\int_0^t \alpha_u\,\mathrm{d}S_{1u}^* = \int_0^t (\mu - r)\alpha_u S_{1u}^*\,\mathrm{d}u + \int_0^t \sigma\alpha_u S_{1u}^*\,\mathrm{d}B_u,$$

for $t \in [0, T]$.

Proof. We show that the stated representation is necessary. For brevity of notation let $V_t = V_t(\varphi)$ and $V_t^* = V_t^*(\varphi) = \mathrm{e}^{-rt}V_t$. Since e^{-rt} has bounded variation, the bracket process $[V, \mathrm{e}^{-r\bullet}]_t$ vanishes. Thus,

$$0 = [V, \mathrm{e}^{-r\bullet}]_t = V_u\mathrm{e}^{-ru}\big|_0^t - \int_0^t V_u\,\mathrm{d}(\mathrm{e}^{-ru}) - \int_0^t \mathrm{e}^{-ru}\,\mathrm{d}V_u.$$

Here, the first term equals $V_t^* - V_0^*$ yielding

$$V_t^* = V_0^* + \int_0^t V_u \, \mathrm{d}\big(\mathrm{e}^{-ru}\big) + \int_0^t \mathrm{e}^{-ru} \, \mathrm{d}V_u.$$

Let us calculate the integrals at the right-hand side. For each $t \in [0, T]$ we have

$$\begin{aligned}\int_0^t \mathrm{e}^{-ru} \, \mathrm{d}V_u &= \int_0^t \mathrm{e}^{-ru} \mu_V(u) \, \mathrm{d}u + \int_0^t \mathrm{e}^{-ru} \sigma_V(u) \, \mathrm{d}B_u \\ &= \int_0^t (\mu \alpha_u S_{1u}^* + r\beta_u) \, \mathrm{d}u + \int_0^t \sigma \alpha_u S_{1u}^* \, \mathrm{d}B_u.\end{aligned}$$

Further, since $V_u = \alpha_u S_{1u} + \beta_u S_{0u}$ and $S_{0u} = \mathrm{e}^{ru}$

$$\begin{aligned}\int_0^t V_u \, d\left(\mathrm{e}^{-ru}\right) &= -\int_0^t V_u r \mathrm{e}^{-ru} \, \mathrm{d}u \\ &= -\int_0^t (r\alpha_u S_{1u}^* + r\beta_u) \, \mathrm{d}u.\end{aligned}$$

Putting things together, we arrive at

$$V_t^* - V_0^* = \int_0^t (\mu - r) \alpha_u S_{1u}^* \, \mathrm{d}u + \int \sigma \alpha_u S_{1u}^* \, \mathrm{d}B_u,$$

for any $t \in [0, T]$, which completes the proof.

7.2 Girsanov's theorem

The Girsanov theorem provides the means to change the real probability measure P in such a way that the discounted stock price process becomes a martingale. Since in the Black–Scholes model the log stock price is driven by a Brownian motion plus a linear drift, the first – and crucial - step is to solve the problem for such a process.

We need some preliminaries. Suppose $L \geq 0$ is a random variable on a probability space $(\Omega, \mathcal{F}, P)$ with $E(L) = 1$. Then we can define a new probability measure Q by using L as the Radon–Nikodym derivative $\frac{\mathrm{d}Q}{\mathrm{d}P}$, i.e.

$$Q(A) = \int_A L \, \mathrm{d}P, \qquad A \in \mathcal{F}. \tag{7.2}$$

Any probability measure defined by a P-density is absolutely continuous with respect to P. If, in addition, L is positive, i.e. $L(\omega) > 0$ for all $\omega \in \Omega$, then Q is equivalent to P.

Lemma 7.2.1 *Suppose that* $X : (\Omega, \mathcal{F}, P) \to (\mathbb{R}, \mathcal{B})$ *follows a* $N(\mu, \sigma^2)$*-distribution,* $\mu \in \mathbb{R}$*,* $\sigma^2 > 0$*. Define*

$$L(\omega) = \exp\left(-\frac{\mu}{\sigma^2}X(\omega) + \frac{\mu^2}{2\sigma^2}\right), \qquad \omega \in \Omega.$$

Then $E(L) = 1$ *and under the probability measure* $\mathrm{d}Q = L\,\mathrm{d}P$*, given by Equation (7.2),* X *follows a* $N(0, \sigma^2)$ *distribution.*

Proof. Clearly, $L > 0$. The proof of $E(L) = 1$ is left to the reader. Let us calculate the characteristic function of X under Q, in order to show that it coincides with the characteristic function of a $N(0, \sigma^2)$ distribution. For any t we have

$$\begin{aligned}
E_Q\left(\mathrm{e}^{itX}\right) &= E_P\left(L\mathrm{e}^{itX}\right) \\
&= E_P \exp\left(-\frac{\mu}{\sigma^2}X + \frac{\mu^2}{2\sigma^2} + itX\right) \\
&= \frac{1}{\sqrt{2\pi}\sigma}\int \exp\left(-\frac{2\mu}{2\sigma^2}x + \frac{\mu^2}{2\sigma^2} + itx - \frac{(x-\mu)^2}{2\sigma^2}\right)\mathrm{d}x \\
&= \frac{1}{\sqrt{2\pi}\sigma}\int \exp\left(-\frac{x^2}{2\sigma^2} + itx\right)\mathrm{d}x \\
&= E_P(\mathrm{e}^{itY}),
\end{aligned}$$

if $Y \sim N(0, \sigma^2)$ under P.

Let $\{B_t : t \in [0, T]\}$ be a standard Brownian motion under P, as appearing in the definition of the stock price process. By Corollary 6.6.8, for any fixed $m \in \mathbb{R}$ the process

$$L_t = \exp\left(-mB_t - \frac{m^2}{2}t\right), \qquad t \in [0, T], \tag{7.3}$$

is a positive martingale and $E(L_t) = 1$ holds true. Thus, we may use L_T as a P-density to define an equivalent probability measure by Equation (7.2).

Lemma 7.2.2 *The probability measure* Q *defined by* $\mathrm{d}Q = L_T\,\mathrm{d}P$*, where* $\{L_t : t \in [0, T]\}$ *is given by* (7.3) *attains the* P*-density* L_t *with respect to the probability space* $(\Omega, \mathcal{F}_t, P)$*.*

Proof. We have to show that

$$Q(A) = \int_A L_t\,\mathrm{d}P \quad \text{for all } A \in \mathcal{F}_t.$$

Let $A \in \mathcal{F}_t$. Then

$$\begin{aligned}
Q(A) &= E(\mathbf{1}_A L_T) \\
&= E(E(\mathbf{1}_A L_T|\mathcal{F}_t)) \\
&= E(\mathbf{1}_A E(L_T|\mathcal{F}_t)) \\
&= E(\mathbf{1}_A L_t),
\end{aligned}$$

which proves the assertion.

Lemma 7.2.3 *Let $\mathcal{A} \subset \mathcal{F}$ be a sub-σ-field. Suppose that X is a random variable with*

$$E\left(\mathrm{e}^{itX}|\mathcal{A}\right) = \mathrm{e}^{-t^2\sigma^2/2}, \quad t \in \mathbb{R}.$$

Then X is independent of $\mathcal{A}$ and $X \sim N(0, \sigma^2)$.

Proof. Let $A \in \mathcal{A}$ and notice that

$$\begin{aligned} E\left(\mathbf{1}_A \mathrm{e}^{itX}\right) &= E\left(\mathbf{1}_A E\left(\mathrm{e}^{itX}|\mathcal{A}\right)\right) \\ &= E\left(\mathbf{1}_A \mathrm{e}^{-t^2\sigma^2/2}\right) \\ &= P(A)\mathrm{e}^{-t^2\sigma^2/2} \end{aligned}$$

for all $t \in \mathbb{R}$. In particular, the characteristic function φ_X of X is given by $\varphi_X(t) = \mathrm{e}^{-t^2\sigma^2/2}$, $t \in \mathbb{R}$, (put $A = \Omega$), such that $X \sim N(0, \sigma^2)$. By assumption, the conditional distribution of X given $A \in \mathcal{A}$ is also $N(0, \sigma^2)$, i.e.

$$P(X \leq x|A) = \Phi\left(\frac{x}{\sigma}\right), \quad x \in \mathbb{R}.$$

It follows that for all $x \in \mathbb{R}$ and $A \in \mathcal{F}$

$$\begin{aligned} P(\{X \leq x\} \cap A) &= P(X \leq x|A) \cdot P(A) \\ &= \Phi\left(\frac{x}{\sigma}\right) P(A). \end{aligned}$$

Since the intervals $(-\infty, x]$, $x \in \mathbb{R}$, generate the Borel σ-field $\mathcal{B}$, we may conclude that X and $\mathcal{A}$ are independent.

In what follows, we make use of the fact that the formula

$$\varphi_X(t) = E\left(\mathrm{e}^{\imath tX}\right) = \mathrm{e}^{\imath t\mu - \sigma^2 t^2/2}, \quad t \in \mathbb{R},$$

for the characteristic function of a random variable $X \sim N(\mu, \sigma^2)$, $\mu \in \mathbb{R}$, $\sigma^2 > 0$, also holds true for complex arguments. This can be seen as follows. The function $\Phi_X(z) = \mathrm{e}^{\imath z\mu - \sigma^2 z^2/2}$, $z \in \mathbb{C}$, is analytic, agrees with φ_X for real numbers and all derivatives of Φ_X and φ_X exist. Hence, all algebraic and absolute moments

$$\alpha_k = EX^k, \quad \beta_k = E|X|^k, \quad k = 0, 1, 2, \ldots$$

exist and Φ_X admits the series expansion

$$\Phi_X(z) = \sum_{k=0}^{\infty} \frac{\imath^k \alpha_k}{k!} z^k, \quad z \in \mathbb{C}.$$

It also holds true that the series

$$\sum_{k=0}^{\infty} \beta_k \frac{z^k}{k!}, \quad z \in \mathbb{C},$$

converges. To verify this, it suffices to establish the following upper bound for β_{2k-1}, $k \in \mathbb{N}$:

$$\frac{\beta_{2k-1}}{(2k-1)!} \leq \frac{1}{2}\left(\frac{\alpha_{2k}}{(2k)!}(2k) + \frac{\alpha_{2k-2}}{(2k-2)!}\right).$$

But this follows from the inequality

$$\left|x^{2k-1}\right| \leq \frac{1}{2}\left(x^{2k} + x^{2k-2}\right), \quad k = 1, 2, \ldots,$$

which can be shown by induction and implies

$$\beta_{2k-1} = E\left|X^{2k-1}\right| \leq \frac{1}{2}\left(E\left(X^{2k}\right) + E\left(X^{2k-2}\right)\right) = \frac{1}{2}\left(\alpha_{2k} + \alpha_{2k-2}\right).$$

It follows that for any $A > 0$

$$\begin{aligned}
\sum_{k=0}^{\infty} \frac{\beta_k}{k!}|y|^k &= \sum_{k=0}^{\infty} \frac{|y|^k}{k!} E|X|^k \\
&\geq \sum_{k=0}^{\infty} \frac{|y|^k}{k!} \int_{-A}^{A} |x|^k \, \mathrm{d}F_X(x) \\
&= \int_{-A}^{A} \sum_{k=0}^{\infty} \frac{|y|^k}{k!} |x|^k \, \mathrm{d}F_X(x) \\
&= \int_{-A}^{A} \mathrm{e}^{|xy|} \, \mathrm{d}F_X(x).
\end{aligned}$$

Therefore, the integral $\int_{-\infty}^{\infty} \mathrm{e}^{|xy|} \, \mathrm{d}F_X(x)$ exists, which implies that the integral

$$\int_{-\infty}^{\infty} \mathrm{e}^{\imath z x} \, \mathrm{d}F_X(x), \quad z \in \mathbb{C},$$

converges and agrees with $\varphi(z)$ if z is real. But then it must agree with $\varphi(z)$ for all $z \in \mathbb{C}$, such that

$$E\left(\mathrm{e}^{\imath z X}\right) = \mathrm{e}^{\imath z \mu - \sigma^2 z^2/2}, \quad z \in \mathbb{C}.$$

We are now in a position to prove the following version of Girsanov's theorem.

Theorem 7.2.4 (GIRSANOV) *Let $\{B_t : t \in [0, T]\}$ be a standard Brownian motion under P. If Q is defined by $\mathrm{d}Q = L_T \, \mathrm{d}P$ with L_T as in Equation* (7.3)*, then*

$$B_t' = B_t + mt, \quad t \in [0, T],$$

is a standard Brownian motion under Q.

Proof.
We have, since $B'_t - B'_s$ is independent from $\mathcal{F}_s$ and $A \in \mathcal{F}_s$,

$$\begin{aligned} E\left(\mathbf{1}_A e^{\imath u(B'_t - B'_s)} L_t\right) &= E\left(\mathbf{1}_A e^{\imath u(B'_t - B'_s)} L_t L_s^{-1} L_s\right) \\ &= E\left(\mathbf{1}_A L_s E\left(e^{\imath u(B'_t - B'_s)} L_t L_s^{-1} | \mathcal{F}_s\right)\right) \\ &= E\left(\mathbf{1}_A L_s\right) E\left(e^{\imath u(B'_t - B'_s)} L_t L_s^{-1}\right). \end{aligned}$$

Let us calculate the second factor

$$\begin{aligned} E\left(e^{\imath u(B'_t - B'_s)} L_t L_s^{-1}\right) &= E\left(\exp\left\{\imath u(B_t - B_s) + \imath u m(t-s) - m(B_t - B_s) - \frac{m^2}{2}(t-s)\right\}\right) \\ &= E\left(\exp\left\{(\imath u - m)(B_t - B_s)\right\} \exp\left\{\imath u m(t-s) - \frac{m^2}{2}(t-s)\right\}\right). \end{aligned}$$

We aim at calculating the first factor via the characteristic function of $B_t - B_s$, which is given by $u \mapsto e^{-u^2(t-s)/2}$ for $u \in \mathbb{C}$.

$$\begin{aligned} E\exp\left\{(\imath u - m)(B_t - B_s)\right\} &= E\exp\left\{\imath(u + im)(B_t - B_s)\right\} \\ &= e^{-(u+\imath m)^2(t-s)/2}. \end{aligned}$$

Hence,

$$\begin{aligned} E\left(e^{\imath u(B'_t - B'_s)} L_t L_s^{-1}\right) &= \exp\left\{-(u + \imath m)^2(t-s)/2 + \imath u m(t-s) - \frac{m^2}{2}(t-s)\right\} \\ &= \exp\left\{(t-s)\left[-\frac{u^2}{2} - \imath u m + \frac{m^2}{2} + \imath u m - \frac{m^2}{2}\right]\right\} \\ &= \exp\left\{-u^2(t-s)/2\right\}, \end{aligned}$$

which completes the proof.

7.3 Equivalent martingale measure

We are now in a position to determine an equivalent martingale measure P^* such that the discounted stock price process is a P^*-martingale. The discounted stock price is

$$S^*_{1t} = S_{1t} e^{-rt} = S_{10} \exp\left((\mu - r)t - \frac{\sigma^2}{2} t + \sigma B_t\right). \tag{7.4}$$

The term $-\frac{\sigma^2}{2} t$ is the right trend we have to take into account to ensure that the geometric Brownian motion $\exp\left(\sigma W_t - \frac{\sigma^2}{2} t\right)$ inherits the martingale property from W_t. Obviously, we may write S^*_{1t} in this form, if we put

$$W_t = \frac{\mu - r}{\sigma} t + B_t.$$

W_t is not a martingale under P, but, by virtue of Girsanov's theorem, there exists an equivalent martingale measure P^* given by

$$P^*(A) = \int_A L_T \,\mathrm{d}P, \quad A \in \mathcal{F},$$

where

$$L_T = \exp\left(-\Theta B_T - \frac{\Theta^2}{2}T\right), \quad \Theta = \frac{\mu - r}{\sigma},$$

such that W_t is a martingale under P^*. Θ is called the **market price of risk**. The stock price S_{1t} is given by

$$S_{1t} = S_{10} \exp\left(\left(\mu - \frac{\sigma^2}{2}\right)t + \sigma B_t\right)$$

where B_t is a standard Brownian motion under P. In terms of $W_t = \frac{\mu - r}{\sigma}t + B_t$, we have

$$S_{1t} = S_{10} \exp\left(\left(r - \frac{\sigma^2}{2}\right)t + \sigma W_t\right),$$

where W_t is a standard Brownian motion under P^*. This means that for calculations concerning S_{1t} under P^* one may simply replace μ by the riskless interest rate r.

The discounted price process $S_{1t}^* = \exp\left(-\frac{\sigma^2}{2}t + \sigma W_t\right)$ can also be written in the form

$$S_{1t}^* = \int_0^t \sigma S_{1u}^* \,\mathrm{d}W_u, \quad t \in [0, T].$$

Next, consider the discounted value process. We have

$$V_t^* = V_0 + \int_0^t \alpha_u \sigma S_{1u}^* \,\mathrm{d}W_u, \quad t \in [0, T],$$

provided the Itô integral $\int_0^t \alpha_u \sigma S_{1u}^* \,\mathrm{d}W_u$ is defined. Let us restrict the class of trading strategies to ensure that this is the case. This means, in the following we only consider self-financing trading strategy satisfying

$$E^*\left(\int_0^T (\alpha_u S_{1u}^*)^2 \,\mathrm{d}u\right) < \infty.$$

7.4 Arbitrage-free pricing and hedging claims

We may now use the results obtained in the previous sections to price and hedge contingent claims in the Black–Scholes framework.

Let $C : (\Omega, \mathcal{F}, P) \to (\mathbb{R}, \mathcal{B})$ be a claim providing the payoff C at time T. The arbitrage-free time 0 price is

$$\pi(C) = E^*\left(\mathrm{e}^{-rT}C\right) = \int \mathrm{e}^{-rT}C(\omega)\,\mathrm{d}P(\omega).$$

In particular, consider a European call option $C = (S_T - K)^+$. We have to calculate

$$E^*\left(\mathrm{e}^{-rT}C\right) = E^*\left(\mathrm{e}^{-rT}\max(S_T - K, 0)\right),$$

where $\log S_T$ is log normal under P^*,

$$\log S_T \overset{P^*}{\sim} \left(r - \frac{\sigma^2}{2}\right)T + \sigma W_T \sim N\left(\left(r - \frac{\sigma^2}{2}\right)T, \sigma^2 T\right).$$

We are led to the calculations provided in detail in Section 1.5.5.

Consider a claim, i.e. a $\mathcal{F}_T$-measurable random variable. Notice that, by $\mathcal{F}_T$-measurability, the payment at maturity may depend on the whole trajectory of the price process thus including path-dependent derivatives.

Definition 7.4.1 *A claim is called* **replicable (attainable)** *if there is a proper trading strategy, $\varphi = \{\varphi_t : t \in [0, T]\}$, called a* **hedge**, *such that*

$$V_T(\varphi) = C.$$

Suppose we are given a claim C and have a predictable process $\{\alpha_t\}$ at our disposal such that

$$C^* = \mathrm{e}^{-rT}C = \alpha_0 + \int_0^T \alpha_u\,\mathrm{d}S^*_{1u}.$$

Then, one may define a self-financing trading strategy by choosing β_t such that $\varphi_t = (\alpha_t, \beta_t)$ satisfies

$$V_t^*(\varphi) = \alpha_t S^*_{1t} + \beta_t \mathrm{e}^{-rt} = \alpha_0 + \int_0^t \alpha_u\,\mathrm{d}S^*_{1u}.$$

This means, the hedge is financed via the bank account. The hedge replicates C, i.e. $V_T(\varphi) = C$.

Let P^* be the equivalent martingale measure and assume that $E\int_0^T \alpha_t^2\,\mathrm{d}t < \infty$. Then $\int_0^t \alpha_u\,\mathrm{d}S^*_{1u}$ is a P^*-martingale and we obtain

$$E^*\left(V_T^*(\varphi)\right) = \alpha_0.$$

Consequently, α_0 is the arbitrage-free price.

The following theorem provides a way to calculate the value process without requiring an explicit hedging strategy.

Theorem 7.4.2 *Let C_T be a claim with $E^*(C_T^2) < \infty$. If C_T is replicable, then its time t value is given by*

$$V_t = E^*\left(\mathrm{e}^{-r(T-t)} C_T | \mathcal{F}_t\right)$$

and, in particular, $V_0 = E^*\left(\mathrm{e}^{-rT} C_T\right)$ represents the fair price.

Proof. Let $\{\alpha_t\}$ be number of stocks of a replicating hedge such that

$$C_T^* = \alpha_0 + \int_0^T \alpha_u \, \mathrm{d}S_{1u}^*,$$

and let $\varphi_t = (\alpha_t, \beta_t)$ be the associated self-financing trading strategy constructed above. Then the corresponding value process $V_t^* = \alpha_0 + \int_0^t \alpha_u \, \mathrm{d}S_{1u}^*$, $t \in [0, T]$, is a P^*-martingale, since the stochastic integral has that property. This means,

$$V_t^* = E^*(V_T^* | \mathcal{F}_t) = E^*(C_T^* | \mathcal{F}_t)$$

holds true for all $t \in [0, T]$. Since $C_T^* = \mathrm{e}^{-rT} C_T$ and $V_t = \mathrm{e}^{rt} V_t^*$, we obtain

$$V_t = E^*(\mathrm{e}^{-r(T-t)} C_T | \mathcal{F}_t), \quad t \in [0, T].$$

Our next aim is to show that the Black–Scholes model is complete, i.e. any $\mathcal{F}_t$-measurable claim can be replicated. The proof is based on the following representation theorem for martingales in continuous time.

Theorem 7.4.3 (Representation Theorem) *Let $M = \{M_t : t \in [0, T]\}$ be a right-continuous P^*-martingale. Then there exists a $\left(\mathcal{B}_{[0,T]} \otimes \mathcal{F}_T\right)$-measurable adapted process $\eta = \{\eta_t : t \in [0, T]\}$, such that*

$$\int_0^T \eta_t^2 \, \mathrm{d}t < \infty$$

P^*-a.s. and

$$M_t = M_0 + \int_0^t \eta_s \, \mathrm{d}B_s^*, \quad t \in [0, T].$$

Theorem 7.4.4 (Completeness) *Let C be a $\mathcal{F}_T$-measurable random variable with $E^*|C| < \infty$. Then there exists a replicating hedge $\varphi = \{\varphi_t : t \in [0, T]\}$, i.e.*

$$V_T(\varphi) = C.$$

In addition, $V_t^*(\varphi) = E^*(C | \mathcal{F}_t)$, $t \in [0, T]$.

Proof. We have already shown that any replicating hedge satisfies

$$V_t^* = E^*(C^* | \mathcal{F}_t), \quad t \in [0, T].$$

This means that the right-hand side of this equation yields the discounted value process. Thus, we consider the P^*-martingale

$$M_t = E^*(C^*|\mathcal{F}_t), \quad t \in [0, T].$$

We may assume that M_t is continuous. The representation theorem yields the existence of an adapted process $\{\eta_t\}$ with $\int_0^T \eta_t^2 \, dt < \infty$, P^*-a.s., such that

$$M_t = M_0 + \int_0^t \eta_s \, dB_s^*.$$

To determine a self-financing replicating trading strategy, we make the following ansatz. If $\varphi_t = (\alpha_t, \beta_t)$ is such a self-financing hedge, then

$$V_t^* = V_0 + \int_0^t \alpha_u \, dS_{1u}^*$$

follows. Since

$$S_{1t}^* = \int_0^t \sigma S_{1u}^* \, dB_u^*$$

is an Itô process, we have

$$\int_0^t \alpha_u \, dS_{1u}^* = \int_0^t \alpha_u \sigma S_{1u}^* \, dB_u^*.$$

Let us equate $M_t - M_0 = \int_0^t \eta_u \, dB_u^*$ and $V_t^* - V_0 = \int_0^t \alpha_u \sigma S_{1u}^* \, dB_u^*$ and solve for α_u. A solution is given by

$$\alpha_t = \frac{\eta_t}{\sigma S_{1t}^*}, \quad t \in [0, T].$$

Put $\beta_t = M_t - \alpha_t S_{1t}^*$. Notice that both α_t and β_t are adapted. Now $\varphi_t = (\alpha_t, \beta_t)$ satisfies

$$V_t^*(\varphi) = \alpha_t S_{1t}^* + \beta_t = M_t, \quad t \in [0, T].$$

It remains to verify, that φ_t replicates C, is proper and self-financing. It is trivial to note that φ_t replicates C, since

$$V_T^*(\varphi) = M_T = E^*(C^*|\mathcal{F}_T) = C^*.$$

Further, φ_t is self-financing, if

$$V_t^*(\varphi) = V_0(\varphi) + \int_0^t \alpha_u \, dS_{1u}^*, \quad t \in [0, T].$$

Let us first check that the Itô integral $\int_0^t \alpha_u \, \mathrm{d}S_{1u}^*$ is well defined. We verify the sufficient condition

$$\int_0^T \alpha_u^2 \, du < \infty, \qquad P\text{-a.s.}$$

Since $S_{1t}^* > 0$, P^*-a.s., for all $t \in [0, T]$, we have $\inf_{t\in[0,T]} |S_{1t}^*| > 0$, since S_{1t}^* is a discounted stock price process. Hence

$$\begin{aligned}\int_0^T \alpha_u^2 \, du &= \int_0^T \left(\frac{\eta_u}{\sigma S_{1u}^*}\right)^2 du \\ &\leq \left(\sigma \cdot \inf_{t\in[0,T]} |S_{1t}^*|\right)^{-2} \int_0^T \eta_u^2 \, du < \infty,\end{aligned}$$

P^*-a.s. φ_t is self-financing, since $V_0(\varphi) = M_0$ and $\eta_u = \alpha_u \sigma S_{1u}^*$, such that

$$\begin{aligned}V_t^*(\varphi) &= M_t \\ &= V_0 + \int_0^t \eta_u \, \mathrm{d}B_u^* \\ &= V_0 + \int_0^t \alpha_u \sigma S_{1u}^* \, \mathrm{d}B_u^* \\ &= V_0 + \int_0^t \alpha_u \, \mathrm{d}S_{1u}^*,\end{aligned}$$

which ensures the self-financing property by Proposition 7.1.5. Finally, φ_t is admissible, since

$$\int_0^T (\alpha_u S_{1u}^*)^2 \, du = \int_0^T \frac{\eta_u^2}{\sigma^2} \, du < \infty.$$

7.5 The delta hedge

Let us assume that the value process of a derivative C,

$$V_t = \mathrm{e}^{rt} E^* \left(C\mathrm{e}^{-rT} | \mathcal{F}_t\right),$$

is a twice continuously differentiable function of t and S_{1t}, i.e.

$$V_t = V(t, S_{1t}), \quad t \in [0, T],$$

for some C^2 function $V : [0, T] \times \mathbb{R} \to \mathbb{R}$, $(t, x) \mapsto V(t, x)$, for $(t, x) \in [0, T] \times \mathbb{R}$. Denote the partial derivatives of V by $\frac{\partial V}{\partial t}$, $\frac{\partial V}{\partial x}$, etc. Let $\{\varphi_t\}$ be a self-financing trading strategy. Then, by Definition 7.1.3,

$$\begin{aligned} V_t(\varphi) &= V_0(\varphi) + \int_0^t \alpha_u \, \mathrm{d}S_{1u} + \int_0^t \beta_u \, \mathrm{d}S_{0u} \\ &= V_0(\varphi) + \int_0^t \alpha_u \, \mathrm{d}S_{1u} + \int_0^t \beta_u r \mathrm{e}^{ru} \, \mathrm{d}u. \end{aligned}$$

An application of Itô's formula to $V(t, S_{1t})$ yields the representation

$$\begin{aligned} V_t = V_0 &+ \int_0^t \left[\frac{\partial V}{\partial t}(u, S_{1u}) + \frac{\partial V}{\partial x}(u, S_{1u})\mu S_{1u} + \frac{1}{2}\frac{\partial^2 V}{\partial x^2}(u, S_{1u})\sigma^2 S_{1u}^2 \right] \mathrm{d}u \\ &+ \int_0^t \frac{\partial V}{\partial x}(u, S_{1u})\sigma S_{1u} \, \mathrm{d}B_u. \end{aligned}$$

Since $\mathrm{d}S_{1t} = \mu S_{1t} \, \mathrm{d}t + \sigma S_{1t} \, \mathrm{d}B_t$, the last expression can be written as

$$V_t = V_0 + \int_0^t \frac{\partial V}{\partial x}(u, S_{1u}) \, \mathrm{d}S_{1u} + \int_0^t \left[\frac{\partial V}{\partial t}(u, S_{1u}) + \frac{1}{2}\frac{\partial^2 V}{\partial x^2}(u, S_{1u})\sigma^2 S_{1u}^2 \right] \mathrm{d}u.$$

Equating the integrand leads to the **delta hedge**

$$\begin{aligned} \alpha_t &= \frac{\partial V}{\partial x}(t, S_{1t}), \\ \beta_t &= \left[\frac{\partial V}{\partial t}(t, S_{1t}) + \frac{1}{2}\frac{\partial^2 V}{\partial x^2}(t, S_{1t})\sigma^2 S_{1t}^2 \right] \Big/ \left(r\mathrm{e}^{rt} \right). \end{aligned}$$

If we plug these formulas in $V_t = \alpha_t S_{1t} + \beta_t S_{0t}$, we obtain

$$rV_t = r\frac{\partial V}{\partial x}(t, S_{1t})S_{1t} + \frac{\partial V}{\partial t}(t, S_{1t}) + \frac{1}{2}\frac{\partial^2 V}{\partial x^2}(t, S_{1t})\sigma^2 S_{1t}^2.$$

This means that the value process satisfies a partial differential equation.

7.6 Time-dependent volatility

So far we have assumed that the volatility of the stock price is a non random constant. However, real data from financial markets show that this is an unrealistic assumption. Let us now assume that the volatility is a function of t. Thus, taking the stochastic differential equation as a starting point, let us assume that the stock price S_{1t} follows the model

$$\mathrm{d}S_{1t} = \mu_t S_{1t} \, \mathrm{d}t + \sigma_t S_{1t} \, \mathrm{d}B_t,$$

for some deterministic function $\sigma : [0, T] \to [0, \infty)$ satisfying

$$\int_0^T \sigma_u^2 \, du < \infty,$$

ensuring that integrals such as $\int_0^t \sigma_u \, dB_u$, $t \in [0, T]$, are well defined, and an adapted drift process μ_t satisfying

$$\int_0^T |\mu_t| \, dt < \infty,$$

almost surely.

Solving the above equation means that we are seeking a solution of the integral equation

$$S_{1t} = \int_0^t \mu S_{1u} \, du + \int_0^t \sigma_t \, dB_t, \qquad t \in [0, T].$$

We have shown in Example 6.6.9 that the generalized Brownian motion solves that equation,

$$S_{1t} = S_{10} \exp\left(\int_0^t \left(\mu - \frac{\sigma_u^2}{2} \right) du + \int_0^t \sigma_u \, dB_u \right),$$

for $t \in [0, T]$, where S_{10} denotes the initial value. Here $\int_0^t \sigma_u \, dB_u$, $t \in [0, T]$, is a martingale under P. Notice that we may write

$$S_{1t} = S_{10} \exp\left(\left(\mu - \frac{\overline{\sigma^2}(0, t)}{2} \right) t + \int_0^t \sigma_u \, dB_u \right),$$

where for $0 \le s \le t \le T$

$$\overline{\sigma^2}(s, t) = \frac{1}{t - s} \int_s^t \sigma_u^2 \, du.$$

Now, the discounted price process is given by

$$S_t^* = S_{10} \exp\left(\left(\mu - r - \frac{\overline{\sigma^2}(0, t)}{2} \right) t + \int_0^t \sigma_u \, dB_u \right),$$

which has exactly the same form as for the case of constant volatility, see Equation (7.4), with σ^2 replaced by $\overline{\sigma^2}(0, t)$ and σB_u by $\int_0^t \sigma_u \, dB_u$. It follows that the price of a European call option within this model is given by the Black–Scholes formula with σ replaced by $\sqrt{\overline{\sigma^2}(0, T)}$.

The function σ_t can be inferred from implied volatilities as follows. Suppose we have determined at time t implied volatility $\sigma_{\text{imp}}(T)$ for maturity T by equating the Black–Scholes formula and option prices for a fixed strike price and maturity T. Then, one solves

$$\frac{1}{T - t} \int_t^T \sigma_u^2 \, du = \sigma_{\text{imp}}^2(T)$$

by differentiating both sides, which leads to the formula

$$\sigma_T^2 = 2\sigma_{\text{imp}}^2(T) \left(\sigma_{\text{imp}}^2 \right)'(T)(T - t) + \sigma_{\text{imp}}^2(T),$$

which holds true for any maturity T, thus revealing the functional form of time-dependent volatility.

7.7 The generalized Black–Scholes model

Let us now assume that the volatility σ_t is an adapted process, which is more realistic than working with a deterministic function. Again taking the stochastic differential equation as a starting point, let us assume that the stock price S_{1t} follows the model

$$dS_{1t} = \mu_t S_{1t}\,dt + \sigma_t S_{1t}\,dB_t, \tag{7.5}$$

for some (possibly random but adapted) function $\sigma : [0, T] \to [0, \infty)$ satisfying

$$\int_0^T \sigma_u^2\,du < \infty,$$

almost surely, ensuring that integrals such as $\int_0^t \sigma_u\,dB_u$, $t \in [0, T]$, are well defined and an adapted drift process μ_t satisfying

$$\int_0^T |\mu_t|\,dt < \infty,$$

almost surely. Recall our interpretation of the differential Equation (7.5): From t to $t + dt$, dt small, the stock price changes by the linear drift part $\mu_t S_{1t} dt$, which is known at time t, and is affected by a random disturbance $\sigma_t S_{1t}\sqrt{dt}Z$, with $Z \sim N(0, 1)$, which is proportional to level S_{1t}. In other words, Equation (7.5) models the relative change dS_{1t}/S_{1t} by

$$\frac{dS_{1t}}{S_{1t}} \approx \mu_t dt + \sqrt{dt}\sigma_t Z,$$

where μ_t and σ_t are known at time t.

The bank account (money-market) is modeled by a locally riskless bond that is driven by a riskless rate (short rate, instantaneous rate) r_t,

$$S_{0t} = r_t dt, \qquad t \in [0, T],$$

where

$$\int_0^T |r_u|\,du < \infty,$$

almost surely.

The explicit solutions of the above equations are as follows. By virtue of Example 6.6.9, the generalized Brownian motion solves the above stochastic differential equation,

$$S_{1t} = S_{10} \exp\left(\int_0^t \left(\mu_u - \frac{\sigma_u^2}{2} \right) du + \int_0^t \sigma_u\,dB_u \right),$$

for $t \in [0, T]$, where S_{10} denotes the initial stock price at time 0. For the bank account we obtain the explicit solution

$$S_{0t} = S_{00} \exp\left(\int_0^t r_u \, \mathrm{d}u\right), \qquad t \in [0, T].$$

where we may assume that $S_{00} = 1$.

To proceed, we need a general version of Girsanov's theorem.

Theorem 7.7.1 (GIRSANOV'S THEOREM)
Let $\{B_t\}$ be a Brownian motion under the real probability measure P and $\{\gamma_t\}$ be a $\mathcal{F}_t$-adapted process satisfying Novikov's condition

$$E\left[\exp\left(\frac{1}{2}\int_0^T \gamma_t^2 \, \mathrm{d}t\right)\right] < \infty.$$

Define

$$L_t = \exp\left(-\int_0^t \gamma_u \, \mathrm{d}B_u - \frac{1}{2}\int_0^t \gamma_u^2 \, \mathrm{d}u\right)$$

and let P^* be the probability measure defined by

$$P^*(A) = \int_A L_T(\omega) \, \mathrm{d}P(\omega), \qquad t \in [0, T].$$

Then, under the probability measure P^*, the process

$$B_t^* = B_t + \int_0^t \gamma_u \, \mathrm{d}u, \qquad t \in [0, T],$$

is a standard Brownian motion.

We have to determine an equivalent martingale measure such that the discounted stock price is a martingale. Let us make the ansatz

$$B_t^* = B_t + \int_0^t \gamma_u \, \mathrm{d}u, \qquad t \in [0, T].$$

The discounted stock price process is given by

$$\begin{aligned} S_{1t}^* &= S_{1t} \exp\left(-\int_0^t r_u \, \mathrm{d}u\right) \\ &= S_{10} \exp\left\{\int_0^t \left(\mu_u - r_u - \frac{\sigma_u^2}{2}\right) \mathrm{d}u + \int_0^t \sigma_u \, \mathrm{d}B_u\right\}, \end{aligned}$$

which is a solution of the stochastic differential equation

$$\mathrm{d}S_{1t}^* = (\mu_t - r_t) S_{1t}^* \, \mathrm{d}t + \sigma_t S_{1t}^* \, \mathrm{d}B_t,$$

that is,

$$S_{1t}^* = \int_0^t (\mu_u - r_u) S_{1u}^* \, du + \int_0^t \sigma_u S_{1u}^* \, dB_u.$$

Noting that B_t^* is an Itô process with drift γ_t and volatility 1, an application of the rule of integration given in Theorem 6.6.5 shows that

$$\int_0^t \sigma_u S_{1u}^* \, dB_u^* = \int_0^t \sigma_u \gamma_u S_{1u}^* \, du + \int_0^t \sigma_u S_{1u}^* \, dB_u.$$

Hence, we can rewrite S_{1t}^* in terms of B_t^*,

$$S_{1t}^* = \int_0^t (\mu_u - r_u - \gamma_u \sigma_u) \, S_{1u}^* \, du + \int_0^t \sigma_u S_{1u}^* \, dB_t^*.$$

We can eliminate the trend if we put

$$\gamma_t = \frac{\mu_t - r_t}{\sigma_t}, \qquad t \in [0, T],$$

such that S_{1t}^* is a martingale if B_t^* is a martingale. But having defined γ_t, this can be achieved by the probability measure P^* given in Girsanov's theorem. γ_t is the *market price of risk*.

One can now easily see that under P^* the undiscounted stock price S_{1t},

$$S_{1t} = S_{10} + \int_0^t \mu_u S_u \, du + \int_0^t \sigma_u S_u \, dB_t,$$

has a mean rate of return equal to the riskless interest rate. Indeed, the substitution

$$\int_0^t \sigma_u S_u \, dB_t = \int_0^t \sigma_u S_{1u} \, dB_u^* - \int_0^t \sigma_u \gamma_u S_{1u} \, du,$$

gives

$$\begin{aligned} S_{1t} &= S_{10} + \int_0^t (\mu_u + \sigma_u \gamma_u) S_u \, du + \int_0^t \sigma_u S_u \, dB_t \\ &= S_{10} + \int_0^t r_u S_u \, du + \int_0^t \sigma_u S_u \, dB_t. \end{aligned}$$

Notice that this change of measure from the real probability P to the risk-neutral one, P^*, only changes the drift but not the volatility.

7.8 Notes and further reading

To some extent, our presentation is inspired by the concise treatment of Williams (2006) and can be supplemented by the treatments of Etheridge (2002) and Shiryaev (1999). The classic reference for the representation theorem for square integrable martingales is Kunita and Watanabe (1967); also see (Karatzas and Shreve, 1991, Ch. 3, Th. 4.15). Further extensions can be found in Jacod and Shiryaev (2003). The elementary proof based on analytic characteristic function arguments of Girsanov's theorem is taken from Nualart (2011). Proofs for the general case presented in Theorem 7.7.1 can be found in many books, e.g. Revuz and Yor (1999). More on (analytic) characteristic functions can be found in Lukacs (1970). For a nice derivation of the general case we refer to Shreve (2004). The connection to partial differential equations is discussed in various books, e.g. Shreve (2004).

References

Etheridge A. (2002) *A Course in Financial Calculus*. Cambridge University Press, Cambridge.

Jacod J. and Shiryaev A.N. (2003) *Limit Theorems for Stochastic Processes*. vol. 288 of *Grundlehren der Mathematischen Wissenschaften [Fundamental Principles of Mathematical Sciences]* 2nd edn. Springer-Verlag, Berlin.

Karatzas I. and Shreve S.E. (1991) *Brownian Motion and Stochastic Calculus*. vol. 113 of *Graduate Texts in Mathematics* second edn. Springer-Verlag, New York.

Kunita H. and Watanabe S. (1967) On square integrable martingales. *Nagoya Math. J.* **30**, 209–245.

Lukacs E. (1970) *Characteristic Functions*. Hafner Publishing Co., New York. 2nd edition, revised and enlarged.

Nualart D. (2011) *Stochastic Calculus*. Lecture Notes.

Revuz D. and Yor M. (1999) *Continuous Martingales and Brownian Motion*. vol. 293 of *Grundlehren der Mathematischen Wissenschaften [Fundamental Principles of Mathematical Sciences]* 3rd edn. Springer-Verlag, Berlin.

Shiryaev A.N. (1999) *Essentials of Stochastic Finance: Facts, Models, Theory*. vol. 3 of *Advanced Series on Statistical Science & Applied Probability*. World Scientific Publishing Co. Inc., River Edge, NJ. Translated from the Russian manuscript by N. Kruzhilin.

Shreve S.E. (2004) *Stochastic Calculus for Finance. II: Continuous-time Models*. Springer Finance. Springer-Verlag, New York.

Williams R.J. (2006) *Introduction to the Mathematics of Finance*. vol. 72 of *Graduate Studies in Mathematics*. American Mathematical Society, Providence, RI.

8

Limit theory for discrete-time processes

We have already mentioned that limit theorems play a crucial role in the analysis of financial data, since often the knowledge about the underlying data generating processes is not sufficient to formulate appropriate parametric models. Further, the analysis of the method of estimation and inference of those parametric models also makes use of nonparametric or model-free concepts and results of the asymptotic distribution theory.

Our exposition starts with a law of large numbers for correlated time series justifying the common averaging procedures used in the construction of estimation methods, which rely directly or indirectly on that basic result. Whereas the law of large numbers asserts that

$$\overline{X}_T = \frac{1}{T}\sum_{t=1}^{T} X_t \stackrel{P}{\to} E(X_1),$$

as $T \to \infty$, even for dependent time series, the central limit theorem provides approximations of the scaled error $\sqrt{T}(\overline{X}_T - E(X_1))$. We discuss a general central limit theorem that asserts that

$$\sqrt{T}(\overline{X}_T - E(X_1)) \stackrel{d}{\to} N(0, \eta^2),$$

for some $\eta \in (0, \infty)$. Those two results ensure that those two fundamental probabilistic results are also valid for correlated series under weak regularity conditions.

Regression is a fundamental tool of statistical data analysis. In financial applications, the regressors are typically random. Thus, we discuss multiple linear regression with stochastic regressors under general conditions that also cover the case that one regresses a time series on its lagged values. Along our way, we introduce and apply some useful limit theorems for martingale differences.

Financial Statistics and Mathematical Finance: Methods, Models and Applications, First Edition. Ansgar Steland.

Then we proceed to nonparametric density estimation and nonparametric regression. For quite a long time, nonparametric methods have played a minor role in empirical work in finance, but the situation has changed drastically and nonparametrics is now ubiquituous. Thus, we provide detailed derivations of the main results.

The class of linear processes provides the right framework to work with ARMA(p, q)-processes. The *Beveridge–Nelson decomposition* provides an elegant tool to derive the central limit theorem for linear processes using rather elementary arguments. For this reason, a separate section is devoted to an exposition of this nice and useful result.

We then discuss α-mixing processes in some detail. Although the derivation of the related calculus is a little involved, measuring the degree of dependence by α-mixing coefficients is very intuitive and allows powerful results to be established without the need to assume that the underlying time is a linear process or even belongs to a parametric class.

The chapter closes with a discussion of the asymptotics of the sample autocovariances and the commonly used Newey–West estimator for the long-run variance parameter that appears in many asymptotic formula.

8.1 Limit theorems for correlated time series

Whereas martingale differences are uncorrelated, time series such as ARMA(p, q) processes have non vanishing autocovariances for lags $|h| > 1$. We will see that arithmetic averages still converge to their means, provided the autocovariances die out sufficiently fast, as the lag increases.

To understand the effect of correlations on probabilistic calculations, which is usually substantial and far from being negligible, let us consider the case of a stationary Gaussian time series

$$\mu = E(X_t), \qquad \sigma^2 = \text{Var}\,(X_t)$$

with autocovariances

$$\gamma(h) = E(X_t - \mu)(X_{t+|h|} - \mu), \qquad h \in \mathbb{Z}.$$

We assume that we can observe the process until time T leading to the sample $X_1, \ldots, X_T$. It is natural to estimate the unknown expectation μ by

$$\widehat{\mu}_T = \frac{1}{T} \sum_{t=1}^{T} X_t.$$

By stationarity, $E(\widehat{\mu}_T) = \mu$. But what about the variance? The summands X_t are not independent. So let us calculate $\text{Var}\,(\widehat{\mu}_T)$ explicitly.

$$\begin{aligned}\text{Var}\,(\widehat{\mu}_T) &= E(\widehat{\mu}_T - \mu)^2 \\ &= \frac{1}{T^2} \sum_{t=1}^{T} \sum_{s=1}^{T} \text{Cov}(X_s, X_t).\end{aligned}$$

The double sum sums up all elements of the symmetric $(T \times T)$-dimensional matrix with elements $\text{Cov}(X_s, X_t)$. We may also sum over all diagonals leading to the formula

$$\begin{aligned}\text{Var}\,(\widehat{\mu}_T) &= \frac{1}{T^2}\left\{2\sum_{k=1}^{T-1}(T-k)\gamma(k) + T\gamma(0)\right\}\\ &= \frac{1}{T}\sum_{|k|<T}\left(1-\frac{|k|}{T}\right)\gamma(k).\end{aligned}$$

An alternative expression is

$$\eta_T^2 = \text{Var}\,(\widehat{\mu}_T) = \frac{1}{T}\,\text{Var}\,(X_1) + \frac{2}{T^2}\sum_{k=1}^{T-1}(T-k)\gamma(k).$$

This is an exact calculation and we are led to the following exact distributional result.

Proposition 8.1.1 *For a Gaussian time series $\{X_t\}$ with autocovariances $\gamma(h)$ we have for all $T \in \mathbb{N}$*

$$\sqrt{T}(\widehat{\mu}_T - \mu) \sim N\left(0, \eta_T^2\right),$$

where

$$\eta_T^2 = \sum_{|k|<T}\left(1-\frac{|k|}{T}\right)\gamma(k).$$

The above proposition has a direct consequence: $\widehat{\mu}_T = \frac{1}{T}\sum_{t=1}^{T} X_t$ converges in probability to its expectation μ, i.e. the weak law of large numbers

$$\widehat{\mu}_T \xrightarrow{P} \mu,$$

as $T \to \infty$, holds with convergence rate $T^{1/2}$. Two questions arise:

(i) Does the the convergence also hold when the time series in non-Gaussian?

(ii) Does the distribution of $\sqrt{T}(\widehat{\mu}_T - \mu)$ converge to a normal law, at least under appropriate conditions on the dependencies?

The first question can be answered rather easily under the weak assumption that the autocorrelations tend to 0, as the lag increases. But the validity of the central limit theorem is much more involved and requires stronger assumptions on the dependencies of the time series. We will study several central limit theorems working under different types of assumptions, which will turn out to be very useful to address different model classes for dependent processes in discrete time.

Recall the notion of a Cesaro sum. If $\{\alpha_n : n \in \mathbb{N}\}$ is a sequence of real numbers with $\alpha_n \to \alpha \in \mathbb{R}$, as $n \to \infty$, then the Cesaro averages $\frac{1}{n}\sum_{i=1}^{n} \alpha_i$ converge as well to α, as $n \to \infty$. Put

$$\overline{X}_T = \frac{1}{T}\sum_{t=1}^{T} X_t, \qquad T \geq 1.$$

Theorem 8.1.2 (Law of Large Numbers)
Let $\{X_t\}$ be a weakly stationary time series with mean $\mu = E(X_1)$ and autocovariances $\gamma(k)$, $k \in \mathbb{Z}$.

(i) If $\lim_{k\to\infty} \gamma(k) = 0$, then

$$\text{Var}\,(\overline{X}_T) \to 0, \qquad T \to \infty,$$

such that

$$\overline{X}_T \stackrel{P,L_2}{\to} \mu,$$

as $T \to \infty$.

(ii) If

$$\sum_{k\in\mathbb{Z}} |\gamma(k)| < \infty,$$

then

$$\text{Var}\,(\sqrt{T}\,\overline{X}_T) = TE(\overline{X}_T - \mu)^2 \to \sum_{k\in\mathbb{Z}} \gamma(k),$$

as $T \to \infty$.

Proof.

(i) Notice that

$$\gamma(k) - \frac{|k|}{T}\gamma(k) \leq \max\{0, \gamma(k)\}.$$

Hence, the Cesaro average $\frac{1}{T}\sum_{k=1}^{T}\left(1 - \frac{|k|}{T}\right)\gamma(k)$ converges to 0, if $T \to \infty$, which in turn implies that

$$\lim_{T\to\infty} \text{Var}\,(\overline{X}_T) = 0.$$

Now the result follows from the Chebychev inequality: For any $\varepsilon > 0$,

$$P(|\overline{X}_T - \mu| > \varepsilon) \leq \frac{\text{Var}\,(\overline{X}_T)}{\varepsilon^2} \to 0,$$

as $T \to \infty$.

(ii) Next consider

$$T\ \mathrm{Var}\,(\overline{X}_T) = \sum_{|k|<T}\left(1-\frac{|k|}{T}\right)\gamma(k) = \int f_T(k)\,\mathrm{d}\nu(k),$$

where ν denotes the counting measure on $\mathbb{Z}$ and

$$f_T(k) = \left(1-\frac{|k|}{T}\right)\gamma(k)\mathbf{1}(|k|<T), \qquad k\in\mathbb{Z}.$$

f_T attains the majorant $f(k) = |\gamma(k)|$, $k\in\mathbb{Z}$, which is ν-integrable by assumption, since $\int |f|\mathrm{d}\nu = \sum_{k\in\mathbb{Z}}|\gamma(k)| < \infty$. Further, f_T converges pointwise. Indeed, for arbitrary but fixed $k\in\mathbb{Z}$,

$$f_T(k) = \left(1-\frac{|k|}{T}\right)\gamma(k)\mathbf{1}(|k|<T) \to \gamma(k) =: f(k).$$

Now by dominated convergence,

$$T\ \mathrm{Var}\,(\overline{X}_T) = \int f_T(k)\,\mathrm{d}\nu(k) \to \int f(k)\,\mathrm{d}\nu(k) = \sum_{k\in\mathbb{Z}}|\gamma(k)|,$$

as $T\to\infty$, which completes the proof.

Remark 8.1.3 *Notice that assertion (i) tells us that* Var $(\overline{X}_T)$ *converges to* 0*, which suffices to obtain the weak law of large numbers. Part (ii) of Theorem 8.1.2 is stronger and asserts the convergence of* T Var $(\overline{X}_T)$*, which coincides with* Var $(\sqrt{T}\overline{X}_T)$*, thus providing the existence of the asymptotic variance of the statistic* $\sqrt{T}(\widehat{\mu}_T-\mu)$*, which we expect to be normally distributed in large samples under certain conditions on the correlations.*

The limit

$$\eta^2 = \sum_{h\in\mathbb{Z}}\gamma(h) = \gamma(0) + 2\sum_{h=0}^{\infty}\gamma(h) \tag{8.1}$$

of η_T^2, that is the the asymptotic variance of the scaled sample mean, is called the **long-run variance**. Let us briefly consider an example.

Example 8.1.4 *Let* $\{X_t\}$ *be an* AR(1) *process with* AR *parameter* ρ *satisfying* $|\rho|<1$*, i.e.*

$$X_t - \mu = \rho(X_{t-1}-\mu) + \epsilon_t,$$

where the innovations are i.i.d. $N(0,\sigma^2)$*. Since*

$$\gamma(h) = \rho^{|h|}\frac{\sigma^2}{1-\rho^2},$$

the long-run variance is given by the limit of

$$\eta_T^2 = \frac{\sigma^2}{1-\rho^2} + 2\frac{\sigma^2}{1-\rho^2}\sum_{h=1}^{T-1}\rho^h = \frac{\sigma^2}{1-\rho^2} + 2\frac{\sigma^2}{1-\rho^2}\frac{1-\rho^T}{1-\rho}.$$

The above law of large numbers tells us that the temporal mean

$$\overline{X}_T(\omega) = \frac{1}{T}\sum_{t=1}^{T} X_t(\omega)$$

converges to the expectation, that is the (integral) weighted average over all ω

$$EX_t = \int X_t(\omega)\,\mathrm{d}P(\omega).$$

Such results are called **ergodic theorems** and a process $\{X_t\}$ satisfying an ergodic theorem is called **ergodic**. The weak law of large numbers can be weakened as follows.

Theorem 8.1.5 (ERGODIC THEOREM)
Any weakly stationary process $\{X_t\}$ is ergodic, that is

$$\overline{X}_T \stackrel{P}{\to} E(X_1),$$

as $T \to \infty$.

Suppose we are given a strictly stationary process $\{X_t : t \in \mathbb{Z}\}$ with $E(X_1) = 0$. Consider the natural filtration $\mathcal{F}_t = \sigma(X_s : s \leq t)$. Recall that conditional expectation $E(X_t|\mathcal{F}_s)$, $s \leq t$, is the L_2-optimal prediction for X_t and $\|E(X_t|\mathcal{F}_s)\|_2$, which usually decreases as $|s-t|$ increases, measures how well we can predict X_t using a measurable function of $X_s, X_{s-1}, \ldots$. As an example, let us consider a linear process

$$X_t = \sum_{i=0}^{\infty} \theta_i \epsilon_{t-i}, \qquad t \in \mathbb{Z},$$

with i.i.d.$(0,\sigma^2)$-error terms $\{\epsilon_t\}$, $\sigma > 0$. Then the optimal predictor of X_t given $X_s, X_{s-1}, \ldots$ is

$$E(X_t|\mathcal{F}_s) = \sum_{i=t-s}^{\infty} \theta_i \epsilon_{t-i}.$$

Its norm

$$\|E(X_t|\mathcal{F}_s)\|_2 = \sigma\sqrt{\sum_{i=t-s}^{\infty}\theta_i^2}.$$

converges to 0, as $t - s \to \infty$. The following general central limit theorem establishes the asymptotic normality of $\frac{1}{\sqrt{T}}\sum_{t=1}^{T} X_t$, provided the $\|E(X_t|\mathcal{F}_s)\|_2$, $s = t-1, t-2, \ldots$ converge to zero sufficiently fast such that the associated series converges. As for linear processes, such a condition is often relatively easy to verify for time-series models induced by i.i.d. innovation processes. We shall also use it to derive central limit theorems for mixing processes and the sample autocovariance function.

As a preparation, we need the following lemma that establishes a relationship between the L_2 norm of a conditional expectation $E(X|Y)$ and the covariance $\mathrm{Cov}(X, Y)$.

Lemma 8.1.6 *Let X be a random variable with $E(X) = 0$ and $\mathcal{A}$ a σ-field. Then*

$$\|E(X|\mathcal{A})\|_{L_2} = \sup\{E(XY) : Y \ \mathcal{A}\text{-measurable}, \ \|Y\|_{L_2} = 1\}.$$

Proof. Let Y be $\mathcal{A}$-measurable with $\|Y\|_{L_2} = 1$. Then the Cauchy–Schwarz inequality yields

$$\begin{aligned} E(XY) &\leq E(E(XY|\mathcal{A})) \\ &= E(YE(X|\mathcal{A})) \\ &\leq \|Y\|_{L_2} \|E(X|\mathcal{A})\|_{L_2} \\ &= \|E(X|\mathcal{A})\|_{L_2} \end{aligned}$$

with equality if and only if Y and $E(X|\mathcal{A})$ are linear dependent, this means

$$Y = \frac{E(X|\mathcal{A})}{\|E(X|\mathcal{A})\|_{L_2}}.$$

Theorem 8.1.7 *Let $\{X_t\}$ be a strictly stationary series with $E(X_1) = 0$, which satisfies the ergodic theorem. Suppose that*

$$\sum_{n=1}^{\infty} \|E(X_0|\mathcal{F}_{-n})\|_2 < \infty.$$

Then

$$\frac{1}{\sqrt{T}} \sum_{t=1}^{T} X_t \xrightarrow{d} N(0, \sigma^2),$$

as $T \to \infty$, where

$$\sigma^2 = E\left(X_1^2\right) + 2 \sum_{h=1}^{\infty} E(X_1 X_{1+h})$$

converges absolutely.

Proof. Define

$$\begin{aligned} Z_t &= \sum_{j=0}^{\infty} E(X_{t+j}|\mathcal{F}_{t-1}), \\ Y_t &= \sum_{j=0}^{\infty} \left[E(X_{t+j}|\mathcal{F}_t) - E(X_{t+j}|\mathcal{F}_{t-1})\right]. \end{aligned}$$

We shall verify:

(i) The above series Z_t and Y_t converge in L_2.

(ii) The decomposition

$$X_t = Y_t + Z_t - Z_{t+1}$$

holds true a.s. for all t.

(iii) We have

$$\frac{Z_1 - Z_{T+1}}{\sqrt{T}} \stackrel{P}{\to} 0,$$

as $T \to \infty$.

(iv) $\{Y_t\}$ is a L_2-martingale difference sequence to which we can apply Theorem 8.3.8.

(v) The series σ^2 converges absolutely.

To check (i) it suffices to show that

$$\sum_{j=0}^{\infty} \|E(X_{t+j}|\mathcal{F}_{t-1})\|_2 < \infty.$$

By stationarity of $\{X_t\}$, we have using Lemma 8.1.6

$$\begin{aligned}
\|E(X_{t+j}|\mathcal{F}_{t-1})\|_2 &= \sup\{E(X_{t+j}Z) : Z \text{ is } \mathcal{F}_{t-1}\text{-measurable}, \|Z\|_2 = 1\} \\
&= \sup\{E(X_{t+j} f(X_{t-1}, \ldots)) : f \text{ measurable}, \|f(X_{t-1}, \ldots)\|_2 = 1\} \\
&= \sup\{E(X_0 f(X_{-j-1}, \ldots)) : f \text{ measurable}, \|f(X_{-j-1}, \ldots)\|_2 = 1\} \\
&= \|E(X_0|\mathcal{F}_{-j-1})\|_2,
\end{aligned}$$

for $j = 0, 1, \ldots$ It follows that

$$\begin{aligned}
\sum_{j=0}^{\infty} \|E(X_{t+j}|\mathcal{F}_{t-1})\|_2 &= \sum_{j=0}^{\infty} \|E(X_0|\mathcal{F}_{-j-1})\|_2 \\
&= \sum_{j=1}^{\infty} \|E(X_0|\mathcal{F}_{-j})\|_2 < \infty,
\end{aligned}$$

by assumption.

To show (ii), notice that by definition of Y_t and Z_t the terms $E(X_{t+j}|\mathcal{F}_{t-1})$ cancel when taking their sum, such that

$$\begin{aligned} Z_t + Y_t &= \sum_{j=0}^{\infty} E(X_{t+j}|\mathcal{F}_t) \\ &= E(X_t|\mathcal{F}_t) + \sum_{j=1}^{\infty} E(X_{t+j}|\mathcal{F}_t) \\ &= X_t + \sum_{j=0}^{\infty} E(X_{t+j+1}|\mathcal{F}_t) \\ &= X_t + Z_{t+1}, \end{aligned}$$

a.s.

Let us now verify (iii). First, we need to show the strict stationarity of $\{Z_t\}$, which is technically involved. Here are the details of this step. The proof is based on the fact that $E(X_s|\mathcal{F}_t)$ is shift invariant. Clearly, we can write

$$E(X_s|\mathcal{F}_t) = f(X_t, X_{t-1}, \ldots)$$

for some measurable function f. We claim that $f(X_{t+h}, X_{t+h-1}, \ldots)$ is a version of $E(X_{s+h}|\mathcal{F}_{t+h})$ for any $h \in \mathbb{Z}$. This follows, if we verify that

$$\int_A f(X_{t+h}, X_{t+h-1}, \ldots)\,\mathrm{d}P = \int_A X_{s+h}\,\mathrm{d}P,$$

for all $A \in \mathcal{F}_{t+h}$. Fix such an A. Then $A = \{(X_{t+h}, X_{t+h-1}, \ldots) \in B\}$ for a measurable set $B \subset \mathbb{R}^{\mathbb{N}_0}$. We have

$$\begin{aligned} \int_A f(X_{t+h}, X_{t+h-1}, \ldots)\,\mathrm{d}P &= \int_{\{(X_{t+h}, X_{t+h-1},\ldots)\in B\}} f(X_{t+h}, X_{t+h-1}, \ldots)\,\mathrm{d}P \\ &= \int_{\{(X_t, X_{t-1},\ldots)\in B\}} f(X_t, X_{t-1}, \ldots)\,\mathrm{d}P \\ &= \int_{\{(X_t, X_{t-1},\ldots)\in B\}} E(X_s|\mathcal{F}_t)\,\mathrm{d}P \\ &= \int_{\{(X_t, X_{t-1},\ldots)\in B\}} X_s\,\mathrm{d}P \\ &= \int_{\{(X_{t+h}, X_{t+h-1},\ldots)\in B\}} X_{s+h}\,\mathrm{d}P \\ &= \int_A X_{s+h}\,\mathrm{d}P. \end{aligned}$$

But this implies

$$E(X_s|\mathcal{F}_t) \stackrel{a.s.}{=} f(X_t, X_{t-1}, \ldots) \stackrel{d}{=} f(X_{t+h}, X_{t+h-1}, \ldots) \stackrel{a.s.}{=} E(X_{s+h}|\mathcal{F}_{t+h}).$$

Further, for fixed $t_1 < \cdots < t_k$, $k \in \mathbb{N}$, there are measurable functions $f_1, \ldots, f_k$, such that

$$\left(E(X_{t_1+j}|\mathcal{F}_{t_1}), \ldots, E(X_{t_k+j}|\mathcal{F}_{t_k})\right) \overset{a.s.}{=} \left(f_1(X_{t_1+j}, \ldots), \cdots, f_k(X_{t_k+j}, \ldots)\right).$$

where the right-hand side is equal in distribution to $\left(f_1(X_{t_1+j+h}, \ldots), \ldots, f_k(X_{t_k+j+h}, \ldots)\right)$, which is a.s. equal to $\left(E(X_{t_1+j+h}|\mathcal{F}_{t_1+h}), \ldots, E(X_{t_k+j+h}|\mathcal{F}_{t_k+h})\right)$.

We may now conclude that $\{Z_t\}$ is strictly stationary. Indeed, there exist measurable functions $f_1, \ldots, f_k$ such that

$$\begin{aligned}
(Z_{t_1}, \ldots, Z_{t_k}) &= \sum_{j=0}^{\infty} \left(E(X_{t_1+j}|\mathcal{F}_{t_1-1}), \ldots, E(X_{t_k+j}|\mathcal{F}_{t_k-1})\right) \\
&\overset{a.s.}{=} \sum_{j=1}^{\infty} \left(f_1(X_{t_1+j-1}, \ldots), \ldots, f_k(X_{t_k+j-1}, \ldots)\right) \\
&\overset{d}{=} \sum_{j=1}^{\infty} \left(f_1(X_{t_1+h+j-1}, \ldots), \ldots, f_k(X_{t_k+h+j-1}, \ldots)\right) \\
&\overset{a.s.}{=} \sum_{j=0}^{\infty} \left(E(X_{t_1+h+j}|\mathcal{F}_{t_1+h-1}), \ldots, E(X_{t_k+h+j}|\mathcal{F}_{t_k+h-1})\right) \\
&= \sum \left(Z_{t_1+h}, \ldots, Z_{t_k+h}\right).
\end{aligned}$$

This establishes the strict stationarity of $\{Z_t\}$. Now it easily follows that

$$\begin{aligned}
\operatorname{Var}\left(\frac{Z_1 - Z_{T+1}}{\sqrt{T}}\right) &= \frac{1}{T} E(Z_1 - Z_{T+1})^2 \\
&\leq \frac{2}{T}(EZ_1^2 + EZ_{T+1}^2) \\
&= \frac{4EZ_1^2}{T},
\end{aligned}$$

which converges to 0, as $T \to \infty$.

To verify (iv), notice that

$$\begin{aligned}
E(Y_t|\mathcal{F}_{t-1}) &= \sum_{j=0}^{\infty} E\left(E(X_{t+j}|\mathcal{F}_t) - E(X_{t+j}|\mathcal{F}_{t-1})|\mathcal{F}_{t-1}\right) \\
&= \sum_{j=0}^{\infty} (E(X_{t+j}|\mathcal{F}_{t-1}) - E(X_{t+j}|\mathcal{F}_{t-1})),
\end{aligned}$$

which vanishes a.s.

The proof can now be completed as follows. $\{Y_t\}$ is a strictly stationary L_2-martingale difference sequence and therefore satisfies the central limit theorem. The decomposition

$$X_t = Y_t + Z_t - Z_{t+1}$$

yields

$$\frac{1}{\sqrt{T}}\sum_{t=1}^{T} X_t = \frac{1}{\sqrt{T}}\sum_{t=1}^{T} Y_t + R_T,$$

where the remainder term

$$R_T = \frac{Z_1 - Z_{T+1}}{\sqrt{T}}$$

converges to 0 in L_2. This implies that, first,

$$\lim_{T\to\infty} P\left(\frac{1}{\sqrt{T}}\sum_{t=1}^{T} X_t \le x\right) = \lim_{T\to\infty} P\left(\frac{1}{\sqrt{T}}\sum_{t=1}^{T} Y_t \le x\right) = \Phi\left(\frac{x}{\sigma}\right),$$

for all $x \in \mathbb{R}$, for some $\sigma \ge 0$, by an application of Slutzky's lemma, and, secondly,

$$\lim_{n\to\infty} \operatorname{Var}\left(\frac{1}{\sqrt{T}}\sum_{t=1}^{T} X_t\right) = \lim_{n\to\infty} \operatorname{Var}\left(\frac{1}{\sqrt{T}}\sum_{t=1}^{T} Y_t\right).$$

Hence,

$$\sigma^2 = E\left(X_1^2\right) + 2\sum_{h=0}^{\infty} E(X_1 X_{1+h}),$$

provided the series at the right-hand side converges. To show the latter notice that

$$E\left(\frac{1}{\sqrt{T}}\sum_{t=1}^{T} X_t\right)^2 = E\left(X_1^2\right) + 2\sum_{h=1}^{T-1}\left(1 - \frac{h}{T}\right) E(X_1 X_{1+h}).$$

We have

$$\begin{aligned}
|E(X_1 X_{1+h})| &= E(X_0 E(X_h|\mathcal{F}_0)) \\
&\le E(|X_0 E(X_h|\mathcal{F}_0)|) \\
&\le \|X_0\|_2 \|E(X_h|\mathcal{F}_0)\|_2 \\
&= \|X_0\|_2 \|E(X_0|\mathcal{F}_{-h})\|_2.
\end{aligned}$$

Hence,

$$\sum_{h=1}^{\infty} |E(X_1 X_{1+h})| \le \|X_0\|_2 \sum_{h=1}^{\infty} \|E(X_0|\mathcal{F}_{-h})\|_2 < \infty.$$

8.2 A regression model for financial time series

The classic multiple linear regression model regresses a response variable Y on p explanatory variables (regressors) $x_1, \ldots, x_p$ by assuming that

$$Y = \beta_0 + \beta_1 x_1 + \cdots + \beta_p x_p + \epsilon$$

for unknown regression coefficients $\beta_0, \ldots, \beta_p$ and a mean zero error term ϵ. Both the regressors and the coefficients are assumed to be non random. In order to estimate the above model, one collects T observation vectors $(Y_t, x_{t1}, \ldots, x_{tp})$, $t = 1, \ldots, T$. The corresponding model equations are

$$Y_t = \beta_0 + \beta_1 x_{t1} + \cdots + \beta_p x_{tp} + \epsilon_t, \; t = 1, \ldots, T.$$

The sampling mechanism yielding the sample of size T determines the stochastic relationships between the random variables $Y_t, x_{t1}, \ldots, x_{tp}, \epsilon_t$, $t = 1, \ldots, T$. The classic regression model assumes that the error terms are i.i.d. and follow a normal distribution. In this form, the regression model is not applicable to financial data sets. First, one wants to analyze random regressors such as returns. Secondly, lagged values, Y_{t-j} for $j \in \mathbb{N}$, often appear as explanatory variables. Lastly, the errors ϵ_t are usually neither independent nor Gaussian. For these reasons, we consider a general regression model with stochastic regressors,

$$Y_t = X_t'\beta + \epsilon_t, \; t = 1, \ldots, T, \tag{8.2}$$

where X_t are d-dimensional random vectors, $\beta \in \mathbb{R}^d$ is an unknown but fixed coefficient vector and ϵ_t are stochastic error terms.

Concerning the regressors, we make the following assumption.

(R1) $\{X_t : t \in \mathbb{N}\}$ is weakly stationary, i.e. $E(X_t) = E(X_1)$ and $E(X_t X_t') = E(X_1, X_1')$ for all t as well as $E(X_{tj}^2) < \infty$ for all j and t.

To proceed, we introduce a filtration $\mathcal{F}_t$, $t \geq 1$. Its purpose is to collect the information at time t on which the regressors X_t may depend. We assume that

(F1) X_{t-j} is $\mathcal{F}_t$ – measurable, $j \geq 0$.

(F2) Y_{t-j} is $\mathcal{F}_t$ – measurable, $j \geq 1$.

Obviously, one may use $\mathcal{F}_t = \sigma(X_{t-j}, Y_{t-1-j} : j \geq 0)$, $t \in \mathbb{N}$, but sometimes $\mathcal{F}_t$ is given such that one has to check that (F1) and (F2) are satisfied.

The assumptions on the error terms are as follows.

(E1) $E(\epsilon_t|\mathcal{F}_t) = 0 \, a.s.$, for all $t \in \mathbb{N}$.

(E2) $E(\epsilon_t^2|\mathcal{F}_t) = \sigma^2 \; a.s., t \in \mathbb{N}$, for some constant $\sigma^2 \in (0, \infty)$.

In the regression model (8.2) the ϵ_t describe the random fluctuation around the predictor $X_t'\beta$. Thus, Equation (E1) assumes that their conditional mean is zero given the regressors, i.e. when the predictor is fixed. Assumption (E2) tells us that the conditional dispersion of the errors is constant as time proceeds. We shall later discuss how one can relax that condition.

The following observations are simple but crucial: ϵ_t is not $\mathcal{F}_t$-measurable; otherwise

$$\epsilon_t^2 = E\left(\epsilon_t^2|\mathcal{F}_t\right) = \sigma^2$$

holds true for all t, such that ϵ_t^2 is a constant time series that has marginal variance $\sigma^2 = E(\epsilon_t^2) = 0$, a contradiction. This fact also implies that $Y_t = X_t'\beta + \epsilon_t$ is not $\mathcal{F}_t$-measurable, since otherwise $\epsilon_t = Y_t - X_t'\beta$ would have that property. But all lagged values ϵ_{t-j}, $j \geq 1$, are $\mathcal{F}_t$-measurable, as one can easily check.

Also, notice that Assumption (E1) implies that the $\epsilon_t' s$ are uncorrelated, since for $j \geq 1$

$$E(\epsilon_{t-j}\epsilon_t) = E(E(\epsilon_{t-j}\epsilon_t|\mathcal{F}_t))$$
$$= E(\epsilon_{t-j}E(\epsilon_t|\mathcal{F}_t)) = 0,$$

since ϵ_{t-j} is $\mathcal{F}_t$-measurable and (E1) holds true. However, $\{\epsilon_t\}$ is not a martingale difference sequence with respect to the filtration $\{\mathcal{F}_t\}$, since we assume $E(\epsilon_t|\mathcal{F}_t) = 0$ and not $E(\epsilon_t|\mathcal{F}_{t-1}) = 0$. But it is possible to enlarge the filtration in such a way that $\{\epsilon_t\}$ becomes a martingale difference, which will allow us to apply powerful limit theorems to construct asymptotic hypothesis tests and confidence intervals for the regression model with stochastic regressors. For that purpose define

$$\mathcal{F}_t^* = \sigma(\mathcal{F}_t \cup \sigma(Y_t)), \quad t \in \mathbb{N}.$$

Then the following properties hold true

(i) Y_t is $\mathcal{F}_t^*$-measurable.

(ii) $\mathcal{F}_{t-1}^* = \sigma(\mathcal{F}_{t-1} \cup \sigma(Y_{t-1})) \subset \mathcal{F}_t$, since $\mathcal{F}_{t-1} \subset \mathcal{F}_t$ and $\sigma(Y_{t-1}) \subset \mathcal{F}_t$.

(iii) Using (ii) we may calculate conditional expectations such as $E(Z|\mathcal{F}_{t-1}^*)$ by using the rule

$$E(Z|\mathcal{F}_{t-1}^*) = E(E(Z|\mathcal{F}_t)|\mathcal{F}_{t-1}^*).$$

Property (iii) now immediately implies

$$E(\epsilon_t|\mathcal{F}_{t-1}^*) = E(E(\epsilon_t|\mathcal{F}_t)|\mathcal{F}_{t-1}^*) = 0,$$

which verifies that ϵ_t is a $\mathcal{F}_t^*$-martingale difference sequence.

Example 8.2.1 *The celebrated* **capital asset pricing model (CAPM)** *fits the linear regression framework. Assume we are given an economy (market) with N financial assets and a bank where one can lend and borrow money at the same risk-free rate r. Let us denote by R_{tj} the log-return of asset j at time t, $j = 1, \ldots, N$, $t = 1, \ldots, T$. It is assumed that there exists a market portfolio with log-returns $R_t^{(M)}$. If R denotes the log-return of a risky asset, the difference*

$$R_e = R - r$$

is called the **excess return**. *The Sharpe–Lintner version of the CAPM now states that the excess returns $R_{ti} - r$ satisfy a linear regression model of the form*

$$R_{ti} - r = \beta_0 + \beta_j\left(R_t^{(M)} - r\right) + \epsilon_{ti}, \qquad t = 1, \ldots, T, \; i = 1, \ldots, N.$$

The regression coefficients β_j are the **beta factors (investment betas)** *of the assets. They measure the sensitivity of the asset returns when the market moves. Assets with a beta smaller than one are less sensitive to market fluctuations, while assets with a beta factor larger than one are riskier than the market portfolio. The CAPM theory asserts that $\beta_0 = 0$ and that the innovations are uncorrelated and normally distributed. In practice, often an index is used as* *a proxy for the market portfolio. Further, one has to fix a riskless interest rate, e.g. the interest rate of a treasury bond.*

8.2.1 Least squares estimation

The most popular approach to estimate a regression model is the method of least squares where the function

$$Q_T(\beta) = \sum_{t=1}^{T} (Y_t - X_t'\beta)^2, \quad \beta \in \mathbb{R}^d,$$

is minimized. It is easy to check that any solution $\widehat{\beta}_T \in \mathbb{R}^d$ satisfies the normal equations

$$\mathbf{X}_T'\mathbf{X}_T\widehat{\beta}_T = \mathbf{X}_T'\mathbf{Y},$$

where $\mathbf{X}_T$ is the $T \times d$ random matrix

$$\mathbf{X}_T = (X_1, \ldots, X_T)'$$

and

$$\mathbf{Y} = (Y_1, \ldots, Y_T)'.$$

Notice that

$$\frac{1}{T}\mathbf{X}_T'\mathbf{X}_T = \frac{1}{T}\sum_{t=1}^{T} X_t X_t'$$

is the average of the rank 1 matrices $X_t X_t'$. We may solve the normal equations for $\widehat{\beta}_T$, if $\mathbf{X}_T'\mathbf{X}_T$ is almost surely invertible, at least for large T. Thus, we impose the following assumption that ensures that property.

(R2) $\frac{1}{T}\mathbf{X}_T'\mathbf{X}_T \to \Sigma_{\mathbf{X}}$ in probability, as $T \to \infty$, for some non random and positive definite matrix $\Sigma_{\mathbf{X}}$.

In some cases, one can even guarantee a.s. convergence, and then the following derivations are easier to understand, since we can then find some set $A \subset \Omega$, such that $\frac{1}{T}\mathbf{X}_T(\omega)'\mathbf{X}_T(\omega) \to \Sigma_{\mathbf{X}}$, as $T \to \infty$, holds true for all $\omega \in A$, and the remaining $\omega \in A^c$ play no role. Thus, let us assume that a.s. convergence holds true. For large enough T we obtain the explicit formula

$$\widehat{\beta}_T = (\mathbf{X}_T'\mathbf{X}_T)^{-1}\mathbf{X}_T'\mathbf{Y}.$$

We are now going to derive a formula that plays a key role in what follows. Using $\mathbf{Y} = \mathbf{X}\boldsymbol{\beta} + \boldsymbol{\epsilon}$ we obtain

$$\begin{aligned}
\widehat{\boldsymbol{\beta}}_T &= (\mathbf{X}_T'\mathbf{X}_T)^{-1}\mathbf{X}_T'\mathbf{Y} \\
&= (\mathbf{X}_T'\mathbf{X}_T)^{-1}\mathbf{X}_T'(\mathbf{X}_T\boldsymbol{\beta} + \boldsymbol{\epsilon}) \\
&= \boldsymbol{\beta} + (\mathbf{X}_T'\mathbf{X}_T)^{-1}\mathbf{X}_T'\boldsymbol{\epsilon},
\end{aligned}$$

yielding

$$\widehat{\boldsymbol{\beta}}_T - \boldsymbol{\beta} = (\mathbf{X}_T'\mathbf{X}_T)^{-1} \sum_{t=1}^{T} X_t \epsilon_t,$$

or equivalently

$$\sqrt{T}(\widehat{\boldsymbol{\beta}}_T - \boldsymbol{\beta}) = \left(\frac{1}{T} \mathbf{X}_T'\mathbf{X}_T \right)^{-1} \frac{1}{\sqrt{T}} \sum_{t=1}^{T} X_t \epsilon_t.$$

Assumption (R2) ensures that the first factor of that matrix product, $(T^{-1}\mathbf{X}_T'\mathbf{X}_T)^{-1}$, converges in probability to $\Sigma_{\mathbf{X}}^{-1}$, as $T \to \infty$. If we know in addition that

$$\frac{1}{\sqrt{T}} \sum_{t=1}^{T} X_t \epsilon_t \underset{T\to\infty}{\overset{d}{\longrightarrow}} N(\mathbf{0}, \mathbf{S}),$$

as $T \to \infty$, holds true for some matrix $\mathbf{S}$, we may conclude that $\widehat{\boldsymbol{\beta}}_T$ satisfies a central limit theorem, since then

$$\sqrt{T}\left(\widehat{\boldsymbol{\beta}}_T - \boldsymbol{\beta}\right) \underset{T\to\infty}{\overset{d}{\longrightarrow}} N(\mathbf{0}, \Sigma_X^{-1}\mathbf{S}(\Sigma_X^{-1})'),$$

as $T \to \infty$. This means, given the regressors satisfy (R2), we have to find conditions on the errors ϵ_t such that the process $\xi_t = X_t \epsilon_t$, $t \in \mathbb{N}$, satisfies a central limit theorem.

The following simple lemma is crucial for what follows.

Lemma 8.2.2 *Given assumptions (E1), (E2), (F1), (F2) and (R1), the process $\xi_t = X_t \epsilon_t$, $t \in \mathbb{N}$, is a $\mathcal{F}_t^*$- martingale difference sequence with*

$$\mathrm{Var}\,(\xi_t) = \sigma^2 \Sigma_{\mathbf{X}}, \quad \Sigma_{\mathbf{X}} = E(\mathbf{X}_t \mathbf{X}_t')$$

and

$$\mathrm{Cov}(\xi_s, \xi_t) = 0, \quad s < t.$$

Proof. $X_t, \epsilon_t \in L_2$ implies that $\xi_t = X_t \epsilon_t \in L_1$. Using $E(\epsilon_t | \mathcal{F}_t) = 0$ we have

$$\begin{aligned} E(X_t \epsilon_t | \mathcal{F}_{t-1}^*) &= E(E(X_t \epsilon_t | \mathcal{F}_t) | \mathcal{F}_{t-1}^*) \\ &= E(X_t E(\epsilon_t | \mathcal{F}_t) | \mathcal{F}_{t-1}^*) = 0, \end{aligned}$$

and

$$\begin{aligned} \mathrm{Var}\,(X_t \epsilon_t) &= E(\epsilon_t^2 X_t X_t') \\ &= E(E(\epsilon_t^2 X_t X_t' | \mathcal{F}_t)) \\ &= E(X_t X_t' E(\epsilon_t^2 | \mathcal{F}_t)) \\ &= \sigma^2 E(X_t X_t') = \sigma^2 \Sigma_{\mathbf{X}}, \end{aligned}$$

where the last equality follows from (R1).

8.3 Limit theorems for martingale difference

We shall now collect some basic results for martingale difference arrays, which generalize the notion of a martingale difference sequence.

Definition 8.3.1 *An array $X_{nk} : (\Omega, \mathcal{F}, P) \to (\mathbb{R}, \mathcal{B})$, $1 \le k \le r_n, n \ge 1$, of random variables and random vectors, respectively, is called a martingale difference array with respect to an non decreasing family $\{\mathcal{F}_{nk} : 1 \le k \le r_n, n \ge 1\}$ of σ- fields, if for each $n \in \mathbb{N}$ the sequence*

$$X_{n1}, \dots, X_{nrn}$$

is a martingale difference sequence with respect to the filtration

$$\mathcal{F}_{n1} \subset \cdots \subset \mathcal{F}_{nr},$$

that is

(i) $\mathcal{F}_{n0} = \{\emptyset, \Omega\}$, $\mathcal{F}_{nk} \subset \mathcal{F}_{n,k+1} \subset \cdots$ *for* $1 \le k \le r_n, n \ge 1$.

(ii) X_{nk} *is* $\mathcal{F}_{nk}$*-measurable with* $E|X_{nk}| < \infty$ *for* $1 \le k \le r_n, n \ge 1$, *and*

(iii) $E(X_{nk}|\mathcal{F}_{n,k-1}) = 0$ *for* $1 \le k \le r_n, n \ge 1$.

Martingale difference arrays satisfy a weak law of large numbers under fairly general conditions.

Theorem 8.3.2 *Let $\{X_{Tt}\} = \{X_{Tt} : 1 \le t \le T, T \ge 1\}$ be a $\{\mathcal{F}_{T,t}\}$-martingale difference array. Suppose that one of the following two conditions is satisfied.*

(i) *There exists some $\sigma^2 \in [0, \infty)$ such that*

$$\frac{1}{T} \sum_{t=1}^{T} \text{Var}\,(X_{Tt}) \to \sigma^2,$$

as $T \to \infty$.

(ii) *There exists some $\sigma^2 \in [0, \infty)$ such that*

$$\frac{1}{T} \sum_{t=1}^{T} \sigma_{Tt}^2 \to \sigma^2, \qquad \text{in } L_1,$$

as $T \to \infty$, where $\sigma_{Tt}^2 = E(X_{Tt}^2|\mathcal{F}_{T,t-1})$.

Then the weak law of large numbers holds true,

$$\frac{1}{T} \sum_{t=1}^{T} X_{Tt} \xrightarrow{P} 0,$$

as $T \to \infty$.

Proof. The proof relies on Chebychev's inequality. We have

$$P\left(\left|\frac{1}{T}\sum_{t=1}^{T} X_{Tt}\right| > \varepsilon\right) \leq \frac{E\left(\frac{1}{T}\sum_{t=1}^{T} X_{Tt}\right)^2}{\varepsilon^2}$$
$$\leq \frac{\frac{1}{T^2}\sum_{t=1}^{T} E(X_{Tt}^2)}{\varepsilon^2},$$

where we used the fact martingale differences are centered and uncorrelated. If (ii) is satisfied, we obtain

$$\frac{1}{T^2}\sum_{t=1}^{T} E(X_{Tt}^2) = \frac{1}{T}E\left[\frac{1}{T}\sum_{t=1}^{T} E(X_{Tt}^2|\mathcal{F}_{T,t-1})\right]$$
$$= \frac{1}{T}E\left(\frac{1}{T}\sum_{t=1}^{T}\sigma_{Tt}^2\right) \leq \frac{c}{T},$$

since $T^{-1}\sum_{t=1}^{T}\sigma_{Tt}^2 \to \sigma^2$ in L_1 means that $a_T = E\left(T^{-1}\sum_{t=1}^{T}\sigma_{Tt}^2\right) \to \sigma^2$, as $T \to \infty$, such that $\{a_T\}$ is bounded by some constant c. The sufficiency of (i) is left to the reader.

Let us illustrate Theorem 8.3.2 by the problem to estimate mean and volatility of log returns.

Example 8.3.3 *Suppose $R_1, R_2, \ldots$ are log returns of an asset. To estimate mean and volatility based on $R_1, \ldots, R_T$, let us consider the following basic model,*

$$R_t = \mu + \epsilon_t, \quad t \in \mathbb{N},$$

where $\mu \in \mathbb{R}$ is a constant and $\{\epsilon_t\}$ is a martingale difference array w.r.t. the natural filtration $\mathcal{F}_t = \sigma(\epsilon_1, \ldots, \epsilon_t)$ with

$$E\left(\epsilon_t^2|\mathcal{F}_{t-1}\right) = \sigma^2, \ E\left(\epsilon_t^4|\mathcal{F}_{t-1}\right) = \gamma_4 < \infty,$$

a.s., for all t, for constants γ_4 and σ^2. Clearly, $E(R_t|\mathcal{F}_{t-1}) = \mu + E(\epsilon_t|\mathcal{F}_{t-1}) = \mu$ and thus $E(R_t) = \mu$ as well. It follows that

$$\widehat{\mu}_T = \frac{1}{T}\sum_{t=1}^{T} R_t$$

is an unbiased estimator for μ. Since $\widehat{\mu}_T - \mu = \frac{1}{T}\sum_{t=1}^{T}\epsilon_t$, Theorem 8.3.2 yields

$$\widehat{\mu}_T - \mu \stackrel{P}{\to} 0 \quad \Leftrightarrow \quad \widehat{\mu}_T \stackrel{P}{\to} \mu,$$

as $T \to \infty$, such that $\widehat{\mu}_T$ is weakly consistent for μ.

In order to estimate σ^2, let us first consider the case that μ is known. Since then $(R_t - \mu)^2 = \epsilon_t^2$ such that $E(R_t - \mu)^2 = \sigma^2$, it is natural to estimate σ^2 by

$$\widetilde{\sigma}_T^2 = \frac{1}{T}\sum_{t=1}^{T}(R_t - \mu)^2.$$

We claim that $\{\epsilon_t^2 - \sigma^2 : t \geq 1\}$, *is a martingale difference sequence. Indeed,*

$$E|\epsilon_t^2 - \sigma^2| \leq E\epsilon_t^2 + \sigma^2 < \infty$$

and

$$E\left(\epsilon_t^2 - \sigma^2 | \mathcal{F}_{t-1}\right) = E\left(\epsilon_t^2 | \mathcal{F}_{t-1}\right) - \sigma^2 = 0,$$

a.s., by assumption. In order to apply Theorem 8.3.2, notice that

$$\begin{aligned} \frac{1}{T}\sum_{t=1}^{T} E[(\epsilon_t^2 - \sigma^2)^2 | \mathcal{F}_{t-1}] &= \frac{1}{T}\sum_{t=1}^{T} E\left(\epsilon_t^4 - 2\epsilon_t^2\sigma^2 + \sigma^4 | \mathcal{F}_{t-1}\right) \\ &= \gamma_4 - \sigma^4 < \infty. \end{aligned}$$

If follows that $\widetilde{\sigma}_T^2 - \sigma^2 \stackrel{P}{\to} 0$, *as* $T \to \infty$. *If* μ *is unknown, one replaces* μ *by* $\widehat{\mu}_T$ *and considers*

$$\widehat{\sigma}_T^2 = \frac{1}{T}\sum_{t=1}^{T}(R_t - \widehat{\mu}_T)^2.$$

Consistency of $\widehat{\sigma}_T^2$ *follows now easily by observing that*

$$\begin{aligned} \widehat{\sigma}_T^2 &= \frac{1}{T}\sum_{t=1}^{T}(R_t - \mu + \mu - \widehat{\mu}_T)^2 \\ &= \frac{1}{T}\sum_{t=1}^{T}(R_t - \mu)^2 + 2(\mu - \widehat{\mu}_T)\frac{1}{T}\sum_{t=1}^{T}(R_t - \mu) + (\mu - \widehat{\mu}_T)^2. \end{aligned}$$

The first term is $\widetilde{\sigma}_T^2$. *The second one converges to* 0 *in probability, since* $\widehat{\mu}_T - \mu \stackrel{P}{\to} 0$ *and* $\frac{1}{T}\sum_{t=1}^{T}(R_t - \mu) \stackrel{P}{\to} 0$, *as* $T \to \infty$. *The last term,* $(\mu - \widehat{\mu}_T)^2$, *converges in probability to* 0 *as well. By virtue of Slutzky's lemma, we may conclude that* $\widehat{\sigma}_T^2 \stackrel{P}{\to} \sigma^2$, *as* $T \to \infty$.

Let us continue our study of the general regression model. The next proposition asserts that the martingale difference property is preserved, if each observation is multiplied with a function of past values.

Proposition 8.3.4 *Let* $\{X_t : t \geq 0\}$ *be a* $\mathcal{F}_t$*-martingale difference array with* $X_t \in L_p$ *for all* t *and let* g *be a function on* $\mathbb{R}^\infty$ *such that*

$$Y_{t-1} = g(X_{t-1}, X_{t-2}, \ldots) \in L_q \text{ for all } t,$$

where $\frac{1}{p} + \frac{1}{q} = 1,\ p, q \in [1, \infty]$. *Then* $\{X_t Y_{t-1} : t \geq 1\}$ *is a martingale difference sequence. Further,* $\{X_t : t \geq 1\}$ *and* $\{Y_{t-1} : t \geq 1\}$ *are uncorrelated.*

Proof. Hölder's inequality yields

$$E|X_t Y_{t-1}| \leq \|X_t\|_p \, \|Y_{t-1}\|_q < \infty.$$

Further, $E(X_t Y_{t-1}|\mathcal{F}_{t-1}) = Y_{t-1}E(X_t|\mathcal{F}_{t-1}) = 0$, a.s., and

$$\text{Cov}(X_t, Y_{t-1}) = E(X_t Y_{t-1}) = E(Y_{t-1}E(X_t|\mathcal{F}_{t-1})) = 0.$$

Combining Theorem 8.3.2 and Proposition 8.3.4 yields the following interesting result on the martingale transform.

Theorem 8.3.5 *Let $\{X_t\}$ be a $\mathcal{F}_t$-martingale and φ_t $\mathcal{F}_t$-predictable. If*

$$\frac{1}{T}\sum_{t=1}^{T}\varphi_t^2 E[(\Delta X_t)^2|\mathcal{F}_{t-1}] \stackrel{P}{\to} \sigma^2,$$

as $T \to \infty$, for some $\sigma^2 \in (0, \infty)$, then

$$\frac{1}{T}\int_0^T \varphi_t \mathrm{d}X_t \underset{T\to\infty}{\stackrel{P}{\longrightarrow}} 0.$$

Proof. Notice that

$$\frac{1}{T}\int_0^T \varphi_t \mathrm{d}X_t = \frac{1}{T}\sum_{t=1}^{T}\varphi_t\xi_t = \frac{1}{T}\sum_{t=1}^{T}\widetilde{\varphi}_{t-1}\xi_t,$$

where $\xi_t = \Delta X_t = X_t - X_{t-1}$ is a $\mathcal{F}_t$-martingale difference sequence and $\widetilde{\varphi}_{t-1} = \varphi_t$ is $\mathcal{F}_{t-1}$-measurable. Hence, $\widetilde{\varphi}_{t-1}\xi_t$ is a martingale difference sequence and the assertion follows from Theorem 8.3.2.

Under fairly general conditions, martingale differences satisfy the central limit theorem. For an elaborated proof consult the references.

Theorem 8.3.6 *Suppose $\{X_{Tt} : 1 \le t \le T, T \ge 1\}$ is a martingale difference array with $\sigma_{Tt}^2 = \text{Var}(X_{Tt}) < \infty$ for all t, T and*

$$\text{Var}\left(\sum_{t=1}^{T} X_{Tt}\right) = \sum_{t=1}^{T}\sigma_{Tt}^2 = 1.$$

If

(i) $\sum_{t=1}^{T} X_{Tt}^2 \stackrel{P}{\to} 1$, *as $T \to \infty$, and*

(ii) $\max_{1\le t\le T} |X_{Tt}| \stackrel{P}{\to} 0$, *as $T \to \infty$,*

then

$$\sum_{t=1}^{T} X_{Tt} \stackrel{d}{\to} N(0, 1),$$

as $T \to \infty$.

Remark 8.3.7 *There are various versions of the central limit theorem for martingale differences. The sufficient conditions given in Theorem 8.3.6 will allow us to establish asymptotic normality of statistics relevant for financial data analysis by relatively elementary arguments. In particular, instead of (ii) one can verify the Linderberg condition*

$$\sum_{i=1}^{T} E(\xi_{Ti}^2 \mathbf{1}(|\xi_{Ti}| > \varepsilon)|\mathcal{F}_{T,i-1}) \to 0, \tag{8.3}$$

for any $\varepsilon > 0$. However, some alternative versions frequently used in financial statistics are presented in Appendix B.

For strictly stationary sequences of martingale sequences, one has the following simplified result.

Theorem 8.3.8 *Let $\{X_t : t \in \mathbb{N}\}$ be a strictly stationary sequence of L_2-martingale differences, that is*

$$\sigma^2 = E(X_t^2) < \infty$$

and

$$E(X_t|\mathcal{F}_{t-1}) = 0,$$

hold for all t, where $\mathcal{F}_t = \sigma(X_s : s \leq t)$, for $t \in \mathbb{N}$, and $\mathcal{F}_0 = \{\emptyset, \Omega\}$. Then

$$\frac{1}{\sqrt{T}} \sum_{t=1}^{T} X_t \xrightarrow{d} N(0, \sigma^2),$$

as $T \to \infty$.

Proof. Define the random variables

$$\xi_t = E(X_t^2|\mathcal{F}_{t-1}), \qquad t \geq 1,$$

and notice that $\xi_t = f(X_{t-1}, X_{t-2}, \ldots)$ for some Borel-measurable function f defined on $\mathbb{R}^{\mathbb{N}_0}$. $\{\xi_t\}$ is a strictly stationary process and therefore satisfies the ergodic theorem. Hence,

$$\frac{1}{T} \sum_{t=1}^{T} E(X_t^2|\mathcal{F}_{t-1}) = \frac{1}{T} \sum_{t=1}^{T} \xi_t \xrightarrow{P} \sigma^2,$$

as $T \to \infty$. Further, by strict stationarity

$$\frac{1}{T} \sum_{t=1}^{T} E\left(X_t^2 \mathbf{1}_{\{|X_t| > \varepsilon\sqrt{T}\}}\right) = E\left(X_1^2 \mathbf{1}_{\{|X_1| > \varepsilon\sqrt{T}\}}\right),$$

which converges to 0, if $T \to \infty$, by dominated convergence. This verifies condition (8.3).

To illustrate how to apply Theorem 8.3.6, we show that the asymptotics leading to the statistical tests and confidence intervals for the mean of (log) returns remains valid, when the returns form a martingale difference sequence.

Example 8.3.9 *At the beginning of the present chapter, we reviewed the central limit theorem for i.i.d. returns* $R_1, \dots, R_T$ *with mean* μ *and volatility* σ*. We will now show that under the assumptions of Example 8.3.3*

$$\sqrt{T}(\widehat{\mu}_T - \mu) = \frac{1}{\sqrt{T}} \sum_{t=1}^{T} (R_t - \mu) \underset{T \to \infty}{\overset{d}{\to}} N(0, \sigma^2) \tag{8.4}$$

still holds true. For that purpose put

$$X_{Tt} = \frac{R_t - \mu}{\sqrt{T} \cdot \sigma}, \quad 1 \le t \le T, T \in \mathbb{N}.$$

Then $\frac{\sqrt{T}(\widehat{\mu}_T - \mu)}{\sigma} = \sum_{t=1}^{T} X_{Tt}$*. Obviously,* $\operatorname{Var}\left(\sum_{t=1}^{T} X_{Tt}\right) = 1$*. To show condition (ii) of Theorem 8.3.6, notice that according to Example 8.3.3*

$$\sum_{t=1}^{T} X_{Tt}^2 = \frac{1}{T\sigma^2} \sum_{t=1}^{T} \epsilon_t^2 \to 1,$$

as $T \to \infty$*.*

Condition (ii) follows from the union bound and Markov's inequality. We have

$$\begin{aligned} P\left(\max_{1 \le t \le T} |R_t - \mu| > \varepsilon\sqrt{T}\sigma \right) &= P\left(\bigcup_{t=1}^{T} \{|\epsilon_t| > \varepsilon\sqrt{T}\sigma\} \right) \\ &\le \sum_{t=1}^{T} P(|\epsilon_t| > \varepsilon\sqrt{T}\sigma) \\ &\le \sum_{t=1}^{T} \frac{E|\epsilon_t|^4}{\varepsilon^4 T^2 \sigma^4} \\ &= \frac{E(\epsilon_1^4)}{T\varepsilon^4\sigma^4} \to 0, \end{aligned}$$

as $T \to \infty$, which completes the proof of Equation (8.4).

8.4 Asymptotics

Before proceeding, let us take a closer look at Assumption (R2), which requires that

$$\frac{1}{T} \sum_{t=1}^{T} X_t X_t' \overset{P}{\to} \Sigma_X, \quad T \to \infty.$$

First, recall that convergence in probability of a sequence of random matrices $\{A, A_T\}$ means $P(\|A_T - A\| > \varepsilon) \to 0$, $T \to \infty$, for any $\varepsilon > 0$, where $\|\cdot\|$ is an arbitrary matrix norm. But this is equivalent to element-wise convergence in probability. Thus, for ease of presentation, we confine ourselves to dimension $d = 1$, where one has to show

$$\frac{1}{T}\sum_{t=1}^{T} X_t^2 \xrightarrow{P} \sigma^2 = E\left(X_1^2\right) \in (0, \infty), \tag{8.5}$$

as $T \to \infty$, which is obviously true for i.i.d. regressors. Under Assumption (E2) $\{X_t^2 - \sigma^2 : t \geq 1\}$ is a $\mathcal{F}_t$-martingale difference sequence, and Example 8.3.3 told us that now again Equation (8.5) holds true.

A further important case is the autoregressive model of order 1 with martingale difference errors, i.e.

$$X_t = \rho\, X_{t-1} + \epsilon_t,\ t \geq 1, \tag{8.6}$$

where $\{\epsilon_t : t \in \mathbb{Z}\}$ is a $\mathcal{F}_t$-martingale difference sequence and $|\rho| < 1$. We have seen in Example 3.4.10 that the above equations have a stationary solution, namely

$$X_t = \sum_{i=0}^{\infty} \rho^i \epsilon_{t-i}, \qquad t \in \mathbb{Z}.$$

Thus, if we take the random starting value $X_0 = \sum_{i=0}^{\infty} \rho^i \epsilon_{-i}$ and put

$$X_t = \sum_{i=0}^{\infty} \rho^i \epsilon_{t-i}, \qquad t \geq 1,$$

we obtain a solution $\{X_t : t \in \mathbb{N}_0\}$ satisfying the weak stationarity conditions

$$E\left(X_t^i\right) = E\left(X_1^i\right) \text{ for all } t \text{ and } i = 1, 2. \tag{8.7}$$

Proposition 8.4.1 *Suppose model (8.6) holds true. Then*

$$\frac{1}{T}\sum_{t=1}^{T} X_t^2 \xrightarrow{P} \frac{\sigma^2}{1-\rho^2},$$

as $T \to \infty$, provided $E(\epsilon_t^2|\mathcal{F}_{t-1}) = \sigma^2 > 0$ and $E(\epsilon_t^4|\mathcal{F}_{t-1}) = \gamma_4$ with constants $\sigma^2, \gamma_4 < \infty$ for all t.

Proof. We have

$$X_t^2 = \rho^2 X_{t-1}^2 + 2\rho X_{t-1}\epsilon_t + \epsilon_t^2,$$

leading to

$$
\begin{aligned}
\frac{1}{T}\sum_{t=1}^{T} X_t^2 &= \frac{\rho^2}{T}\sum_{t=1}^{T} X_{t-1}^2 + \frac{2\rho}{T}\sum_{t=1}^{T} X_{t-1}\epsilon_t + \frac{1}{T}\sum_{t=1}^{T}\epsilon_t^2 \\
&= \rho^2\left(\frac{1}{T}\sum_{t=1}^{T} X_t^2\right) + \rho^2\left(\frac{X_0^2}{T} - \frac{X_T^2}{T}\right) + \frac{2\rho}{T}\sum_{t=1}^{T} X_{t-1}\epsilon_t + \frac{1}{T}\sum_{t=1}^{T}\epsilon_t^2 \\
&= \rho^2\left(\frac{1}{T}\sum_{t=1}^{T} X_t^2\right) + A_T + B_T + C_T.
\end{aligned}
$$

We arrive at

$$
(1-\rho^2)\frac{1}{T}\sum_{t=1}^{T} X_t^2 = A_T + B_T + C_T.
$$

Clearly, for $i = 0, \ldots, T$

$$
P\left(\rho^2\left|\frac{X_i^2}{T}\right| > \varepsilon\right) \le \rho^2 \frac{EX_0^2}{T\varepsilon} \to 0,
$$

as $T \to \infty$, which implies $A_T \xrightarrow{P} 0$, as $T \to \infty$. Further, by Proposition 8.3.4, $\epsilon_t X_{t-1}$ is a $\mathcal{F}_t$-martingale difference sequence. Its martingal variance is

$$
\begin{aligned}
\text{Var}\,(\epsilon_t X_{t-1}) &= E(\epsilon_t^2 X_{t-1}^2) \\
&= E(X_{t-1}^2 E(\epsilon_t^2|\mathcal{F}_{t-1})) \\
&= \sigma^2 E(X_{t-1}^2) \\
&= \sigma^2 E(X_1^2) > 0,
\end{aligned}
$$

due to Equation (8.7). Thus, condition (i) of Theorem 8.3.2 is satisfied, such that $B_T \xrightarrow{P} 0$, as $T \to \infty$, follows. Finally, consider C_T. The random variables $\epsilon_t^2 - \sigma^2,\ t \ge 1$, are martingale difference with variance

$$
\text{Var}\,(\epsilon_t^2 - \sigma^2) = E(\epsilon_t^2 - \sigma^2)^2 = \gamma_4 - \sigma^4 \in [0, \infty).
$$

Again, Theorem 8.3.2 can be applied, yielding $C_T \xrightarrow{P} \sigma^2$, as $T \to \infty$.

In order to establish a central limit theorem for the least squares estimator $\widehat{\boldsymbol{\beta}}_T$, the assumptions (R1) and (R2) have to be strenghtened a little.

Theorem 8.4.2 *If in addition to assumptions (R1), (R2), (F1), (F2), (E1), (E2) the condition*

(SM) *there exists a* $\delta > 0$ *such that* $E|X_t\epsilon_t|^{2+\delta} = E|X_1\epsilon_1|^{2+\delta} < \infty$ *for all* t *and* $E(X_t^4\epsilon_t^2) = E(X_1^4\epsilon_1^2)$ *for all* t, *holds true, then the least squares estimator* $\widehat{\boldsymbol{\beta}}_T$ *is asymptotically normal,*

$$\sqrt{T}(\widehat{\boldsymbol{\beta}}_T - \boldsymbol{\beta}) \underset{T\to\infty}{\overset{d}{\to}} N(0, \sigma^2 \Sigma_{\mathbf{X}}),$$

as $T \to \infty$.

Proof. In dimension $d = 1$ we have

$$\sqrt{T}(\widehat{\boldsymbol{\beta}}_T - \boldsymbol{\beta}) = \left(\frac{1}{T} \sum_{t=1}^{T} X_t^2 \right)^{-1} \frac{1}{\sqrt{T}} \sum_{t=1}^{T} X_t \epsilon_t.$$

It suffices to show that we may apply Theorem 8.3.6 to the array of random variables

$$X_{Tt} = \frac{X_t \epsilon_t}{\sqrt{T}\sigma\sqrt{\Sigma_x}}, \quad 1 \le t \le T, T \ge 1,$$

where $\Sigma_x = \Sigma_{\mathbf{X}}$ signifies that we now work in dimension $d = 1$. We have shown in Lemma 8.2.2 that $X_t \epsilon_t$ is a $\mathcal{F}_t^*$-martingale difference sequence. It is easily verified that this implies that X_{Tt} is a $\mathcal{F}_t^*$-martingale difference array. This particulary yields $\operatorname{Var}\left(\sum_{t=1}^T X_{Tt}\right) = 1$ for all T. We will now show that

(i) $\frac{1}{T} \sum_{t=1}^T X_t^2 \epsilon_t^2 \overset{P}{\to} \sigma^2 \Sigma_X$, $T \to \infty$, and

(ii) $P\left(\max_{1 \le t \le T} |X_t \epsilon_t| > \varepsilon\eta\sqrt{T}\right) \to 0$, $T \to \infty$, where $\eta = \sigma\sqrt{\Sigma_x}$.

Let us first verify (ii). As in Example 8.3.9, we apply the union bound and Markov's inequality, but this time using the function $x \mapsto x^{2+\delta}$.

$$\begin{aligned} P\left(\max_{1 \le t \le T} |X_t \epsilon_t| > \varepsilon\eta\sqrt{T}\right) &= P\left(\bigcup_{t=1}^{T} \{|X_t \epsilon_t| > \varepsilon\eta\sqrt{T}\}\right) \\ &\le \sum_{t=1}^{T} P\left(|X_t \epsilon_t| > \varepsilon\eta\sqrt{T}\right) \\ &\le \sum_{t=1}^{T} \frac{E|X_t \epsilon_t|^{2+\delta}}{(\varepsilon\eta)^{2+\delta} T^{1+\delta/2}} \\ &= \frac{E|X_1 \epsilon_1|^{2+\delta}}{(\varepsilon\eta)^{2+\delta} T^{\delta/2}} \to 0, \end{aligned}$$

as $T \to \infty$, which shows (ii). The verification of (i) is more involved. Observe that

$$X_t^2 \epsilon_t^2 - \sigma^2 \Sigma_X = X_t^2 (\epsilon_t^2 - \sigma^2) + \sigma^2 (X_t^2 - \Sigma_X),$$

which implies the decomposition

$$\frac{1}{T} \sum_{t=1}^{T} (X_t^2 \epsilon_t^2 - \sigma^2 \Sigma_X) = \frac{1}{T} \sum_{t=1}^{T} X_t^2 (\epsilon_t^2 - \sigma^2) + \sigma^2 \frac{1}{T} \sum_{t=1}^{T} (X_t^2 - \Sigma_X).$$

By Assumption (R2), the second term on the right side converges to 0 in probability, as $T \to \infty$. We intend to apply the law of large numbers for martingale differences to the first term. Using (R1) and (SM), we have

$$\begin{aligned} E|X_t^2(\epsilon_t^2 - \sigma^2)| &\le E|X_t^2\epsilon_t^2| \,+\, E|X_t^2|\sigma^2 \\ &\le \sqrt{EX_t^4}\sqrt{E\epsilon_t^4} + E|X_t^2|\sigma^2 < \infty. \end{aligned}$$

Further,

$$E\big(X_t^2(\epsilon_t^2 - \sigma^2)|\mathcal{F}_{t-1}^*\big) = E\big(X_t^2 \cdot E(\epsilon_t^2 - \sigma^2|\mathcal{F}_{t-1})|\mathcal{F}_{t-1}^*\big) = 0,$$

by virtue of Assumption (E2), such that $X_t^2(\epsilon_t^2 - \sigma^2) \ge 1$, are $\mathcal{F}_t$-martingale differences. Now we check that

$$\frac{1}{T}\sum_{t=1}^{T} X_t^4(\epsilon_t^2 - \sigma^2)^2 \underset{T\to\infty}{\longrightarrow} c < \infty \text{ in } L_1,$$

for some constant c. Using (SM) we have

$$E\left(\frac{1}{T}\sum_{t=1}^{T} E\big(X_t^4(\epsilon_t^2 - \sigma^2)^2|\mathcal{F}_{t-1}^*\big)\right) = E\big(X_1^4(\epsilon_1^2 - \sigma^2)^2\big). \tag{8.8}$$

Now

$$\begin{aligned} E\big(X_1^4(\epsilon_1^2 - \sigma^2)^2\big) &= E\big(X_1^4\epsilon_1^4 - 2X_1^4\epsilon_1^2\sigma^2 + \sigma^4 X_1^4\big) \\ &= \sigma^2 E\big(X_1^4\epsilon_1^4\big) - 2\sigma^4 E\big(X_1^4\big) + \sigma^4 E\big(X_1^4\big), \end{aligned}$$

since $E\big(X_1^4\epsilon_1^2\big) = E\big(X_1^4 E(\epsilon_1^2|\mathcal{F}_1)\big) = \sigma^2 E\big(X_1^4\big)$.

Here, we used the $\mathcal{F}_1$-measureability of X_1, cf. (F1). We see that the left-hand side of Equation (8.8) converges to $c = E\big(X_1^4\epsilon_1^4 - \sigma^4 E\big(X_1^4\big) < \infty$. Thus, (i) is shown, which completes the proof.

8.5 Density estimation and nonparametric regression

We have already discussed the Rosenblatt–Parzen kernel density estimator in Chapter 1 for univariate samples. But in financial applications one typically has to deal with the multivariate case, for instance when analyzing the log returns of a portfolio. We have already seen that the bias is an issue in nonparametric estimation, both in its application in practice and in the derivation of analytic results. To simplify presentation, the analytic treatment focuses on the i.i.d. setting. However, the results are worked in such a way that they can be extended to dependent time series with relative ease. We shall discuss this issue in some detail also in Chapter 9, where the generalization to local polynomial estimation is given.

8.5.1 Multivariate density estimation

Having already discussed univariate nonparametric density estimation in Chapter 1, we shall now extend the method to the multivariate case, which is of practical interest as well as needed to understand nonparameteric regression. Let

$$X_t = (X_{t1}, \dots, X_{td}), \qquad t = 1, \dots, T,$$

be i.i.d. random vectors of dimension $d \in \mathbb{N}$ with common density function f. Let $L : \mathbb{R} \to [0, \infty)$ be an univariate kernel satisfying

$$\int L(z)\,\mathrm{d}z = 1, \int zL(z)\mathrm{d}z = 0, \tag{8.9}$$

$$L_2 = \int L(z)^2\,\mathrm{d}z < \infty, \quad \widetilde{L}_2 = \int z^2 L(z)\,\mathrm{d}z < \infty. \tag{8.10}$$

Let $h_1, \dots, h_d > 0$ be bandwidths, which we use to obtain a scaled smoothing kernel $h_j^{-1} L(x/h_j)$, $x \in \mathbb{R}$, for dimension $j = 1, \dots, d$. It is worth mentioning that one could also use a different smoothing kernel, L_j, for each dimension j, without any changes in the derivation except the extra indices. Thus, to simplify the exposition, we confine ourselves to the case $L_j = L$ for $j = 1, \dots, d$.

These d scaled univariate kernels are now combined to the corresponding scaled product kernel

$$\frac{1}{h_1 \cdots h_d} K\left(\frac{x}{h}\right) = \frac{1}{h_1 \cdots h_d} \prod_{j=1}^{d} L\left(\frac{x_j}{h_j}\right), \qquad x = (x_1, \dots, x_d) \in \mathbb{R}^d. \tag{8.11}$$

At first glance the notation on the left-hand side looks like a shortcut, but when putting $h = (h_1, \dots, h_d)$ and defining x/h for a vector $x \in \mathbb{R}^d$ as

$$(x_1/h_1, \dots, x_d/h_d),$$

then the left-hand side of Equation (8.11) is well defined, if we define $K : \mathbb{R}^d \to \mathbb{R}$ by

$$K(x_1, \dots, x_d) = \prod_{j=1}^{d} L(x_j), \qquad (x_1, \dots, x_d) \in \mathbb{R}^d.$$

Notice that the product kernel defines a d-dimensional distribution with independent coordinates that are marginally distributed as $h_j Z_j$, if $Z_j \sim L$, $j = 1, \dots, d$.

The multivariate kernel density estimator of f is now defined by

$$\widehat{f}_T(x) = \frac{1}{Th_1 \cdots h_d} \sum_{t=1}^{T} K\left(\frac{x - X_t}{h}\right), \qquad x \in \mathbb{R}^d.$$

When using the same bandwidth $h_j = h$, $j = 1, \dots, d$, for each dimension, the formula simplifies, but in practice the d variables often have different scales, which requires us to select the bandwidths separately in order to take this into account.

As for the univariate setting, it is interesting to study bias, variance and MSE of that estimator.

Proposition 8.5.1 *Assume that the partial derivatives of f up to the order three exist and are uniformly bounded. Then*

$$\mathrm{Bias}(\widehat{f}_T(x); f(x)) = \frac{\widetilde{L}_2}{2}\sum_{j=1}^{d} h_j^2 \frac{\partial^2 f}{\partial x_j^2} + O\left(\sum_{j=1}^{d} h_j^3\right), \tag{8.12}$$

$$\mathrm{Var}\,(\widehat{f}_T(x)) = \frac{1}{Th_1\cdots h_d}\left\{L_2^d f(x) + O\left(\sum_{j=1}^{d} h_j^2\right)\right\} \tag{8.13}$$

and

$$\mathrm{MSE}(\widehat{f}_T(x); f(x)) = O\left(\left(\sum_{j=1}^{d} h_j^2\right)^2 + \frac{1}{Th_1\cdots h_d}\right).$$

Proof. Let us calculate the bias of the kernel density estimator. We have

$$\begin{aligned} E(\widehat{f}_T(x)) - f(x) &= E\left(\frac{1}{h_1\cdots h_d}K\left(\frac{X_1 - x}{h}\right) - f(x)\right)\\ &= \frac{1}{h_1\cdots h_d}\int K\left(\frac{z-x}{h}\right) f(z)\,\mathrm{d}z - f(x)\\ &= \int K(u)[f(x+hu) - f(x)]\,\mathrm{d}u, \end{aligned}$$

by the d substitutions $u_j = (z_j - x_j)/h_j$, $j = 1, \ldots, d$, where here and in the following $\mathrm{d}u = \mathrm{d}u_1 \ldots \mathrm{d}u_d$. A Taylor expansion up to the terms of order three shows that the last expression of the above display can be written as

$$\int K(u)\left[\sum_{j=1}^{d}\frac{\partial f(x)}{\partial x_j}h_j u_j + \frac{1}{2}\sum_{j,k=1}^{d}\frac{\partial^2 f(x)}{\partial x_j \partial x_k}h_j h_k u_j u_k + O\Big(\sum_{j=1}^{d} h_j^3\Big)\right]\mathrm{d}u,$$

where we used the estimate $h_j h_k h_l \le \max\{h_j, h_k, h_l\}^3 = \max\left\{h_j^3, h_k^3, h_l^3\right\} \le \sum_{j=1}^{d} h_j^3$ to handle the terms involving derivatives of the order three. Observe that for all j

$$\int u_j K(u)\,\mathrm{d}u = \prod_{k\neq j}\int L(u_k)\,\mathrm{d}u_k \cdot \int u_j L(u_j)\,\mathrm{d}u_j = 0$$

and for $j \neq k$

$$\int u_j u_k K(u)\,\mathrm{d}u = \prod_{l\neq j,k}\int L(u_l)\,\mathrm{d}u_l \cdot \int u_j L(u_j)\,\mathrm{d}u_j \int u_k L(u_k)\,\mathrm{d}u_k = 0$$

as well as

$$\int u_j^2 K(u)\,\mathrm{d}u = \prod_{l \neq j} \int L(u_l)\,\mathrm{d}u_l \int u_j^2 L(u_j)\,\mathrm{d}u_j = \widetilde{L}_2.$$

Putting things together, we arrive at

$$E(\widehat{f}_T(x)) - f(x) = \left\{ \frac{\widetilde{L}_2}{2} \sum_{j=1}^d h_j^2 \frac{\partial^2 f}{\partial x_j^2} + O\left(\sum_{j=1}^d h_j^3 \right) \right\}.$$

Noticing that, by independence of the summands of $\widehat{f}_T(x)$,

$$\begin{aligned}(Th_1 \cdots h_d) \operatorname{Var}(\widehat{f}_T(x)) &= \frac{1}{h_1 \cdots h_d} \operatorname{Var} K\left(\frac{X_1 - x}{h} \right) \\ &\leq EK^2\left(\frac{X_1 - x}{h} \right),\end{aligned}$$

where

$$\begin{aligned}EK^2\left(\frac{X_1 - x}{h} \right) &= \int K^2\left(\frac{z - x}{h} \right) f(z)\,\mathrm{d}z \\ &= h_1 \cdots h_d \int K^2(u) f(x + hu)\,\mathrm{d}u \\ &= h_1 \cdots h_d \int K^2(u) \left[f(x) + \nabla f(x)'(hu) + \frac{1}{2}(hu)' Df(x^*)(hu) \right] \mathrm{d}u \\ &= h_1 \cdots h_d \biggl(f(x) \int K^2(u)\,\mathrm{d}u + \int K^2(u) \nabla f(x)' hu\,\mathrm{d}u \\ &\quad + \frac{1}{2} \int K^2(u)(hu)' Df(x^*)(hu)\mathrm{d}u \biggr),\end{aligned}$$

with $\int K^2(u)\,\mathrm{d}u = L_2^d$, for some x^* between x and $x + hu$, where $\nabla f(x)$ denotes the gradient of f at x and $Df(x^*)$ is the matrix of second partial derivatives of f evaluated at x^*. Denote by $\|x\| = \sum_{i=1}^d |x_i|$ the l_1 norm of a vector $x = (x_1, \ldots, x_d)' \in \mathbb{R}^d$. The associated matrix norm will be denoted by $\| \bullet \|$ as well. Since $\|Df(x^*)\| \leq c < \infty$ by assumptions (8.9) and (8.10), the inequality $|x'Ax| \leq \|A\| \|x\|^2$ leads to

$$\begin{aligned}\int |K^2(u)(hu)' Df(x^*)(hu)|\,\mathrm{d}u &\leq c \int K^2(u) \left(\sum_{j=1}^d |h_j||u_j| \right)^2 \mathrm{d}u \\ &\leq 2dc \int K^2(u) \sum_{j=1}^d h_j^2 u_j^2\,\mathrm{d}u.\end{aligned}$$

Here, we used the inequality $\left(\sum_{j=1}^d \alpha_j \right)^2 = O\left(\sum_{j=1}^d \alpha_j^2 \right)$ for real numbers $\alpha_1, \ldots, \alpha_d$. Indeed, $\alpha_j \alpha_k \leq \alpha_j^2 + \alpha_k^2$ leads to $\left(\sum_{j=1}^d \alpha_j \right)^2 = \sum_{j,k=1}^d \alpha_j \alpha_k \leq 2d \sum_{j=1}^d \alpha_j^2$. Since, in addition, $\sum_{j=1}^d \alpha_j^2 \beta_j^2 \leq \sum_{j=1}^d \alpha_j^2 (\beta_1^2 + \cdots + \beta_d^2) \leq \left(\sum_{j=1}^d \alpha_j^2 \right)\left(\sum_{j=1}^d \beta_j^2 \right)$ for real numbers

$\alpha_1, \ldots, \alpha_d, \beta_1, \ldots, \beta_d$, we can proceed as follows.

$$\begin{aligned}\int |K^2(u)(hu)' Df(x^*)(hu)|\,\mathrm{d}u &\leq 2dc \int K^2(u) \left(\sum_{j=1}^d h_j^2\right)\left(\sum_{j=1}^d u_j^2\right) \mathrm{d}u \\ &= 2dc \int \left(K^2(u) \sum_{j=1}^d u_j^2\right) \mathrm{d}u \sum_{j=1}^d h_j^2 \\ &= 2dc \sum_{j=1}^d \int K^2(u) u_j^2 \,\mathrm{d}u \sum_{j=1}^d h_j^2 \\ &= O\left(\sum_{j=1}^d h_j^2\right),\end{aligned}$$

since

$$\int u_j^2 K^2(u)\,\mathrm{d}u = \int u_j^2 \prod_{k=1}^d L(u_k)\,\mathrm{d}u = \prod_{k \neq j} \int L(u_k)\,\mathrm{d}u_j \int u_j^2 L(u_j)\,\mathrm{d}u_j < \infty.$$

Further,

$$\begin{aligned}\left|\int K^2(u) \nabla f(x)' hu \,\mathrm{d}u\right| &\leq \int K^2(u) \|\nabla f(x)\| \, \|hu\| \,\mathrm{d}u \\ &\leq \sqrt{\int K^2(u) \|\nabla f(x)\|^2 \,\mathrm{d}u} \sqrt{\int K^2(u) \|hu\|^2 \,\mathrm{d}u} \\ &= O\left(\int K^2(u) \sum_{j=1}^d h_j^2 u_j^2 \,\mathrm{d}u\right) \\ &= O\left(\sum_{j=1}^d h_j^2\right).\end{aligned}$$

Hence, using the formula $\mathrm{Var}\,(Z) = E(Z^2) - (E(Z))^2$ for any random variable Z with $EZ^2 < \infty$ leads to

$$\mathrm{Var}\, K\left(\frac{X_1 - x}{h}\right) \leq EK^2\left(\frac{X_1 - x}{h}\right) = O\left(\sum_{j=1}^d h_j^2\right).$$

Noting that $\int K^2(u)\,\mathrm{d}u f(x) = L_2^d f(x)$, we can conclude that

$$(Th_1 \cdots h_d)\, \mathrm{Var}\,(\widehat{f}_T(x)) = L_2^d f(x) + O\left(\sum_{j=1}^d h_j^2\right),$$

which proves Equation (8.13). The formula for the MSE now follows easily, since $\mathrm{MSE}(\widehat{f}_T(x); f(x)) = \mathrm{Bias}(\widehat{f}_T(x); f(x))^2 + \mathrm{Var}\,(\widehat{f}_T(x))$.

Having explicit formulas for the bias and variance, we can formulate explicit conditions for consistency of the kernel density estimator in d dimensions and determine the optimal bandwidth that should balance bias and variance.

Corollary 8.5.2 *The kernel density estimator $\widehat{f}_T(x)$ is consistent, if*

$$\max\{h_1, \dots, h_d\} \to 0, \qquad Th_1 \cdots h_d \to \infty.$$

The optimal bandwidths $h_1^*, \dots, h_d^*$ are such that $h_j^4 = O((Th_j)^{-1})$ leading to

$$h_j = c_j T^{-1/(d+4)}, \qquad j = 1, \dots, d.$$

It is remarkable that the kernel density is consistent under such general conditions; it is capable of recovering the arbitrary dependence between the coordinates of the X_t. This works, although we used, for simplicity, the product kernel corresponding to independence across components.

For what follows, it is worth mentioning the following simple fact. If one uses the optimal bandwidth choices $h_j = c_j T^{-1/(d+4)}$, then

$$\sqrt{Th_1 \cdots h_d h_j^2} = \sqrt{Th_1 \cdots h_d h_j^4} = \sqrt{C_j T^{1-d/(d+4)-4/(d+4)}} = \sqrt{C_j}, \tag{8.14}$$

where $C_j = \prod_{k=1, k \neq j}^{d} c_k c_j^5$, is constant, and therefore $\sqrt{Th_1 \cdots h_d} \sum_{j=1}^{d} h_j^2 \phi_j$ is also constant for arbitrary given numbers $\phi_1, \dots, \phi_d$, but $\sqrt{Th_1 \cdots h_d h_j^2 h^\gamma} \to 0$ and $\sqrt{Th_1 \cdots h_d} \sum_{j=1}^{d} h_j^2 h^\gamma \to 0$, as $T \to \infty$, for any $\gamma > 0$. We obtain the following assertion on the rescaled bias.

Remark 8.5.3 *When using the optimal bandwidths $h_j = c_j T^{-1/(d+4)}$, $j = 1, \dots, d$, then*

$$\sqrt{Th_1 \cdots h_d} \, \text{Bias}(\widehat{f}_T(x)) = \frac{\widetilde{L}_2}{2} \sum_{j=1}^{d} \sqrt{C_j} \frac{\partial^2 f}{\partial x_j^2} + o(1).$$

Indeed,

$$\begin{aligned}
\sqrt{Th_1 \cdots h_d} \, \text{Bias}(\widehat{f}_T(x)) &= \frac{\widetilde{L}_2}{2} \sum_{j=1}^{d} \sqrt{Th_1 \cdots h_d h_j^2} \frac{\partial^2 f}{\partial x_j^2} + o\left(\sum_{j=1}^{d} \sqrt{Th_1 \cdots h_d h_j^3} \right) \\
&= \frac{\widetilde{L}_2}{2} \sum_{j=1}^{d} \sqrt{C_j} \frac{\partial^2 f}{\partial x_j^2} + o(1).
\end{aligned}$$

Let us now study conditions under which the kernel density estimator is asymptotically normal. We will apply the following version of the central limit theorem for row-wise i.i.d. arrays of random variables.

Theorem 8.5.4 *Let $\{\xi_{Tt} : 1 \le t \le T, T \in \mathbb{N}\}$ be an array of random variables such that*

$$\xi_{T1}, \dots, \xi_{TT}$$

are i.i.d. for each $T \in \mathbb{N}$. If

$$\text{Var}\ (\xi_{T1}) \to \sigma^2 \in (0, \infty),$$

as $T \to \infty$, then

$$\frac{1}{\sqrt{T}} \sum_{t=1}^{T} [\xi_{Tt} - E(\xi_{Tt})] \stackrel{d}{\to} N(0, \sigma^2),$$

as $T \to \infty$.

Theorem 8.5.5 *Let $X_1, \ldots, X_T$ be i.i.d. random vectors taking values in $\mathbb{R}^d$ according to a common probability density $f(x)$. Suppose that f has compact support* supp(f) *or K is compactly supported and bounded. Assume that $f(x)$ has partial derivatives up to the order three with bounded derivatives of order three and fix $x \in supp(f)$. If*

$$h_j \to 0,\ j = 1, \ldots, d, \quad Th_1 \cdots h_d \to \infty$$

and

$$(Th_1 \cdots h_d) \sum_{j=1}^{d} h_j^6 \to 0, \tag{8.15}$$

as $T \to \infty$, then

$$\sqrt{Th_1 \cdots h_d} \left[\widehat{f}_T(x) - f(x) - \frac{\widetilde{L}_2}{2} \sum_{j=1}^{d} h_j^2 \frac{\partial^2 f(x)}{\partial x_j^2} \right] \stackrel{d}{\to} N(0, L_2^d f(x)),$$

as $T \to \infty$.

Proof. Since we already know that $\widehat{f}_T(x)$ is a biased estimator for $f(x)$, we consider the decomposition

$$\widehat{f}_T(x) - f(x) = \left(\widehat{f}_T(x) - E\widehat{f}_T(x) \right) + \left(E\widehat{f}_T(x) - f(x) \right).$$

The first step is to show

$$\sqrt{Th_1 \cdots h_d} \left[\widehat{f}_T(x) - E\widehat{f}_T(x) \right] \stackrel{d}{\to} N(0, L_2^d f(x)), \tag{8.16}$$

as $T \to \infty$. Notice that

$$\sqrt{Th_1 \cdots h_d} \left[\widehat{f}_T(x) - E\widehat{f}_T(x) \right] = \frac{1}{\sqrt{T}} \sum_{t=1}^{T} [\xi_{Tt} - E\xi_{Tt}],$$

with

$$\xi_{Tt} = \frac{1}{\sqrt{h_1 \cdots h_d}} K\left(\frac{x - X_t}{h} \right). \quad t = 1, \ldots, T.$$

Let us check whether $\mathrm{Var}\,(\xi_{T1})$ converges. We have $x = (x_1, \ldots, x_d)$

$$\begin{aligned}
E\,(\xi_{T1})^2 &= \frac{1}{h_1 \cdots h_d} \int \cdots \int \prod_{j=1}^{d} L\left(\frac{x_j - z_j}{h_j}\right)^2 f(z_1, \ldots, z_d) \prod_{j=1}^{d} \mathrm{d}z_j \\
&= \int \cdots \int \prod_{j=1}^{d} L(u_j)^2 f(x_1 + h_1 u_1, \ldots, x_d + h_d u_d) \prod_{j=1}^{d} \mathrm{d}u_j \\
&= \int \cdots \int \prod_{j=1}^{d} L(u_j)^2 \left[f(x_1, \ldots, x_d) + O\left(\sum_{j=1}^{d} h_j\right)\right] \prod_{j=1}^{d} \mathrm{d}u_j \\
&\to L_2^d f(x),
\end{aligned}$$

as $T \to \infty$, where we used the Taylor expansion

$$f(x + hu) = f(x) + \nabla f(x^*)'(hu) = f(x) + O\left(\sum_{j=1}^{d} h_j\right)$$

for some x^* between x and $x + hu$. Further, a similar argument shows that

$$E(\xi_{T1}) = O(\sqrt{h_1 \cdots h_d}),$$

and therefore

$$\mathrm{Var}\,(\xi_{T1}) \to L_2^d f(x),$$

as $T \to \infty$. Hence we can apply Theorem 8.5.4 and obtain Equation (8.16).

It remains to study the behavior of the bias term under the stated conditions. Since we have to scale $\widehat{f}_T(x) - f(x)$ with the factor $\sqrt{Th_1 \cdots h_d}$, we have to take into account our findings leading to and summarized in Remark 8.5.3. First, consider the scaled second term of the bias. We have

$$\begin{aligned}
\sqrt{Th_1 \cdots h_d} \sum_{j=1}^{d} h_j^3 &= \sum_{j=1}^{d} \sqrt{Th_1 \cdots h_d h_j^6} \\
&\le \sum_{j=1}^{d} \sqrt{Th_1 \cdots h_d \sum_{j=1}^{d} h_j^6} \\
&= d \sqrt{Th_1 \cdots h_d \sum_{j=1}^{d} h_j^6}
\end{aligned}$$

and analogously, with $\phi_j = \frac{\partial^2 f}{\partial x_j^2}$, for the first term one gets

$$\sqrt{Th_1 \cdots h_d} \sum_{j=1}^{d} h_j^2 \phi_j \le \sqrt{CTh_1 \cdots h_d \sum_{j=1}^{d} h_j^4},$$

where $C = d^2 \max_{1 \leq j \leq d} \phi_j^2$. The assumption $Th_1 \cdots h_d \sum_{j=1}^d h_j^6 = o(1)$ now ensures that the second term of $\sqrt{Th_1 \cdots h_d}(E\widehat{f}_T(x) - f(x))$ converges to 0, as $T \to \infty$, and allows for the optimal bandwidth choices so that the first term converges to a constant, see Remark 8.5.3. However, under the more general conditions stated in the present theorem, that term is subtracted from the $\sqrt{Th_1 \cdots h_d}[\widehat{f}_T(x) - f(x)]$, i.e. with

$$b_h(x) = \frac{\widetilde{L}_2}{2} \sum_{j=1}^d h_j^2 \frac{\partial^2 f(x)}{\partial x_j^2}$$

one considers $\sqrt{Th_1 \cdots h_d}[\widehat{f}_T(x) - f(x) - b_h(x)]$ instead. The corresponding decomposition is

$$\begin{aligned} \sqrt{Th_1 \cdots h_d}\left[\widehat{f}_T(x) - f(x) - b_h(x)\right] &= \sqrt{Th_1 \cdots h_d}\left[\widehat{f}_T(x) - E\widehat{f}_T(x)\right] \\ &\quad + \sqrt{Th_1 \cdots h_d}\left[E\widehat{f}_T(x) - f(x) - b_h(x)\right], \end{aligned}$$

where now

$$\sqrt{Th_1 \cdots h_d}\left[E\widehat{f}_T(x) - f(x) - b_h(x)\right] = O\left(\sqrt{Th_1 \cdots h_d}\sum_{j=1}^d h_j^3\right) = o(1).$$

as shown above. Now an application of Slutzky's lemma establishes the result,

$$\begin{aligned} \sqrt{Th_1 \cdots h_d}\left[\widehat{f}_T(x) - f(x) - b_h(x)\right] &= \sqrt{Th_1 \cdots h_d}\left[\widehat{f}_T(x) - E\widehat{f}_T(x)\right] + o(1) \\ &\stackrel{d}{\to} N(0, L_2^d f(x)), \end{aligned}$$

as $T \to \infty$, which completes the proof.

8.5.2 Nonparametric regression

Assume that (Y, X), (Y_t, X_t), $t = 1, \ldots, T$, are i.i.d. random vectors where Y is univariate and $X = (X_1, \ldots, X_d)'$ a d-dimensional random vector, distributed according to a joint probability density $f_{(Y,X)}(y, x)$ with marginal densities $f_Y(y)$ and $f_X(x)$, respectively. It is assumed that

$$Y_t = m(X_t) + \epsilon_t, \qquad t = 1, \ldots, T, \tag{8.17}$$

for some smooth function $m(x)$ and i.i.d. error terms $\{\epsilon_t\}$ with $E(\epsilon_t|X_t) = 0$ for all $t = 1, \ldots, T$. The extension to dependent processes will become clear in Section 9.2. This implies that $m(X_t)$ is the conditional mean of Y given X,

$$m(X_t) = E(Y_t|X_t),$$

almost surely. Further, we allow for **conditional heteroscedasticity** given X_t and assume that

$$E\left(\epsilon_t^2|X = X_t\right) = \sigma^2(X_t)$$

for some continuous function σ^2 defined on the range of the regressors, such that

$$\int K^2(x)\sigma^2(x) f_X(x)\,dx < \infty.$$

It is well known that

$$E(Y|X=x) = \int y f_{Y|X=x}(y)\,dy,$$

where

$$f_{Y|X=x}(y) = \begin{cases} \frac{f_{(Y,X)}(y,x)}{f_X(x)}, & y \in \mathbb{R}, x \in \{f_X \neq 0\}, \\ f_Y(y), & y \in \mathbb{R}, x \in \{f_X = 0\}. \end{cases}$$

Since $P(X \in \{f_X \neq 0\}) = 0$, it is common to ignore the latter case and assume without loss of generality that $f_X > 0$. In order to estimate $E(Y|X=x)$, it suffices to estimate the conditional density $f_{Y|X=x}(y)$. But that problem can be reduced to the problem to estimate the joint $(d+1)$-dimensional density $f_{(Y,X)}(y,x)$ and the d-dimensional density $f_X(x)$, which we studied in the previous subsection. Given such estimates $\widehat{f}_T(y,x)$ and $\widehat{f}_T(x)$, we can use

$$\widehat{f}_T(y|x) = \widehat{f}_{Y|X=x}(y) = \frac{\widehat{f}_T(y,x)}{\widehat{f}_T(x)}$$

to estimate the conditional density. In what follows, we use the kernel density estimator.

Select a smoothing kernel G for the y-coordinate satisfying assumptions (8.9) and (8.10) and a bandwidth $h_0 > 0$. Define the scaled product kernel

$$\frac{1}{h_0 \cdots h_d} K\left(\frac{x}{h}\right) G\left(\frac{y}{h_0}\right) = \frac{1}{h_0 \cdots h_d} \prod_{j=1}^{d} L\left(\frac{x_j}{h_j}\right) G\left(\frac{y}{h_0}\right), \qquad x \in \mathbb{R}^d, y \in \mathbb{R},$$

where again $x/h = (x_1/h_1, \ldots, x_d/h_d)$. The corresponding estimator of the joint density of (Y, X) is then given by

$$\widehat{f}_T(y,x) = \frac{1}{Th_0 \cdots h_d} \sum_{t=1}^{T} K\left(\frac{x - X_t}{h}\right) G\left(\frac{y - Y_t}{h_0}\right), \qquad x \in \mathbb{R}^d, y \in \mathbb{R}.$$

Let us calculate the associated conditional mean

$$\int y \widehat{f}_T(y|x)\,dy = \frac{1}{\widehat{f}_T(x)} \int y \widehat{f}_T(y,x)\,dy.$$

We have

$$\int y \widehat{f}_T(y,x)\,dy = \frac{1}{Th_0 \cdots h_d} \sum_{t=1}^{T} K\left(\frac{x - X_t}{h}\right) \int yG\left(\frac{y - Y_t}{h_0}\right) dy$$

where

$$\begin{aligned}\int yG\left(\frac{y-Y_t}{h_0}\right)\mathrm{d}y &= h_0\int (Y_t+h_0u)G(u)\,\mathrm{d}u\\ &= h_0Y_t + h_0^2\int uG(u)\,\mathrm{d}u\\ &= h_0Y_t.\end{aligned}$$

Hence, we obtain

$$\int y\widehat{f}_T(y,x)\,\mathrm{d}y = \frac{1}{Th_1\cdots h_d}\sum_{t=1}^{T}K\left(\frac{x-X_t}{h}\right)Y_t.$$

These considerations lead to the following definition.

Definition 8.5.6 *The* **Nadaraya–Watson estimator** *for the conditional mean $m(x) = E(Y|X=x)$ is defined as*

$$\widehat{m}_T(x) = \frac{\sum_{t=1}^{T}K\left(\frac{x-X_t}{h}\right)Y_t}{\sum_{t=1}^{T}K\left(\frac{x-X_t}{h}\right)}, \qquad x\in\mathbb{R}^d.$$

Notice that $\widehat{m}_T(x) = \widehat{m}_T(x; Y_1, \ldots, Y_T)$ is a weighted average of $Y_1, \ldots, Y_T$ with weights summing up to 1. Hence, for fixed x and any constant a, we have the rule of calculation

$$\widehat{m}_T(x; Y_1, \ldots, Y_T) - a = \widehat{m}_T(x; Y_1 - a, \ldots, Y_T - a). \tag{8.18}$$

Our next goal is to study the behavior of the difference $\widehat{m}_T(x) - m(x)$. The basic identity used for that purpose is

$$\widehat{m}_T(x) - m(x) = \frac{[\widehat{m}_T(x) - m(x)]\widehat{f}_T(x)}{\widehat{f}_T(x)} = \frac{\widehat{g}_T(x)}{\widehat{f}_T(x)}. \tag{8.19}$$

If, for example, we are in a position to establish the convergence in distribution of a scaled version of the numerator, perhaps after some bias correction as was necessary in the case of the kernel density estimator, the asymptotic normality of $\widehat{m}_T$ will follow quite easily. Consider

$$\widehat{g}_T(x) = [\widehat{m}_T(x) - m(x)]\widehat{f}_T(x) = \frac{1}{Th_1\ldots h_d}\sum_{t=1}^{T}K\left(\frac{x-X_t}{h}\right)(Y_t - m(x)).$$

The model Equation (8.17), which implies that $Y_t - m(X_t) = \epsilon_t$, leads us to the decomposition

$$\widehat{g}_T(x) = \widetilde{g}_T(x) + \overline{g}_T(x),$$

where

$$\widetilde{g}_T(x) = \frac{1}{Th_1 \ldots h_d} \sum_{t=1}^{T} K\left(\frac{x - X_t}{h}\right) \epsilon_t,$$

$$\overline{g}_T(x) = \frac{1}{Th_1 \ldots h_d} \sum_{t=1}^{T} K\left(\frac{x - X_t}{h}\right) [m(X_t) - m(x)].$$

The strategy is now to study both terms separately. The following theorem shows that $\overline{g}_T(x)$ converges in L_2 to a deterministic expression involving the bandwidths.

Theorem 8.5.7 *Suppose that the third-order partial derivatives of f and m exist and are bounded, and that K is bounded, symmetric and is compactly supported.*

(i) We have

$$E\overline{g}_T(x) = \widetilde{L}_2 \sum_{j=1}^{d} h_j^2 b_j(m; f) + O\left(\sum_{j=1}^{d} h_j^3\right),$$

where

$$b_j(m; f) = \frac{\partial m(x)}{\partial x_j} \frac{\partial f(x)}{\partial x_j} + \frac{1}{2} \frac{\partial^2 m(x)}{\partial x_j^2} f(x). \tag{8.20}$$

(ii) For i.i.d. random vectors (Y_t, X_t), $t = 1, \ldots, T$,

$$\operatorname{Var}\left(\overline{g}_T(x)\right) = O\left(\frac{1}{Th_1 \cdots h_d} \sum_{j=1}^{d} h_j^2\right).$$

Proof. For brevity of notation put $f = f_X$.

(i) We have

$$\begin{aligned} E\overline{g}_T(x) &= \frac{1}{h_1 \ldots h_d} \int K\left(\frac{x - z}{h}\right) [m(z) - m(x)] f(z)\, dz \\ &= \int K(u)[m(x + hu) - m(x)] f(x + hu)\, du \end{aligned}$$

Taylor expansions show that the above expression is equal to

$$\int K(u) \left[\nabla m(x)(hu) + \frac{1}{2}(hu)' Dm(x)(hu) + r_m\right] \left[f(x) + \nabla f(x)(hu) + r_f\right] du,$$

which in turn simplifies to

$$\int K(u) \left[\frac{1}{2}(hu)' Dm(x)(hu) + r_m\right] \left[\nabla f(x)(hu) + r_f\right] du,$$

since

$$f(x)\nabla m(x)\int uK(u)\,\mathrm{d}u = 0$$

by Assumption (8.9). Here, by virtue of our assumptions on the higher-order partial derivatives of m and f and since integration is over a compact interval,

$$r_m = \frac{1}{3!}\sum_{i,j,k=1}^{d}\frac{\partial^3 m(x^*)}{\partial x_i \partial x_j \partial x_k} h_i h_j h_k u_i u_j u_k = O\left(\sum_{j=1}^{d} h_j^3\right), \tag{8.21}$$

$$r_f = \frac{1}{2}(hu)' Df(x^+)(hu) = O\left(\sum_{j=1}^{d} h_j^2\right), \tag{8.22}$$

and x^* and x^+ lie between x and $x + hu$. Now expand the product $[\cdots][\cdots]$. We claim that

$$E\overline{g}_T(x) = \int K(u)\left[\nabla m(x)(hu)\nabla f(x)(hu) + \frac{1}{2}(hu)' Dm(x)(hu) f(x)\right]\mathrm{d}u$$
$$+ O\left(\sum_{j=1}^{d} h_j^3\right),$$

where the first integral, denoted by I_T in what follows, is of the order $O\big(\sum_{j=1}^{d} h_j^2\big)$. Precisely, we have

$$\begin{aligned} I_T &= \int K(u)\sum_{j,k=1}^{d}\left[\frac{\partial m(x)}{\partial x_j}\frac{\partial f(x)}{\partial x_k} + \frac{1}{2}\frac{\partial^2 m(x)}{\partial x_j \partial x_k} f(x)\right] h_j h_k u_j u_k\,\mathrm{d}u \\ &= \int K(u)\sum_{j=1}^{d}\left[\frac{\partial m(x)}{\partial x_j}\frac{\partial f(x)}{\partial x_j} + \frac{1}{2}\frac{\partial^2 m(x)}{\partial x_j^2} f(x)\right] h_j^2 u_j^2\,\mathrm{d}u \\ &= \widetilde{L}_2 \sum_{j=1}^{d} h_j^2 b_j(m; f), \end{aligned}$$

where $b_j(m; f)$ is given in Equation (8.20). The other terms indeed vanish such as

$$\int K(u)\nabla m(x)'(hu) f(x)\,\mathrm{d}u = f(x)\nabla m(x)' h\left(\int K(u) u_j\,\mathrm{d}u\right)_{j=1}^{d} = 0,$$

are of the order $O\left(\sum_{j=1}^{d} h_j^3\right)$. For example,

$$
\begin{aligned}
J_h &= \int K(u)\left[\frac{1}{2}(hu)' Dm(x)(hu)\nabla f(x)(hu)\right] du \\
&= \frac{1}{2}\int K(u) \sum_{i,j,k=1}^{d} \frac{\partial^2 m(x)}{\partial x_j \partial x_k}\frac{\partial f(x)}{\partial x_i} h_i h_j h_k u_i u_j u_k \, du \\
&= \frac{1}{2}\int K(u) \sum_{j=1}^{d} \frac{\partial^2 m(x)}{\partial x_j^2}\frac{\partial f(x)}{\partial x_j} h_j^3 u_j^3 \, du \\
&= O\left(\sum_{j=1}^{d} h_j^3\right).
\end{aligned}
$$

(ii) By virtue of the i.i.d. assumption and the usual substitutions $u_j = (z_j - x_j)/h_j$, $j = 1, \dots, d$,

$$
\begin{aligned}
\mathrm{Var}\,(\overline{g}_T) &= \frac{1}{Th_1^2 \cdots h_d^2} \mathrm{Var}\left(K\left(\frac{x - X_t}{h}\right)(m(X_T) - m(x))\right) \\
&\leq \frac{1}{Th_1^2 \cdots h_d^2} E\left(K\left(\frac{x - X_t}{h}\right)^2 (m(X_T) - m(x))^2\right) \\
&= \frac{1}{Th_1 \cdots h_d}\int K^2(u)[m(x + hu) - m(x)]^2 f(x + hu)\, du.
\end{aligned}
$$

The latter expression can now be treated similarly as in (i), yielding the estimate

$$
\mathrm{Var}\,(\overline{g}_T) = \frac{1}{Th_1 \cdots h_d} O\left(\sum_{j=1}^{d} h_j^2\right).
$$

Notice that the proof of Theorem 8.5.7 does not assume that $m(x)$ is the conditional mean $E(Y|X = x)$. Indeed, the result can be extended to more general functions leading to the following weak law of large numbers for smoothing averages.

Theorem 8.5.8 (Weak Law of Large Numbers for Kernel Smoothers)
Suppose $X_t \overset{i.i.d.}{\sim} f(x)$ and let $\psi(z, x)$ be a function defined on $X(\Omega) \times X(\Omega)$ with $E|\psi(X_1, x)|^2 < \infty$. If the third-order partial derivatives of f and ψ exist and are bounded, then

$$
\begin{aligned}
&\frac{1}{Th_1 \cdots h_d}\sum_{t=1}^{T} K\left(\frac{X_t - x}{h}\right)[\psi(X_t, x) - \psi(x, x)] \\
&\quad = \int K\left(\frac{z - x}{h}\right)[\psi(z, x) - \psi(x, x)](z)\, dz + o_P(1), \qquad (8.23)
\end{aligned}
$$

and

$$\int K\left(\frac{z-x}{h}\right)[\psi(z,x)-\psi(x,x)](z)\,\mathrm{d}z = o(1).$$

Theorem 8.5.9 *Assume that the third-order partial derivatives of f_X and m exist and are bounded, and that K is bounded and compactly supported. If*

$$\max_{j=1,\ldots,d} h_j \to 0, \qquad Th_1\cdots h_d \to \infty$$

and

$$(Th_1\cdots h_d)\sum_{j=1}^{d} h_j^6 \to 0, \tag{8.24}$$

as $T\to\infty$, then

$$\sqrt{Th_1\cdots h_d}\left(\widehat{m}_T(x)-m(x)-\widetilde{L}_2\sum_{j=1}^{d}h_j^2 b_j(m;f)\right)\xrightarrow{d} N\left(0, L_2^d\sigma^2(x)/f_X(x)\right),$$

as $T\to\infty$, for any x such that $f_X(x)>0$.

Proof. Recalling the decomposition (8.19) we see that

$$\widehat{f}_T(x)\sqrt{Th_1\cdots h_d}\left(\widehat{m}_T(x)-m(x)-L_2\sum_{j=1}^{d}h_j^2 b_j(m;f)\right) = \widehat{f}_T(x)(A_T(x)+B_T(x)),$$

with

$$A_T(x)=\sqrt{Th_1\cdots h_d}\,\widetilde{g}_T(x),$$

$$B_T(x)=\sqrt{Th_1\cdots h_d}\left(\overline{g}_T(x)-\widetilde{L}_2\sum_{j=1}^{d}h_j^2 b_j(m;f)\right).$$

According to our results of the previous proposition,

$$EB_T(x)=O\left(\sqrt{Th_1\cdots h_d}\sum_{j=1}^{d}h_j^3\right),$$

and

$$\operatorname{Var} B_T(x)=O\left(\sum_{j=1}^{d}h_j^2\right).$$

Both right-hand sides converge to 0, as $T\to\infty$, by virtue of Assumption (8.24), as shown in the proof of Theorem 8.5.5. This shows that

$$B_T \xrightarrow{L_2,P} 0,$$

as $T \to \infty$. Consider

$$\begin{aligned} A_T(x) &= \frac{1}{\sqrt{Th_1 \cdots h_d}} \sum_{t=1}^{T} K\left(\frac{x - X_t}{h}\right) \epsilon_t \\ &= \frac{1}{\sqrt{T}} \sum_{t=1}^{T} \xi_{Tt}, \end{aligned}$$

where

$$\xi_{Tt} = \frac{1}{\sqrt{h_1 \cdots h_d}} K\left(\frac{x - X_t}{h}\right) \epsilon_t,$$

for $t = 1, \ldots, T$ and $T \geq 1$. Clearly, the summands are centered and using $E\left(\epsilon_t^2 | X_t\right) = \sigma^2(X_t)$, we obtain

$$\begin{aligned} \operatorname{Var}(\xi_{Tt}) &= E\left(E\left(\xi_{Tt}^2 | X_t\right)\right) \\ &= E_{X_t}\left(\frac{1}{h_1 \cdots h_d} K^2\left(\frac{x - X_t}{h}\right) \sigma^2(X_t)\right) \\ &= \int \left(\frac{1}{h_1 \cdots h_d} K^2\left(\frac{x - z}{h}\right) \sigma^2(z) f_X(z)\right) \mathrm{d}z \\ &= \int K^2(u) \sigma^2(x + hu) f_X(x + hu) \, \mathrm{d}u. \end{aligned}$$

But the latter expression clearly converges to $L_2^d \sigma^2(x) f_X(x)$. Hence,

$$A_T(x) \xrightarrow{d} N(0, L_2^d \sigma^2(x) f_X(x)),$$

as $T \to \infty$, and, by virtue of Slutzky's theorem,

$$\widehat{f}_T(x) \sqrt{Th_1 \cdots h_d} \left(\widehat{m}_T(x) - m(x) - \widetilde{L}_2 \sum_{j=1}^{d} h_j^2 b_j(m; f) \right) \xrightarrow{d} N(0, L_2^d \sigma^2(x) f_X(x)),$$

as $T \to \infty$, as well. Since under the stated conditions $\widehat{f}_T(x) \xrightarrow{P} f(x)$, as $T \to \infty$, a further application of Slutzky's lemma completes the proof.

8.6 The CLT for linear processes

We have seen that linear processes form an important class of time series. Thus, this section is devoted to a detailed proof of the central limit theorem for such processes. We shall employ the Beveridge–Nelson decomposition, which allows an elegant and transparent proof.

Definition and Theorem 8.6.1 *Let*

$$X_t = \sum_{i=0}^{\infty} \alpha_i \epsilon_{t-i}, \qquad t \in \mathbb{Z},$$

be a linear process with coefficients $\{\alpha_i\}$ *satisfying* $\sum_{i=0}^{\infty} |\alpha_i| < \infty$. *Then the decomposition*

$$X_t = \overline{\alpha}\epsilon_t + \widetilde{\epsilon}_{t-1} - \widetilde{\epsilon}_t, \qquad t \in \mathbb{Z},$$

where $\overline{\alpha} = \sum_{i=0}^{\infty} \alpha_i$ *and*

$$\widetilde{\epsilon}_t = \sum_{i=0}^{\infty} \widetilde{\alpha}_i \epsilon_{t-i}, \quad \widetilde{\alpha}_i = \sum_{k=i+1}^{\infty} \alpha_k,$$

is called the **Beveridge–Nelson decomposition**. *More compactly, the linear process* $X_t = \alpha(L)\epsilon_t$ *can be written as*

$$X_t = \alpha(1)\epsilon_t - (1-L)\widetilde{\epsilon}_t, \quad \widetilde{\epsilon}_t = \widetilde{\alpha}(L)\epsilon_t,$$

where the lag operator $\widetilde{\alpha}(L)$ *is given by the cofficients* $\widetilde{\alpha}_i$.

Proof. Notice that for a linear process

$$\widetilde{\epsilon}_t = \sum_{i=0}^{\infty} \widetilde{\alpha}_i \epsilon_{t-i}$$

with coefficients $\{\widetilde{\alpha}_i\}$ we have

$$\begin{aligned}-\widetilde{\epsilon}_t + \widetilde{\epsilon}_{t-1} = &-(\widetilde{\alpha}_0\epsilon_t + \widetilde{\alpha}_1\epsilon_{t-1} + \widetilde{\alpha}_2\epsilon_{t-1} + \cdots) \\ &+ \widetilde{\alpha}_0\epsilon_{t-1} + \widetilde{\alpha}_1\epsilon_{t-2} + \cdots,\end{aligned}$$

leading to

$$-\widetilde{\epsilon}_t + \widetilde{\epsilon}_{t-1} = -\widetilde{\alpha}_0\epsilon_t + (\widetilde{\alpha}_0 - \widetilde{\alpha}_1)\epsilon_{t-1} + (\widetilde{\alpha}_1 - \widetilde{\alpha}_2)\epsilon_{t-2} + \cdots$$

For the special choice

$$\widetilde{\alpha}_i = \sum_{k>i} \alpha_k$$

we obtain

$$\widetilde{\alpha}_0 = \sum_{k=1}^{\infty} \alpha_k \quad \text{and} \quad \widetilde{\alpha}_i - \widetilde{\alpha}_{i+1} = \left(\sum_{k=i+1}^{\infty} - \sum_{k=i+2}^{\infty} \right) \alpha_k = \alpha_{i+1}.$$

Therefore,

$$-\widetilde{\epsilon}_t + \widetilde{\epsilon}_{t-1} = -\left(\sum_{k=1}^{\infty} \alpha_k \right) \epsilon_t + \sum_{i=1}^{\infty} \alpha_i \epsilon_{t-i}.$$

Adding the term $\alpha(1)\epsilon_t = \left(\sum_{k=0}^{\infty} \alpha_k\right) \epsilon_t$ on both sides now yields the assertion,

$$\alpha(1)\epsilon_t - \widetilde{\epsilon}_t + \widetilde{\epsilon}_{t-1} = \left(\sum_{k=0}^{\infty} \alpha_k - \sum_{k=1}^{\infty} \alpha_k \right) \epsilon_t + \sum_{i=1}^{\infty} \alpha_i \epsilon_{t-i} = X_t.$$

To proceed, we need the following simple facts.

Lemma 8.6.2

(i) If $\sum_{i=0}^{\infty} i|\alpha_i| < \infty$*, then* $\sum_{i=0}^{\infty} i^2\alpha_i^2 < \infty$.

(ii) The process $\{\widetilde{\epsilon}_t\}$ *arising in the Beveridge–Nelson decomposition is weakly stationary (and thus a* L_2 *series), if* $\sum_{i=0}^{\infty} i|\alpha_i| < \infty$.

Proof.

(i) We can find some $i_0 \in \mathbb{N}$ such that $i|\alpha_i| < 1$ for all $i > i_0$. For those i we have $i^2\alpha_i^2 < i|\alpha_i|$. Therefore,

$$\begin{aligned}\sum_{i=0}^{\infty} i^2\alpha_i^2 &= \sum_{i=0}^{i_0} i^2\alpha_i^2 + \sum_{i=i_0+1}^{\infty} i^2\alpha_i^2 \\ &< \sum_{i=0}^{i_0} i^2\alpha_i^2 + \sum_{i=i_0+1}^{\infty} i|\alpha_i|.\end{aligned}$$

The first sum has only a finite number of terms and is therefore finite. The second sum can be bounded by $\sum_{i=0}^{\infty} i|\alpha_i| < \infty$.

(ii) It suffices to show that $\sum_{i=0}^{\infty} \widetilde{\alpha}_i^2 < \infty$, which implies $\sum_{i=0}^{\infty} |\widetilde{\alpha}_i| < \infty$, since then we can apply Theorem A.5.1. We have

$$\begin{aligned}\sum_{i=0}^{\infty} |\widetilde{\alpha}_i| &\leq \sum_{i=0}^{\infty}\sum_{j=i+1}^{\infty} |\alpha_j| \\ &= \sum_{j=1}^{\infty}\sum_{i=0}^{j-1} |\alpha_j| \\ &= \sum_{j=1}^{\infty} j|\alpha_j| < \infty.\end{aligned}$$

Therefore $\{\widetilde{\alpha}_j\}$ is a null sequence such that we can find an integer i_0 so that $|\widetilde{\alpha}_i| \leq 1$ for all $i > i_0$. Arguing as above, we obtain

$$\begin{aligned}\sum_{i=0}^{\infty} \widetilde{\alpha}_i^2 &= \sum_{i=0}^{i_0} \widetilde{\alpha}_i^2 + \sum_{i=i_0+1}^{\infty} \widetilde{\alpha}_i^2 \\ &\leq \sum_{i=0}^{i_0} \widetilde{\alpha}_i^2 + \sum_{i=i_0+1}^{\infty} \widetilde{\alpha}_i^2 \\ &\leq \sum_{i=0}^{i_0} \widetilde{\alpha}_i^2 + \sum_{i=i_0+1}^{\infty} |\widetilde{\alpha}_i| < \infty.\end{aligned}$$

We are now in a position to discuss the central limit theorem.

Theorem 8.6.3 (CLT FOR LINEAR PROCESSES)
Let

$$X_t = \sum_{i=0}^{\infty} \alpha_i \epsilon_{t-i}$$

be a linear process with $\epsilon_t \overset{i.i.d.}{\sim} (0, \sigma^2)$ *for some* $\sigma^2 \in (0, \infty)$. *If*

$$\sum_{i=0}^{\infty} i|\alpha_i| < \infty \qquad \text{and} \qquad \sum_{i=0}^{\infty} \alpha_i \neq 0,$$

then

$$S_T = \frac{1}{\sqrt{T}} \sum_{t=1}^{T} X_t \xrightarrow{d} N\left(0, \sigma_X^2\right),$$

as $T \to \infty$, *where* $\sigma_X^2 = \sigma^2 \left(\sum_{i=0}^{\infty} \alpha_i\right)^2$.

Proof. By virtue of the Beveridge–Nelson decomposition, we obtain

$$S_T = U_T + R_T$$

with

$$U_T = \frac{\overline{\alpha}}{\sqrt{T}} \sum_{t=1}^{T} \epsilon_t,$$

$$R_T = \frac{1}{\sqrt{T}} \sum_{t=1}^{T} (\widetilde{\epsilon}_{t-1} - \widetilde{\epsilon}_t).$$

Clearly,

$$R_T = \frac{1}{\sqrt{T}} (\widetilde{\epsilon}_0 - \widetilde{\epsilon}_T).$$

The classic CLT for i.i.d. random variables $\{\epsilon_t\}$ with $E(\epsilon_1) = 0$ and $E(\epsilon_1^2) < \infty$ provides

$$U_T \xrightarrow{d} N(0, \overline{\alpha}^2 \sigma^2),$$

as $T \to \infty$. It remains to show that

$$R_T \xrightarrow{P} 0,$$

as $T \to \infty$. Notice that

$$|R_T| \leq \frac{1}{\sqrt{T}} |\widetilde{\epsilon}_0 - \widetilde{\epsilon}_T| \leq \frac{2}{\sqrt{T}} \max_{1 \leq t \leq T} |\widetilde{\epsilon}_t|.$$

We shall use the equivalence

$$\max_{1\le t\le T} |\widetilde{\epsilon}_t| > \delta \quad \Leftrightarrow \quad \sum_{t=1}^{T} \widetilde{\epsilon}_t^2 \mathbf{1}(|\widetilde{\epsilon}_t| > \delta) > \delta^2 \tag{8.25}$$

for any $\delta > 0$, which can be shown as follows. First, notice that $\max_{1\le t\le T} |\widetilde{\epsilon}_t| > \delta$ implies that there is some j with $|\widetilde{\epsilon}_j| > \delta$. But then

$$\sum_{t=1}^{T} \widetilde{\epsilon}_t^2 \mathbf{1}(|\widetilde{\epsilon}_t| > \delta) \ge |\widetilde{\epsilon}_j^2| > \delta^2.$$

Vice versa, if $\sum_{t=1}^{T} \widetilde{\epsilon}_t^2 \mathbf{1}(|\widetilde{\epsilon}_t| > \delta) > \delta^2$, then there must be some j with $|\widetilde{\epsilon}_j| > \delta$, since otherwise the sum vanishes. Therefore $\max_{1\le t\le T} |\widetilde{\epsilon}_t| > \delta$, which establishes (8.25). Repeating these arguments with $\widetilde{\epsilon}_t/\sqrt{T}$ instead of $\widetilde{\epsilon}_t$, we arrive at

$$\max_{1\le t\le T} \frac{|\widetilde{\epsilon}_t|}{\sqrt{T}} \xrightarrow{P} 0 \quad \Leftrightarrow \quad \frac{1}{T}\sum_{t=1}^{T} \widetilde{\epsilon}_t^2 \mathbf{1}(|\widetilde{\epsilon}_t| > \delta\sqrt{T}) \xrightarrow{P} 0.$$

Since $\{\epsilon_t\}$ is a strictly stationary L_2-process, we have

$$\begin{aligned} E\left(\frac{1}{T}\sum_{t=1}^{T} \widetilde{\epsilon}_t^2 \mathbf{1}(|\widetilde{\epsilon}_t| > \delta\sqrt{T}\right) &= E\left(\widetilde{\epsilon}_1^2 \mathbf{1}(|\widetilde{\epsilon}_1| > \delta\sqrt{T})\right) \\ &= \int_{-\infty}^{-\delta\sqrt{T}} x^2 \, \mathrm{d}F(x) + \int_{\delta\sqrt{T}}^{\infty} x^2 \, \mathrm{d}F(x) \\ &\to 0, \end{aligned}$$

as $T \to \infty$, if $F(x) = P(\epsilon_1 \le x)$, $x \in \mathbb{R}$. But this shows $R_T \xrightarrow{P} 0$, as $T \to \infty$, which completes the proof.

8.7 Mixing processes

8.7.1 Mixing coefficients

Let $\{X_t\} = \{X_t : t \in \mathbb{Z}\}$ be a time series. Denote current time by t. The question arises to which extent the past represented by previous observations such as X_s, $s < t$, influences the present and future. One approach is to look at the maximal correlation between random variables of the form

$$Y = f(X_s, X_{s-1}, \ldots), \qquad Z = g(X_t, X_{t-1}, \ldots), \tag{8.26}$$

where f and g are measurable functions such that the correlation exists. Clearly, Y is $\mathcal{F}_{-\infty}^{s}$-measurable and Z is $\mathcal{F}_t^{\infty}$-measurable, where for $-\infty \le a \le b \le \infty$

$$\mathcal{F}_a^b = \sigma(X_a, \ldots, X_b). \tag{8.27}$$

Vice versa, any $\mathcal{F}_{-\infty}^{s}$-measurable random variable Y and any $\mathcal{F}_t^{\infty}$-measurable random variable Z can be written in the form (8.26). To turn this rough idea into a definition, let us denote the underlying probability space by $(\Omega, \mathcal{F}, P)$ and, for some sub-σ-field $\mathcal{A}$, let $L_2(\Omega, \mathcal{A}, P)$ be the L_2-space of $\mathcal{A}$-measurable random variables defined on Ω.

Definition 8.7.1 (ρ-MIXING)

(i) Let $\mathcal{A}$ and $\mathcal{B}$ be two sub-σ-fields of $\mathcal{F}$. Then

$$\rho(\mathcal{A}, \mathcal{B}) = \sup\{|\operatorname{Cov}(X, Y)| : X \in L_2(\Omega, \mathcal{A}, P),\ Y \in L_2(\Omega, \mathcal{B}, P)\}$$

is called the ρ**-mixing** *coefficient between $\mathcal{A}$ and $\mathcal{B}$.*

(ii) Let $\{X_t\}$ be a L_2 series, then

$$\rho(k) = \sup_t \rho\left(\mathcal{F}_{-\infty}^{t-k}, \mathcal{F}_t^{\infty}\right), \qquad k \in \mathbb{N}_0,$$

where $\mathcal{F}_a^b$ is defined in Equation (8.27), is called the ρ**-mixing coefficient of** $\{X_t\}$ *for the lag k.*

(iii) A L_2 time series $\{X_t\}$ is called ρ-mixing, if

$$\lim_{k\to\infty} \rho(k) = 0.$$

One can show that

$$\rho(\mathcal{A}, \mathcal{B}) = \sup\{|\operatorname{Cov}(X, Y)|\}$$

where the supremum is taken over all those random variables X and Y that satisfy

$$X \in L_2(\Omega, \mathcal{A}, P),\ Y \in L_2(\Omega, \mathcal{B}, P),\ E(X) = E(Y) = 0,\ \|X\|_{L_2} = \|Y\|_{L_2} = 1.$$

Further, by considering the correlation instead of the covariance, one can drop the constraints on the second moment.

Let $A \in \mathcal{F}_{-\infty}^{t-k}$ be an event depending on the past $\ldots, X_{t-k-1}, X_{t-k}$ up to time $t-k$ and $B \in \mathcal{F}_t^{\infty}$ be an event depending on future values $X_t, X_{t+1}, \ldots$. This means that there is a time lag between these events of k units of time. Time series for which dependencies between lagged events die out for large lags should have the property that A and B are asymptotically independent in the sense that

$$P(A \cap B) \approx P(A)P(B)$$

is valid. In the simplest situation, A and B are independent for large enough k.

Definition 8.7.2 *A time series $\{X_t\}$ is called m-dependent, if $A \in \mathcal{F}_{-\infty}^{t-k}$ and $B \in \mathcal{F}_t^{\infty}$ are independent for all $k > m$.*

In other words, $\{X_t\}$ is m-dependent, if any finite-dimensional marginals $(X_{s_1}, \ldots, X_{s_k})$ and $(X_{t_1}, \ldots, X_{t_l})$, $s_1 < \cdots < s_l \le t_1 < \cdots < t_l$, are independent, if $t_1 - s_l > m$. In particular, a stationary series is 0-dependent, if and only if it is i.i.d.

According to our above discussion, a meaningful approach to measure the departure from the independence is

$$|P(A \cap B) - P(A)P(B)|,$$

giving rise to the following definitions.

Definition 8.7.3 (α-MIXING)

(i) *Let $\mathcal{A}, \mathcal{B} \subset \mathcal{F}$ be sub-σ-fields. Then*

$$\alpha(\mathcal{A}, \mathcal{B}) = \sup\{|P(A \cap B) - P(A)P(B)| : A \in \mathcal{A}, B \in \mathcal{B}\}$$

is called the **α-mixing coefficient**.

(ii) *For a time series $\{X_t\}$*

$$\alpha(k) = \sup_t \alpha\left(\mathcal{F}_{-\infty}^{t-k}, \mathcal{F}_t^{\infty}\right)$$

is called the **strong or α-mixing coefficient of** $\{X_t\}$ **for the lag** $k \in \mathbb{N}_0$.

(iii) *A time series $\{X_t\}$ is called* **strong or α-mixing**, *if*

$$\lim_{k \to \infty} \alpha(k) = 0.$$

Notice that this definition also applies to non stationary time series and, contrary to ρ-mixing, does not require the second moments to exist. Notice that in the latter case ρ-mixing can still be defined, for it addresses only the covariance $\mathrm{Cov}(Y, Z)$ of those random variables Y (measurable w.r.t. $\mathcal{F}_{-\infty}^{t-k}$) and Z (measurable w.r.t $\mathcal{F}_t^{\infty}$), whose second moments are finite. But then it does not make any meaningful assertion on the observations X_t themselves, since one cannot apply the definition with, say, $Y = X_t$ and $Z = X_{t+k}$.

As we shall see, in general one has to make assumptions on the (rate of) decay of the sequence $\{\alpha(k) : k \in \mathbb{N}_0\}$ of α-mixing coefficients. In particular, the mixing property itself is not sufficient to guarantee important statements on sample averages such as the law of large numbers or the central limit theorem.

8.7.2 Inequalities

Let us collect a couple of basic inequalities on the mixing coefficients. In particular, we shall see how one can estimate covariances in terms of the α-mixing coefficient.

First, it is instructive to ρ-mixing is always stronger than α-mixing.

Lemma 8.7.4 *For any sub-σ-fields $\mathcal{A}, \mathcal{B}$ of $\mathcal{F}$*

$$4\alpha(\mathcal{A}, \mathcal{B}) \le \rho(\mathcal{A}, \mathcal{B}).$$

Proof. Let $A \in \mathcal{A}$ and $B \in \mathcal{B}$. Put $X = \mathbf{1}_A$ and $Y = \mathbf{1}_B$. Then $E(X) = P(A)$, $E(Y) = P(B)$ and $E(XY) = P(A \cap B)$, such that

$$|\mathrm{Cov}(X, Y)| = |P(A \cap B) - P(A)P(B)|.$$

Since $\text{Var}(X) = P(A)(1 - P(A)) \leq 1/4$ and $|\text{Cov}(X, Y)| = |\text{Cor}(X, Y)|\sqrt{\text{Var}(X)\,\text{Var}(Y)}$, we obtain the inequality

$$|P(A \cap B) - P(A)P(B)| \leq (1/4)|\text{Cor}(X, Y)|$$

leading to

$$4|P(A \cap B) - P(A)P(B)| \leq \sup\{|\text{Cor}(X, Y)| : X, Y \in L_2\}.$$

Since the right-hand side does not depend on A or B, we may take the sup and obtain $4\alpha(\mathcal{A}, \mathcal{B}) \leq \rho(\mathcal{A}, \mathcal{B})$.

When dealing with α-mixing processes, the following inequality is crucial, whose proof is rather technical but can be found in the literature. It is used to bound the covariance between a $\mathcal{A}$-measurable random variables X and a $\mathcal{B}$-measurable random variable Y, i.e. the absolute of the difference between the cross-moment $E(XY)$ by its value $E(X)E(Y)$ assuming independente, by their corresponding α-mixing coefficient.

Proposition 8.7.5 *Let $p, q, r \in (1, \infty]$ with $\frac{1}{p} + \frac{1}{q} + \frac{1}{r} = 1$. Let X be a $\mathcal{A}$-measurable random variable and Y be a $\mathcal{B}$-measurable random variable with*

$$E|X|^p < \infty \qquad \textit{and} \qquad E|Y|^q < \infty.$$

Then

$$|\text{Cov}(X, Y)| = |E(XY) - E(X)E(Y)| \leq 8\|X\|_p\|Y\|_q\alpha(\mathcal{A}, \mathcal{B})^{1/r}.$$

Here, $x^0 = 1$ if $x > 0$ and $0^0 = 0$.

It is important to note that in the above inequality $\frac{1}{p} + \frac{1}{q}$ must be less than 1 for the bound to be meaningful. Hence, it can not be applied with $p = q = 2$, i.e., if X and Y are known to have finite second moments. But the inequality works when assuming the existence of the absolute moments of order $2 + \delta$ for some arbitrarily small $\delta > 0$. Then $\frac{1}{r} = 1 - \frac{1}{p} - \frac{1}{q} = \frac{2}{2+\delta}$. In this case, Proposition 8.7.5 takes the following form.

Proposition 8.7.6 *Let X be $\mathcal{A}$-measurable and Y be $\mathcal{B}$-measurable. If for some $\delta > 0$*

$$E|X|^{2+\delta} < \infty \qquad \textit{and} \qquad E|Y|^{2+\delta} < \infty,$$

then

$$|\text{Cov}(X, Y)| \leq 8\alpha(\mathcal{A}, \mathcal{B})^{2/(2+\delta)}\|X\|_{2+\delta}\|Y\|_{2+\delta}.$$

If X_t are i.i.d. with $E(X_1) = 0$ and a finite second moment, an elementary calculation shows that the sum $S_n = \sum_{i=1}^n X_i$ satisfies the moment bound

$$E\left(S_n^2\right) = O(n).$$

If even $E(X_1^4) < \infty$ holds, then

$$E\left(S_n^4\right) = O\left(n^2\right).$$

For a mixing process these orders remain true, provided that the mixing coefficients $\alpha(k)$, $k \geq 0$, decay sufficiently fast in the sense that

$$A_r(\delta) = \sum_{i=0}^{\infty} (i+1)^{r/2-1} \alpha(i)^{\delta/(r+\delta)} < \infty,$$

and the series is weakly stationary of order r, i.e. all moments of orders up to $r + \delta$, $\delta > 0$, exist and are time-invariant such that

$$E(X_{t_1} \cdots X_{t_r}) = E(X_{t_1+h} \cdots X_{t_r+h})$$

for all $t_1, \ldots, t_r$ and h. We provide the details for the important cases $r \in \{2, 4\}$.

Theorem 8.7.7

(i) Let $\{X_t\}$ be weakly stationary with $E|X_1|^{2+\delta} < \infty$ for some $\delta > 0$. Then

$$E(S_n^2) \leq 2!8 A_2(\delta) \|X_1\|_{2+\delta}^2.$$

(ii) Let $\{X_t\}$ be weakly stationary of order 4 with $E|X_1|^{4+\delta} < \infty$. Then

$$E(S_n^4) \leq 84! A_4(\delta) \|X_1\|_{4+\delta}^4 n^2.$$

Proof. We have for $r = 2, 4$

$$E(S_n^r) = \sum_{t_1=1}^{n} \cdots \sum_{t_r=1}^{n} E(X_{t_1} \cdots X_{t_r}).$$

Observe that all $r!$ terms $E(X_{t_1} \cdots X_{t_r})$ whose indices are a permutation of the ordered values $t_{(1)} \leq \cdots \leq t_{(r)}$ coincide. By stationarity, $E(X_{t_1} \cdots X_{t_r}) = E(X_0 X_{t_2-t_1} \cdots X_{t_r-t_1})$. Hence,

$$\begin{aligned} E(S_n^r) &= r! \sum_{1 \leq t_1 \leq \cdots \leq t_r \leq n} E(X_{t_1} \cdots X_{t_r}) \\ &= r! n \sum_{0 \leq t_2 \leq \cdots \leq t_r \leq n-1} E(X_0 X_{t_2} \cdots X_{t_{r-1}}) \\ &= r! n \sum_{1 \leq t_1 \leq \cdots \leq t_{r-1} \leq n} E(X_1 X_{t_1} \cdots X_{t_{r-1}}). \end{aligned}$$

For $r = 2$ we have to estimate $\sum_{t_1=1}^{n} |E(X_1 X_{t_1}|$. We claim that the latter is not larger than $8\|X_1\|_{2+\delta}^2 A_2(\delta)$, which establishes Equation (i). To verify this claim notice that $E(X_1) = 0$, such that

$$|E(X_0 X_j)| = |\operatorname{Cov}(X_0, X_j)| \leq 8\|X_1\|_{2+\delta}^2 \alpha(j)^{2/(2+\delta)}.$$

This implies

$$\sum_{t_1=1}^{n} |E(X_1 X_{t_1})| \leq 8\|X_1\|_{2+\delta}^2 \sum_{j=0}^{\infty} \alpha(j)^{2/(2+\delta)}$$
$$= 8\|X_1\|_{2+\delta}^2 A_2(\delta).$$

The case $r = 4$ is more involved and requires us to rearrange the terms appropriately. Let us denote by $\sum_{n,j}^{(h)}$ the sum over all $1 \leq t_1 \leq \cdots \leq t_j \leq n$ such that the maximal difference of successive indices is attained at position h, i.e.

$$r_h = t_h - t_{h-1} = \max\{t_j - t_{j-1}, \ldots, t_2 - t_1, t_1 - t_0\},$$

where we put $t_0 = 0$ for convenience and denote here the maximal difference by r_h. We may now calculate a sum taken over $1 \leq t_1 \leq t_2 \leq t_3 \leq n$ by first summing over the positions $h = 1, 2, 3$ and then over the ordered indices satisfying these constraints. Hence, we consider now the right-hand side of

$$E(S_n^4) = 4!n \sum_{h=1}^{3} \sum_{n,3}^{(h)} E(X_1 X_{t_1} X_{t_2} X_{t_3}).$$

We claim that

$$\sum_{n,3}^{(h)} |E(X_1 X_{t_1} X_{t_2} X_{t_3})| \leq 8\|X_1\|_{4+\delta}^4 \sum_{r=0}^{n-1} C_r \alpha(r)^{\delta/(4+\delta)}.$$

Here, C_r is the number of tuples (t_1, t_2, t_3) with $1 = t_0 \leq t_1 \leq t_2 \leq t_3 \leq n$, $t_h - t_{h-1} = \max_i t_i - t_{i-1} = \ell$ for a fixed value ℓ and h is 1, 2 or 3. Let us denote the corresponding cases by

$$1 \overset{\ell}{-} t_1 t_2 t_3, \quad 1 t_1 \overset{\ell}{-} t_2 t_3, \quad 1 t_1 t_2 \overset{\ell}{-} t_3,$$

where $\overset{\ell}{-}$ indicates the position of the gap of length ℓ. We shall show that in all cases

$$|E(X_1 X_{t_1} X_{t_2} X_{t_3})| \leq 8\alpha(\ell)^{\delta/(4+\delta)} \|X_1\|_{4+\delta}^4$$

and then estimate the number C_r.

Case $1 \overset{\ell}{-} t_1 t_2 t_3$: Apply Proposition 8.7.5 with $p = 4 + \delta$ and $q = (4 + \delta)/3$ such that $1 - 1/p - 1/q = \delta/(4 + \delta)$. We have

$$\begin{aligned} |E(X_1 X_{t_1} X_{t_2} X_{t_3}) - E(X_1) E(X_{t_1} X_{t_2} X_{t_3})| &\leq 8\alpha(\ell)^{\delta/(4+\delta)} \|X_{t_1}\|_{4+\delta} \|X_1 X_{t_2} X_{t_3}\|_{(4+\delta)/3} \\ &\leq 8\alpha(\ell)^{\delta/(4+\delta)} \|X_1\|_{4+\delta}^4, \end{aligned}$$

where the last estimate $\|X_1 X_{t_2} X_{t_3}\|_{(4+\delta)/3} \leq \|X_1\|_{4+\delta} \|X_{t_2}\|_{4+\delta} \|X_{t_3}\|_{4+\delta}$ follows from the generalized Hölder inequality, cf. A.4.1 (iv).

Case $1 t_1 t_2 \overset{\ell}{-} t_3$: This is shown as above.

Case $1 t_1 \overset{\ell}{-} t_2 t_3$: Now chose $p = (4+\delta)/2$ and $q = (4+\delta)/2$ such that $1 - 1/p - 1/q = \delta/(4+\delta)$. This gives

$$\begin{aligned}|E(X_1 X_{t_1} X_{t_2} X_{t_3}) - E(X_1 X_{t_1}) E(X_{t_2} X_{t_3})| \\ &\leq 8\alpha(\ell)^{\delta/(4+\delta)} \|X_1 X_{t_1}\|_{(4+\delta)/2} \|X_{t_2} X_{t_3}\|_{(4+\delta)/2} \\ &\leq 8\alpha(\ell)^{\delta/(4+\delta)} \|X_1\|_{4+\delta}^4,\end{aligned}$$

where we again use the generalized Hölder inequality.

To estimate C_r, observe that the condition $t_h - t_{h-1} = \ell$ fixes t_h at the value $t_{h-1} + \ell$ (ℓ is fixed). Since $t_i - t_{i-1}$ attains its maximum ℓ at $i = h$, each of the other indices t_i, $i \neq h$, attains only the $\ell + 1$ values $t_{i-1}, \ldots, t_{i-1} + \ell$, such that $C_r = \sum_{h=1}^{3} \sum_{n,3}^{(h)} 1 \leq n(\ell + 1)$.

Therefore, we can conculde

$$\begin{aligned}E\left(S_n^4\right) &= 4!n \sum_{h=1}^{3} \sum_{n,3}^{(h)} |E(X_1 X_{t_1} X_{t_2} X_{t_3})| \\ &\leq 4!8n \|X_1\|_{4+\delta} \sum_{\ell=0}^{n-1} C_r \alpha(\ell)^{\delta/(4+\delta)} \\ &\leq 4!8n \|X_1\|_{4+\delta} \sum_{\ell=0}^{\infty} (r+1) \alpha(\ell)^{\delta/(4+\delta)} \\ &\leq 4!8 A_4(\delta) \|X_1\|_{4+\delta}^4,\end{aligned}$$

which completes the proof.

Remark 8.7.8 *The assertion of Theorem 8.7.7 that $E|S_n|^r = O(n^{r/2})$ also holds true for real-valued $r > 2$, if $A_r(\delta) < \infty$ for some $\delta > 0$, cf. Yokoyama (1980).*

Finally, we show how conditional expectations can be bounded by mixing coefficients.

Lemma 8.7.9 *Suppose the time series $\{X_t : t \in \mathbb{Z}\}$ satisfies $E(X_0) = 0$ and $E|X_0|^{2+\delta} < \infty$ for some $\delta > 0$. Let $\mathcal{F}_{-n} = \sigma(X_t : t \leq -n)$. Then*

(i)

$$\|E(X_0|\mathcal{F}_{-n})\|_{L_2} \leq 8\|X_0\|_{2+\delta} \alpha(n)^{\frac{\delta}{2(2+\delta)}}$$

and

(ii)

$$\sum_{n=1}^{\infty} \|E(X_0|\mathcal{F}_{-n})\|_{L_2} < \infty,$$

provided

$$\sum_{n=1}^{\infty} \alpha(n)^{\frac{\delta}{2(2+\delta)}} < \infty.$$

Proof. Observe that for $p = 2$ and $q = 2 + \delta$ we have $1/p - 1/q = \frac{\delta}{2(2+\delta)}$. Hence, for any $\mathcal{F}_{-n}$-measurable Y with $\|Y\|_{L_2} < \infty$, we have

$$|E(X_0 Y)| \le 8\|X_0\|_{2+\delta}\|Y\|_{L_2}\alpha(n)^{\frac{\delta}{2(2+\delta)}}.$$

Lemma 8.1.6 yields

$$\begin{aligned}\|E(X_0|\mathcal{F}_{-n})\|_{L_2} &= \sup\{E(X_0Y) : Y\ \mathcal{F}_{-n}\text{-measurable}, \|Y\|_{L_2} = 1\} \\ &\le 8\|X_0\|_{2+\delta}\alpha(n)^{\frac{\delta}{2(2+\delta)}},\end{aligned}$$

which shows assertion (i). The second claim now follows easily,

$$\sum_{n=1}^{\infty} \|E(X_0|\mathcal{F}_{-n})\|_{L_2} \le 8\|X_0\|_{L_2}\sum_{n=1}^{\infty}\alpha(n)^{\frac{\delta}{2(2+\delta)}}.$$

8.8 Limit theorems for mixing processes

We shall now discuss a strong law of large numbers for α-mixing time series.

Theorem 8.8.1 (STRONG LAW OF LARGE NUMBERS UNDER α-MIXING)
Let $\{X_t\}$ be a weakly stationary process of order 4, which is α-mixing with mixing coefficients satisfying

$$A_4(\delta) = \sum_{i=0}^{\infty}(i+1)\alpha(i)^{\delta/(4+\delta)} < \infty$$

for some $\delta > 0$. Then

$$\frac{1}{T}\sum_{t=1}^{T} X_t \overset{a.s.}{\to} E(X_1),$$

as $T \to \infty$.

Proof. W.l.o.g. assume $E(X_1) = 0$. Put $\overline{X}_T = \frac{1}{T}\sum_{t=1}^{T} X_t$ and $S_T = \sum_{t=1}^{T} X_t$. Consider the event

$$A = \{\omega \in \Omega : \overline{X}_T(\omega) \to 0,\ T \to \infty\}.$$

We shall show $P(A) = 1 \Leftrightarrow P(A^c) = 0$. Observe that

$$\begin{aligned} & \overline{X}_T(\omega) \text{ does not converge} \\ \Leftrightarrow\quad & \exists \varepsilon > 0 : \forall T_0 \in \mathbb{N} : \exists T \ge T_0 : |\overline{X}_T(\omega)| > \varepsilon \\ \Leftrightarrow\quad & A_T = \{|\overline{X}_T| > \varepsilon\} \text{ occurs infinitely often.}\end{aligned}$$

By the lemma of Borel Cantelli,

$$P(A_T \text{ infinitely often}) = P\left(\cap_T \cup_{S \ge T} A_S\right) = 0$$

holds, if

$$\sum_{T=1}^{\infty} P(A_T) < \infty.$$

The generalized Markov inequality yields

$$P(A_T) = P(|\overline{X}_T| > \varepsilon) \leq \frac{ES_T^4}{T^4 \varepsilon^4}.$$

By Theorem 8.7.7 (ii), the right-hand side is $O(T^{-2})$, such that $\sum_{t=1}^{T} P(A_T) < \infty$ follows.

Theorem 8.8.2 (Central Limit Theorem for Mixing Processes) *Let $\{X_t\}$ be a strictly stationary and α-mixing process with*

$$E(X_1) = 0 \qquad \text{and} \qquad E|X_1|^{2+\delta} < \infty$$

for some $\delta > 0$. If

$$\sum_{k=1}^{\infty} \alpha(k)^{\frac{\delta}{2+\delta}} < \infty,$$

then

$$\frac{1}{\sqrt{T}} \sum_{t=1}^{T} X_t \stackrel{d}{\to} N(0, \sigma^2),$$

as $T \to \infty$, where

$$\sigma^2 = E(X_1) + 2 \sum_{h=1}^{\infty} E(X_1 X_{1+h})$$

converges absolutely.

Proof. As a strictly stationary L_2-process, $\{X_t\}$ satisfies the ergodic theorem. By Lemma 8.7.9, the condition on the mixing coefficients ensures that

$$\sum_{n=1}^{\infty} \|E(X_0|\mathcal{F}_{-n})\|_2 < \infty.$$

Hence, we can directly apply Theorem 8.1.7.

We will now establish central limit theorems for the sample autocovariance functions

$$\widehat{\gamma}_T(h) = \frac{1}{T} \sum_{t=1}^{T-h} (X_{t+h} - \overline{X}_T)(X_t - \overline{X}_T)$$

for $0 \le h \le T-1$ and $\widehat{\gamma}_T(h) = 0$, otherwise, and

$$\widetilde{\gamma}_T(h) = \frac{1}{T-h} \sum_{t=1}^{T-h} (X_{t+h} - \mu)(X_t - \mu)$$

for $0 \le h \le T-1$ and again $\widetilde{\gamma}_T(h) = 0$, otherwise.

Theorem 8.8.3 *Let $\{X_t\}$ be a strictly stationary and α-mixing process with $E(X_1) = \mu$ and autocovariance function $\gamma = \gamma_X$. Suppose that there exists some $\delta > 0$ such that the following two conditions are satisfied.*

(i) $\sum_{k=1}^{\infty} \alpha(k)^{\frac{\delta}{2+\delta}} < \infty$.

(ii) $E|X_1|^{4+4\delta} < \infty$.

Then, for any fixed h,

$$\sqrt{T}(\widetilde{\gamma}_T(h) - \gamma(h)) \xrightarrow{d} N(0, \eta_h^2),$$

as $T \to \infty$, where

$$\eta_h^2 = \lim_{T\to\infty} E\left(\frac{1}{\sqrt{T}} \sum_{t=1}^{T} X_t X_{t+h} \right)^2.$$

Proof. W.l.o.g. assume $\mu = 0$. Fix $h \ge 0$ and consider the process

$$\xi_t = X_t X_{t+h}, \qquad t \in \mathbb{N}.$$

Then $E(\xi_t) = \gamma(h)$ for all t. Clearly, $\{\xi_t\}$ is strictly stationary and an L_2-process, since

$$\begin{aligned} E|\xi_t|^{2+\delta} &= E|X_t X_{t+h}|^{2+\delta} \\ &\le \sqrt{E|X_t|^{2(2+\delta)}}\sqrt{E|X_{t+h}|^{2(2+\delta)}} \\ &= E|X_t|^{4+4\delta} < \infty. \end{aligned}$$

Denote the kth α-mixing coefficient of $\{\xi_t\}$ by $\alpha_\xi(k)$. Notice that

$$\mathcal{F}_{t,0} = \sigma(X_s X_{s+h} : s \le t) \subset \sigma(X_s : s \le t+h)$$

as well as

$$\mathcal{F}_{t,k} = \sigma(X_s X_{s+h} : s \ge t+k) \subset \sigma(X_s : s \ge t+k).$$

Hence, for $k > h$ we may estimate $\alpha_\xi(k)$ as

$$\begin{aligned} \alpha_\xi(k) &= \sup_{A\in\mathcal{F}_{t,0}, B\in\mathcal{F}_{t,k}} |P(A\cap B) - P(A)P(B)| \\ &\le \sup_{A\in\sigma(X_s:s\le t+h), B\in\sigma(X_s:s\ge t+k)} |P(A\cap B) - P(A)P(B)| \\ &= \alpha(k-h). \end{aligned}$$

It follows that

$$\sum_{k=1}^{\infty} \alpha_\xi(k)^{\frac{\delta}{2+\delta}} = \sum_{k=1}^{h} \alpha_\xi(k)^{\frac{\delta}{2+\delta}} + \sum_{k=h+1}^{\infty} \alpha_\xi(k)^{\frac{\delta}{2+\delta}}$$

$$= \sum_{k=1}^{h} \alpha_\xi(k)^{\frac{\delta}{2+\delta}} + \sum_{k=1}^{\infty} \alpha(k)^{\frac{\delta}{2+\delta}} < \infty.$$

Consequently, the assertions follow from Theorem 8.8.2.

The following proposition shows that $\widehat{\gamma}_T(h)$ inherits its asymptotic distribution from $\widetilde{\gamma}_T(h)$.

Proposition 8.8.4 *Let $\{X_t\}$ be a weakly stationary series satisfying the central limit theorem,*

$$\frac{1}{\sqrt{T}} \sum_{t=1}^{T} X_t \stackrel{d}{\to} N(0, \sigma^2),$$

as $T \to \infty$. Then

$$\sqrt{T}(\widehat{\gamma}_T(h) - \widetilde{\gamma}_T(h)) \stackrel{P}{\to} 0,$$

as $T \to \infty$.

Proof. A direct calculation yields

$$\widehat{\gamma}_T(h) = \frac{1}{T} \sum_{t=1}^{T} (X_t - \overline{X}_T)(X_{t+h} - \overline{X}_T)$$

$$= \frac{1}{T} \left(\sum_{t=1}^{T-h} X_t X_{t+h} - 2\overline{X}_T \sum_{t=1}^{T-h} X_{t+h} + \overline{X}_T^2 \right).$$

Hence,

$$\widehat{\gamma}_T(h) - \widetilde{\gamma}_T(h) = -\overline{X}_T^2 + 2\overline{X}_T \frac{1}{T} \sum_{t=T-h+1}^{T} X_t - \frac{h}{T} \overline{X}_T^2,$$

which eventually leads to

$$\sqrt{T}(\widetilde{\gamma}_T(h) - \widetilde{\gamma}_T(h)) = -[\sqrt{T}\overline{X}_T]\overline{X}_T + 2[\sqrt{T}\overline{X}_T] \frac{1}{T} \sum_{t=T+h+1}^{T} X_t - [\sqrt{T}\overline{X}_T] \frac{h}{T} \overline{X}_T.$$

By assumption, the terms in brackets are asymptotically normal and therefore $O_P(1)$. The remaining factors are $o_P(1)$, since $\overline{X}_T \to 0$, as $T \to \infty$, in probability, as a consequence of the central limit theorem.

As a consequence of Proposition 8.8.4, Theorem 8.8.3 remains true when $\widetilde{\gamma}_T(h)$ is replaced by $\widehat{\gamma}_T(h)$, that is when the time series is centered at the sample average $\overline{X}_T$. Further, one may extend the result to joint convergence in distribution. Assuming the $\{X_t\}$ is a linear process with i.i.d. innovations and coefficients $\{\theta_i\}$ satisfying the conditions of Theorem 8.6.3, one can obtain

$$\sqrt{T}[(\widehat{\rho}_T(1), \ldots, \widehat{\rho}_T(d))' - (\rho(1), \ldots, \rho(d))'] \xrightarrow{d} N(0, W),$$

where $W = (w_{ij})_{i,j}$ is the $(d \times d)$-matrix with entries

$$w_{ij} = \sum_{k=-\infty}^{\infty} [\rho(k+i)\rho(k+j) + \rho(k-i)\rho(k+j) + 2\rho(i)\rho(j)\rho^2(k) \\ - 2\rho(i)\rho(k)\rho(k+j) - 2\rho(j)\rho(k)\rho(k+i)].$$

In particular, for a white-noise process one obtains $W = \text{diag}(1, \ldots, 1)$, which justifies the asymptotic hypothesis test and confidence interval discussed in Section 3.3.1.

It is worth mentioning that the assumption of a finite moment of order $4 + \delta$ is essential. Indeed, for heavy-tailed distributions the sample autocovariances and autocorrelations, respectively, may exhibit a completely different behavior.

Having consistent estimators of the autocovariances at our disposal, it seems natural to use them in order to estimate the (asymptotic) variance of the scaled sample mean,

$$\Sigma_T = \text{Var}\left(\frac{1}{\sqrt{T}}\sum_{t=1}^{T} X_t\right) = \Gamma_0 + \sum_{j=1}^{T-1} \frac{T-j}{T}(\Gamma_j + \Gamma_j'),$$

where $\{X_t : t = 1, 2, \ldots\}$ is a stationary d-dimensional time series with mean μ and autocovariance matrices

$$\Gamma_j = E(X_1 - \mu)(X_{1+j} - \mu)', \qquad j = 1, 2, \ldots.$$

Generalizing from the univariate setting, see Equation (8.1),

$$\Sigma = \lim_{T\to\infty} \Sigma_T = \Gamma_0 + \sum_{j=1}^{\infty}(\Gamma_j + \Gamma_j')$$

is called the **long-run variance**, provided it exists.

In applications, the d-dimensional time series $\{X_t\}$ for which one has to estimate Σ is often given by

$$X_t = h(Z_t; \vartheta),$$

where $\{Z_t\}$ is another strictly stationary l-dimensional time series, $\vartheta \in \Theta$ a parameter and $h(z, \vartheta)$ a known function defined on $\mathbb{R}^l \times \Theta$ such that

$$E|h(Z_1; \vartheta)|^{4+\delta} < \infty,$$

for some $\delta > 0$.

For brevity of notation, put

$$h_t(\vartheta) = h(Z_t; \vartheta), \qquad t \in \mathbb{N},$$

and assume that there is some $\vartheta_0 \in \Theta$, the *true* parameter, such that

$$Eh_t(\vartheta_0) = 0, \qquad \text{for all } t,$$

holds true.

The **Newey–West estimator** for Σ is defined by

$$\widehat{\Sigma}_T = \widehat{\Gamma}_{T0} + \sum_{j=1}^{m} w(j,m)(\widehat{\Gamma}_{Tj} + \widehat{\Gamma}'_{Tj}), \tag{8.28}$$

where $0 \le m < T$ is the **lag truncation parameter** and

$$\widehat{\Gamma}_{Tj} = \frac{1}{T} \sum_{t=j+1}^{T} \widehat{h}_t(\vartheta_0)\widehat{h}_{t-j}(\vartheta_0)'.$$

For simplicity of presentation, we omit the parameter in our notation, since it will be fixed at $\vartheta = \vartheta_0$. The lag truncation parameter is chosen as a function of the sample size, $m = m_T$, and we shall see that it has to be of smaller order than T.

The above estimator relies on estimates of the random variables h_t. We consider the following two cases:

(i) If $h_t(\vartheta) = h(Z_t)$ does not depend on ϑ, then we simply use $\widehat{h}_t = h(Z_t)$.

(ii) If $h_t(\vartheta) = h(Z_t; \vartheta)$, we assume that we have some consistent estimator $\widehat{\vartheta}_T$ and use

$$\widehat{h}_t = h(Z_t; \widehat{\vartheta}_T), \qquad t = 1, 2, \ldots.$$

Notice that the estimator $\widehat{\Sigma}_T$ uses only the first m diagonals of the sample autocovariance matrix of $\widehat{h}_1, \ldots, \widehat{h}_T$, which are additionally weighted using the **Bartlett weights**

$$w(j,m) = 1 - \frac{j}{m+1}, \qquad j = 0, \ldots, m.$$

Those weights satisfy the following regularity conditions

(i) $w(j,m) \le C_w$ for all $0 \le j \le m$, $m \in \mathbb{N}$, for some constant $0 < C_w < \infty$.

(ii) For each $0 \le j \le m$ it holds $w(j,m) \to 1$, as $m \to \infty$.

Also observe that

$$w(j,m) = k_B\left(\frac{j}{h_T}\right), \qquad j = 0, \ldots, m, m \ge 1,$$

if we put $h_T = m + 1$ and introduce the Bartlett kernel function

$$k_B(x) = (1 - |x|)\mathbf{1}(|x| \le 1), \qquad x \in \mathbb{R}.$$

One can define other weighting schemes by selecting other kernel functions. Here is a list of the most commonly used kernels.

(i) **Truncated:** $k_{TR}(x) = \mathbf{1}(|x| \le 1)$, $x \in \mathbb{R}$.

(ii) **Parzen:**

$$k_P(x) = \begin{cases} 1 - 6x^2 + 6|x|^3, & 0 \le |x| \le 1/2, \\ 2(1 - |x|)^3, & 1/2 \le |x| \le 1, \\ 0, & \text{otherwise.} \end{cases}$$

(iii) **Tukey–Hanning:**

$$k_{TH}(x) = \begin{cases} (1 + \cos(\pi x))/2, & |x| \le 1, \\ 0, & \text{otherwise.} \end{cases}$$

(iv) **Quadratic Spectral:**

$$k_{QS}(x) = \frac{25}{12\pi^2 x^2} \left(\frac{\sin(6\pi x/5)}{6\pi x/5} - \cos(6\pi x/5) \right).$$

To motivate an assumption on the estimates $\widehat{h}_t$, suppose that $h(z; \vartheta)$ is twice differentiable with

$$\sup_t \|\nabla h(Z_t; \vartheta_0)\| = O_P(1), \quad \sup_t \|Dh(Z_t; \vartheta_0)\| = O_P(1).$$

Then a Taylor expansion leads to

$$\widehat{h}_t - h_t = \nabla h(Z_t; \vartheta_0)(\widehat{\vartheta}_T - \vartheta_0) + O_P(\|\widehat{\vartheta}_T - \vartheta_0\|).$$

If $\widehat{\vartheta}_T$ is $\sqrt{T}$-consistent, say, such that

$$\sqrt{T}(\widehat{\vartheta}_T - \vartheta_0) = O_P(1),$$

then the estimates $\widehat{h}_t$ are uniformly consistent with rate $\sqrt{T}$ in the sense that

$$\max_{t \le T} \sqrt{T} |\widehat{h}_t - h_t| = O_P(1).$$

To derive the consistency of the Newey–West estimator, we need the following maximal inequality.

Proposition 8.8.5 (Maximal Inequality for α-Mixing Series)
Let $\{\xi_t : t \in \mathbb{N}\}$ be a mean zero α-mixing time series such that there is some $\delta > 0$ with

$$E|\xi_t|^{2+\delta} < \infty$$

for all $t \ge 1$ and mixing coefficients $\{\alpha(k) : k \in \mathbb{N}\}$ satisfying

$$\sum_{k=1}^{\infty} \alpha(k)^{\frac{2}{2+\delta}} < \infty.$$

Then,

$$P\left(\max_{1\le t\le T}\left|\sum_{t=1}^{T}\xi_t\right|>\varepsilon\right)\le\frac{4\sum_{t=1}^{T}E\xi_t^2+c(\alpha,\delta)\sum_{t=1}^{T}\left(E|\xi_t|^{2+\delta}\right)^{\frac{\delta}{2+\delta}}}{\varepsilon^2},$$

where

$$c(\alpha,\delta)=16\left[(4\delta^{-1}+2)\sum_{k=1}^{\infty}(k+1)^{\frac{2}{\delta}}\alpha(k)^{\frac{\delta}{2+\delta}}\right].$$

Theorem 8.8.6 (Consistency of the Newey–West Estimator)
Let $\{Z_t\}$ be a strictly stationary α-mixing time series with mixing coefficients satisfying

$$\sum_{k=1}^{\infty}\alpha(k)^{\frac{\delta}{2+\delta}}<\infty.$$

Further, let $h_t(\vartheta)=h(Z_t;\vartheta)$, $t\ge 1$, for some function $h(z,\vartheta)$ satisfying

$$Eh(Z_1;\vartheta_0)=0 \qquad \text{and} \qquad E|h(Z_1;\vartheta_0)|^{4+4\delta}<\infty.$$

Let $\widehat{h}_t$ be uniformly consistent estimates for h_t, that is

$$\max_{t\le T}|\widehat{h}_t-h_t|=o_P(1). \tag{8.29}$$

Then the Newey–West estimator $\widehat{\Sigma}_T$ for Σ_T is consistent,

$$\widehat{\Sigma}_T-\Sigma_T\xrightarrow{P}0,$$

as $T\to\infty$, provided the lag truncation rule satisfies

$$\frac{m^2}{T}=o(1),$$

as $T\to\infty$.

Proof. For simplicity of notation put $h_t=h_t(\vartheta_0)$. Notice that $\widehat{\Sigma}_T$ can be written as

$$\widehat{\Sigma}_T=\frac{1}{T}\sum_{t=1}^{T}\widehat{h}_t\widehat{h}_t'+\frac{2}{T}\sum_{j=1}^{m}\sum_{t=j+1}^{T}w(j,m)\widehat{h}_t\widehat{h}_{t-j}'.$$

We shall show consistency of

$$\breve{\Sigma}_T=\frac{1}{T}\sum_{t=1}^{T}h_th_t'+\frac{2}{T}\sum_{j=1}^{m}\sum_{t=j+1}^{T}w(j,m)h_th_{t-j}',$$

which uses the the random variables h_t instead of their estimates $\widehat{h}_t$, and then argue that the h_t can be replaced by their estimates. Consider the following corresponding truncated version of Σ_T,

$$\widetilde{\Sigma}_T = \frac{1}{T}\sum_{t=1}^{T} E(h_t h_t') + \frac{2}{T}\sum_{j=1}^{m}\sum_{t=j+1}^{T} w(j,m)E(h_t h_{t-j}')$$

and notice that

$$\widetilde{\Sigma}_T = E(\breve{\Sigma}_T).$$

The theorem follows, if we verify the following assertions.

(i) $\|\Sigma_T - \widetilde{\Sigma}_T\| = o(1)$.

(ii) $\|\breve{\Sigma}_T - \widetilde{\Sigma}_T\| = O(m^2/T)$.

(iii) $\|\widehat{\Sigma}_T - \breve{\Sigma}_T\| = o_P(1)$.

In what follows, $\| \bullet \|$ denotes the maximum matrix norm. Let us start with (i) and estimate the error $\|\Sigma_T - \widetilde{\Sigma}_T\|$. We have

$$\|\Sigma_T - \widetilde{\Sigma}_T\| \le \frac{2}{T}\sum_{j=1}^{m}\sum_{t=j+1}^{T} |w(j,m)-1| \|E(h_t h_{t-j}')\| + \frac{2}{T}\sum_{j=m+1}^{T}\sum_{t=j+1}^{T} \|E(h_t h_{t-j}')\|.$$

The α-mixing condition ensures that the second term can be estimated by

$$2C\max_{N\le T}\sum_{j=1}^{N}\frac{N-j}{N}\alpha(j)^{\frac{2}{2+\delta}} = O\left(\sum_{j=m+1}^{\infty}\alpha(j)^{\frac{2}{2+\delta}}\right) = o(1),$$

as $m \to \infty$, since $\sum_{j=1}^{\infty}\alpha(j)^{\frac{2}{2+\delta}} < \infty$, and by $Eh_t = 0$ and Proposition 8.7.6 the element (k,l), $h_{tk}h_{t-j,l}$, of the random matrix $h_t h_{t-j}'$ satisfies

$$|E(h_{tk}h_{t-j,l})| \le C\alpha(j)^{\frac{2}{2+\delta}}$$

for some constant $0 < C < \infty$ not depending on $k, l \in \{1, \dots, d\}$. Using similar arguments, the first term can be bounded by

$$B_m = c\sum_{j=1}^{m}|w(j,m)-1|\alpha(j)^{\frac{2}{2+\delta}}$$

for some constant $0 < c < \infty$. Define the sequence of functions

$$f_n(j) = |w(j,n)-1|\alpha(j)^{\frac{2}{2+\delta}}\mathbf{1}(j \le n), \qquad j \in \mathbb{N}_0,$$

for $n \in \mathbb{N}$. Then

$$B_m = c\int f_m(x)\,d\nu(x),$$

where $\nu(x)$ denotes the counting measure on $\mathbb{N}_0$. Now $B_m = o(1)$ follows by dominated convergence, since clearly $f_m(x) \to 0$, as $m \to \infty$, for each fixed $x \in \mathbb{N}$, and

$$f_m(x) \le g(x) = 2\alpha(x)^{\frac{2}{2+\delta}}$$

where

$$\int g(x)\,\mathrm{d}\nu(x) = 2\sum_{k=1}^{\infty} \alpha(x)^{\frac{2}{2+\delta}} < \infty.$$

To show (ii) notice that

$$\breve{\Sigma}_T - \widetilde{\Sigma}_T = R_{T1} + R_{T2},$$

where

$$R_{T1} = \frac{1}{T}\sum_{t=1}^{T}(h_t h_t' - E(h_t h_t')),$$

$$R_{T2} = \frac{2}{T}\sum_{j=1}^{m} w(j,m) \sum_{t=j+1}^{T} [h_t h_{t-j}' - E(h_t h_{t-j}')]$$

$$\le \frac{2C_w}{T}\sum_{j=1}^{m}\left|\sum_{t=j+1}^{T} [h_t h_{t-j}' - E(h_t h_{t-j}')]\right|.$$

We show that $R_{T2} = O(m^2/T)$ by applying the maximal inequality given in Proposition 8.8.5. Fix $1 \le j \le T-1$ and consider the sequence

$$U_{tj} = h_t h_{t-j}' - E(h_t h_{t-j}'), \qquad t \ge 1.$$

Denoting the (k,l)th element of U_{tj} by $U_{tj}^{(k,l)}$, $1 \le k, l \le d$, the union bound together with Proposition 8.8.5 now allows us to estimate R_{T2} as follows.

$$P(|R_{T2}| > \varepsilon) \le \sum_{k,l=1}^{d}\sum_{j=1}^{m} P\left(\max_{1\le N\le T}\frac{1}{N}\left|\sum_{t=j+1}^{N} U_t^{(k,l)}\right| > \frac{\varepsilon}{C_w m}\right)$$

$$\le \sum_{k,l=1}^{d}\sum_{j=1}^{m} C_w^2 m^2 \frac{4TE|U_1^{(k,l)}|^{4+2\delta} + c(\alpha,\delta)T(E|U_1^{(k,l)}|^{4+2\delta})^{\frac{2}{2+\delta}}}{\varepsilon^2 T^2}$$

$$= O\left(\frac{m^2}{T}\right).$$

We can replace the h_ts by their estimates $\widetilde{h}_t$, since Equation (8.29) implies that

$$\widehat{h}_t = h_t + \frac{\sqrt{T}(\widehat{h}_t - h_t)}{\sqrt{T}} = h_t + O_P(1/\sqrt{T}),$$

uniformly in $t \le T$, which leads to

$$\begin{aligned}\widehat{h}_t \widehat{h}'_{t-j} &= [h_t + O_P(1/\sqrt{T})][h'_{t-j} + O_P(1/\sqrt{T})] \\ &= h_t h'_{t-j} + O_P(1/\sqrt{T}),\end{aligned}$$

where the $O_P(1/\sqrt{T})$ term is uniform in $j \le T-1$ and $t \le T$, since the fourth absolute moment of the h_ts is uniformly bounded by assumption. We obtain

$$\frac{1}{T}\sum_{t=1}^{T} \widehat{h}_t \widehat{h}'_{t-j} = \frac{1}{T}\sum_{t=1}^{T} h_t h'_{t-j} + O_P(1/\sqrt{T}),$$

which establishes (iii). Now the assertion follows easily.

8.9 Notes and further reading

An exposition of the asymptotic distribution theory for time series with a focus on parametric approaches can also be found in Brockwell and Davis (1991). A nice and elaborated proof of Theorem 8.3.6, the central limit theorem for martingale differences, which is omitted here, can be found in the monograph (Davidson, 1994, p. 383), where it is shown by verifying the conditions of a fundamental theorem due to McLeish (1974). The expositions on mixing processes draws from Durrett (1996), Bosq (1998) and Doukhan (1994). It is worth mentioning that the condition $\sum_{n=1}^{\infty} \|E(X_0|\mathcal{F}_{-n})\|_2 < \infty$ in Theorem 8.1.7 can be weakened to $\sum_{n=1}^{\infty} \|E(X_n|\mathcal{F}_0) - E(X_n|\mathcal{F}_{-1}\|_2 < \infty$, see (Hall and Heyde, 1980, Theorem 5.3) and the discussion in Durrett (1996), resulting in the weaker condition $\sum_i |\theta_i| < \infty$ for Theorem 8.6.3, the central limit theorems for linear processes, see (Hannan, 1970, Theorem 11). The method of proof can be generalized and is discussed in the review paper Merlevède et al. (2006). For an advanced probabilistic monograph on limit theorems for semimartinagles the reader is referred to Jacod and Shiryaev (2003). An exposition of the asymptotics of linear processes can be found in Phillips and Solo (1992). Taking an econometric point of view, the asymptotic theory for the multiple linear regression model from with stochastic regressors can be found in Davidson (2000). For nonparametric density estimation, nonparametric regression and its applications, we refer to Silverman (1986), Härdle (1990), Fan and Gijbels (1996), Fan and Yao (2003), Li and Racine (2007) and Franke et al. (2008). Further econometric monographs related to the material of this chapter are White (2001) and Tanaka (1996). The maximal inequality in Theorem 8.8.5 is due to (Peligrad, 1999, Corollary 2.4). For a general result on the almost sure convergence of Bartlett's estimator, see Berkes et al. (2005).

References

Berkes I., Horváth L., Kokoszka P. and Shao Q.M. (2005) Almost sure convergence of the Bartlett estimator. *Period. Math. Hungar.* **51**(1), 11–25.

Bosq D. (1998) *Nonparametric Statistics for Stochastic Processes*. vol. 110 of *Lecture Notes in Statistics* 2nd edn. Springer-Verlag, New York. Estimation and prediction.

Brockwell P.J. and Davis R.A. (1991) *Time Series: Theory and Methods*. Springer Series in Statistics 2nd edn. Springer-Verlag, New York.

Davidson J. (1994) *Stochastic Limit Theory: An Introduction for Econometricians*. Advanced Texts in Econometrics. The Clarendon Press Oxford University Press, New York.

Davidson J. (2000) *Econometric Theory*. Blackwell Publishing.

Doukhan P. (1994) *Mixing*. vol. 85 of *Lecture Notes in Statistics: Properties and Examples*. Springer-Verlag, New York.

Durrett R. (1996) *Probability: Theory and Examples*. 2nd edn. Duxbury Press, Belmont, CA.

Fan J. and Gijbels I. (1996) *Local Polynomial Modelling and its Applications*. vol. 66 of *Monographs on Statistics and Applied Probability*. Chapman & Hall, London.

Fan J. and Yao Q. (2003) *Nonlinear Time Series: Nonparametric and Parametric Methods*. Springer Series in Statistics. Springer-Verlag, New York.

Franke J., Härdle W.K. and Hafner C.M. (2008) *Statistics of Financial Markets: An Introduction*. Universitext 2nd edn. Springer-Verlag, Berlin.

Hall P. and Heyde C.C. (1980) *Martingale Limit Theory and its Application*. Academic Press Inc. [Harcourt Brace Jovanovich Publishers], New York. Probability and Mathematical Statistics.

Hannan E.J. (1970) *Multiple Time Series*. John Wiley and Sons, Inc., New York-London-Sydney.

Härdle W. (1990) *Applied Nonparametric Regression*. vol. 19 of *Econometric Society Monographs*. Cambridge University Press, Cambridge.

Jacod J. and Shiryaev A.N. (2003) *Limit Theorems for Stochastic Processes*. vol. 288 of *Grundlehren der Mathematischen Wissenschaften [Fundamental Principles of Mathematical Sciences]* 2nd edn. Springer-Verlag, Berlin.

Li Q. and Racine J.S. (2007) *Nonparametric Econometrics: Theory and Practice*. Princeton University Press, Princeton, NJ.

McLeish D.L. (1974) Dependent central limit theorems and invariance principles. *Ann. Probability* **2**, 620–628.

Merlevède F., Peligrad M. and Utev S. (2006) Recent advances in invariance principles for stationary sequences. *Probab. Surv.* **3**, 1–36 (electronic).

Peligrad M. (1999) Convergence of stopped sums of weakly dependent random variables. *Electron. J. Probab.* **4**, no. 13, 13 pp. (electronic).

Phillips P.C.B. and Solo V. (1992) Asymptotics for linear processes. *Ann. Statist.* **20**(2), 971–1001.

Silverman B.W. (1986) *Density Estimation for Statistics and Data Analysis: Monographs on Statistics and Applied Probability*. Chapman & Hall, London.

Tanaka K. (1996) *Time Series Analysis*. Wiley Series in Probability and Statistics: Probability and Statistics. John Wiley & Sons Inc., New York. Nonstationary and noninvertible distribution theory, A Wiley-Interscience Publication.

White H. (2001) *Asymptotic Theory for Econometricians*. Academic Press, San Diego, CA.

Yokoyama R. (1980) Moment bounds for stationary mixing sequences. *Z. Wahrsch. Verw. Gebiete* **52**(1), 45–57.

9

Special topics

This chapter is devoted to some special topics. First, we discuss the copula approach to the modeling and analysis of high-dimensional multivariate distributions, which has found widespread applications in finance. A prominent application is the pricing of collaterized debt obligations, and thus we take the opportunity to discuss how copulas became part of the 2008 financial crisis.

Nonparametric methods have become an integral part of the analysis of financial data analysis as well. In many areas, the local polynomial approach, particularly the local linear estimator, provides a good balance between simplicity and accuracy. Thus, we study in greater detail the corresponding asymptotic theory complementing and extending the discussions of previous chapters.

Lastly, we discuss some selected change-point methods. Those methods assume that the time series of interest may have structural breaks (change-points) where the distribution changes. Such a setting is quite realistic for financial data according to the large number of potential factors that may have impact on prices, returns, risk measures, indices or other random quantities of interest: Unexpected news such as profit warnings, mergers, fusions, political events, weather extremes, etc. Events like those may lead to changes of the distribution, which call for immediate action such as portfolio updates, hedges, risk selling or closing of positions. Therefore, the analysis of time series in order to test for such changes as well as the application of monitoring procedures to detect them quickly have received considerable interest.

9.1 Copulas – and the 2008 financial crisis

Copulas have become a common approach to model multivariate distributions of random vectors, e.g. the joint default times of credits. They allow us to separate the problem to specify or estimate marginal distributions and the problem to handle dependencies between the coordinates. Many copula models are actually parsimonious parametric models with only

Financial Statistics and Mathematical Finance: Methods, Models and Applications, First Edition. Ansgar Steland.

a few parameters or even only one parameter. In this way, intractable problems such as the calculation of complex risk measures of hundreds or thousands of financial instruments can be melted down to quite simple formulas and procedures. That problem is of particular concern when dealing with debts. The default events are often dependent, particularly when a crisis hits markets and one default induces the next one. Usually, there is not enough historical data for each credit and obligor, respectively, to estimate default probabilities from them. Copula models allow us to describe the dependence by a few parameters, which in some cases can be linked to available market information.

The price one has to pay for that parsimonity is that the structural shape of the dependence is fixed for a given copula. Of course, a one-dimensional model for a, say, 1000-dimensional distribution represents only a tiny subset of all possible distributions. But if the selected copula model is wrong and does not fit reality, risk measures and prices calculated from that model can be highly misleading.

This is what actually happened before and during the 2008 financial crisis. They were more and more used to price collateralized default obligations as well as credit default swaps. The Li model became the industry standard in this field and was heavily used by rating agencies and banks. Li's model relies on a simple Gaussian copula model, too simplistic for the problem at hand. The default probabilities calculated from this model were too optimistic.

After an introduction to copulas and their basic properties, we give a brief outline of the major reasons leading to the financial turmoil before discussing the famous Li model in some detail.

9.1.1 Copulas

Let $X = (X_1, \ldots, X_d)$ be a random vector of dimension $d \in \mathbb{N}$, where d can be very large. Its distribution is determined by the multivariate distribution function

$$F(x_1, \ldots, x_d) = P(X_1 \le x_1, \ldots, X_d \le x_d), \qquad x_1, \ldots, x_d \in \mathbb{R}.$$

The basic idea of the copula approach is based on the following observation: Let $C(u_1, \ldots, u_d)$ be a distribution function of a random vector $(U_1, \ldots, U_d)$ with uniform marginals, that is $U_i \sim U[0, 1]$, $i = 1, \ldots, d$. Then

$$C(F_1(x_1), \ldots, F_d(x_d)), \qquad x_1, \ldots, x_d \in \mathbb{R}, \tag{9.1}$$

defines a distribution function on $\mathbb{R}^d$ with marginals $F_1, \ldots, F_d$. Let us verify the latter fact for the first coordinate. We have $\lim_{x \to \infty} F_j(x) = 1$, $j = 1, \ldots, d$, and

$$\lim_{u_2, \ldots, u_d \to \infty} C(u_1, \ldots, u_d) = C(u_1, 1, \ldots, 1) = u_1, \qquad u_1 \in [0, 1],$$

leading to

$$\lim_{x_2, \ldots, x_d \to \infty} C(F_1(x_1), \ldots, F_d(x_d)) = F_1(x_1), \qquad x_1 \in \mathbb{R}.$$

Definition 9.1.1 (COPULA)
A **copula** *is a distribution function* $C : [0, 1]^d \to [0, 1]$ *of a d-dimensional random vector* $(U_1, \ldots, U_d)$ *with* $U_i \sim U(0, 1)$, $i = 1, \ldots, d$.

Let $X \sim F$ be a random variable and define for $x \in \mathbb{R}$ and $a > 0$

$$F(x, a) = P(X \leq x) + aP(X = x).$$

Let V be a random variable with $V \sim U(0, 1)$. Then

$$U = F(X, V) = F(X-) + V[F(X) - F(X-)]$$

is called a **distributional transform** or **probability integral transform**. Notice that for a continuous distribution function F, $F(x, a) = F(x)$. In this case the distributional transform is simply given by $U = F(X)$. The distributional transform satisfies

$$U \sim U(0, 1) \qquad \text{and} \qquad X = F^{-1}(U) \text{ a.s..}$$

Let us check the latter fact. By definition

$$F(X-) \leq U = F(X, V) \leq F(X)$$

and $P(U = F(X-)) = P(V = 0)$. Since $F^{-1}(u) = x$ for all $u \in (F(x-), F(x)]$, $F^{-1}(U) = X$ a.s. follows.

The following celebrated theorem of Sklar asserts that any distribution function on $\mathbb{R}^d$ can be represented in the form (9.1) and provides the basis of the copula approach.

Theorem 9.1.2 (SKLAR'S THEOREM, 1959)

(i) Let $F : \mathbb{R}^d \to [0, 1]$ be a d-dimensional distribution function and F_i, $i = 1, \ldots, d$, be the associated marginal distribution functions. Then there exists a d-dimensional distribution function $C : [0, 1]^d \to [0, 1]$ such that

$$F(x_1, \ldots, x_d) = C(F_1(x_1), \ldots, F_d(x_d)),$$

for all $(x_1, \ldots, x_d)' \in \mathbb{R}^d$.

(ii) If $C : [0, 1]^d \to [0, 1]$ is a copula and $F_1, \ldots, F_d$ are univariate distribution functions, then

$$C(F_1(x_1), \ldots, F_d(x_d)), \qquad (x_1, \ldots, x_d) \in \mathbb{R}^d,$$

defines a d-dimensional distribution function with marginals $F_1, \ldots, F_d$.

Proof. We show (i). Let $(X_1, \ldots, X_d)$ be a random vector with distribution function F. For a random variable $V \sim U(0, 1)$ independent of X consider the distributional transforms $U_i = F_i(X_i, V_i)$ leading to the a.s. representations $X_i = F_i^{-1}(U_i), i = 1, \ldots d$. Now the copula

$$C(u_1, \ldots, u_d) = P(U_1 \leq u_1, \ldots, U_d \leq u_d), \qquad u_1, \ldots, u_d \in [0, 1],$$

satisfies

$$\begin{aligned} F(x_1, \ldots, x_d) &= P(X_1 \leq x_1, \ldots, X_d \leq x_d) \\ &= P(F_1(U_1) \leq x_1, \ldots, F_d(U_d) \leq x_d) \\ &= P\big(U_1 \leq F_1^{-1}(x_1), \ldots, U_d \leq F_d^{-1}(x_d)\big), \end{aligned}$$

for all $x_1, \ldots, x_d \in \mathbb{R}$.

If a multivariate distribution is expressed by a copula C and the marginal distribution functions $F_1, \ldots, F_d$, one can simulate random vectors $(X_1, \ldots, X_d)$ having the distribution function $F(x_1, \ldots, x_d) = C(F_1(x_1), \ldots, F_d(x_d))$ as follows.

1. Draw a random vector $(U_1, \ldots, U_d) \sim C$ from the copula C.
2. Apply coordinate-wise the quantile functions F_i^{-1} to obain the random vector
$$(X_1, \ldots, X_d) = \left(F_1^{-1}(U_1), \ldots, F_d^{-1}(U_d)\right),$$
which is distributed according to F.
3. Repeat steps 1. and 2. n times to obtain a random sample $X^{(1)}, \ldots, X^{(n)} \sim F$ of d-dimensional random vectors.

Notice that, if the d-dimensional distribution function F in Sklar's theorem has continuous marginals $F_1, \ldots, F_d$, then the copula satisfies
$$C(u_1, \ldots, u_d) = F\left(F_1^{-1}(u_1), \ldots, F_d^{-1}(u_d)\right),$$
for every $u = (u_1, \ldots, u_d)' \in \mathbb{R}^d$. Further, we have the representation
$$C(u_1, \ldots, u_d) = P(F_1(X_1) \le u_1, \ldots, F_d(X_d) \le u_d),$$
for every $u = (u_1, \ldots, u_d)' \in \mathbb{R}^d$.

Here is a list of some frequently used copulas.

Example 9.1.3 (COPULAS)

(i) **Independence:** *Suppose $U_1, \ldots, U_d \sim U(0,1)$ are independent. The corresponding independence copula is given by*
$$P(U_1 \le u_1, \ldots, U_d \le u_d) = \prod_{i=1}^d P(U_i \le u_i) = \prod_{i=1}^d u_i,$$
for $u_1, \ldots, u_d$.

(ii) **Perfect correlation:** *If $U_1, \ldots, U_d \sim U(0,1)$ are perfectly correlated such that $U_1 = \cdots = U_d$, then the associated copula is*
$$C_u(u_1, \ldots, u_d) = P(U_1 \le \min(u_1, \ldots, u_d)) = \min(u_1, \ldots, u_d),$$
for $u_1, \ldots, u_d \in [0,1]$, called the **upper Fréchet copula**.

(iii) **Mixtures of copulas***: If $C_0(u_1, \ldots, u_d)$ is an arbitrary copula (e.g. the independence copula) and $\rho > 0$ a mixing coefficient, then we obtain a new copula by mixing it with the copula corresponding to perfect correlation,*
$$C(u_1, \ldots, u_d) = (1-\rho)C_0(u_1, \ldots, u_d) + \rho \min(u_1, \ldots, u_d),$$
for $u_1, \ldots, u_d \in [0,1]$.

(iv) **Bivariate Gaussian copula:** *For $d = 2$ one may consider the bivariate normal distribution with means zero, unit variances and correlation coefficient $\rho \in (-1, 1)$, that is*

$$C(u_1, u_2) = \int_{-\infty}^{\Phi^{-1}(u_1)} \int_{-\infty}^{\Phi^{-1}(u_2)} \frac{1}{2\pi(1-\rho^2)^{1/2}} \left(-\frac{s^2 - 2\rho st + t^2}{2(1-\rho^2)} \right) \mathrm{d}s\, \mathrm{d}t,$$

for $u_1, u_2 \in [0, 1]$. Notice that this is a one-parameter family.

(v) **Bivariate t-copula:**

$$C(u_1, u_2) = \int_{-\infty}^{F_\nu^{-1}(u_1)} \int_{-\infty}^{F_\nu^{-1}(u_2)} \frac{1}{2\pi(1-\rho^2)^{1/2}} \left(1 + \frac{s^2 - 2\rho st + t^2}{\nu(1-\rho^2)} \right)^{-(\nu+2)/2} \mathrm{d}s\, \mathrm{d}t,$$

for $u_1, u_2 \in [0, 1]$, where $F_\nu^{-1}(p)$ denotes the quantile function of the t-distribution with ν degrees of freedom and $\rho \in (-1, 1)$ is the coefficient of correlation.

(vi) **Gaussian copula:** *The bivariate Gaussian copula easily extends to dimension d. In general,*

$$C(u_1, \ldots, u_d) = \Phi_\Sigma(\Phi^{-1}(u_1), \ldots, \Phi^{-1}(u_d)), \qquad u_1, \ldots, u_d \in [0, 1],$$

is called a Gaussian copula, where Φ_Σ is the distribution function of a multivariate normal distribution with mean zero and a covariance matrix Σ, and $\Phi^{-1}$is the quantile function of the $N(0, 1)$ distribution. For $d = 2$ and $\Sigma = \begin{pmatrix} 1 & \rho \\ \rho & 1 \end{pmatrix}$ one obtains the bivariate Gaussian copula.

(vii) **Extreme value copuals:** *Such copulas for bivariate distributions have the form*

$$C(u_1, u_2) = \exp\left[\log(xy) A\left(\frac{\log(u_1)}{\log(u_1 u_2)} \right) \right], \qquad u_1, u_2 \in [0, 1],$$

where $A : [0, 1] \to [1/2, 1]$, is a convex function and $A(t)$ satisfies the constraints $\max(t, 1-t) \leq A(t) \leq 1$ for all $t \in [0, 1]$.

(viii) **Kimeldorf and Sampson copula:**

$$C(u_1, u_2) = \left(u_1^{-\gamma} + u_2^{-\gamma} - 1 \right)^{1/\gamma}, \qquad u_1, u_2 \in [0, 1],$$

for some $0 \leq \gamma < \infty$.

(ix) **Archimedean copulas:** *A copula is Archimedean, if it is of the form*

$$C(u_1, \ldots, u_d) = \psi(\psi^{-1}(u_1) + \cdots + \psi^{-1}(u_d))$$

for some function ψ *that is called a* **generator**. *To ensure that C is a copula, the generator has to be d-monotone on* $[0, \infty)$*, i.e.*

$$(-1)^k \psi^{(k)}(x) \geq 0,$$

for all $x \geq 0$ *and* $k = 0, 1, \ldots, d-2$*, and* $(-1)^{d-2}\psi^{(d-2)}(x)$ *is nonincreasing and convex. Frequently used generators and their inverse are as follows.*

– **Ali–Mikhail–Haq:** *Generator*

$$\psi(t;\theta) = \frac{1-\theta}{\exp(t)-\theta}, \qquad \theta \in [0, 1),$$

with inverse

$$\psi^{-1}(t;\theta) = \log \frac{1-\theta+\theta t}{t}.$$

– **Clayton:** *Generator*

$$\psi(t;\theta) = (1+t)^{-1/\theta}, \qquad \theta \in (0, \infty),$$

with inverse

$$\psi^{-1}(t;\theta) = t^{-\theta} - 1.$$

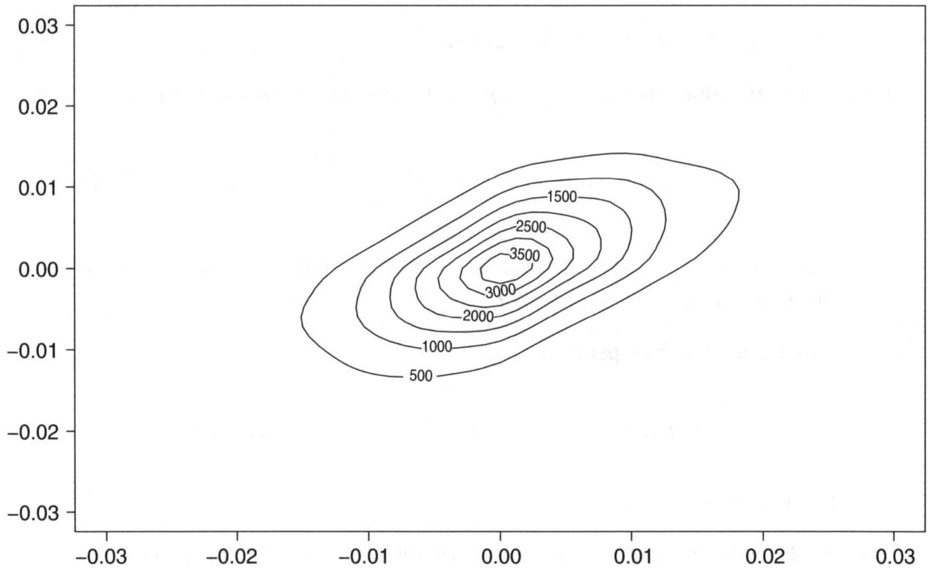

Figure 9.1 Contour plot for the distribution of DAX and FTSE daily log returns using the Gaussian copula and kernel estimates for the marginals.

– **Frank:** *Generator*

$$\psi(t;\theta) = -\frac{\log(1-(1-\exp(-\theta))\exp(-t))}{\theta}, \qquad \theta \in (0,\infty),$$

with inverse

$$\psi^{-1}(t;\theta) = -\log\frac{\exp(-\theta t)-1}{\exp(-\theta)-1}.$$

Figure 9.1 and Figure 9.2 show a Gaussian copula fitted by maximum likelihood to daily log returns of of DAX and FTSE with marginals estimated nonparametrically by a kernel density estimator yielding a copula-based estimator of the bivariate distribution.

A nonparametric estimator for the copula function $C(u_1, \ldots, u_d)$ based on a sample $X_1, \ldots, X_T$ of d-dimensional random vectors,

$$X_t = (X_{t1}, \ldots, X_{td})', \qquad t = 1, \ldots, T,$$

with common distribution function $F(x_1, \ldots, x_d) = C(F_1(x_1), \ldots, F_d(x_d))$ can be derived as follows. For simplicity of the exposition, let us assume that the marginal distribution functions $F_1, \ldots, F_d$ are continuous. Recall that

$$C(u_1, \ldots, u_d) = P(F_1(X_1) \le u_1, \ldots, F_d(X_d) \le u_d),$$

for $u_1, \ldots, u_d \in [0, 1]$. If we knew the transforms $U_{ij} = F_j(X_{ij})$, a natural estimator for the copula is

$$V_T(u_1, \ldots, u_d) = \frac{1}{T}\sum_{t=1}^{T} \mathbf{1}(U_{t1} \le u_1, \ldots, U_{td} \le u_d), \qquad u_1, \ldots, u_d \in [0, 1].$$

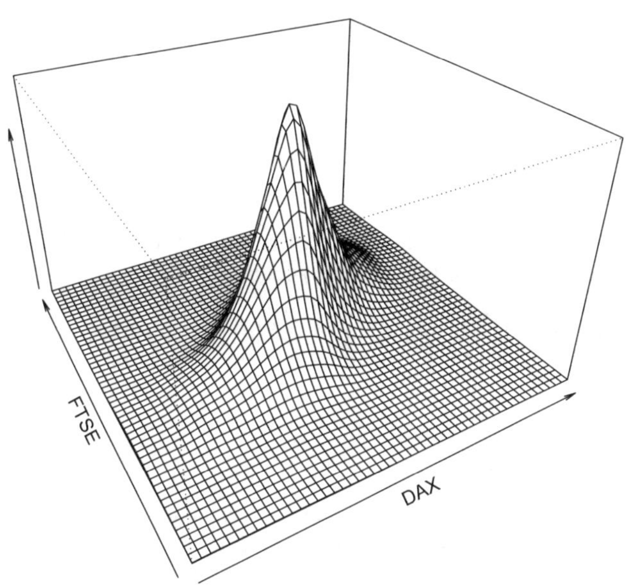

Figure 9.2 3D plot of the joint density of the DAX and FTSE daily log returns constructed from the Gaussian copula and kernel estimates for the marginals.

Since the F_j are unknown, we now replace them by their nonparametric estimators

$$\widehat{F}_{Tj}(x) = \frac{1}{T}\sum_{t=1}^{T} \mathbf{1}(X_{tj} \le x), \qquad x \in \mathbb{R},$$

leading to the empirical counterparts of $U_{tj} = F_j(X_{tj})$ given by

$$\widehat{U}_{tj} = \widehat{F}_{Tj}(X_{tj}), \qquad j = 1, \ldots, d;\ t = 1, \ldots, T.$$

Notice that $T\widehat{U}_{tj}$ is the sum of all $X_{t'j}, t' = 1, \ldots, T$, with $X_{t'j} \le X_j$, i.e. the rank R_{tj} of X_{tj} in the sample $X_{1j}, \ldots, X_{Tj}$ of the jth coordinates, and the ranks are a permutation of the numbers $1, \ldots, T$, almost surely. Now the unknown copula function is estimated nonparametrically by

$$\widehat{C}_T(u_1, \ldots, u_d) = \frac{1}{T}\sum_{i=1}^{T} \mathbf{1}(\widehat{F}_{T1}(X_{i1}) \le u_1, \ldots, \widehat{F}_{Td}(X_{id}) \le u_d), \qquad u_1, \ldots, u_d \in [0, 1].$$

Since $\widehat{F}_{T1}(X_{i1}) \le u_1 \Leftrightarrow X_{i1} \le \widehat{F}_{T1}^{-1}(u_1)$, this estimator can also be written as

$$\widehat{C}_T(u_1, \ldots, u_d) = \frac{1}{T}\sum_{i=1}^{n} \mathbf{1}\big(X_{i1} \le \widehat{F}_{Td}^{-1}(u_1), \ldots, X_{id} \le \widehat{F}_{T1}^{-1}(u_d)\big).$$

The corresponding empirical process

$$\sqrt{T}[\widehat{C}_T(u_1, \ldots, u_d) - C(u_1, \ldots, u_d)], \qquad u_1, \ldots, u_d \in [0, 1],$$

is known to converge weakly to the Gaussian process

$$\begin{aligned} G_C(u_1, \ldots, u_d) &= B_C(u_1, \ldots, u_d) \\ &\quad - \sum_{j=1}^{d} \frac{\partial C}{\partial u_j}(u_1, \ldots, u_d) B_C(u_1, \ldots, u_{j-1}, 1, u_{j+1}, \ldots, u_d), \end{aligned}$$

$u_1, \ldots, u_d \in [0, 1]$, as $T \to \infty$, where B_C is a Brownian bridge on $[0, 1]^d$ with covariance function $\mathrm{Cov}(B_C(u_1, \ldots, u_d), B_C(u_1', \ldots, u_d'))$ equal to

$$C(\min(u_1, u_1'), \ldots, \min(u_d, u_d')) - C(u_1, \ldots, u_d)C(u_1', \ldots, u_d'),$$

for $u_1, \ldots, u_d, u_1', \ldots, u_d' \in [0, 1]$.

9.1.2 The financial crisis

From the mid-1990s to 2006 US American house prices increased each year across the whole country, forming a substantial bubble. During that time the interest rates were low, which spurred increases in mortgage financing, of course, and house prices, and also encouraged financial institutions to construct instruments enhancing yields. The bubble was inflated by a rapid rise of lending to subprime borrowers, which started in 2000. *Adjustable Rate Mortgages*

(ARMs) were invented for low-income not credit-worthy people, which would otherwise be excluded from the mortgage market, and speculators. The ARMs usually offered low teaser-rates in the first few years and required no down-payment, for example by offering a second mortgage contract for the down-payment. Sometimes, the borrowers were even allowed to postpone some of the interest due and add it to the principal. The logic behind these deals was the expected constant rise of home prices, which would allow the borrowers (or speculators) to re-finance after a few years by a new mortgage with teaser-rate or sell the house at a higher price. But the rate of the US house price rise began to decline after April 2005, such that the possibility to re-finance early was pushed into the future. Many subprime borrowers ended up with substantially higher mortage costs. As a consequence, the delinquency rate among home owners of subprime ARMs increased until 2006 to 10.09% - compared to 2.27% for prime fixed-rate mortgages.

The rise of subprime lending was enhanced by new financial innovations, which transformed mortgages into a standardized financial instrument, a process called *securitization*. Two government-sponsored enterprises, which were rated like AAA bonds, heavily engaged in mortgage lending, Fannie Mae and Freddie Mac, developed *mortgage-backed securities* (MBS) in the 1970s, which pooled geographically diversified mortgages into packages and added some guarantees, making them marketable. In this way, Fannie Mae and Freddie Mac were able to buy loans across the country from local banks and sell the associated risk to the risk-eager investors. The local banks were then in a position to expand their mortgage lending to customers. The basic and fascinating idea of MBS is that the mortgages are repackaged in tranches with different risk profiles. The funds available to the mortgage market can be increased substantially in this way. Since a country-wide series of defaults had never occurred before, these MBS were regarded as rather safe investments. Both enterprises were regulated and only allowed to deal with *conforming* loans of borrowers with a credit score above a certain threshold. But private firms developed MBS backed by subprime mortgages, which offer higher yields than standard mortgages.

It is worth looking at how MBS are constructed. A MBS transaction has two sides linked by cash flows. The asset side is the underlying reference portfolio and the liability side consists of securities issued by an issuer, often a *special purpose vehicle* (SPV). SPVs are off-balance sheet legal entities created by the owner of the pool in order to insulate investors from the credit risk of the originator, usually a bank. The originator sells assets to a SPV that then issues structured notes backed by the portfolio. The tranches are called senior, mezzanine, mezzanine junior and equity, to which percentages of losses are assigned. Assume for example that these percentages are 3%, 4%, 5%, and 85%. The equity tranche has to cover the first 3% of a loss, the mezzanine junior tranche the next 4%, and so forth. When the portolio suffers a loss of, e.g., 5%, by a defaulting debt, then the equity tranche has to bear 3% and the mezzanine junior tranche 2%. The other tranches are not affected. In this way, holders of a senior tranche take less risk than holders of the equity tranche, for example. Further, the senior tranche has preferred claims on the returns generated by the mortgages; they are paid first. Once the senior holders are paid, the mezzanine holders are paid next, and so on. The equity tranche receives what is left. In this way, the riskier tranches are *subordinated* to the senior tranche. The higher the subordination, the safer the senior tranche. MBS achieve *credit enhancement* by *over-collaterization*. This means that the face value of the mortgage assets is higher than the face value of the re-packaged securities. The over-collaterized part is the equity tranche. In the above example, 3% of the mortgage payments can default before the higher tranches suffer any loss.

The re-packaging into tranches does not reduce the risk of the underlying portfolio, it is only rearranged. The senior tranche is eligible to high investment grade credit ratings, since they are more or less insulated from the default risk. But the lower tranches are substantially more risky and suffer from losses rather quickly.

Since around 2000, commercial and investment banks constructed new financial instruments to securitize subprime mortgages, which boomed and fueled the bubble substantially: *collateralized debt obligations* (CDOs) package securities backed by mortgages together with other assets. CDO issuers purchase different MBS tranches and pool them with other asset-backed securities such as credit card loans, auto loans or student loans. Whereas the assets of a MBS consist of actual mortgage payments, the assets of a CDO are the securities that collect these mortgage payments. Clearly, now the risks an investor is taking are less transparent. CDOs were issued in great quantities in 2006 and 2007. Actually, there were not enough ABS traded, so many CDOs were backed on synthetic ABS. Before the financial crisis, the issuers worked directly with rating agencies to structure CDOs tranches and purchased credit default swaps or credit insurances in such a way that they received high ratings for their CDOs. Actually, the senior tranches of those CDOs were mispriced: The investment banks purchased BBB mortgage-backed securities with high yields by issuing AAA-rated CDO bonds paying lower yields. By simply re-packaging cashflows, it was possible to generate a positive net present value. But when in 2007 waves of CDO downratings hit the market on a massive scale, since the CDOs were mispriced by the too-simplistic Li model, previously high-rated tranches became exposed to severe losses. Now, the holders of the CDOs came into trouble. Investors such as pensions funds that can invest only a certain amount of money in bonds below triple-A rating had pressure to sell their holdings at any price.

There was a general positive feedback mechanism in effect: When prices increase constantly during a bubble, the net worth of a bank increases when the positions are mark-to-market. This decreases their leverage. Hence, they aim at making use of their new surplus capital, i.e. expand their balance-sheet, since it is unprofitable for a bank to be under-leveraged when prices are rising. This fueled further demand for mortgage-related products. The supply of subprime assets adjusted to this continuously increasing demand. When the crisis hit, the leverage sharply increased, the institutions lacked liquidity and had many securities on bad loans in their books, which were no longer tradeable.

Banks held part of the MBS, CDOs and other debts in *structured investment vehicles* (SIVs), which are off-balance sheet SPV created by banks to hold such assets, in order to leverage their positions more than they could on their balance-sheets due to capital requirements of regulators. To finance the SIV's positions, the SPV issued *asset-backed commercial papers* (ABCPs) as liabilities. These instruments mostly had short-term maturities less than two weeks and had to roll over constantly. However, when the crisis hit, many banks put their SIVs on the balance-sheet again.

Overnight repurchase agreements (repo loans) became a popular instrument for short-term borrowing for investment banks. Here a bank takes its assets as collateral in an overnight loan with another bank. In this way, the financial institutions were linked together so that when one bank got into trouble, the problem spreads to the other institutions. It is estimated that overnight repos grew as a share of the total assets from 12% to 25%. In 2007 37% of Lehman Brother's liabilities were collateralized borrowing and 22% short positions, and these short-term instruments were used to finance long-term assets. The drying up of these short-term liquidity funding, especially ABSPs and repo loans, as a consequence of the mistrust among the banks has been an important element in the financial crisis.

A natural idea to enhance mortgage-backed securities and CDOs is credit insurance. In the 1970s, mono-line insurers such as Ambac with strong credit ratings started to back municipal bond issues. By providing default insurance, the municipals were able to borrow at AAA rates. Since the default probabilities of municipal bonds were overestimated, the mono-liners took advantage of ratings arbitrage. This business model was expanded into the housing market by selling credit default swaps to insure holders of CDOs and other mortgage-backed securities.

Of course, the extensive trading of a certain financial instrument backed on large portfolios of credits requires pricing formulas to determine their fair value. David X. Li used the Gaussian copula to model default correlation and proposed to use spreads of credit default swaps in order to calibrate them to the market. As mentioned above, this became the standard in the financial industry and was also used by the rating agencies. When the crisis evolved, the CDS spreads increased rapidly due to the increased default risks of the underlying obligors, which in turn led to falling values of CDO tranches. The rating agencies adapted their ratings of CDOs to reality. In June 2007, Moody's downgraded the ratings of subprime MBS backed by residential mortgages worth 5 billion US dollars and started to review 184 CDO tranches for downgrade. Standard & Poors placed MBS worth 7.3 billion US dollars on downgrade watch.

9.1.3 Models for credit defaults and CDOs

Let us start with some elementary calculus. Fix a time horizon T and denote by A and B the events that two credits, obligors or firms, default. The corresponding probabilities of a default before time T are $p_A = P(A) = E(\mathbf{1}_A)$ and $p_B = P(B) = E(\mathbf{1}_B)$. The joint default is the event $A \cap B$ that occurs with probability $p_{AB} = P(A \cap B)$. When the defaults are independent, $p_{AB} = p_A p_B$ holds true. Otherwise, the defaults are correlated and it makes sense to calculate the conditional probabilities

$$p_{A|B} = \frac{p_{AB}}{p_B}, \qquad p_{B|A} = \frac{p_{AB}}{p_A},$$

and the coefficient of correlation $\rho_{AB} = \mathrm{Cor}(\mathbf{1}_A, \mathbf{1}_B)$, which is easily seen to be

$$\rho_{AB} = \frac{p_{AB} - p_A p_B}{\sqrt{p_A(1-p_A)p_B(1-p_B)}}.$$

Suppose we are given n credits from n obligors with face values $N_1, \ldots, N_n$. Let $r_1, \ldots, r_n$ be the recovery rates, i.e. if the credit i defaults, one still gets the payment $r_i N_i$, $i = 1, \ldots, n$. In what follows, the recovery rates are assumed deterministic.

Let

$$\tau_1, \ldots, \tau_n : (\Omega, \mathcal{F}, Q) \to [0, \infty)$$

be the default time of n obligors (borrowers), which may be individuals or firms, and Q be a fixed pricing probability measure. Further, denote by F and S, respectively, the joint distribution and survival functions, i.e.

$$F(x_1, \ldots, x_n) = Q(\tau_1 \le x_1, \ldots, \tau_n \le x_n)$$

and

$$S(x_1, \ldots, x_n) = Q(\tau_1 > x_1, \ldots, \tau_n > x_n)$$

for $(x_1, \ldots, x_n) \in \mathbb{R}^n$. By virtue of Sklar's theorem, there exists a copula C such that

$$F(x_1, \ldots, x_n) = C(F_1(x_1), \ldots, F_n(x_n)),$$

where $F_1, \ldots, F_n$ are the marginal distribution functions of the default times $\tau_1, \ldots, \tau_n$. Vice versa, we obtain a model for the joint distribution of defaults, if we select a copula and marginals $F_1, \ldots, F_n$.

As CDOs are often based on a large number of credits, there is usually a lack of historical data, which would allow us to estimate (joint) default probabilities. A further issue is dimensionality. n may be very large, which may even preclude the estimation of correlations in practice. Recall that there exists $n(n-1)/2$ pairwise correlations. For example, for $n = 100$, one has to estimate 4590 correlations.

In asset-based or structural models, the underlying obligor's value $V_i(t)$ are modelled. A default is triggered if $V_i(t) < K_i$ for the first time, such that in such a model

$$\tau_i = \inf\{t \geq 0 : V_i(t) < K_i\},$$

where K_i is a transition barrier, $i = 1, \ldots, n$. In the same vein, downgradings in ratings can be modeled by selecting such barriers. Correlated defaults are then a consequence of the dependence structure of the values $V_1(t), \ldots, V_n(t)$.

In **intensity-based models**, one assumes the existence of d non-negative functions $\lambda_1, \ldots, \lambda_d : [0, \infty) \to [0, \infty)$ such that the survival probability that the ith obligor (or credit) survives t is given by

$$p_i(t) = P(\tau_i > t) = \exp\left(-\int_0^t \lambda_i(u)\, du\right), \qquad t \in [0, \infty).$$

Notice that $p_i(t)$ is nonincreasing and the corresponding distribution functions are

$$F_i(t) = 1 - p_i(t) = 1 - \exp\left(-\int_0^t \lambda_i(u)\, du\right), \qquad t \in [0, \infty).$$

We claim that the default times can be assumed to be given by

$$\tau_i = \inf\{t \geq 0 : p_i(t) \leq U_i\} \tag{9.2}$$

for random variables $U_1, \ldots, U_n$, which are uniformly distributed on the unit interval $[0, 1]$. Indeed, recall that for any number $p \in [0, 1]$ and any random variable $U \sim U(0, 1)$, we have $P(U < p) = p$. Hence, for τ_i as in Equation (9.2),

$$\begin{aligned} P(\tau_i > t) &= P(p_i(s) > U_i, s \in [0, t]) \\ &= P(p_i(t) > U_i) \\ &= p_i(t), \end{aligned}$$

since $p_i(t)$ is nonincreasing. Hence,

$$P(\tau_i \leq t) = F_i(t), \qquad t \in \mathbb{R},$$

for $i = 1, \ldots, d$. Actually, one may replace the U_i by

$$F_i(\tau_i), \qquad i = 1, \ldots, d,$$

since those random variables are uniformly distributed on [0, 1]. Dependence between defaults is now introduced by allowing for dependent $U_1, \ldots, U_n$. The Li model assumes a Gaussian copula. That means that the joint distribution function of the default times is modelled by a Gaussian copula C_g, i.e.

$$P(\tau_1 \leq x_1, \ldots, \tau_d \leq x_d) = C_g(F_1(x_1), \ldots, F_d(x_d)), \qquad x_1, \ldots, x_d \in \mathbb{R}. \tag{9.3}$$

The d loans or bonds of the underlying portfolio are assumed to be issued by d companies (or obligors) with asset value Z_i. The one-factor approach assumes now that

$$V_i = \rho F + \sqrt{1-\rho^2} G_i, \tag{9.4}$$

where $F, G_1, \ldots, G_d$ are independent standard normal Gaussian random variables under the fixed pricing measure Q, F is a market factor common to all obligors and $\rho \in [0, 1]$ determines the exposure to the factor F. For $\rho = 0$ the defaults are independent, whereas $\rho = 1$ corresponds to the comonotonic case. The random variables G_i represent the idiosyncratic risks of the obligors. Notice that $\mathrm{Cov}(V_i, V_j) = \rho$, if $i \neq j$, and $\mathrm{Var}\,(V_i) = 1$. Consequently, the random vector $(V_1, \ldots, V_n)$ is multivariate normally distributed with mean zero and a covariance matrix $\Sigma(\rho)$ with diagonal elements 1 and off-diagonal elements equal to ρ. This means, ρ is the between-asset correlation.

Recall that $\tau_i \sim F_i$, such that $F_i(\tau_i) \sim U(0, 1)$. In Li's model it is now assumed that the standard normal random variables V_i are related to the default times τ_i by

$$V_i = \Phi^{-1}(F_i(\tau_i)), \qquad i = 1, \ldots, d. \tag{9.5}$$

Noting that $V_i = \Phi^{-1}(F_i(\tau_i)) \Leftrightarrow F_i(\tau_i) = \Phi(V_i)$, we obtain

$$\begin{aligned} P(\tau_1 \leq x_1, \ldots, \tau_d \leq x_d) &= P(F_1(\tau_1) \leq F_1(x_1), \ldots, F_d(\tau_d) \leq F_d(x_d)) \\ &= P(V_1 \leq \Phi^{-1}(F_1(x_1)), \ldots, V_d \leq \Phi^{-1}(F_d(x_d))) \\ &= C_{\Sigma(\rho)}(F_1(x_1), \ldots, F_d(x_d)). \end{aligned}$$

This means that the joint distribution of the default times is given by (9.3) with $C_g = C_{\Sigma(\rho)}$. Notice that for this argument the validity of Equation (9.5) is not needed. It is sufficient to assume that

$$(V_1, \ldots, V_n) \stackrel{d}{=} (\Phi^{-1}(F_1(\tau_1)), \ldots, \Phi^{-1}(F_n(\tau_n))).$$

The conditional default probability at time t given Fcan be calculated as follows.

$$\begin{aligned} p_i(t|V) &= Q(\tau_i \leq t|F) \\ &= Q(F_i^{-1}(\Phi(V_i)) \leq t|F) \\ &= Q(V_i \leq \Phi^{-1}(F_i(t))|F). \end{aligned}$$

Now use the model Equation (9.4) to conclude that

$$\begin{aligned} p_i(t|V) &= Q(\rho F + \sqrt{1-\rho^2}G_i \leq \Phi^{-1}(F_i(t))|V) \\ &= Q\left(G_i \leq \frac{\Phi^{-1}(F_i(t)) - \rho F}{\sqrt{1-\rho^2}}\right) \\ &= \Phi\left(\frac{\Phi^{-1}(F_i(t)) - \rho F}{\sqrt{1-\rho^2}}\right). \end{aligned}$$

To price CDOs using the above model, the correlation ρ needs to be estimated. A common approach is calibration. Here, the parameter is selected such that the model price matches the market prices of credit default swaps.

The simplicity of a Gaussian copula model made it very attractive for the financial industry and it was quickly adopted and became a de facto standard. Having only one parameter, it can be easily calibrated to market prices, allows us to price large portfolios and enables fast computations. But it has a couple of severe drawbacks. The modeling of (joint) defaults is inadequate. Defaults tend to occur in clusters. If one company defaults, it is likely that other companies default as well, for example since they have the same business, or a similar exposure to external risks that caused the default.

Even worse, under the Gaussian copula, the defaults become independent as the size of default increases, in the sense that the upper-tail dependence vanishes, which is defined as follows. Let T_1 and T_2 be default times with marginal distribution functions F_1 and F_2, respectively. Provided the limit exists,

$$d = \lim_{p \to 1-} P\big(T_2 > F^{-1}(q)|T_1 > F^{-1}(q)\big)$$

is called the **coefficient of upper tail dependence**. T_1 and T_2 are called **asymptotically independent in the upper tail**, if $d = 0$. The coefficient d depends only on the copula. For a Gaussian copula with correlation less than one, one can show that $d = 0$. This means that under the Gaussian copula, extreme events look independent, whatever the value of the coefficient of correlation.

9.2 Local linear nonparametric regression

Let us consider the following general framework for nonparametric regression, which extends the setup studied in previous chapters. It is assumed that $\{(Y_t, X_t) : t \in \mathbb{N}\}$ is a discrete-time process satisfying

$$\begin{aligned} Y_t &= m(X_t) + \epsilon_t, \\ \epsilon_t &= \sigma(X_t)\xi_t, \end{aligned}$$

where Y_t is an univariate response variable, X_t d-dimensional regressors, $d \in \mathbb{N}$, $\{\xi_t\}$ is an innovation process with

$$E(\xi_t|X_t) = 0 \quad \text{and} \quad E\big(\xi_t^2|X_t\big) = 1,$$

$\sigma : \mathbb{R} \to (0, \infty)$ and $m : \mathbb{R} \to \mathbb{R}$ functions, $\sigma(x)$ continuous and $m(x)$ a function whose partial derivatives of order $p + 1$ for some $p \in \mathbb{N}$ exist and are bounded. Let us introduce the filtration

$$\mathcal{F}_0 = \{\emptyset, \Omega\}, \ \mathcal{F}_t = \sigma(Y_{s-1}, X_s : s \leq t), \qquad t \geq 0$$

and notice that

$$E(Y_t | X_t = x) = m(x) + \sigma(x) E(\xi_t | X_t = x) = m(x)$$

as well as

$$\mathrm{Var}\,(Y_t | X_t = x) = \sigma^2(x) E\left(\xi_t^2 | X_t = x\right) = \sigma^2(x),$$

almost surely. This means, $m(x)$ is the regression function and $\sigma(x)$ the conditional volatility.

9.2.1 Applications in finance: Estimation of martingale measures and Itô diffusions

The above framework is general enough to cover many important estimation problems arising in finance: The estimation of risk-neutral densities (equivalent martingale measures), estimation of nonparametric models for autoregressive conditional heteroscedasticity and the estimation of discretely sampled diffusion processes.

Example 9.2.1 (Estimation of the Risk-Neutral Density)
Recall from Chapter 1, Section 1.5.8, that the risk-neutral density $\varphi_T^*(x)$ of the stock price S_T at time T is related to the arbitrage-free price $C_e(K)$ of a European call on the stock with strike price K and expiration date T via

$$\frac{\partial^2 C_e(K)}{\partial K^2} = \varphi_T^*(K).$$

Given a sample $C_1, \ldots, C_n$ of option prices with maturities $K_1, \ldots, K_n$, write

$$C_i = m(K_i) + \epsilon_i, \qquad i = 1, \ldots, n,$$

where $m(K_i) = E(C_i | K_i)$ and $\epsilon_i = C_i - m(K_i)$. Then we are led to the problem to estimate the second derivative $m^{(2)}(x)$ from the sample $(C_1, K_1), \ldots, (C_n, K_n)$.

Example 9.2.2 (Nonparametric ARCH Model)
Let $R_t = \log(P_t / P_{t-1})$, $t = 1, \ldots, T$, denote the log returns of price process $\{P_t\}$ that is observed at T equidistant time instants. Let us assume that $\{R_t\}$ form a stationary time series with mean zero and assume a nonparametric ARCH *model of order $p \in \mathbb{N}$,*

$$R_t = \sigma\left(R_{t-1}^2, \ldots, R_{t-p}^2\right) \xi_t, \qquad t = 1, 2, \ldots,$$

where ξ_t are random variables, independent of

$$\mathcal{F}_{t-1} = \sigma(R_{t-1}, \ldots, R_{t-p}),$$

with $E(\xi_t) = 0$ and $E(\xi_t^2) = 1$, for all t, and $\sigma(x_1, \ldots, x_p)$ is a smooth function defined on $[0, \infty) \times \cdots \times [0, \infty)$. Note that $\sigma(R_{t-1}, \ldots, R_{t-p})$ is the conditional volatility given the *past* $\mathcal{F}_{t-1} = \sigma(R_{t-1}, R_{t-2}, \ldots)$. *Consider*

$$R_t^2 = \sigma^2(R_{t-1}, \ldots, R_{t-p})\xi_t^2, \qquad t = 1, 2, \ldots.$$

Then

$$E\left(R_t^2|\mathcal{F}_{t-1}\right) = \sigma^2(R_{t-1}, \ldots, R_{t-p}),$$

and we may estimate $\sigma^2(x_1, \ldots, x_p)$ *by using the model*

$$R_t^2 = \sigma^2(R_{t-1}, \ldots, R_{t-p}) + \epsilon_t, \qquad t = 1, \ldots, T,$$

where $\epsilon_t = R_t^2 - \sigma^2(R_{t-1}, \ldots, R_{t-p})$ *have conditional mean* 0.

Example 9.2.3 (NONPARAMETRIC ESTIMATION OF ITÔ DIFFUSIONS)
Let $\{X(t) : t \geq 0\}$ *be a stationary ergodic Itô diffusion, that is a stationary ergodic solution of the stochastic differential equation*

$$\mathrm{d}X(t) = \mu(X(t))\mathrm{d}t + \sigma(X(t))\mathrm{d}B_t,$$

where $X(0)$ *is distributed according to the stationary distribution* μ. *Recall Section 6.7 for a discussion of this class of processes, sufficient conditions for stationary and ergodic solutions and many examples. The Euler approximation scheme with time step* $\Delta > 0$,

$$X_t = X(t\Delta), \qquad t = 1, 2, \ldots,$$

leads to the equations

$$Y_t = \mu(X_{t-1})\Delta + \sigma(X_{t-1})\sqrt{\Delta}\epsilon_t, \qquad t = 1, 2, \ldots,$$

where

$$Y_t = X_t - X_{t-1}, \qquad t = 1, 2, \ldots$$

is a stationary time series.

9.2.2 Method and asymptotics

Let us first assume that X_t takes values in $\mathbb{R}$. We aim at estimating $m(x)$ for some fixed x. Let us approximate $m(x)$ by its Taylor polynomial around x,

$$m(z) \approx \sum_{k=0}^{p} \frac{m^{(k)}(x)}{k!}(z - x)^k.$$

Fix some univariate smoothing kernel K and a bandwidth $h > 0$. The **local polynomial estimator** of $m(x)$ and its derivatives $m'(x), \ldots, m^{(p)}(x)$ are defined by

$$\widehat{m}_T(x) = \widehat{\delta}_{T0}(x),$$
$$\widehat{m}_T^{(k)}(x) = k!\widehat{\delta}_{Tk}(x), \quad k = 1, \ldots, p,$$

where $\widehat{\delta}_T(x) = (\widehat{\delta}_{T0}, \ldots, \widehat{\delta}_{Tp})'$ with

$$\widehat{\delta}_T(x) = \operatorname{argmin}_{(\delta_0,\ldots,\delta_p)\in\mathbb{R}^{p+1}} \sum_{t=1}^{T} \left(Y_t - \sum_{j=0}^{p} \delta_j (X_t - x)^j \right)^2 K\left(\frac{X_t - x}{h}\right).$$

Define

$$X = \begin{pmatrix} 1 & X_1 - x & (X_1 - x)^2 & \cdots & (X_1 - x)^p \\ \vdots & \vdots & \vdots & & \vdots \\ 1 & X_T - x & (X_T - x)^2 & \cdots & (X_T - x)^p \end{pmatrix},$$
$$Y = (Y_1, \ldots, Y_T)',$$
$$W = \operatorname{diag}\left(K\left(\frac{X_1 - x}{h}\right), \ldots, K\left(\frac{X_T - x}{h}\right)\right).$$

Then $\widehat{\delta}_T(x)$ is a solution of the linear equations

$$(X'WX)\widehat{\delta}_T(x) = X'WY$$

and allows the explicit representation

$$\widehat{\delta}_T(x) = (X'WX)^{-1}X'WY.$$

Notice that $X'WX = (a_{ij})_{i,j}$ is a Vandermondsche matrix with entries $a_{ij} = \sum_{t=1}^{T} K\left(\frac{X_t - x}{h}\right)(X_t - x)^{i+j}$, $1 \le i, j \le p+1$. Thus, it is almost surely regular, since the X_t are continuously distributed.

In what follows, we confine ourselves to the local linear estimator given by $p = 1$, but allow for the multivariate case, that is $d > 1$. To define the smoothing weights, we shall again use the product kernel $K(u) = \prod_{j=1}^{d} L(u_j)$, $u = (u_1, \ldots, u_d) \in \mathbb{R}^d$, see Equation (8.11), for some univariate smoothing kernel L satisfying Assumptions (8.9) and (8.10).

Assume that the regression function $m : \mathbb{R}^d \to \mathbb{R}$ is of the class C_b^3 and consider the Taylor expansion

$$m(X_t) = m(x) + \nabla m(x)(X_t - x) + \frac{1}{2}(X_t - x)'Dm(x)(X_t - x) + o(\|X_t - x\|^2),$$

where $\nabla m(x)$ denotes the gradient and $Dm(x)$ the matrix of second-order partial derivatives. Observe that

$$m(X_t) = \begin{pmatrix} 1 \\ X_t - x \end{pmatrix}' \vartheta(x) + \frac{1}{2}(X_t - x)'Dm(x)(X_t - x) + o(\|X_t - x\|^2),$$

where

$$\vartheta(x) = \begin{pmatrix} m(x) \\ \nabla m(x) \end{pmatrix}.$$

The **local linear estimator** of $\vartheta(x)$ is defined by

$$\widehat{\vartheta}_T(x) = \operatorname{argmin}_{\vartheta(x)\in\mathbb{R}^2} \sum_{t=1}^{T} K\left(\frac{X_t - x}{h}\right)\left(Y_t - \begin{pmatrix} 1 \\ X_t - x \end{pmatrix}' \vartheta(x)\right)^2$$

and attains the explicit representation

$$\widehat{\vartheta}_T(x) = \left[\frac{1}{Th}\sum_{t=1}^{T} K\left(\frac{X_t - x}{h}\right)\begin{pmatrix} 1 \\ X_t - x \end{pmatrix}(1, X_t - x)\right]^{-1} \times \frac{1}{Th}\sum_{t=1}^{T} K\left(\frac{X_t - x}{h}\right)\begin{pmatrix} 1 \\ X_t - x \end{pmatrix} Y_t,$$

It turns out that the matrix in brackets converges to a regular matrix, in probability, if we multiply it with the diagonal matrix $Q_h(x)$ with entries $1, h_1^{-2}, \ldots, h_d^{-2}$. Recall the formula $Q_h \sum_{t=1}^{T} w_t a_t a_t' = \sum_{t=1}^{T} w_t a_t (Q_h a_t)'$ for arbitrary scalars w_t and vectors a_t, $t = 1, \ldots, T$. This leads to

$$\widehat{\vartheta}_T(x) = A_T^{-1}(x) \cdot \frac{1}{Th}\sum_{t=1}^{T} K\left(\frac{X_t - x}{h}\right)\begin{pmatrix} 1 \\ \frac{X_t - x}{h^2} \end{pmatrix} Y_t, \tag{9.6}$$

$$A_T(x) = \frac{1}{Th}\sum_{t=1}^{T} K\left(\frac{X_t - x}{h}\right)\begin{pmatrix} 1 \\ \frac{X_t - x}{h^2} \end{pmatrix}(1, X_t - x) \tag{9.7}$$

$$= \frac{1}{Th}\sum_{t=1}^{T} K\left(\frac{X_t - x}{h}\right)\begin{pmatrix} 1 & (X_t - x)' \\ \frac{X_t - x}{h^2} & \frac{(X_t - x)(X_t - x)'}{h^2} \end{pmatrix}, \tag{9.8}$$

which will be the starting point for the asymptotic analysis. Here and in the following h^2 is understood element-wise such that

$$\frac{X_t - x}{h^2} = \left(\frac{X_{ti} - x_i}{h_i^2}\right)_{i=1}^{d}.$$

It turns out that the convergence rate of the intercept differs from the rate for the slopes. Precisely,

(i) $\widehat{\vartheta}_{T1}(x)$ is $\sqrt{Th_1 \cdots h_d}$-consistent for $\delta_1(x) = m(x)$; whereas

(ii) $\widehat{\vartheta}_{T2}(x)$ is $\sqrt{Th_1 \cdots h_d}\,\operatorname{diag}(h_1, \ldots, h_d)$-consistent.

Here, $\widehat{\vartheta}_T = (\widehat{\vartheta}_{T1}, \widehat{\vartheta}_{T2}')'$, that is $\widehat{\vartheta}_{T1}$ estimates the intercept, whereas $\widehat{\vartheta}_{T2}$ estimates the slopes. The right scaling matrix is therefore the diagonal matrix with those elements,

$$H_T = \operatorname{diag}\left(\sqrt{Th_1 \cdots h_d}, \sqrt{Th_1 \cdots h_d}\,h_1, \ldots, \sqrt{Th_1 \cdots h_d}\,h_d\right).$$

We are now in a position to state and prove the following result on the asymptotic normality of the local linear estimator. The theorem works under the weak assumption that the central limit theorem for smoothing averages holds true.

Theorem 9.2.4 (CENTRAL LIMIT THEOREM FOR LOCAL LINEAR ESTIMATION) *Suppose that* $\{(Y_t, X_t) : t = 1, 2, \dots, \}$ *is a stationary and ergodic* L_2 *time series such that the random vectors*

$$\begin{pmatrix} 1 \\ \frac{X_t - x}{h^2} \end{pmatrix} \epsilon_t, \qquad t = 1, 2, \dots,$$

where $\epsilon_t = Y_t - E(Y_t|X_t)$*, satisfy a central limit theorem in the sense that*

$$U_T(x) = \frac{1}{\sqrt{Th_1 \cdots h_d}} \sum_{t=1}^{T} K\left(\frac{X_t - x}{h}\right) \begin{pmatrix} 1 \\ \frac{X_t - x}{h} \end{pmatrix} \epsilon_t \xrightarrow{d} N(0, V), \tag{9.9}$$

as $T \to \infty$*, for some matrix* V*, and*

$$\max_j h_j = o(1) \qquad \text{and} \qquad Th_1 \cdots h_d \sum_{j=1}^{d} h_j^4 = o(1).$$

Suppose that the third-order partial derivatives of f *exist and are bounded and that* f *has compact support or* K *is bounded and compactly supported. Then*

$$H_T \left(\widehat{\vartheta}_T - \vartheta(x) - \begin{pmatrix} b_h(x) \\ 0 \end{pmatrix} \right) \xrightarrow{d} N(0, A^{-1}(x)V(x)(A^{-1})'),$$

as $T \to \infty$*, where*

$$b_h(x) = \sum_{i=1}^{d} h_i^2 m_{ii}^{(2)}(x) f(x) \widetilde{L}_2,$$

$$A^{-1} = \begin{pmatrix} 1/f(x) & 0 \\ -\nabla f(x)/f^2(x) & \mathbb{I}/(\widetilde{L}_2 f(x)) \end{pmatrix}. \tag{9.10}$$

and

$$\Sigma = \begin{pmatrix} \widetilde{L}_2^d \sigma^2(x)/f(x) & 0 \\ 0 & \mathbb{I}_d \widetilde{L}_2^{d-1} \widetilde{L}_{22}/(\widetilde{L}_2^2 f(x)) \end{pmatrix},$$

with $\widetilde{L}_{22} = \int u^2 L^2(u)\, du$.

Proof. First, notice that the weak law of large numbers for kernel smoothers, that is assertion (8.23) of Theorem 8.5.8, which easily extends to the multivariate case, holds true for the functions $\psi(X_t, x) = X_t - x$ and $\psi(X_t, x) = (X_t - x)(X_t - x)'$, since $\{X_t\}$ is ergodic by assumption. In the proof we shall frequently make use of this fact that implies that

$$\frac{1}{Th}\sum_{t=1}^{T} K\left(\frac{X_t - x}{h}\right)\psi(X_t, x) = \int \frac{1}{h} K\left(\frac{z - x}{h}\right)\psi(z, x) f(z)\,\mathrm{d}z + o_P(1).$$

Next, observe that straightforward algebra of Equation (9.6) shows that

$$H_T(\widehat{\vartheta}_T(x) - \vartheta(x)) = A_T^{-1}(x)\{U_T(x) + H_T B_T(x) + H_T R_T(x)\}, \tag{9.11}$$

where $U_T(x)$ is defined in Equation (9.9) and and

$$B_T(x) = \frac{1}{Th_1 \cdots h_d}\sum_{t=1}^{T} K\left(\frac{X_t - x}{h}\right)\begin{pmatrix} 1 \\ \frac{X_t - x}{h^2}\end{pmatrix}\frac{1}{2}(X_t - x)' Dm(x)(X_t - x)$$

$$R_T(x) = \frac{1}{Th_1 \cdots h_d}\sum_{t=1}^{T} K\left(\frac{X_t - x}{h}\right)\begin{pmatrix} 1 \\ \frac{X_t - x}{h^2}\end{pmatrix} o(\|X_t - x\|^2).$$

We shall show that

(i) $A_T(x) \stackrel{P}{\to} A(x)$, as $T \to \infty$, for some invertible matrix $A(x)$.

(ii) $H_T B_T(x) = \begin{pmatrix} H_{T1}(b_h(x) + o(1) + o_P(1)) \\ H_{T2}\max_j h_j + o_P(1)\end{pmatrix}$, where

$$H_T = \begin{pmatrix} H_{T1} & 0 \\ 0 & H_{T2}\end{pmatrix}$$

with $H_{T1} = \sqrt{Th_1 \cdots h_d}$ and $H_{T2} = \sqrt{Th_1 \cdots h_d}\,\mathrm{diag}(h_1, \ldots, h_d)$.

(iii) $H_T R_T(x) = o_P(1)$.

Noting that $R_T(x)$ is, of course, of smaller order than $B_T(x)$, (iii) follows easily and thus we omit the details. Notice that (ii) and (iii) imply that

$$H_T\left(\widehat{\vartheta}_T(x) - \vartheta(x) - \begin{pmatrix} b_h(x) \\ 0 \end{pmatrix}\right) = A_T^{-1}(x) U_T(x) + o_P(1),$$

as $T \to \infty$, which together with (i) and Equation (9.9) then yields the assertion, by virtue of Slutzky's lemma. Let us now start with (i). We have to show that the sequence, $A_T(x)$, $T \geq 1$, converges to some regular matrix $A(x)$. Let us partition $A_T(x)$ as

$$A_T(x) = \begin{pmatrix} \alpha_T(x) & \beta_T(x) \\ \gamma_T(x) & \delta_T(x)\end{pmatrix},$$

with

$$\alpha_T(x) = \frac{1}{Th_1 \cdots h_d} \sum_{t=1}^{T} K\left(\frac{X_t - x}{h}\right),$$

$$\beta_T(x) = \frac{1}{Th_1 \cdots h_d} \sum_{t=1}^{T} K\left(\frac{X_t - x}{h}\right)(X_t - x)',$$

$$\gamma_T(x) = \frac{1}{Th_1 \cdots h_d} \sum_{t=1}^{T} K\left(\frac{X_t - x}{h}\right)\frac{(X_t - x)}{h^2},$$

$$\delta_T(x) = \frac{1}{Th_1 \cdots h_d} \sum_{t=1}^{T} K\left(\frac{X_t - x}{h}\right)\frac{(X_t - x)(X_t - x)'}{h^2}.$$

Clearly, $\alpha_T(x) \to f(x)$, as $T \to \infty$, in probability. Let us consider the element (i, j) of $\delta_T(x)$. The usual substitutions $u_l = (z_l - x_l)/h_l$, $l = 1, \ldots, d$, lead to

$$\begin{aligned}
(\delta_T(x))_{i,j} &= \frac{1}{Th_1 \cdots h_d} \sum_{t=1}^{T} K\left(\frac{X_t - x}{h}\right)\frac{X_{ti} - x_i}{h_i}\frac{X_{tj} - x_j}{h_j} \\
&= \int \frac{1}{\prod_{j=1}^{d} h_j} K\left(\frac{z - x}{h}\right)\frac{z_i - x_i}{h_i}\frac{z_j - x_j}{h_j} f(z)\,\mathrm{d}z + o_P(1) \\
&= \int K(u)u_i u_j f(x + hu)\,\mathrm{d}u + o_P(1) \\
&= \int K(u)u_i u_j [f(x) + \nabla f(x)hu + (1/2)(hu)'Df(x)(hu) + r_f]\,\mathrm{d}u + o_P(1).
\end{aligned}$$

Here and throughout the proof, we make use of the third-order Taylor expansion

$$f(x + hu) = f(x) + \nabla f(x)hu + (hu)'Df(x)(hu) + r_f \tag{9.12}$$

with remainder $r_f = O(\sum_{j=1}^{d} h_j^3)$, see the formulas and arguments provided for Equation (8.21). Recall that

$$\int u_i u_j K(u)\,\mathrm{d}u = \begin{cases} \widetilde{L}_2, & i = j, \\ 0, & i \neq j. \end{cases}$$

Further

$$\begin{aligned}
\int K(u)u_i u_j \nabla f(x)hu\,\mathrm{d}u &= \sum_{k=1}^{d} h_k \frac{\partial f(x)}{\partial x_k} \int K(u)u_i u_j u_k\,\mathrm{d}u \\
&= h_i \frac{\partial f(x)}{\partial x_i} \int K(u)u_i^2 u_j\,\mathrm{d}u + h_j \frac{\partial f(x)}{\partial x_j} \int K(u)u_j^2 u_i\,\mathrm{d}u \\
&= \begin{cases} 0, & i \neq j, \\ O(\max\{h_i, h_j\}), & i = h. \end{cases}
\end{aligned}$$

Since $f(x) \int K(u)u_i u_j \, du = f(x)\widetilde{L}_2 \delta_{ij}$ as well, where $\delta_{ij} = \mathbf{1}(i = j)$ denotes the Kronecker symbol, we arrive at

$$(\delta_T(x))_{i,j} = f(x)\widetilde{L}_2 + o(1).$$

Hence,

$$\delta_T(x) = \mathbb{I}_d f(x)\widetilde{L}_2 + o(1) + o_P(1),$$

as $T \to \infty$. Consider now the kth element of $\gamma_T(x)$. We have

$$\begin{aligned}
(\gamma_T(x))_k &= \frac{1}{Th_1 \cdots h_d} \sum_{t=1}^{T} K\left(\frac{X_t - x}{h}\right) \frac{X_{tk} - x_k}{h_k^2} \\
&= \frac{1}{h_k} \int \frac{1}{\prod_{j=1}^{d} h_j} K\left(\frac{z - x}{h}\right) \frac{z_{tk} - x_k}{h_k} f(z) \, dz + o_P(1) \\
&= \frac{1}{h_k} \int u_k K(u)[f(x) + \nabla f(x)hu + (1/2)(hu)' Df(x)(hu) + r_f] \, du + o_P(1) \\
&= \frac{1}{h_k} \int u_k K(u)[\nabla f(x)hu + (1/2)(hu)' Df(x)(hu) + r_f] \, du + o_P(1),
\end{aligned}$$

since $\int u_k K(u) \, du = 0$, which implies that the term involving $f(x)$ vanishes. Further

$$O\left(\frac{1}{h_k} \int |u_k K(u)| \, du \, r_f\right) = O\left(\sum_{j=1}^{d} h_j^2\right) = o(1).$$

It remains to analyze the terms involving the linear and quadratic terms of the Taylor expansion. Using

$$\begin{aligned}
\frac{1}{h_k} \int u_k K(u) \nabla f(x) hu \, du &= \sum_{j=1}^{d} \frac{h_j}{h_k} \frac{\partial f(x)}{\partial x_j} \int u_k u_j K(u) \, du \\
&= \frac{\partial f(x)}{\partial x_k} \int u_k^2 K(u) \, du \\
&= \frac{\partial f(x)}{\partial x_k} \widetilde{L}_2
\end{aligned}$$

and

$$\frac{1}{2h_k} \int u_k K(u)(hu)' Df(x)(hu) \, du = o(1),$$

we obtain

$$\gamma_T(x) = \widetilde{L}_2 \nabla f(x) + o(1) + o_P(1),$$

as $T \to \infty$. Noting that $\beta_T(x) = h^2 \gamma_T(x)'$, we obtain

$$\beta_T(x) = o(1) + o_P(1),$$

This establishes part (i) of the proof, namely

$$A_T \xrightarrow{P} A = \begin{pmatrix} f(x) & 0 \\ \widetilde{L}_2 \nabla f(x) & \mathbb{I}_d f(x) \widetilde{L}_2 \end{pmatrix}.$$

Obviously, A is regular and its inverse is given by Equation (9.10). Let us now verify (ii). To do so, consider the first component of $H_T B_h(x)$. We have, using the usual substitutions

$$\begin{aligned}
(H_T B_T(x))_1 &= \sqrt{Th_1 \cdots h_d} \frac{1}{2T} \sum_{t=1}^{T} \frac{1}{\prod_{j=1}^{d} h_j} K\left(\frac{X_t - x}{h}\right) (X_t - x)' Dm(x)(X_t - x) \\
&= \sqrt{Th_1 \cdots h_d} \sum_{i,j=1}^{d} m_{ij}^{(2)}(x) \frac{1}{T} \sum_{t=1}^{T} \frac{1}{\prod_{k \neq i,j}^{d} h_k} K\left(\frac{X_t - x}{h}\right) \frac{X_{ti} - x_i}{h_i} \frac{X_{tj} - x_i}{h_j}, \\
&= \sqrt{Th_1 \cdots h_d} \sum_{i,j=1}^{d} \left\{ m_{ij}^{(2)}(x) h_i h_j \int K(u) u_i u_j f(x + hu) \, \mathrm{d}u + o_P(1) \right\},
\end{aligned}$$

where $Dm(x) = \left(m_{ij}^{(2)}(x)\right)_{i,j}$, since

$$\begin{aligned}
&\int \frac{1}{\prod_{k \neq i,j}^{d} h_k} K\left(\frac{z - x}{h}\right) \frac{z_i - x_i}{h_i} \frac{z_j - x_j}{h_j} f(z) \, \mathrm{d}z \\
&= h_i h_j \int u_i u_j K(u) f(x + hu) \, \mathrm{d}u.
\end{aligned}$$

Again, plugging in the Taylor expansion (9.12) and noting that the leading term matters, we obtain

$$\begin{aligned}
(H_T B_T(x))_1 &= \sqrt{Th_1 \cdots h_d} \sum_{i,j=1}^{d} m_{ij}^{(2)}(x) \left\{ h_i h_j f(x) \int u_i u_j K(u) \, \mathrm{d}u + o_P(1) \right\} \\
&= \sqrt{Th_1 \cdots h_d} \sum_{i=1}^{d} h_i^2 m_{ii}^{(2)}(x) \left\{ f(x) \widetilde{L}_2 + o_P(1) \right\}.
\end{aligned}$$

Consider now the kth component of $H_T B_T(x)$ for $k > 1$. We have

$$\begin{aligned}
&\sqrt{Th_1 \cdots h_d} h_k \frac{1}{T} \sum_{t=1}^{T} \frac{1}{\prod_{l=1}^{d} h_l} K\left(\frac{X_t - x}{h}\right) \frac{X_{tk} - x_k}{h_k^2} \frac{1}{2} \sum_{i,j=1}^{d} (X_{tj} - x_j) m_{ij}^{(2)}(x)(X_{ti} - x_i) \\
&= \sqrt{Th_1 \cdots h_d} \frac{1}{2} \sum_{i,j=1}^{d} h_i h_j \left\{ \int \frac{1}{\prod_{l=1j}^{d} h_l} K\left(\frac{z - x}{h}\right) \frac{z_k - x_k}{h_k} \frac{z_i - x_i}{h_i} \frac{z_j - x_j}{h_j} f(z) \, \mathrm{d}z + o_P(1) \right\} \\
&= \sqrt{Th_1 \cdots h_d} \frac{1}{2} \sum_{i,j=1}^{d} \left\{ h_i h_j \int K(u) u_i u_j u_k \left[f(x) + \nabla f(x) hu + O\left(\sum_l h_l^2\right)\right] \mathrm{d}u + o_P(1) \right\} \\
&= \sqrt{Th_1 \cdots h_d} \frac{1}{2} \sum_{i,j=1}^{d} \left\{ h_i h_j \int K(u) u_i u_j u_k \left[\nabla f(x) hu + O\left(\sum_l h_l^2\right)\right] \mathrm{d}u + o_P(1) \right\}.
\end{aligned}$$

As a preparation for the final estimates of the terms involving the linear term $\nabla f(x)hu$, observe the following fact. If $d = 1$, then $i = j = k$, yielding $\int u_i u_j u_k K(u)\,\mathrm{d}u = \int t^3 L(t)\,\mathrm{d}t = 0$, since $\int_{-K}^{0} t^3 L(t)\,\mathrm{d}t = -\int_0^K z^3 L(t)\,\mathrm{d}t$ for any $K > 0$. Otherwise, if $i \neq j$, either i or j differs from k and we can first integrate w.r.t. that index yielding $\int u_i u_j u_k K(u)\,\mathrm{d}u = 0$, and if $i = j$ we can find some $1 \leq \nu \leq d$ with $\nu \neq i = j$ and argue the same way. We obtain

$$
\begin{aligned}
h_i h_j & \int u_i u_j u_k K(u) \left(\sum_{l=1}^{d} \frac{\partial f(x)}{\partial x_l} h_l u_l \right) du \\
&= h_i h_j \sum_{l=1}^{d} h_l \frac{\partial f(x)}{\partial x_l} \int u_i u_j u_k K(u) u_l \,\mathrm{d}u \\
&= h_i h_j \sum_{l \in \{i,j,k\}} h_l \frac{\partial f(x)}{\partial x_l} \int u_i u_j u_k u_l K(u)\,\mathrm{d}u \\
&= O\left(h_i h_j \max\{h_i, h_j, h_k\}\right) \\
&= O\left(\max\left\{h_i^3, h_j^3\right\}\right).
\end{aligned}
$$

Summarizing the above arguments, we may conclude that

$$
(H_T B_T(x))_k = \sqrt{Th_1 \cdots h_d} \frac{1}{2} \sum_{i,j=1}^{d} \{o(1) + o_P(1)\} = \sqrt{Th_1 \cdots h_d}(0 + o_P(1)),
$$

since

$$
O\left(\sqrt{Th_1 \cdots h_d} \int |u_i u_j u_k K(u)|\,\mathrm{d}u \sum_{j=1}^{d} h_j^2 \right) = O\left(\sqrt{Th_1 \cdots h_d \sum_{j=1}^{d} h_j^4} \right),
$$

which completes the proof.

Theorem 9.2.4 provides the right asymptotics of the local linear estimator for *any* time series that satisfies the smoothing central limit theorem (9.9).

Theorem 9.2.5 *Assume that $\epsilon_t = Y_t - E(Y_t|X_t)$ forms a series of i.i.d. random variables with $E(\epsilon_1^2) < \infty$. Then Equation (9.9) holds true with*

$$
V = \sigma^2(x) f^2(x) \begin{pmatrix} L_2^d & \mathbf{1}'\widetilde{L}_2 L_2^{d-1} \\ \mathbf{1}\widetilde{L}_2 L_2^{d-1} & W \end{pmatrix},
$$

where W is a d-dimensional matrix with diagonal elements equal to $\widetilde{L}_2 L_2^{d-1}$ and off-diagonal elements $L_2^{d-1} \left(\int u L^2(u)\,\mathrm{d}u\right)^2$.

Proof. It is easy to verify that Theorem 8.5.4 also holds true for random vectors. Hence, it suffices to show that

$$
V_T = \operatorname{Var}\left(\frac{1}{\sqrt{Th_1 \cdots h_d}} \sum_{t=1}^{T} K\left(\frac{X_t - x}{h}\right) \begin{pmatrix} 1 \\ \frac{X_t - x}{h} \end{pmatrix} \epsilon_t \right)
$$

converges to V, as $T \to \infty$. We have

$$
\begin{aligned}
V_T &= \frac{1}{Th_1 \cdots h_d} \sum_{t=1}^{T} E\left(K^2\left(\frac{X_t - x}{h}\right) \begin{pmatrix} 1 \\ \frac{X_t - x}{h} \end{pmatrix} \left(1, \frac{(X_t - x)'}{h}\right) \epsilon_t \right) \\
&= \frac{1}{h_1 \cdots h_d} E_{X_1} E\left(K^2\left(\frac{X_1 - x}{h}\right) \begin{pmatrix} 1 \\ \frac{X_1 - x}{h} \end{pmatrix} \left(1, \frac{(X_1 - x)'}{h}\right) E(\epsilon_1^2 | X_1) \right) \\
&= \frac{1}{h_1 \cdots h_d} \int K^2\left(\frac{z - x}{h}\right) \begin{pmatrix} 1 \\ \frac{z - x}{h} \end{pmatrix} \left(1, \frac{(z - x)'}{h}\right) \sigma^2(z) f(z)\, \mathrm{d}z \\
&= \int K^2(u) \begin{pmatrix} 1 \\ u \end{pmatrix} (1, u')(\sigma^2 f)(x + hu)\, \mathrm{d}u.
\end{aligned}
$$

Noting that $K^2(u) = \prod_{j=1}^{d} L^2(u_j)$, $u = (u_1, \ldots, u_d)$, we have

$$
\begin{aligned}
\int K^2(u)(\sigma^2 f)(x + hu)\, \mathrm{d}u &= \int \prod_{j=1}^{d} L^2(u_j)(\sigma^2 f)(x + hu)\, \mathrm{d}u_1 \cdots \mathrm{d}u_d \\
&= (\sigma^2 f)(x) \left(\int L^2(t)\, \mathrm{d}t \right)^d + o(1),
\end{aligned}
$$

by d iterated applications of Bochner's lemma. Indeed, notice that $\int K^2(u)(\sigma^2 f)(x + hu)\, \mathrm{d}u$ equals

$$
\int \prod_{j=2}^{d} L(u_j) \left[\int L^2(u_1)(\sigma^2 f)(x_1 + h_1 u_1, x_2 + h_2 u_2, \ldots, u_d + h_d u_d)\, \mathrm{d}u_1 \right] \mathrm{d}u_2 \cdots \mathrm{d}u_d,
$$

where the expression in brackets converges to $(\sigma^2 f)(x_1, x_2 + h_2 u_2, \ldots, u_d + h_d u_d)$, as $h_1 \to 0$, and so forth. Further, for $k = 1, \ldots, d$

$$
\begin{aligned}
\int K^2(u) u_k (\sigma^2 f)(x + hu)\, \mathrm{d}u &= (\sigma^2 f)(x) \int u_k L^2(u_k)\, \mathrm{d}u_k \prod_{j \neq k} \int L^2(u_j)\, \mathrm{d}u_j + o(1) \\
&= (\sigma^2 f)(x) \int t L^2(t)\, \mathrm{d}t \left(\int L^2(t)\, \mathrm{d}t \right)^{d-1} + o(1),
\end{aligned}
$$

as well as

$$
\begin{aligned}
&\int K^2(u) u_i u_j (\sigma^2 f)(x + hu)\, \mathrm{d}u \\
&\quad = \begin{cases} (\sigma^2 f)(x) \int u_i L^2(u_i)\, \mathrm{d}u \int u_j L^2(u_j) \left(\int L^2(t)\, \mathrm{d}t\right)^{d-2}, & i \neq j, \\ (\sigma^2 f)(x) \left(\int L^2(t)\, \mathrm{d}t\right)^{d-1} \int t^2 L(t)\, \mathrm{d}t, & i = j. \end{cases}
\end{aligned}
$$

Recalling that $\widetilde{L}_2 = \int t^2 L(t)\, \mathrm{d}t$ and $L_2 = \int L^2(t)\, \mathrm{d}t$, the assertion follows now easily.

9.3 Change-point detection and monitoring

Since the true finite-dimensional distributions of time series as arising in finance are unknown, inferential procedures have to rely on large sample asymptotics, both for parametric and nonparametric approaches. The common setting studied extensively throughout this book is to assume that we are given the first T observations $X_1, \dots, X_T$ of an infinite one- or two-sided time series

$$X_1, X_2, \dots \qquad \text{or} \qquad \dots, X_{-1}, X_0, X_1, \dots$$

of random variables are at our disposal, where T is large. Then, a statistic $U_T(X_1, \dots, X_T)$ is calculated, for instance to conduct a statistical test, and its distribution is approximated by applying a central limit theorem in order to determine quantities of interest such as critical values, confidence intervals or risk measures.

Since usually the nth observation X_n is available to us with negligible delay, this means that in the classical setting one waits T units of time before one applies an inferential procedure once at time T to make a decision. However, often the same decision could be made much earlier and many financial tasks such as risk monitoring of traders, portfolio management for clients, liquidity management or the handling of risks by initiating hedges, selling risks or closing positions, demand for pseudo-continuous actions taking place on a sufficiently small time scale.

In such situations, monitoring procedures are in order, which analyze a data stream $X_1, X_2, \dots$ sequentially. At each time t the available sample $X_1, \dots, X_t$ of size t is analyzed, usually by calculating a so-called control statistic, and, depending on the observed value of the control statistic, one decides to either continue monitoring (no action) or give a signal to trigger a measurement. Here an analyst, trader, portfolio or risk manager is only involved when such a sequential monitoring procedure gives a signal, in order to decide how to react, but can focus on other issues as long as no signal is observed. In this way, the sequential approach is very attractive, since it allows for automatic monitoring of a huge number of financial instruments, positions or portfolios.

A further important aspect of financial data has to be taken into account. Real financial time series are often non-stationary, for they represent quantitative measures of economic relationships and entities. At best, one can assume that they are stationary over short time intervals. In this case the time series $X_1, X_2, \dots$ can be decomposed in blocks

$$X_1, \dots, X_{q_1-1}, \quad X_{q_1}, \dots, X_{q_2-1}, \quad X_{q_2}, \dots, X_{q_3-1}, \dots$$

in such a way that the observations in each block are weakly or strictly stationary, but any subset containing observations from two or more blocks leads to a non-stationary series. In this situation, $q_1, q_2, \dots, q_L$ are called **change-points**. L may be infinite, but usually one assumes only a finite number of change-points.

The change of the marginal and/or finite-dimensional distribution or induced functionals such as autocovariances can have many specific forms. The basic change-in-mean model assumes that the expectation changes at a change-point such that the mean function $m(t) = E(X_t)$, $t \geq 0$, is a piecewise linear function.

$$m(t) = \sum_{i=0}^{\infty} m_i \mathbf{1}_{[q_i, q_{i+1})}(t), \tag{9.13}$$

where $m_0, m_1, m_2, \ldots$ are real numbers and $q_0 = 0$. Here, we ignore the X_t with $t < 0$, since this part of the time series simplifies the mathematical treatment, but cannot be observed in practice. It is natural to embed the case of no change into the model. This can be achieved by allowing for $m_t = m_0$ for all $t \in \mathbb{N}$ and $q_j \in \mathbb{N}_0$ for all j.

It is worth mentioning that various other change-point models can be transformed to the change-in-mean model. For example, if $\{Z_t\}$ is a mean zero time series, a change-point model for the lag h autocovariance,

$$\gamma_Z(h) = E(Z_t Z_{t+h}) = \begin{cases} \gamma_0(h), & t < q, \\ \gamma_0(h) + \Delta(h), & t \geq q, \end{cases}$$

for some autocovariance function $\gamma_0(h)$ and a function $\Delta(h) \neq 0$ such that $(\gamma_0 + \Delta)(h)$ is an autocovariance function as well, can be analyzed by the above change-in-mean model, if one defines

$$X_t = Z_t Z_{t+h}, \qquad t \geq 1,$$

since then $E(X_t) = \gamma_Z(h)$ for all t.

Returning to the basic change-in-mean model (9.14), notice that the model can also be interpreted as follows when $m_1 \neq m_0$: After the (first) change-point $q = q_1$ the mean function is a piece-wise linear function. However, often it is more realistic to assume that the mean function belongs to a more general class of functions. The change-point setting can be subsumed under the following model for the mean function $m(t) = E(Y_t)$,

$$m(t) = m_0 \mathbf{1}_{[0,q)}(t) + m^*(t)\mathbf{1}_{[q,\infty)}(t), \tag{9.14}$$

for some function m^* satisfying $m^*(q+) \neq m_0$.

9.3.1 Offline detection

Suppose we are given a sample $X_1, \ldots, X_T$ and aim at testing whether there is a change-point where the (marginal) distribution of the observation changes. Before focussing on the change-in-mean problem, let us briefly consider the general case. Let us assume that

$$\begin{aligned} X_t &\sim f_0(x), & t &= 1, \ldots, k-1, \\ X_t &\sim f_1(x), & t &= k, \ldots, T, \end{aligned}$$

for two different densities. In what follows, we assume that f_0 and f_1 are Lebesgue densities, but the results can be easily extended to the general case of densities that are dominated by a measure μ. As before, k is assumed to be an unknown, fixed and nonrandom change-point. The **likelihood ratio statistic** for that problem is given by

$$\Lambda_T = \Lambda_T(k) = \frac{\prod_{t=1}^{k-1} f_0(X_t) \prod_{t=k}^{T} f_1(X_t)}{\prod_{t=1}^{T} f_0(X_t)} = \prod_{t=k}^{T} \frac{f_1(X_t)}{f_0(X_t)}$$

on the set $A = \{\prod_{t=1}^{T} f_0(X_t) \neq 0\}$ and $\Lambda_T = 0$ on A^c. Notice that $P_0(A) = 1$, where P_0 indicates that the probability is calculated assuming that $X_t \overset{i.i.d.}{\sim} f_0$ for all t. The likelihood-ratio (LR) test rejects the null hypothesis

$$H_0 : k = \infty \quad \text{(no change)}$$

in favor of the alternative hypothesis of a change at time $k > 0$, if Λ_T attains large values, that is the corresponding statistical test is given by

$$\phi_{LR} = \mathbf{1}(\Lambda_T > c)$$

for some critical value c. The question arises whether the test ϕ_{LR} is optimal in the sense of maximizing the power $P_k(\phi = 1)$, where P_k indicates that the probability is calculated assuming the change occurs at $k \in \{1, \dots, T\}$, among all statistical tests ϕ operating at the same significance level $P_0(\phi = 1)$.

We need the following change-of-measure lemma.

Lemma 9.3.1 *Let A be a measurable set. Then*

$$E_0(\Lambda_T \mathbf{1}_A) = E_k(\mathbf{1}_A) = P_k(A)$$

and

$$E_0(\Lambda_T Y) = E_k(Y),$$

for any random variable Y that is measurable with respect to $\sigma(X_1, \dots, X_T)$, where E_0 indicates that the expectation is calculated assuming that $X_1, \dots, X_T \sim f_0$.

Proof. We have

$$\begin{aligned} E_0(\Lambda_T \mathbf{1}_A) &= \int_A \prod_{t=k}^{T} \frac{f_1(x_t)}{f_0(x_t)} \prod_{t=1}^{T} f_0(x_t)\, \mathrm{d}x_1 \cdots \mathrm{d}x_T \\ &= \int_A \prod_{t=1}^{k-1} f_0(x_t) \prod_{t=k}^{T} f_1(x_t)\, \mathrm{d}x_1 \cdots \mathrm{d}x_t \\ &= E_k(\mathbf{1}_A) \\ &= P_k(A). \end{aligned}$$

Similarly,

$$E_0(\Lambda_T Y) = \int \prod_{t=1}^{k-1} f_0(x_t) \prod_{t=k}^{T} f_1(x_t) Y\, \mathrm{d}x_1 \cdots \mathrm{d}x_T = E_k(Y).$$

Here is an elegant and simple proof of the optimality of the LR test.

Theorem 9.3.2 *Let $\delta : (\Omega, \mathcal{F}, P) \to ([0, 1], \mathcal{B}_{[0,1]})$ be an arbitrary randomized statistical test. If ϕ_{LR} and δ operate on the same significance level, that is*

$$P_0(\Lambda_T > c) = P_0(\delta = 1) = \alpha,$$

then ϕ_{LR} is more powerful than δ in the sense that

$$P_k(\Lambda_T > c) \geq P_k(\delta = 1).$$

Proof. Notice that for all $x, y \in \mathbb{R}$

$$(x - y)[\mathbf{1}(x > y) - \delta] \geq 0$$

holds, since δ attains values in $[0, 1]$. Apply that inequality with $x = \Lambda_T$ and $y = c$ to obtain

$$(\Lambda_T - c)\mathbf{1}(\Lambda_T > c) \geq (\Lambda_T - c)\delta. \tag{9.15}$$

But this implies

$$E_0((\Lambda_T - c)\mathbf{1}(\Lambda_T > c)) \geq E_0(\Lambda_T\delta) - E_0(c\delta),$$

that is

$$E_0(\Lambda_T\mathbf{1}(\Lambda_T > c)) - cP_0(\Lambda_T > c) \geq E_0(\Lambda_T\delta) - cP_0(\delta = 1).$$

By virtue of the preceding lemma

$$E_0(\Lambda_T\delta) = E_k(\delta) = P_k(\delta = 1)$$

as well as

$$E_0(\Lambda_T\mathbf{1}(\Lambda_T > c)) = E_k(\phi_{LR}) = P_k(\phi_{LR} = 1).$$

Hence, Equation (9.15) is equivalent to

$$P_k(\phi_{LR} = 1) - cP_0(\Lambda_T > c) \geq P_k(\delta = 1) - cP_0(\delta = 1),$$

which in turn is equivalent to

$$P_k(\phi_{LR} = 1) \geq P_k(\delta = 1),$$

since $P_0(\phi_{LR} = 1) = P_0(\delta = 1)$ by assumption.

The LR test requires to know the pre- and after-change distributions f_0 and f_1, which is rarely the case, since otherwise one can not calculate the test statistic. A common approach is to investigate for a specific problem the LR test statistic assuming that the observations follow a certain idealized model such as the normal law. Then one tries to establish appropriate limit theorems that are also valid for non-normal observations. This allows us to make approximative probabilistic calculations, for example in order to obtain critical values or make assertions on the power. In a last step, one investigates whether such approximations can also be derived for dependent time series, perhaps after some appropriate modifications, which is of particular concern in the analysis of financial data.

In what follows, we shall study the change-in-mean problem and tentatively assume that the X_t are independent normal variables with means $\mu_t = E(X_t)$ and common variance $\sigma^2 \in (0, \infty)$. Then, the change-in-mean change-point testing problem addresses the no-change null hypothesis

$$H_0 : \mu_1 = \cdots = \mu_T,$$

which we aim at testing against the composite alternative hypothesis

$$H_1 : \mu_1 = \cdots = \mu_{k-1} \neq \mu_k = \cdots = \mu_T \quad \text{for some } 1 < k < T.$$

If there is a change, it occurs at some fixed k. Thus, consider the testing problem

$$H_0 : \text{no change} \qquad \text{against} \qquad H_1^{(k)} : \text{change at time } k.$$

The associated likelihood ratio statistic is given by

$$\Lambda_k = \frac{L_k(X_1, \ldots, X_T)}{L_1(X_1, \ldots, X_T)},$$

where

$$L_k(X_1, \ldots, X_T) = \sup_{\mu_1, \mu_2 \in \mathbb{R}} \prod_{t=1}^{k-1} \varphi(X_t - \mu_1) \prod_{t=k}^{T} \varphi(X_t - \mu_2)$$

is the likelihood of the alternative, that is the joint density of $X_1, \ldots, X_T$ maximized over the pre- and after-change means μ_1, μ_2, and

$$L_1(X_1, \ldots, X_T) = \sup_{\mu \in \mathbb{R}} \prod_{t=1}^{T} \varphi(X_t - \mu)$$

the corresponding likelihood for the null hypothesis. Here and in what follows, $\varphi(x) = \varphi_{(\mu,\sigma)}(x)$ denotes the density of the $N(0, \sigma^2)$-distribution.

Recall that products of the form

$$\prod_{t=a}^{b} \varphi(X_t - \mu) = \left(\frac{1}{\sqrt{2\pi\sigma^2}} \right)^{(b-a+1)} \exp\left\{ -\frac{1}{2\sigma^2} \sum_{t=a}^{b} (X_t - \mu)^2 \right\},$$

for $1 \le a \le b \le T$ and $\mu \in \mathbb{R}$, are maximized (in μ) by minimizing $\sum_{t=a}^{b} (X_t - \mu)^2$, the least squares criterion, leading to the unique maximizer $\frac{1}{b-a+1} \sum_{t=a}^{b} X_t$. Hence, with the notations

$$\overline{X}_k = \frac{1}{k} \sum_{t=1}^{k} X_t, \qquad \overline{X}_{(k+1):T} = \frac{1}{T-k} \sum_{t=k+1}^{T} X_t,$$

we obtain the formula

$$\Lambda_k(X_1, \ldots, X_T) = \frac{\prod_{t=1}^{k} \varphi(X_t - \overline{X}_k) \prod_{t=k+1}^{T} \varphi(X_t - \overline{X}_{(k+1):T})}{\prod_{t=1}^{T} \varphi(X_t - \overline{X}_T)}.$$

To proceed, we shall consider the log likelihood ratio statistic

$$l_k(X_1, \ldots, X_T) = \log \Lambda_k(X_1, \ldots, X_T).$$

Lemma 9.3.3 *The log likelihood ratio statistic allows the representations*

$$l_k(X_1, \ldots, X_T) = \frac{1}{2\sigma^2} \frac{T}{(T-k)k} \left(S_k - \frac{k}{T} S_T \right)^2$$

and

$$l_k(X_1, \dots, X_T) = \frac{1}{2\sigma^2} \frac{T}{(T-k)k} \left(\sum_{t=1}^{k} (X_t - \overline{X}_T)^2 \right)^2,$$

where

$$S_k = \sum_{t=1}^{k} X_t, \qquad \text{and} \qquad S_{(k+1):T} = \sum_{t=k+1}^{T} X_t.$$

Proof. The calculation is a slightly involved chain of simple manipulations. First, notice that by taking the log, the factors involving $\sqrt{2\pi\sigma}$ cancel, such that $l_k(X_1, \dots, X_T)$ equals

$$\frac{1}{2\sigma^2} \left\{ -\sum_{t=1}^{k} (X_t - \overline{X}_k)^2 - \sum_{t=k+1}^{T} (X_t - \overline{X}_{(k+1):T})^2 + \sum_{t=1}^{T} (X_t - \overline{X}_T)^2 \right\}. \tag{9.16}$$

Using simple identities such as

$$2S_{(k+1):T}\overline{X}_{(k+1):T} = \frac{2}{T-k} S_{(k+1):T}^2 \quad \text{and} \quad k\overline{X}_k^2 = \frac{1}{k} S_k^2,$$

and collecting terms, we see that the expression (9.16) is equal to

$$\frac{1}{2\sigma^2} \left\{ \left(\frac{2}{k} - \frac{1}{k} \right) S_k^2 + \left(\frac{2}{T-k} - \frac{1}{T-k} \right) S_{(k+1):T}^2 - \frac{1}{T} S_T^2 \right\}.$$

Now use $S_{(k+1):T}^2 = (S_T - S_k)^2 = S_T^2 - 2S_k S_T + S_k^2$ and collect terms to arrive at

$$\begin{aligned} l_k(X_1, \dots, X_T) &= \frac{1}{2\sigma^2} \left\{ \left(\frac{1}{k} + \frac{1}{T-k} \right) S_k^2 - \left(\frac{1}{T} - \frac{1}{T-k} \right) S_T^2 - \frac{2}{T-k} S_k S_T \right\} \\ &= \frac{1}{2\sigma^2} \left\{ \frac{T}{(T-k)k} S_k^2 - \frac{k}{(T-k)k} S_T^2 - \frac{2}{T-k} S_k S_t \right\} \\ &= \frac{1}{2\sigma^2} \frac{T}{(T-k)k} \left\{ S_k^2 - \frac{k}{T} S_T^2 - 2\frac{k}{T} S_k S_T \right\}. \end{aligned}$$

Hence,

$$l_k(X_1, \dots, X_T) = \frac{1}{2\sigma^2} \frac{T}{(T-k)k} \left(S_k - \frac{k}{T} S_T \right)^2.$$

The above facts lead us to the **maximally selected weighted CUSUM statistic**

$$C_T^w = \max_{1 \le k \le T} \sqrt{\frac{T}{k(T-k)}} \sum_{t=1}^{k} (X_t - \overline{X}_T),$$

or the unweighted version

$$C_T = \max_{1 \le k \le T} \sum_{t=1}^{k} (X_t - \overline{X}_T).$$

Both statistics can also be used when the observations are non-normal. Indeed, the limiting law turns out to be a functional of Brownian motion, for *any* time series $\{X_t\}$ that satisfies a FCLT. As a preparation, let us consider the following fact.

Lemma 9.3.4 *Suppose that $\{X_t\}$ satisfies a FCLT in the sense that*

$$S_T(u) = \frac{1}{\sqrt{T}} \sum_{t=1}^{\lfloor Tu \rfloor} X_t, \qquad u \in [0, 1],$$

converges weakly to $\eta B(u)$, as $T \to \infty$, for some constant $\eta \in (0, \infty)$ and standard Brownian motion B. Then

$$B_T(u) = \frac{1}{\sqrt{T}} \sum_{t=1}^{\lfloor Tu \rfloor} (X_t - \overline{X}_T) \Rightarrow \eta B^0(u),$$

as $T \to \infty$, where

$$B^0(t) = B(t) - tB(1), \qquad t \in [0, 1],$$

is a Brownian bridge.

Proof. Notice that

$$\begin{aligned} \frac{1}{\sqrt{T}} \sum_{t=1}^{\lfloor Tu \rfloor} (X_t - \overline{X}_T) &= \frac{1}{\sqrt{T}} \sum_{t=1}^{\lfloor Tu \rfloor} X_t - \frac{\lfloor Tu \rfloor}{T} \frac{1}{\sqrt{T}} \sum_{t=1}^{T} X_t \\ &= S_T(u) - \frac{\lfloor Tu \rfloor}{T} S_T(1). \end{aligned}$$

But the latter expression converges weakly to $\eta B(u) - u\eta B(u) = \eta B^0(u)$, by an application of the continuous mapping theorem.

From Lemma 9.3.4 it is only a small step to a general limit theorem for the (weighted) CUSUM statistic.

Theorem 9.3.5 *Suppose that $\{X_t\}$ satisfies a FCLT in the sense that*

$$S_T(u) = \frac{1}{\sqrt{T}} \sum_{t=1}^{\lfloor Tu \rfloor} X_t, \qquad u \in [0, 1],$$

converges weakly to $\eta B(u)$, as $T \to \infty$, for some constant $\eta \in (0, \infty)$ and standard Brownian motion B. Then,

$$C_T^w \Rightarrow \sup_{u \in [0,1]} \frac{\eta B^0(u)}{\sqrt{u(1-u)}},$$

and

$$C_T \Rightarrow \sup_{u \in [0,1]} \eta B^0(u),$$

as $T \to \infty$.

Proof. We show the result for C_T^w. Notice that $C_T^w(u) = 0$ for $u \in [0, 1/T)$. Hence,

$$\begin{aligned} C_T^w &= \max_{1 \le k \le T} \sqrt{\frac{T}{k(T-k)}} \sum_{t=1}^{k} (X_t - \overline{X}_T) \\ &= \sup_{u \in [1/T, 1]} \sqrt{\frac{T}{\lfloor Tu \rfloor (T - \lfloor Tu \rfloor)}} \sum_{t=1}^{\lfloor Tu \rfloor} (X_t - \overline{X}_T) \\ &= \sup_{u \in [1/T, 1]} \frac{B_T^0(u)}{g_T(u)}, \end{aligned}$$

where

$$g_T(u) = \sqrt{\frac{\lfloor Tu \rfloor}{T} \frac{T - \lfloor Tu \rfloor}{T}}$$

and

$$B_T^0(u) = \frac{1}{\sqrt{T}} \sum_{t=1}^{\lfloor Tu \rfloor} (X_t - \overline{X}_T)$$

is the process we studied in Lemma 9.3.4. We have $B_T^0 \Rightarrow \eta B^0$, as $T \to \infty$, and

$$\lim_{T \to \infty} \sup_{u \in [0,1]} |g_T(u) - g(u)| = 0,$$

where

$$g(u) = \sqrt{u(1-u)}, \qquad u \in [0, 1],$$

is a deterministic function. Hence, $(B_T^0, g_T) \Rightarrow (\eta B^0, g)$, as $T \to \infty$, jointly in the product space $D([0, 1]; \mathbb{R}) \otimes D([0, 1]; \mathbb{R})$, and the continuous mapping theorem yields

$$\frac{B_T^0}{g_T} \Rightarrow \frac{\eta B^0}{g},$$

as $T \to \infty$, in $D([0, 1]; \mathbb{R})$, since $B^0(0) = B^0(1) = 0$, a.s., such that $\eta B^0/g$ is well defined for $u \in \{0, 1\}$. A further application of the continuous mapping theorem shows that

$$\sup_{u \in [0,1]} \frac{B_T^0(u)}{g_T(u)} \Rightarrow \sup_{s \in [0,1]} \frac{\eta B^0(u)}{g(u)},$$

as $T \to \infty$, which completes the proof.

In practice, the parameter η, the long-run variance, is unknown. However, it can be estimated from $X_1, \ldots, X_T$ by the Newey–West estimator $\widehat{\eta}_T$, see Equation (8.28). Assuming that the assumptions of Theorem 8.8.6 are satisfied, one may divide the statistics C_T and C_T^w, respectively, by $\widehat{\eta}_T$, and then use the above limit processes with $\eta = 1$. In other words,

$$\frac{C_T^w}{\widehat{\eta}_T} \Rightarrow \sup_{u \in [0,1]} \frac{B^0(u)}{\sqrt{u(1-u)}},$$

and

$$\frac{C_T}{\widehat{\eta}_T} \Rightarrow \sup_{u\in[0,1]} B^0(u),$$

as $T \to \infty$.

Let us now study the above tests under the alternative hypothesis that there is a change-point $q = \lfloor T\vartheta \rfloor$ where the mean changes from μ to $\mu + \Delta$ for some constant $\Delta \neq 0$. Now the model for the observations $X_1, X_2, \ldots$ is

$$X_t = \mu + \Delta \mathbf{1}_{\{q,q+1,\ldots\}}(t) + \epsilon_t, \qquad t = 1, 2, \ldots, \tag{9.17}$$

where $\{\epsilon_t\}$ is a mean zero time series satisfying a FCLT. We have seen that the process $B_T(u)$ defined in Lemma 9.3.4 governs the statistics C_T and C_T^w. The following proposition reveals its behavior under the above alternative model.

Proposition 9.3.6 *Suppose that $\{\epsilon_t\}$ satisfies a FCLT in the sense that*

$$S_T^\epsilon(u) = \frac{1}{\sqrt{T}} \sum_{t=1}^{\lfloor Tu \rfloor} \epsilon_t, \qquad u \in [0, 1],$$

converges weakly to $\eta B(u)$, as $T \to \infty$, for some constant $\eta \in (0, \infty)$ and standard Brownian motion B. Then under the model (9.17), the process

$$B_T(u) = \frac{1}{\sqrt{T}} \sum_{t=1}^{\lfloor Tu \rfloor} (X_t - \overline{X}_T), \qquad u \in [0, 1],$$

satisfies

$$B_T(u) - \delta_T(u) \Rightarrow B^0(u),$$

as $T \to \infty$, where B^0 is a Brownian bridge motion and the drift

$$\delta_T(u) = \Delta \frac{\lfloor Tu \rfloor - \lfloor T\vartheta \rfloor + 1}{\sqrt{T}} \mathbf{1}(\lfloor Tu \rfloor > \lfloor T\vartheta \rfloor)$$

converges to zero, if $u < \vartheta$, and diverges to $\operatorname{sign}(\Delta)\infty$, *if $u \geq \vartheta$, as $T \to \infty$.*

Proof. Notice that

$$\overline{X}_T = \overline{\epsilon}_T + \Delta \frac{T - q + 1}{T}$$

and

$$X_t - \overline{X}_T = \begin{cases} \epsilon_t - \overline{\epsilon}_T - \Delta \frac{T-q+1}{T}, & t < q, \\ \epsilon_t - \overline{\epsilon}_T + \Delta(t - q + 1) - \Delta \frac{T-q+1}{T}, & t \geq q. \end{cases}$$

Define

$$B_T^\epsilon(u) = \frac{1}{\sqrt{T}} \sum_{t=1}^{\lfloor Tu \rfloor} (\epsilon_t - \overline{\epsilon}_T), \qquad u \in [0, 1].$$

We have

$$B_T(u) = B_T^{\epsilon}(u) + \frac{1}{\sqrt{T}} \sum_{t=1}^{\lfloor Tu \rfloor} \left[\Delta(t-q+1)\mathbf{1}_{\{q,q+1,\ldots\}}(t) - \frac{T-q+1}{T}\Delta \right]$$
$$= B_T^{\epsilon}(u) + \Delta \frac{\lfloor Tu \rfloor - q + 1}{\sqrt{T}} \mathbf{1}(\lfloor Tu \rfloor \geq \lfloor T\vartheta \rfloor) - \frac{T-q+1}{\sqrt{T}}\Delta$$
$$= B_T^{\epsilon}(u) + \delta_T(u) + o(1).$$

Hence,

$$B_T(u) - \delta_T(u) = B_T^{\epsilon}(u) + o(1) \Rightarrow \eta B(u),$$

as $T \to \infty$, since $S_T^{\epsilon} \Rightarrow \eta B$, as $T \to \infty$, implies $B_T^{\epsilon} \Rightarrow \eta B^0$, as $T \to \infty$, by virtue of Lemma 9.3.4.

Proposition 9.3.6 shows that the process B_T before the change, that is $\{B_T(u) : u < \frac{\lfloor Tu \rfloor}{T}\}$ converges to the corresponding part $\{\eta B_T^0 : u \in [0, \vartheta]\}$ of the Brownian bridge. But behind the change-point the process $B_T(u)$ diverges such that the maximally selected CUSUM statistic diverges as well.

9.3.2 Online detection

We have seen that offline detection aims at detecting a change-point by analyzing a sample of size T once. We have already argued that for many problems arising in finance **sequential monitoring**, also called **online detection** or **monitoring** is the method of choice, provided the observations arrive sequentially, thus forming a data stream, and the observation X_t at time t is available to us with no or a negligible delay. A *sequential* method analyzes at time t the available sample $X_1, \ldots, X_t$ of size t in order to generate a decision. It is common practice to design such a decision rule (detector) in the following way. One calculates a *control statistic* $U_t = U_t(X_1, \ldots, X_t)$ taking values in the real numbers. Then a signal is given indicating that the data provide empirical evidence in favor of a change-point, if the control statistics U_t attains values in a critical set A; otherwise monitoring continues by proceeding to the next time instant. One may either consider monitoring schemes that may last forever or schemes that stop at the latest when reaching a time horizon T defining the maximal sample size. In financial applications, the latter approach is often more natural, since basic assumptions and models are checked and updated from time to time, usually on a regular basis, for example on a quarterly or yearly basis. In mathematical terms, the first time point $t \leq T$ when $U_t \in A$ is given by the stopping time,

$$R_T = \inf\{k \leq t \leq T : U_t \in A\}.$$

In what follows, we assume $T < \infty$.

As a large class of control statistics, let us consider weighted averages of the form

$$\widehat{m}_t = \sum_{i=1}^{t} K\left(\frac{t-i}{h}\right) Y_i, \qquad t = 1, 2, \ldots,$$

for some smoothing kernel $K(x)$ and a bandwidth $h > 0$. Notice the similarity to the Nadaraya–Watson estimator. However, in order to detect a change in the mean, it is not necessary to estimate the mean consistently. Thus, we do not norm the weights. Further, assuming that the observations $Y_1, Y_2, \ldots$ arrive sequentially and are observed at fixed equidistant time instants, which after rescaling time can be assumed to be the natural numbers, we assume that the bandwidth tends to infinity as the time horizon T approaches infinity. Precisely, we assume that $h = h_T$ with

$$\lim_{T\to\infty} \frac{T}{h_T} = \zeta$$

for some $\zeta \in [1, \infty)$. Anticipating the $\sqrt{T}$-convergence rate of $\widehat{m}_T$, let us consider the stopping time R_T with $U_t = \frac{1}{\sqrt{T}}\widehat{m}_t$, that is

$$R_T = \inf\left\{k \le t \le T : \frac{1}{\sqrt{T}}\frac{1}{\sqrt{T}}\widehat{m}_t > c\right\}.$$

Let us define the corresponding **sequential kernel-weighted partial sum process** by

$$M_T(s) = \sum_{t=1}^{\lfloor Ts \rfloor} K\left(\frac{\lfloor Ts \rfloor - t}{h}\right) Y_t, \qquad s \in [0, 1].$$

Then,

$$R_T = T \inf\left\{s \in [0, 1] : \frac{1}{\sqrt{T}} M_T(s) > c\right\}.$$

In order to proceed, let us slightly modify the change-in-mean model (9.14). Let us assume that a time horizon $T \in \mathbb{N}$ is given and the first T observations used for monitoring depend on T, that is $Y_t = Y_{Tt}$. This means that we assume that we sequentially observe $Y_{T1}, \ldots, Y_{TT}$ satisfying the model

$$Y_{Tt} = \frac{1}{\sqrt{T}} m_0\left(\frac{t-q}{T}\right) \mathbf{1}(t \ge q) + \epsilon_t, \qquad t = 1, \ldots, T, \tag{9.18}$$

for some function $m_0 : \mathbb{R} \to [0, \infty)$, the change-point q, and a strictly stationary mean zero process $\{\epsilon_t\}$.

If the function m_0 satisfies for some $t^* > 0$ the condition

$$m_0(s) = 0, \quad s \le 0, \quad \text{and} \quad m_0(s) > 0 \quad \text{for } s \in (0, t^*), \tag{9.19}$$

then q is a change-point. Before the change the observations have mean zero, whereas after the change the mean function is induced by the strictly positive function $m_0(t)$, $t > 0$. Model (9.18) is a local alternative model: As T gets larger, the problem to detect the change becomes harder, since the mean tends to zero. In particular, if m_0 is bounded in the supnorm, $E(Y_{Tt}) = O(1/\sqrt{T})$ for $t \ge q$.

The following result provides the distribution under the no-change null hypothesis.

Theorem 9.3.7 *Assume that* $\{\epsilon_t\}$ *satisfies* $E|\epsilon_1|^{r+\delta} < \infty$ *for* $r \geq 4, \delta > 0$, *and is* α*-mixing with mixing coefficients* $\alpha(k)$, $k \in \mathbb{N}$, *satisfying*

$$\alpha(k) = O(k^{-\beta}) \quad \text{for some} \quad \beta > \frac{r(r+\delta)}{2\delta}.$$

K *is assumed to be a bounded and Lipschitz continuous kernel. Then, under the no-change null hypothesis* $H_0 : m_0 = 0$, *the following assertions hold true.*

(i) For all $0 \leq s, t \leq 1$ *the limit*

$$C(r, s) = \lim_{T\to\infty} \frac{1}{T} \sum_{i=1}^{\lfloor Tr \rfloor} \sum_{j=1}^{\lfloor Ts \rfloor} K\left(\frac{\lfloor Ts \rfloor - i}{h}\right) K\left(\frac{\lfloor Tr \rfloor - j}{h}\right) r_0(|i - j|) \tag{9.20}$$

exists, where $r_0(k) = E(\epsilon_1 \epsilon_{1+k})$, $k \in \mathbb{N}$, *is the autocovariance function of* $\{\epsilon_t\}$.

(ii) The empirical process $\{M_T(s) : s \in [0, 1]\}$ *converges weakly,*

$$M_T \Rightarrow M,$$

as $T \to \infty$, *where* M *is a mean zero Gaussian process with covariance function (9.20).*

(iii) The stopping time R_T *satisfies the central limit theorem*

$$\frac{R_T}{T} \xrightarrow{d} \inf\{s \in [0, 1] : M(s) > c\},$$

as $T \to \infty$.

Let us now study the behavior when the change-in-mean condition (9.19) holds. In what follows, the change-point q is allowed to be a function of T, that is $q = q_T$.

Proposition 9.3.8 *Suppose that, in addition to the assumption of Theorem 9.3.7, the change-in-mean condition (9.19) holds. If the product* Km_0 *is integrable, then*

$$\frac{1}{\sqrt{T}} M_T(s) \Rightarrow M^1(s) = M(s) + \int_{\lim q/T}^{s} K(\zeta(s - r)) m_0(r - \lim q/T)\, dr,$$

as $T \to \infty$. *Consequently,*

$$\frac{R_T}{T} \xrightarrow{d} \inf\{s \in [0, 1] : M^1(s) > c\},$$

as $T \to \infty$.

Proof. The result follows by an application of Slutzky's lemma using the fact that

$$\frac{1}{\sqrt{T}} M_T(s) = \frac{1}{\sqrt{T}} M_T^0(s) + m_T(s),$$

where

$$M_T^0(s) = \frac{1}{\sqrt{T}} \sum_{t=1}^{\lfloor Ts \rfloor} K\left(\frac{\lfloor Ts \rfloor - t}{h}\right) \epsilon_t,$$

$$m_T(s) = \frac{1}{T} \sum_{t=1}^{\lfloor Ts \rfloor} K\left(\frac{\lfloor Ts \rfloor - t}{h}\right) m_0\left(\frac{t-q}{T}\right) \mathbf{1}(t \geq q).$$

Clearly, $M_T^0 \Rightarrow M$, as $T \to \infty$, by Theorem 9.3.7. Further, by virtue of the Lipschitz continuity of K and m_0,

$$\begin{aligned} m_T(s) &= \frac{1}{T} \sum_{t=1}^{\lfloor Ts \rfloor} K\left(\frac{T}{h}\frac{\lfloor Ts \rfloor - t}{T}\right) m_0\left(\frac{t-q}{T}\right) \mathbf{1}(t \geq q) \\ &= \frac{1}{T} \sum_{t=q}^{\lfloor Ts \rfloor} K\left(\zeta\left(s - \frac{t}{T}\right)\right) m_0\left(\frac{t}{T} - \lim \frac{q}{T}\right) + o(1) \\ &= \int_{\lim q/T}^{s} K(\zeta(s-r)) m_0(r - \lim q/T)\, dr + o(1), \end{aligned}$$

as $T \to \infty$. The result for R_T/T follows by an application of the continuous mapping theorem, see Theorem B.2.1 and Example B.2.2 (iv).

Remark 9.3.9 *Observe that the limit process M^1 depends on the function m_0 determining the mean function of the process after the change only through the deterministic drift function*

$$\mu(s; K) = \int_{\lim q/T}^{s} K(\zeta(s-r)) m_0(r - \lim q/T)\, dr,$$

which depends on the kernel K. The natural question arises how one should select the kernel in order to optimize the procedure in terms of the delay $R_T - q$ of the procedure in some reasonable sense. Within the following modified framework, one can find optimal kernels. Consider the model

$$Y_{Tt} = \frac{1}{\sqrt{T}} m_0\left(\frac{t-q}{T}\right) \mathbf{1}(t \geq q) + \epsilon_t, \qquad t = 1, \ldots, T,$$

with $\lim q/T = 0$. *If one uses the control statistic $U_t = \widehat{m}_t/h$ and considers the stopping time*

$$R_\infty = \inf\{t \geq 0 : \widehat{m}_t/h > c\},$$

then the normed delay $\rho_h = \frac{R_T - q}{h}$ converges almost surely to

$$\rho_0 = \inf\{s > 0 : \widetilde{\mu}(s; K) > c\},$$

as $h \to \infty$, for some function $\widetilde{\mu}(s; K)$ similarly defined as $\mu(s; K)$. Now the optimal kernel can be determined under rather weak conditions for a large class of functions m_0. For details we refer to the references.

9.4 Unit roots and random walk

Let us tentatively assume that $u_1, u_2, \ldots$ are i.i.d. random variables with mean 0 and common variance $\sigma^2 = E(u_1^2) \in (0, \infty)$. We shall relax that assumption below substantially. Recalling Bachelier's historic approach to model stock prices, let us consider the pure random walk

$$Y_t = \sum_{i=0}^{t} u_i, \qquad t = 0, 1, \ldots$$

with starting value $Y_0 = u_0$. The above equation for Y_t can be written as a recursion yielding

$$Y_t = Y_{t-1} + u_t, \qquad t = 1, 2, \ldots$$

with starting value $Y_0 = u_0$. Obviously, this recursion is a member of the family

$$Y_t = \rho Y_{t-1} + u_t, \qquad t = 1, 2, \ldots \tag{9.21}$$

of recursions parameterized by $\rho \in (-1, 1]$. $\rho = 1$ reproduces the pure random walk above, whereas for $|\rho| < 1$ we obtain an AR(1) process with start at $Y_0 = u_0$. In Chapter 3, we discussed the question whether there exist stationary solutions of the Equations (9.21). If u_{-n}, $n = 0, 1, \ldots$, are i.i.d. copies of u_1, and $Y_0 = \sum_{i=0}^{\infty} \rho^i u_{-i}$, then

$$Y_t = \sum_{i=0}^{\infty} \rho^i u_{t-i}, \qquad t \in \mathbb{N}_0$$

is a stationary (and causal) solution of the AR(1) equations.

Whether or not we are in the stationary or random-walk regime can be inferred from the characteristic polynomial

$$\varphi(z) = 1 - \rho z, \qquad z \in \mathbb{C},$$

which has exactly one root given by $z = 1/\rho$. That root lies outside the unit disc $\{z \in \mathbb{C} : |z| \le 1\}$ if and only if $|\rho| < 1$. For $\rho = 1$ the characteristic polynomial has its root on the unit circle. Such a root is called a **unit root**. In other words, model (9.21) corresponds to a random walk if and only if the characteristic polynomial has a unit root.

As we shall in the next subsections, the stochastic behavior of basic statistics change completely when the underlying time series is a stationary AR(1) or a non-stationary random walk with independent or serially dependent innovations. For many series representing log prices or interest rates, the random-walk assumption can be justified by economic theory and statistical evidence, and the statistical task is then to test whether that assumption can be confirmed by empirical data.

This gives rise to the following two testing problems within the AR(1) framework discussed here.

Stationarity tests: Here, one tests the null hypothesis of stationarity

$$H_0 : |\rho| < 1$$

against the alternative hypothesis of a unit root (random-walk hypothesis)

$$H_1 : \rho = 1.$$

Unit root tests: A unit root test reverses the null hypothesis and tests

$$H_0 : \rho = 1$$

against the alternative stationarity hypothesis

$$H_1 : |\rho| < 1.$$

We shall later extend this class of testing problems.

Unit root tests also have interesting applications in trading. Here is an example.

Example 9.4.1 (PAIRS TRADING)
Assume that $\{X_t\}$ and $\{Y_t\}$ are two random walks, for example the prices of two similar assets or commodities. Some assets and commodities, respectively, are exchangeable. For example, usually one may substitute white pepper by black pepper and vice versa. But then their prices should be quite similar in such a way that the differences fluctuate around zero in a stationary way. This suggests the following trading strategy: If the price of white pepper is much smaller than the price of black pepper, one enters a long position in white pepper and a short position in black pepper. One waits until white pepper has become more expensive than black pepper and then reverts the positions. This strategy may work, if the difference

$$D_t = Y_t - X_t, \qquad t = 1, 2, \ldots,$$

is indeed a stationary process with mean zero and not a random walk. In the latter case, the probability that the price difference exceeds a given threshold is much higher than for a stationary process. To check that assumption behind the rationale of pairs trading, one may assume that D_t follows the model (9.21) and apply a unit root test. If the test rejects the random-walk null hypothesis, one may enter a pairs trading strategy. Even more interesting would be the application of a procedure that continuously monitors the price processes and gives a signal if the series provide evidence in favor of the stationarity assumption. Then one can automatically enter a pairs trading strategy. We shall study such procedures in Section 9.4.4.

The mathematical phenomenon behind Example 9.4.1 is called **cointegration** and will be defined in Section 9.4.2 in a more general framework.

9.4.1 The OLS estimator in the stationary AR(1) model

In model (9.21) we assumed i.i.d. innovations u_t. To study inferential procedures to decide in favor of the unit root hypothesis or the stationarity hypothesis, let us now generalize model (9.21) to the case of martingale difference errors. Thus, we assume

$$Y_t = \rho Y_{t-1} + u_t, \qquad t = 1, 2, \ldots, \tag{9.22}$$

where $\{u_t\}$ is a martingale difference sequence with respect to the filtration $\mathcal{F}_t = \sigma(u_s : s \leq t)$, i.e.

$$E(u_t | \mathcal{F}_{t-1}) = 0, \qquad t = 1, 2, \ldots.$$

In addition, we assume that there are finite constants σ^2 and γ_4 such that

$$E\left(u_t^2 | \mathcal{F}_{t-1}\right) = \sigma^2 \qquad \text{and} \qquad E\left(u_t^4 | \mathcal{F}_{t-1}\right) = \gamma_4,$$

a.s. These assumptions will allow us to apply the limit theorems for martingale difference sequences and arrays, respectively, discussed in Chapter 3.

Given $Y_1, \ldots, Y_T$, the ordinary least squares (OLS) estimator $\widehat{\rho}_T$ of ρ defined by

$$\widehat{\rho}_T = \operatorname{argmin}_{\rho \in \mathbb{R}} \sum_{t=1}^{T} (Y_t - \rho Y_{t-1})^2.$$

A straightforward calculation shows that it is given by

$$\widehat{\rho}_T = \frac{\sum_{t=1}^{T} Y_t Y_{t-1}}{\sum_{t=1}^{T} Y_{t-1}^2}.$$

It coincides with the ML estimate for i.i.d. normal errors, and in this case the denominator $\sum_t Y_{t-1}^2$ turns out to be the Fisher information.

Let us first study the asymptotic distribution of $\widehat{\rho}_T$ when $|\rho| < 1$. The following theorem shows that $\widehat{\rho}_T$ converges to ρ in probability at the usual convergence rate $1/\sqrt{T}$, and is asymptotically normal. It appears as a special case of the least squares estimation theory for multiple regression models given in Section 8.2, but we will give a direct proof to highlight in the next subsection the completely different stochastic behavior of certain statistics for integrated time series.

Theorem 9.4.2 *Suppose that $\{u_t\}$ is a strictly stationary martingale difference sequence with $E(u_t^2|\mathcal{F}_{t-1}) = \sigma^2$ and $E(u_t^4|\mathcal{F}_{t-1}) = \gamma_4 < \infty$. It holds that*

(i)

$$\frac{1}{T}\sum_{t=1}^{T} Y_{t-1}^2 \stackrel{P}{\to} \frac{\sigma^2}{1-\rho^2}, \tag{9.23}$$

(ii)

$$\frac{1}{\sqrt{T}}\sum_{t=1}^{T} Y_{t-1}u_t \stackrel{d}{\to} N(0, \sigma^4/(1-\rho^2)), \tag{9.24}$$

as $T \to \infty$.

This implies that

$$\sqrt{T}(\widehat{\rho}_T - \rho) \stackrel{d}{\to} N\big(0, (1-\rho^2)\big),$$

as $T \to \infty$, where $\sigma^2 = E(\epsilon_1^2)$.

Proof. Notice that

$$\widehat{\rho}_T - \rho = \frac{\sum_{t=1}^{T}\left(Y_t Y_{t-1} - \rho Y_{t-1}^2\right)}{\sum_{t=1}^{T} Y_{t-1}^2} = \frac{\sum_{t=1}^{T} \epsilon_t Y_{t-1}}{\sum_{t=1}^{T} Y_{t-1}^2},$$

which leads to the representation

$$\sqrt{T}(\widehat{\rho}_T - \rho) = \frac{T^{-1/2}\sum_{t=1}^{T} Y_{t-1}u_t}{T^{-1}\sum_{t=1}^{T} Y_{t-1}^2}.$$

We shall first treat the denominator. Assertion (i) is simply Proposition 8.4.1, which assumes $E(u_t^2|\mathcal{F}_{t-1}) = \sigma^2$ and $E(u_t^4|\mathcal{F}_{t-1}) = \gamma_4 < \infty$. Next, put

$$\xi_t = Y_{t-1}u_t, \qquad t = 1, 2, \ldots,$$

and let us show that $\{\xi_t\}$ is a strictly stationary martingale difference sequence with respect to the filtration $\mathcal{F}_t = \sigma(u_s : s \leq t)$. Recall that the stationary solution of the AR(1) equations is given by the linear process

$$Y_t = \sum_{i=0}^{\infty} \rho^i u_{t-i}.$$

It follows that $\xi_t = Y_{t-1}u_t$ is of the form $f(u_t, u_{t-1}, \ldots)$ for some Borel measurable function $f : \mathbb{R}^\infty \to \mathbb{R}$. Hence, ξ_t is strictly stationary. Since $E(Y_{t-1}^2) = \sigma^2 \sum_{i=0}^{\infty} \rho^{2i} < \infty$, we have

$$E|\xi_t| \leq \sqrt{E(Y_{t-1}^2)E(u_t^2)} < \infty \qquad \text{for all } t.$$

Further,

$$E(\xi_t|\mathcal{F}_{t-1}) = E(Y_{t-1}u_t|Y_{t-1}) = Y_{t-1}E(u_t|\mathcal{F}_{t-1}) = 0,$$

a.s., for all t. It follows that $\{\xi_t\}$ is a strictly stationary martingale difference sequence. To show the central limit theorem for $\frac{1}{\sqrt{T}}\sum_{t=1}^{T} \xi_t$ consider for $\varepsilon > 0$ the Lindeberg condition,

$$L_T(\varepsilon) = \frac{1}{T}\sum_{t=1}^{T} E\big(E(Y_{t-1}^2 u_t^2 \mathbf{1}(Y_{t-1}u_t > \varepsilon) \mid \mathcal{F}_{t-1})\big),$$

which takes the form

$$\frac{1}{T}\sum_{t=1}^{T} \int y^2 \int u_t^2(\omega)\mathbf{1}(|yu_t(\omega)| > \sqrt{T}\varepsilon)\, \mathrm{d}P_{u_t|Y_{t-1}=y}(\omega)\, \mathrm{d}P_{Y_{t-1}}(y)$$

Clearly, the indicator converges to 0, as $T \to \infty$, for any fixed ω and y. Since, additionally, the integrand is bounded by the integrable r.v. u_t^2, which implies that $E(u_T^2|\mathcal{F}_{t-1}) < \infty$, a.s., the conditional dominated convergence theorem ensures that for all fixed t

$$\int u_t^2(\omega)\mathbf{1}(|yu_t(\omega)| > \sqrt{T}\varepsilon)\, \mathrm{d}P_{u_t|Y_{t-1}=y}(\omega) \to 0,$$

as $T \to \infty$, and a further application of dominated convergence shows that for each t

$$E\big(Y_{t-1}^2 u_t^2 \mathbf{1}(Y_{t-1}u_t > \sqrt{T}\varepsilon)\big) \to 0,$$

as $T \to \infty$. But by strict stationarity these summands do not depend on t, such that

$$L_T(\varepsilon) \to 0, \qquad T \to \infty,$$

follows. It remains to study the convergence of

$$\frac{1}{T}\sum_{t=1}^{T} E\left(Y_{t-1}^2 u_t^2 | \mathcal{F}_{t-1}\right).$$

Since $E(u_t^2|\mathcal{F}_{t-1}) = \sigma^2$, we obtain

$$\frac{1}{T}\sum_{t=1}^{T} E\left(Y_{t-1}^2 u_t^2 | \mathcal{F}_{t-1}\right) = \frac{1}{T}\sum_{t=1}^{T} Y_{t-1}^2 E\left(u_t^2 | \mathcal{F}_{t-1}\right) \to \frac{\sigma^4}{1-\rho^2},$$

as $T \to \infty$. Hence, conditions (i) and (ii) of Theorem B.7.2 are satisfied and we may conclude that

$$\frac{1}{\sqrt{T}}\sum_{t=1}^{T} Y_{t-1}u_t \xrightarrow{d} Z \sim N(0, \sigma^4/(1-\rho^2)),$$

as $T \to \infty$, which shows assertion (ii). Combining this fact with Equation (9.23) and applying Slutzky's lemma, we obtain that

$$\sqrt{T}(\widehat{\rho}_T - \rho) = \frac{T^{-1/2}\sum_{t=1}^{T} Y_{t-1}u_t}{T^{-1}\sum_{t=1}^{T} Y_{t-1}^2} \xrightarrow{d} \left(\frac{\sigma^2}{1-\rho^2}\right)^{-1} Z \sim N(0, (1-\rho^2)),$$

as $T \to \infty$, which completes the proof.

The above result allows us to discuss how to test statistical hypotheses on the AR parameter ρ and how to set up a confidence interval for that parameter. The testing problem

$$H_0 : \rho = \rho_0 \qquad \text{versus} \qquad H_1 : \rho \neq \rho_0$$

for some known $\rho_0 \in (-1, 1)$ can now be treated as follows: For a given significance level $\alpha \in (0, 1)$ reject the null hypothesis, if $|S_T| > z_{1-\alpha/2}$, where

$$S_T = \sqrt{T}\frac{\widehat{\rho}_T - \rho_0}{\sqrt{1-\widehat{\rho}_T^2}},$$

and $z_{1-\alpha/2}$ is the $(1-\alpha/2)$-quantile of the $N(0, 1)$-distribution. A confidence interval with asymptotic coverage probability $1-\alpha$ is given by

$$\left[\widehat{\rho}_T - z_{1-\alpha/2}\frac{\sqrt{1-\widehat{\rho}_T^2}}{\sqrt{T}}, \widehat{\rho}_T + z_{1-\alpha/2}\frac{\sqrt{1-\widehat{\rho}_T^2}}{\sqrt{T}}\right].$$

Notice that the estimator $\widehat{\rho}_T$ and therefore its asymptotic distribution does not depend on the scale of the measurements. Nevertheless, it may be interesting to estimate the dispersion σ of the innovations u_t. It is natural to take the full-sample residuals

$$\widehat{u}_t = Y_t - \widehat{\rho}_T Y_{t-1}, \qquad t = 1, \ldots, T$$

and to estimate σ^2 by

$$\widehat{\sigma}_T^2 = \frac{1}{T}\sum_{t=1}^{T} \widehat{u}_t^2.$$

A direct proof of its consistency is straightforward.

Proposition 9.4.3 *Under the conditions of Theorem 9.4.2,*

$$\widehat{\sigma}_T^2 \stackrel{P}{\to} \sigma^2,$$

as $T \to \infty$.

Proof. Using $\widehat{u}_t = Y_t - \widehat{\rho}_T Y_{t-1} = (\rho - \widehat{\rho}_T)Y_{t-1} + u_t$, we obtain

$$\widehat{u}_t^2 = (\rho - \widehat{\rho}_T)^2 Y_{t-1}^2 + 2(\rho - \widehat{\rho}_T)Y_{t-1}u_t + u_t^2.$$

Clearly, the arithmetic mean of the last term converges to σ^2 a.s., by the strong law of large numbers. Thus, the result follows, if we show that

$$A_T = (\rho - \widehat{\rho}_T)^2 \frac{1}{T}\sum_{t=1}^{T} Y_{t-1}^2 \stackrel{P}{\to} 0, \quad B_T = 2(\rho - \widehat{\rho}_T)\frac{1}{T}\sum_{t=1}^{T} Y_{t-1}u_t \stackrel{P}{\to} 0,$$

as $T \to \infty$. Consider A_T. The asymptotic normality of $\sqrt{T}(\widehat{\rho}_T - \rho)$ implies $\widehat{\rho}_T - \rho \stackrel{P}{\to} 0$, and, in turn, $(\rho - \widehat{\rho}_T)^2 \stackrel{P}{\to} 0$, as $T \to \infty$. Further, in the proof of Theorem 9.4.2 we have shown that $T^{-1}\sum_{t=1}^{T} Y_{t-1}^2$ converges in probability to a finite constant. Hence, $A_T \stackrel{P}{\to} 0$, as $T \to \infty$. The corresponding result for B_T follows from similar arguments.

9.4.2 Nonparametric definitions for the degree of integration

The stationary solution, Y_t, of the AR(1) Equation (9.22) with $|\rho| < 1$ can be represented as a linear process, $Y_t = \sum_{i=0}^{\infty} \rho^i u_{t-i}$, and therefore satisfies a functional central limit theorem. This means that the partial sum process

$$S_T(u) = \frac{1}{\sqrt{T}}\sum_{t=1}^{\lfloor Tu \rfloor} Y_t, \qquad u \in [0, 1], \quad T \geq 1,$$

satisfies

$$S_T(u) \Rightarrow \eta B(u), \tag{9.25}$$

as $T \to \infty$, where B is a standard Brownian motion and $\eta = \sqrt{\eta^2}$. Here

$$\eta^2 = EY_0^2 + 2\sum_{k=0}^{\infty} E(Y_0 Y_k),$$

is the **long-run variance** associated to $\{Y_n\}$. Consistent estimation of η^2 has been discussed in Section 8.8, see Theorem 8.8.6.

If $\rho = 1$, then Y_t is a random walk with increments that are martingale differences and therefore satisfies

$$T^{-1/2} Y_{\lfloor Ts \rfloor} \Rightarrow \sigma B(s), \tag{9.26}$$

as $T \to \infty$, where $Y_{\lfloor Ts \rfloor}$, $s \in [0, 1]$, is the **canonical process** associated to $Y_1, \ldots, Y_T$. In addition, the first-order differences,

$$\Delta Y_t = Y_t - Y_{t-1},$$

are stationary and satisfy a functional central limit, since by telescoping

$$Y_{\lfloor Ts \rfloor} = Y_0 + \sum_{i=0}^{\lfloor Ts \rfloor} \Delta Y_t.$$

Both the FCLT (9.25) as well as the weak convergence (9.26) also hold true, if the summands and increments of the random walk, respectively, are correlated time series.

Let us now suppose that we are given a statistic, say, U_T, which depends on the time-series data $\{Y_t\}$ through the partial sum process S_T, i.e.

$$U_T = U(S_T),$$

for some mapping $U : \mathcal{D} \to \mathbb{R}$ defined on a domain $\mathcal{D}$ of functions such that all trajectories of S_T are elements of $\mathcal{D}$. If the mapping U is smooth in the sense that $U(S_T)$ converges in distribution to $U(\eta B)$, as $T \to \infty$, if S_T converges weakly to ηB, then the asymptotic behavior of the statistic U_T is known for *any* time series $\{Y_t\}$ that satisfies the FCLT (9.25). This means, we get a general limit theorem that holds true for a very rich class of time series. Analogously, the asymptotic distribution of any statistic V_T that depends in such a smooth way on the time-series data via the canonical process, i.e. $V_T = V(T^{-1/2} Y_{\lfloor T \bullet \rfloor})$, is given by the random element $V(\sigma B)$, by virtue of the FCLT (9.26). For various unit root statistics, we encounter that pleasant setting. This fact motivates the following nonparametric definitions.

Definition 9.4.4 (INTEGRATED TIME SERIES)

(i) A stochastic process $\{Y_t\}$ in discrete time is called **integrated of order** 0*, denoted by $Y_t \sim I(0)$, if it satisfies a functional central limit theorem, that is for some constant $\eta \in (0, \infty)$*

$$\frac{1}{\sqrt{T}} \sum_{t=1}^{\lfloor Ts \rfloor} Y_t \Rightarrow \eta B(s),$$

as $T \to \infty$, where B is a standard Brownian motion.

(ii) A stochastic process $\{Y_t\}$ in discrete time is called **integrated of order** 1*, denoted by $Y_t \sim I(1)$, if the canonical process satisfies*

$$T^{-1/2} Y_{\lfloor Ts \rfloor} \Rightarrow \sigma B(s),$$

as $T \to \infty$, and if the first-order differences ΔY_t form a weakly stationary sequence satisfying a functional central limit theorem.

Remark 9.4.5 *It is worth mentioning that in the above definition the notion $I(0)$ covers nonstationary processes as long as they are sufficiently well behaved in the sense that they satisfy a functional central with a Brownian motion limit. However, the models studied by econometricians usually give rise to stationary $I(0)$ series.*

The above definitions allow us to provide a general definition of cointegrated time series.

Definition 9.4.6 *A bivariate time series $\{(Y_t, X_t) : t = 1, 2, \dots\}$ with $Y_t \sim I(1)$ and $X_t \sim I(1)$ is called* **cointegrated**, *if there exists a constant $a \neq 0$ such that $Y_t - aX_t \sim I(0)$. More generally, a d-dimensional time series $\{X_t\}$, $X_t = (X_{t1}, \dots, X_{td})$, with $X_{ti} \sim I(1)$ for all $i = 1, \dots, d$, is called cointegrated, if there is some vector $a \in \mathbb{R}^d$, which is not the null vector, such that $a'X_t \sim I(0)$.*

9.4.3 The Dickey–Fuller test

We are still studying the model (9.22). A widely used statistic to test the unit root null hypothesis $H_0 : \rho = 1$ against the stationarity alternative $H_1 : |\rho| < 1$ is the **Dickey–Fuller statistic**

$$D_T = T(\widehat{\rho}_T - 1).$$

The following theorem provides its asymptotic distribution under the unit root null hypothesis that $Y_t = \sum_{i=1}^t \epsilon_i$ is $I(1)$. It shows that the convergence rate of $\widehat{\rho}_T$ is T, i.e. in the unit root case it converges faster to the true value $\rho_0 = 1$ than under a stationary regime where the convergence rate is $\sqrt{T}$. This means, $\widehat{\rho}_T$ is a *superconsistent* estimator when applied to random-walk series.

Theorem 9.4.7 *Suppose that the partial sum process $S_T(u) = T^{-1/2} \sum_{t=1}^{\lfloor Tu \rfloor} \epsilon_i$, $u \in [0, 1]$, satisfies*

$$S_T \Rightarrow \eta B,$$

as $T \to \infty$, for some constant $\eta \in (0, \infty)$, and that

$$\frac{1}{T} \sum_{t=1}^{T} \epsilon_i^2 \xrightarrow{P} \sigma^2, \qquad T \to \infty.$$

Then,

(i) $\frac{1}{T} \sum_{t=1}^T Y_{t-1}\epsilon_t \xrightarrow{d} \frac{\eta^2}{2} \left(B^2(1) - \frac{\sigma^2}{\eta^2} \right)$, *as $T \to \infty$, and*

(ii) $\frac{1}{T^2} \sum_{t=1}^T Y_{t-1}^2 \xrightarrow{d} \eta^2 \int_0^1 B^2(r)\,\mathrm{d}r$, *as $T \to \infty$,*

and these convergences are jointly. Further, the OLS estimator $\widehat{\rho}_T$ satisfies

$$T(\widehat{\rho}_T - 1) \xrightarrow{d} \frac{1}{2} \frac{B^2(1) - \frac{\sigma^2}{\eta^2}}{\int_0^1 B^2(r)\,\mathrm{d}r},$$

as $T \to \infty$.

Proof. Put $S(T) = \sqrt{T} S_T(1) = \sum_{i=1}^T \epsilon_i$. We first show that

$$\sum_{t=1}^T Y_{t-1}\epsilon_t = \frac{1}{2}\left(S(T)^2 - \sum_{t=1}^T \epsilon_t^2 \right). \tag{9.27}$$

Notice that

$$\sum_{t=1}^T Y_{t-1}\epsilon_t = \sum_{t=1}^T \left(\sum_{j=1}^{t-1} \epsilon_j \right) \epsilon_t = \sum_{1 \le i < j \le T} \epsilon_i \epsilon_j$$

is the sum of all elements above the main diagonal of the symmetric matrix $(\epsilon_i \epsilon_j)_{i,j}$. Thus,

$$\left(\sum_{t=1}^T \epsilon_t \right)^2 = \sum_{i,j=1}^T \epsilon_i \epsilon_j = 2 \sum_{i<j} \epsilon_i \epsilon_j + \sum_{t=1}^T \epsilon_t^2,$$

which implies that

$$\sum_{t=1}^T Y_{t-1}\epsilon_t = \sum_{i<j} \epsilon_i \epsilon_j = \frac{1}{2}\left[\left(\sum_{t=1}^T \epsilon_t \right)^2 - \sum_{t=1}^T \epsilon_t^2 \right] = \frac{1}{2}\left[S(T)^2 - \sum_{t=1}^T \epsilon_t^2 \right],$$

where $S(T)^2 = T S_T^2(1)$ such that

$$\frac{1}{T} \sum_{t=1}^T Y_{t-1}\epsilon_t = \frac{1}{2}\left[S_T^2(1) - \frac{1}{T} \sum_{t=1}^T \epsilon_t^2 \right]. \tag{9.28}$$

Let us now consider

$$\frac{1}{T^2} \sum_{t=1}^T Y_{t-1}^2 = \frac{1}{T} \sum_{t=1}^T \frac{1}{T} \left(\sum_{i=1}^{t-1} \epsilon_i \right)^2.$$

Notice that $1/T = \int_0^1 \mathbf{1}_{[(t-1)/T, t/T)}(r)\,\mathrm{d}r$ and, by definition of the floor function,

$$\sum_{i=1}^{\lfloor Tr \rfloor} \epsilon_i = \sum_{i=1}^{t-1} \epsilon_i, \qquad \text{for } t-1 \le Tr < t \Leftrightarrow r \in [(t-1)/T, t/T),\ t = 1, \ldots, T.$$

We shall now partition the one, $1_{[0,1]} = \sum_{t=1}^T \mathbf{1}_{[(t-1)/T, t/T]}$ and use the fact that

$$\frac{1}{T} \sum_{t=1}^{t-1} \epsilon_i = \sum_{i=1}^{t-1} \int_{\frac{t-1}{T}}^{\frac{t}{T}} \mathrm{d}r = \int_{\frac{t-1}{T}}^{\frac{t}{T}} \sum_{i=1}^{\lfloor Tr \rfloor} \epsilon_i \,\mathrm{d}r.$$

We obtain

$$\begin{aligned}\frac{1}{T^2}\sum_{t=1}^{T} Y_{t-1}^2 &= \frac{1}{T}\sum_{t=1}^{T}\int_0^1 \left(\sum_{i=1}^{\lfloor Tr\rfloor}\epsilon_i\right)^2 \mathbf{1}_{[(t-1)/T,t/T)}(r)\,\mathrm{d}r\\ &= \sum_{t=1}^{T}\int_0^1 S_T^2(r)\mathbf{1}_{[(t-1)/T,t/T)}(r)\,\mathrm{d}r\\ &= \int_0^1 S_T^2(r)\sum_{t=1}^{T}\mathbf{1}_{[(t-1)/T,t/T)}(r)\,\mathrm{d}r.\end{aligned}$$

Hence, we arrive at

$$\frac{1}{T^2}\sum_{t=1}^{T} Y_{t-1}^2 = \int_0^{1-} S_T^2(r)\,\mathrm{d}r. \tag{9.29}$$

To show the joint weak convergence, we consider the bivariate random vector $(T^{-1}\sum_{t=1}^{T} Y_{t-1}\epsilon_t, T^{-2}\sum_{t=1}^{T}\epsilon_t^2)$. Combining Equations (9.28) and (9.29), we may conclude that

$$\begin{aligned}\left(\frac{1}{T}\sum_{t=1}^{T} Y_{t-1}\epsilon_t, \frac{1}{T^2}\sum_{t=1}^{T}\epsilon_t^2\right) &= \left(\frac{1}{2}\left[S_T^2(1) - \frac{1}{T}\sum_{t=1}^{T}\epsilon_t^2\right], \int_0^{1-} S_T^2(r)\,\mathrm{d}r\right)\\ &= \left(\frac{1}{2}[S_T^2(1)-\sigma^2], \int_0^{1-} S_T^2(r)\,\mathrm{d}r\right) + o_P(1)\\ &= \phi(S_T) + o_P(1),\end{aligned}$$

where the functional $\phi : D([0,1];\mathbb{R}) \to \mathbb{R}^2$ is defined by

$$\phi(f) = \left(\frac{1}{2}[f^2(1)-\sigma^2], \int_0^{1-} f^2(r)\,\mathrm{d}r\right), \qquad f \in D([0,1];\mathbb{R}).$$

ϕ is continuous, cf. Example B.2.2. Hence, the result follows from the continuous mapping theorem and we can conclude that

$$\phi(S_T) \Rightarrow \phi(\eta B) = \left(\frac{\eta^2}{2}\left[B^2(1) - \frac{\sigma^2}{\eta^2}\right], \eta^2\int_0^1 B^2(r)\,\mathrm{d}r\right),$$

as $T \to \infty$, which establishes the assertions (i) and (ii). Now the asymptotics for $D_T = T(\widehat{\rho}_T - 1)$ follows by an application of the continuous mapping theorem, since the function $F(x, y) = x/y$ is continuous on $\mathbb{R} \times \mathbb{R} - \{0\}$ and $\int_0^1 B^2(r)\,\mathrm{d}r$ is a.s. positive.

Compare assertions (i) and (ii) of the above theorem with Theorem 9.4.2 that gave the corresponding results for a stationary AR(1) model. For $I(1)$ series the asymptotics for the statistics $\sum_t Y_{t-1}^2$ and $\sum_t Y_{t-1}\epsilon_t$, after appropriate scaling, are completely different, both the convergence rates as well as the type of the limiting distribution.

Observing that the limit process appearing in Theorem 9.4.7 (i) is a linear function of $B^2(1) \sim \chi^2(1)$, we may apply a transformation to the statistic

$$\frac{1}{T}\sum_{t=1}^{T} Y_{t-1}\epsilon_t = \left(\frac{1}{T}\sum_{t} Y_{t-1}^2\right)(\widehat{\rho}_T - 1)$$

to obtain a χ^2-distributed random variable. This leads to the Phillips–Durlauf statistic

$$\frac{2}{\eta^2}\left(\frac{1}{T}\sum_{t=1}^{T} Y_{t-1}^2\right)(\widehat{\rho}_T - 1) + \frac{\sigma^2}{\eta^2} \Rightarrow B^2(1) \sim \chi^2(1),$$

as $T \to \infty$, under the unit root null hypothesis. In this formula we can replace the unknown nuisance parameters η^2 and σ^2 by consistent estimators.

9.4.4 Detecting unit roots and stationarity

The question arises how to monitor a time series in order to detect a change from $I(0)$-stationarity to $I(1)$-nonstationarity in the sense of Definition 9.4.4. To make this idea precise, let us denote the true and deterministic but unknown change-point by q and assume that it is proportional to a time horizon T,

$$q = \lfloor T\vartheta \rfloor$$

for some $\vartheta \in (0, 1)$. Two change-point models are of interest:

I(1) – I(0) Change: It is assumed that the series of pre-change observations

$$\{Y_0, \ldots, Y_{\lfloor T\vartheta \rfloor - 1}\} \sim I(1)$$

is $I(1)$ in the sense of part (ii) of Definition 9.4.4, whereas the subseries of observations

$$\{Y_{\lfloor T\vartheta \rfloor}, \ldots, Y_T\} \sim I(0)$$

is $I(0)$ in the sense of part (i) of that definition.

I(0) – I(1) Change: Here, the first part of the series behaves as a $I(0)$ series

$$\{Y_0, \ldots, Y_{\lfloor T\vartheta \rfloor - 1}\} \sim I(0)$$

and changes its behavior to a $I(1)$ series starting at the change-point q,

$$\{Y_{\lfloor T\vartheta \rfloor}, \ldots, Y_T\} \sim I(1).$$

$I(0)$ **Detection:** In order to detect a change from $I(1)$ to $I(0)$, one may use a sequential version of the KPSS or Dickey–Fuller test statistics. Related to the KPSS test is the following procedure. Define $U_T(s) = 0$ for $s \in [0, 1/T)$ and

$$U_T(s) = \frac{\lfloor Ts \rfloor^{-3} \sum_{i=1}^{\lfloor Ts \rfloor} \left(\sum_{j=1}^{i} Y_j\right)^2 K_h(i - \lfloor Ts \rfloor)}{\lfloor Ts \rfloor^{-2} \sum_{j=1}^{\lfloor Ts \rfloor} Y_j^2}, \quad s \in [1/T, 1]. \tag{9.30}$$

U_T is called a **sequential KPSS process**. Here, K is a kernel with mean zero, a finite non-vanishing second moment and $K_h(\bullet) = K(\bullet/h)/h$ is the scaled version. It is used to ensure that partial sums $\sum_{j=1}^{i} Y_j$ corresponding to distant time points i contribute less to the statistic than partial sums corresponding to time points near the current time instant $t = \lfloor Ts \rfloor$. If K is a kernel with support $[-1, 1]$, only observations located in the time window $[t - h, t]$ are used. Thus, the bandwidth h determines the degree of localization of the procedure. Inference on the degree of integration requires relatively large effective sample sizes and we do not assume that the time scale on which the time series $Y_1, Y_2, \ldots$ is observed tends to zero. Hence, in contrast to the nonparametric smoothing procedures discussed in Section 8.5, we do not assume that the bandwidth tends to zero, as the maximal sample size tends to ∞. Instead, it is assumed that $h = h_T$ is chosen as a function of T in such a way that

$$\lim_{T\to\infty} \frac{T}{h_T} = \zeta \in [1, \infty),$$

This condition ensures that the number of observations effectively used by the procedure tends to ∞, as T gets larger. In practice, one fixes ζ and then selects $h_T = \lfloor T/\zeta \rfloor$.

We stop monitoring and signal evidence for a change from $I(1)$ to $I(0)$, if $U_T(t/T)$ falls below a fixed value for the first time. This means that we consider the stopping time

$$R_T = R_T(c) = \min\{k \le t \le T : U_T(t/T) < c\},$$

where c is some control limit chosen to ensure that the procedure has well-defined statistical properties and k denotes the first time instance where monitoring starts. For the asymptotic results discussed below, it is assumed that

$$k = \lfloor \kappa T \rfloor, \qquad \text{for some } \kappa \in (0, 1). \tag{9.31}$$

The first $k - 1$ observations can be used as a learning sample to estimate other quantities of interest not being related to the monitoring procedure.

A natural approach to specify the control limit c is to ensure that the type I error rate that the procedure stops,

$$\alpha_T = P_0(R_T \le T),$$

where P_0 indicates that the probability is calculated under the null hypothesis that the process $\{Y_t\}$ is $I(1)$, converges to some given nominal significance level $\alpha \in (0, 1)$,

$$\lim_{T\to\infty} \alpha_T = \alpha.$$

The null distribution of the sequential KPSS process is given in the following theorem.

Theorem 9.4.8 *Assume $\{Y_t\}$ is $I(1)$ in the sense of Definition 9.4.4. Then the sequential KPSS process converges weakly,*

$$U_T(s) \Rightarrow \mathcal{U}_1(s) = \frac{\zeta s^{-1} \int_0^s K(\zeta(r - s)) \left[\int_0^r B(t)\, \mathrm{d}t\right]^2 \mathrm{d}r}{\int_0^s B(r)^2\, \mathrm{d}r}, \tag{9.32}$$

in $D[\kappa, 1]$, as $T \to \infty$.

Proof. Notice that

$$U_T(s) = \frac{\left(\frac{T}{\lfloor Ts\rfloor}\right)^3 X_T(s)}{\left(\frac{T}{\lfloor Ts\rfloor}\right)^2 Z_T(s)},$$

if we define

$$X_T(s) = \frac{1}{T^3}\sum_{i=1}^{\lfloor Ts\rfloor}\left(\sum_{j=1}^{i} Y_j\right)^2 K_h(i - \lfloor Ts\rfloor),$$

$$Z_T(s) = \frac{1}{T^2}\sum_{i=1}^{\lfloor Ts\rfloor} Y_i^2 = \int_0^{\lfloor Ts\rfloor/T}\left(\frac{1}{\sqrt{T}}Y_{\lfloor Tr\rfloor}\right)^2 \mathrm{d}r,$$

where

$$S_T(s) = \frac{1}{\sqrt{T}}\sum_{t=1}^{\lfloor Ts\rfloor} Y_t, \qquad s \in [0, 1].$$

If we show that

$$\frac{X_T(s)}{Z_T(s)} \Rightarrow s\mathcal{U}_1(s), \tag{9.33}$$

as $T \to \infty$, then the assertion follows by a straightforward application of Slutzky's lemma, Lemma B.2.4. To show Equation (9.33), we shall approximate jointly numerator and denominator by the right integral functionals of a Brownian motion, the limit process of the partial sum process S_T under the conditions of the theorem. We make use of the Skorohod representation theorem and will assume that, w.l.o.g.,

$$\sup_{u\in[0,1]}\left|\frac{Y_{\lfloor Tu\rfloor}}{\sqrt{T}} - \sigma B(u)\right| \to 0, \tag{9.34}$$

almost surely, as $T \to \infty$. Notice that, for simplicity of presentation, we do not change notation when working with the equivalent versions on the new probability space. Let us represent $X_T(s)$ as an iterated Riemann integral. Use

$$t = \lfloor Tr\rfloor \Leftrightarrow r \in \left[\frac{t}{T}, \frac{t+1}{T}\right)$$

to obtain

$$\frac{1}{T}\left(\frac{1}{T}\sum_{j=1}^{t}\frac{Y_j}{\sqrt{T}}\right)^2 K\left(\frac{t-\lfloor Ts\rfloor}{h}\right)$$
$$= \int \mathbf{1}_{\left[\frac{t}{T},\frac{t+1}{T}\right)}(r)\left(\frac{1}{T}\sum_{j=1}^{\lfloor Tr\rfloor}\frac{Y_j}{\sqrt{T}}\right)^2 K\left(\frac{\lfloor Tr\rfloor - \lfloor Ts\rfloor}{h}\right)\mathrm{d}r$$

In the same vein,

$$j = \lfloor Tu \rfloor \Leftrightarrow u \in \left[\frac{j}{T}, \frac{j+1}{T}\right)$$

gives

$$\frac{1}{T}\sum_{j=1}^{\lfloor Tr \rfloor} \frac{Y_j}{\sqrt{T}} = \sum_{j=1}^{\lfloor Tr \rfloor} \int \mathbf{1}_{\left[\frac{j}{T}, \frac{j+1}{T}\right)}(u) \frac{Y_{\lfloor Tu \rfloor}}{\sqrt{T}} \, \mathrm{d}u.$$

Putting things together, we obtain

$$\begin{aligned} X_T(s) &= \frac{T}{h}\frac{1}{T}\sum_{t=1}^{T} \mathbf{1}(t \leq \lfloor Ts \rfloor) \left(\frac{1}{T}\sum_{j=1}^{t} \frac{Y_j}{\sqrt{T}}\right)^2 K\left(\frac{t - \lfloor Ts \rfloor}{h}\right) \\ &= \frac{T}{h}\sum_{t=1}^{T} \mathbf{1}(t \leq \lfloor Ts \rfloor) \int_{\frac{t}{T}}^{\frac{t+1}{T}} \left(\sum_{j=1}^{\lfloor Tr \rfloor} \int_{\frac{j}{T}}^{\frac{j+1}{T}} \frac{Y_{\lfloor Tu \rfloor}}{\sqrt{T}} \, \mathrm{d}u\right)^2 K\left(\frac{\lfloor Tr \rfloor - \lfloor Ts \rfloor}{h}\right) \mathrm{d}r \\ &= \frac{T}{h}\int_{\frac{1}{T}}^{\frac{\lfloor T(s+1/T) \rfloor}{T}} \left(\int_{\frac{1}{T}}^{\frac{\lfloor T(r+1/T) \rfloor}{T}} \frac{Y_{\lfloor Tu \rfloor}}{\sqrt{T}} \, \mathrm{d}u\right)^2 K\left(\frac{\lfloor Tr \rfloor - \lfloor Ts \rfloor}{h}\right) \mathrm{d}r, \end{aligned}$$

where we used the fact that $\frac{\lfloor Ts \rfloor + 1}{T} = \frac{\lfloor T(s+1/T) \rfloor}{T}$. Notice that Equation (9.34) implies that

$$\int_a^b \frac{Y_{\lfloor Tu \rfloor}}{\sqrt{T}} \, \mathrm{d}u \Rightarrow \int_a^b \sigma B(u) \, \mathrm{d}u,$$

for any $0 \leq a \leq b \leq 1$ as well as

$$I_T(B) = \sup_{a,b \in [0,1], a \leq b} \int_a^b |\sigma B(u)| \, \mathrm{d}u = O_P(1),$$

as $T \to \infty$. Further, we have

$$\begin{aligned} &\sup_{r \in [\kappa,1]} \left| \left(\int_{\frac{1}{T}}^{\frac{\lfloor T(r+1/T) \rfloor}{T}} \frac{Y_{\lfloor Tu \rfloor}}{\sqrt{T}}\right)^2 - \left(\int_{\frac{1}{T}}^{\frac{\lfloor T(r+1/T) \rfloor}{T}} \sigma B(u)\right)^2 \right| \\ &\quad \leq \sup_{r \in [\kappa,1]} \left(\int_0^1 |\sigma B(u)| \, \mathrm{d}u + \int_0^1 \left|\frac{Y_{\lfloor Tu \rfloor}}{\sqrt{T}}\right| \mathrm{d}u\right) \int_0^1 \left|\frac{Y_{\lfloor Tu \rfloor}}{\sqrt{T}} - \sigma B(u)\right| \mathrm{d}u \\ &\leq \sup_{r \in [\kappa,1]} \left(\int_0^1 |\sigma B(u)| \, \mathrm{d}u + \int_0^1 \left|\frac{Y_{\lfloor Tu \rfloor}}{\sqrt{T}}\right| \mathrm{d}u\right) \sup_{u \in [0,1]} \left|\frac{Y_{\lfloor Tu \rfloor}}{\sqrt{T}} - \sigma B(u)\right| \\ &= o_P(1), \end{aligned}$$

as $T \to \infty$, since the expression in parentheses is $O_P(1)$. Combining this fact with the boundedness of K and T/h, we obtain

$$X_T(s) = \frac{T}{h} \int_{\frac{1}{T}}^{\frac{\lfloor T(s+1/T) \rfloor}{T}} \left(\int_{\frac{1}{T}}^{\frac{\lfloor T(r+1/T) \rfloor}{T}} \sigma B(u)\, \mathrm{d}u \right)^2 K\left(\frac{\lfloor Tr \rfloor - \lfloor Ts \rfloor}{h} \right) \mathrm{d}r + o_P(1),$$

as $T \to \infty$, where the $o_P(1)$ term is also uniform in $s \in [\kappa, 1]$. Using $T/h \to \zeta$ and the Lipschitz continuity of K, we may further conclude that

$$X_T(s) = \zeta \int_{\frac{1}{T}}^{\frac{\lfloor T(s+1/T) \rfloor}{T}} \left(\int_{\frac{1}{T}}^{\frac{\lfloor T(r+1/T) \rfloor}{T}} \sigma B(u)\, \mathrm{d}u \right)^2 K(\zeta(r-s))\, \mathrm{d}r + o_P(1),$$

as $T \to \infty$, where again the $o_P(1)$ term is uniform in $s \in [\kappa, 1]$. Replacing the intervals of integration by $(0, s]$ and $(0, r]$, respectively, also leads to an error term of the order $o_P(1)$, uniformly in r, s. For example,

$$\begin{aligned}
\sup_{r \in [\kappa, 1]} & \left| \int_{\frac{1}{T}}^{\frac{\lfloor T(r+1/T) \rfloor}{T}} \sigma B(u)\, \mathrm{d}u - \int_0^r \sigma B(u)\, \mathrm{d}u \right| \\
&\le \frac{1}{T} \left[\sup_{v \in [0,1]} \int_0^v |\sigma B(u)|\, \mathrm{d}u + \sup_{v \in [0,1]} \int_v^1 |\sigma B(u)|\, \mathrm{d}u \right] \\
&= \frac{1}{T} I_T(B) \\
&= o_P(1),
\end{aligned}$$

as $T \to \infty$. The same arguments show that

$$\sup_{s \in [\kappa, 1]} \left| Z_T(s) - \sigma^2 \int_0^s B^2(r)\, \mathrm{d}r \right| = o_P(1),$$

as $T \to \infty$. Putting things together leads to

$$\sup_{s \in [\kappa, 1]} \left\| \begin{pmatrix} X_T(s) \\ Z_T(s) \end{pmatrix} - \begin{pmatrix} \zeta\sigma^2 \int_0^s \left(\int_0^r B(u)\, \mathrm{d}u \right)^2 K(\zeta(r-s))\, \mathrm{d}r \\ \sigma^2 \int_0^s B^2(r)\, \mathrm{d}r \end{pmatrix} \right\| = o_P(1),$$

as $T \to \infty$, where $\| \bullet \|$ denotes an arbitrary vector norm on $\mathbb{R}^2$. But the latter implies the joint weak convergence

$$(X_T, Z_T) \Rightarrow (X, Z),$$

as $T \to \infty$, for the original processes, in the product space $D([0,1]; \mathbb{R}) \times D([0,1]; \mathbb{R})$ (equipped with the product Skorohod toplogy, cf. Appendix B.2), where

$$\begin{aligned}
X(s) &= \zeta\sigma^2 \int_0^s \left(\int_0^r B(u)\, \mathrm{d}u \right)^2 K(\zeta(r-s))\, \mathrm{d}r, \\
Z(s) &= \sigma^2 \int_0^s B^2(r)\, \mathrm{d}r,
\end{aligned}$$

for $s \in [\kappa, 1]$. Next, observe that for any $s \in [\kappa, 1]$

$$P(Z(s) > 0) = 1,$$

since otherwise $B(r) = 0$ for $r \in [\kappa, 1]$, λ-almost everywhere, with positive probability, leading to a contradiction; the variation of Brownian motion on any interval is infinite. Thus, we can apply the continuous mapping theorem, see the discussion in Remark B.2.6, to obtain that

$$U_T = \frac{X_T}{Z_T} \Rightarrow \mathcal{U},$$

as $T \to \infty$, as stated in the theorem.

From the above theorem, we can deduce a central limit theorem for the normed stopping rule

$$\frac{R_T}{T} = \inf\left\{\kappa \leq \frac{t}{T} \leq 1 : U_T(s) < c\right\}.$$

Notice that R_T/T stops later than x, if and only the condition $U_T(s) \geq c$ holds true for all $s \in [\kappa, x]$. This means,

$$\frac{R_T}{T} > x \quad \Leftrightarrow \quad \sup_{s\in[\kappa,x]} U_T(s) \geq c,$$

for all given $x \in \mathbb{R}$. Hence, for the distribution function $F_T(x)$ of R_T/T, we obtain

$$\begin{aligned} F_T(x) &= P(R_T/T \leq x) \\ &= P\left(\sup_{s\in[\kappa,x]} U_T(s) < c\right) \\ &\to P\left(\sup_{s\in[\kappa,x]} \mathcal{U}_1(s) < c\right), \end{aligned}$$

as $T \to \infty$.

As already discussed, the above results allow us to obtain critical values and control limits, respectively, to set up a procedure for testing and monitoring and attaining a given significance level in large samples.

$I(1)$ **Detection:** In order to detect a change from $I(0)$ to $I(1)$, we consider the process

$$\widetilde{U}_T(s) = \frac{1}{Ts_{Tm}^2(s)} \sum_{i=1}^{\lfloor Ts \rfloor} \left(\sum_{j=1}^{i} Y_j \right)^2 K_h(i - \lfloor Ts \rfloor), \qquad s \in [0, 1], \tag{9.35}$$

where

$$s_{Tm}^2(s) = \frac{1}{T} \sum_{i=1}^{\lfloor Ts \rfloor} Y_i^2 + 2 \sum_{k=1}^{m} w(k, m) \frac{1}{T} \sum_{i=1}^{\lfloor Ts \rfloor} Y_i Y_{i+k}, \qquad s \in [0, 1],$$

is the Newey–West estimator as defined in Equation (8.28) and studied in Theorem 8.8.6 of Section 8.8. That estimator is used to eliminate a nuisance parameter appearing otherwise in the limit distribution. The corresponding detector is given by

$$\widetilde{R}_T = \widetilde{R}_T(c) = \min\{k \le n \le T : \widetilde{U}_T(n/T) < c\}.$$

The null distribution of that detector follows from the following result.

Theorem 9.4.9 *Let $\{Y_n\}$ be a strictly stationary α-mixing $I(0)$ process such that $EY_1^{4\nu} < \infty$ with α-mixing coefficients satisfying*

$$\sum_{j=1}^{\infty} j^2 \alpha(j)^{(\nu-1)/\nu} < \infty, \tag{9.36}$$

for some $\nu > 1$. Assume that the lag truncation rule satisfies $m/T^{1/2} = o(1)$. Then

$$\widetilde{U}_T(s) \Rightarrow \widetilde{\mathcal{U}}_2(s) = s^{-1}\zeta \int_0^s B(r)^2 K(\zeta(r-s))\,\mathrm{d}r,$$

in $D[\kappa, 1]$, as $T \to \infty$.

Let us now consider the asymptotics for time series that are *nearly* a random walk. It is common to consider the following mathematical model. Assume that the first T observations we have available for the analysis correspond to the Tth row of a triangular array $\{Y_{Tt} : 1 \le tT, T \in \mathbb{N}\}$ satisfying

$$Y_{T,t+1} = (1 + a/T)Y_{Tt} + u_t, \qquad 1 \le t \le T,\ T \in \mathbb{N}, \tag{9.37}$$

where $a \in \mathbb{R}$ and $\{u_t\}$ is an $I(0)$ process. As $T \to \infty$, the AR coefficient $1 + a/T$ approaches 1 corresponding to the unit root case, although for each fixed T the series $Y_{T1}, \dots, Y_{TT}$ is $I(0)$-stationary. This model is called the **local-to-unity model**.

We shall give a sketch of the arguments leading to the asymptotics of this more involved model. Routine calculus shows that

$$\begin{aligned}
\frac{1}{\sqrt{T}} Y_{T,\lfloor Ts \rfloor} &= \frac{1}{\sqrt{T}} \sum_{t=1}^{\lfloor Ts \rfloor} (1 + a/T)^{\lfloor Ts \rfloor - i} u_t \\
&= \int_0^{\lfloor Ts \rfloor / T} e_T(r; s)\,\mathrm{d}S_T(r),
\end{aligned}$$

where

$$\begin{aligned}
S_T(r) &= \frac{1}{\sqrt{T}} \sum_{t=1}^{\lfloor Ts \rfloor} u_t, \\
e_T(r; s) &= (1 + a/T)^{\lfloor Ts \rfloor - \lfloor Tr \rfloor},
\end{aligned}$$

for $r, s \in [0, 1]$. Noting that $e_T(r; s)$ converges uniformly in $0 \le r \le s \le 1$ to $e(r; s) = e^{a(s-r)}$, and has uniformly bounded variation, one may show that

$$\sup_{s\in[0,1]} \left| \int_0^{\lfloor Ts \rfloor/T} e_T(r; s)\, dS_T(r) - \int_0^s e(r; s)\, dB(r) \right| \to 0,$$

as $T \to \infty$, in probability, leading to the following result.

Proposition 9.4.10 (Convergence to Ornstein–Uhlenbeck Process)
For an array $\{Y_{Tt} : 1 \le t \le T, T \ge 1\}$ satisfying the local-to-unity model (9.37), it holds that

$$\frac{1}{\sqrt{T}} Y_{T,\lfloor Ts \rfloor} = \int_0^{\lfloor Ts \rfloor/T} e_T(r; s)\, dS_T(r) \Rightarrow \mathcal{O}(r; s),$$

as $T \to \infty$, where

$$\mathcal{O}(r; s) = \int_0^s e^{a(s-r)}\, dB(r),$$

is an Ornstein–Uhlenbeck process and B a standard Brownian motion.

Observing that the proof of Theorem 9.4.8 works under the general assumption that the canonical process $T^{-1/2}Y_{T,\lfloor Ts \rfloor}$, $s \in [\kappa, 1]$, converges weakly to an almost surely continuous process, the proof can be easily modified to obtain the asymptotics for data-generating processes leading to more involved limits. Thus, one can establish the following result on the **local-to-unity-asymptotics** of the process

$$U_T(s) = \frac{\lfloor Ts \rfloor^{-3} \sum_{i=1}^{\lfloor Ts \rfloor} \left(\sum_{j=1}^{i} Y_{Tj} \right)^2 K_h(i - \lfloor Ts \rfloor)}{\lfloor Ts \rfloor^{-2} \sum_{j=1}^{\lfloor Ts \rfloor} Y_{Tj}^2}, \quad s \in [1/T, 1]. \tag{9.38}$$

Theorem 9.4.11 *Let $\{Y_{Tt} : 1 \le t \le T, T \in \mathbb{N}\}$ be an array satisfying the local-to-unity model (9.37). Then*

$$U_T(s) \Rightarrow \mathcal{U}_{\mathcal{O}}(s) = \frac{\zeta s^{-1} \int_0^s \left(\int_0^r \mathcal{O}(t; a)\, dt \right)^2 K(\zeta(s-r))\, dr,}{\int_0^s \mathcal{O}(r; a)^2\, dr},$$

in $D[\kappa, 1]$, as $T \to \infty$.

In addition, observe that the process $U_T(s)$ is invariant under changes of the scale. This fact implies that we may also consider strong dependent processes where a different rate of convergence applies. Assume that

$$Y_t = \sum_{i=1}^{t} X_i, \qquad t = 1, 2, \ldots \tag{9.39}$$

with long-memory increments X_t, for example a fractional integrated noise of degree $0 < d < 1/2$ and Hurst index $H = 1/2 + d$, respectively, that is

$$(1 - L)^d X_t \sim \mathrm{WN}(0, \sigma^2), \qquad \sigma^2 \in (0, \infty),$$

see Definition 3.8.3. In what follows, we assume that $\{X_t\}$ satisfies an invariance principle with fractional Brownian motion limit,

$$\frac{1}{T^H}\sum_{t=1}^{\lfloor Ts\rfloor} X_t \Rightarrow c_H B^H(s), \tag{9.40}$$

as $T \to \infty$, for some constant c_H; see Theorem B.7.5 and the references. Consequently, the canonical process with scaling T^{-H} satisfies

$$\frac{1}{T^H} Y_{\lfloor Ts\rfloor} \Rightarrow c_H B^H,$$

as $T \to \infty$. It follows that the process $U_T(s)$ based on the time series (9.39) satisfies

$$U_T(s) \Rightarrow \mathcal{U}_{B^H}(s) = \frac{\zeta s^{-1}\int_0^s \left(\int_0^r B^H(t)\,\mathrm{d}t\right)^2 K(\zeta(s-r))\,\mathrm{d}r,}{\int_0^s B^H(r)^2\,\mathrm{d}r},$$

as $T \to \infty$.

9.5 Notes and further reading

Sklar's theorem can be found in Sklar (1959), also see Rüschendorf (2009). An exposition on correlation and dependencies in risk management as well as related relevant properties of copulas see Embrechts et al. (2002). The brief summary of the financial crisis is mainly based on Baily and Johnson (2008). The application of the Gaussian copula to price CDOs is due to Li (2000). The asymptotics for the nonparametric copula estimator can be found in Fermanian et al. (2004). Local polynomial regression, first studied systematically by Stone (1977), has been extensively studied and applied, see Fan and Gijbels (1996), Fan and Yao (2003) and, for an econometric viewpoint and applications, Li and Racine (2007). Results on minimax estimation can be found in Tsybakov (2009). The asymptotics for mixing processes has been established by Masry and Fan (1997), see also Fan and Yao (2003) and Li and Racine (2007). A thorough exposition on nonparametric statistics for dependent stochastic processes is Bosq (1998). For the methodology of nonparametric econometrics, particularly kernel smoothing methods, we refer to the monograph of Li and Racine (2007). Our presentation draws on those known works and provides a version of this type of asymptotic normality theorems that is new in that shows that the result holds true for *any* stationary and ergodic time series which satisfies a certain smoothing central limit theorem.

For a discussion and applications of the Neyman–Pearson-type result, Theorem 9.3.2, see Vexler and Gurevich (2011). Theorem 9.3.7 and related results can be found in Steland (2004). The kernel optimization theory can be found in Steland (2005). For a thorough exposition on change-point methods for classical problems we refer to the monograph Csörgő and Horváth (1997), the present exposition focuses on sequential procedures for the detection of stationarity and nonstationarity. For asymptotic results on sequential detection of $I(0)$ stationarity and $I(1)$-nonstationarity based on sequential KPSS processes, as discussed here, see Steland (2007a). We also refer to Davies and Krämer (2003). Procedures related to the Dickey–Fuller unit root test statistics are studied in Steland (2007b) and extensions to polynomial trends were investigated by Steland (2008). For methods to detect changes in linear models, we refer to

Ploberger and Krämer (1992) and Hušková et al. (2007), amongst others. For conditions such that Equation (9.40) holds, see Wu and Shao (2006).

References

Baily, Martin Neil L.R.E and Johnson M.S. (2008) The origins of the financial crisis. The Brookings Institution, pp. 1–45.

Bosq D. (1998) *Nonparametric Statistics for Stochastic Processes: Estimation and Prediction*. vol. 110 of *Lecture Notes in Statistics* 2nd edn. Springer-Verlag, New York.

Csörgő M. and Horváth L. (1997) *Limit Theorems in Change-point Analysis*. Wiley Series in Probability and Statistics. John Wiley & Sons Ltd., Chichester. With a foreword by David Kendall.

Davies P.L. and Krämer W. (2003) The Dickey-Fuller test for exponential random walks. *Econometric Theory* **19**(5), 865–883.

Embrechts P., McNeil A.J. and Straumann D. (2002) Correlation and dependence in risk management: properties and pitfalls *Risk Management: Value at Risk and Beyond (Cambridge, 1998)*. Cambridge Univ. Press Cambridge pp. 176–223.

Fan J. and Gijbels I. (1996) *Local Polynomial Modelling and its Applications*. vol. 66 of *Monographs on Statistics and Applied Probability*. Chapman & Hall, London.

Fan J. and Yao Q. (2003) *Nonlinear Time Series: Nonparametric and Parametric Methods*. Springer Series in Statistics. Springer-Verlag, New York.

Fermanian J.D., Radulović D. and Wegkamp M. (2004) Weak convergence of empirical copula processes. *Bernoulli* **10**(5), 847–860.

Hušková M., Prášková Z. and Steinebach J. (2007) On the detection of changes in autoregressive time series. I. Asymptotics. *J. Statist. Plann. Inference* **137**(4), 1243–1259.

Li D.X. (2000) On default correlation: A copula function approach. *The Journal of Fixed Income* **9**, 43–54.

Li Q. and Racine J.S. (2007) *Nonparametric Econometrics*. Princeton University Press, Princeton, NJ. Theory and practice.

Masry E. and Fan J. (1997) Local polynomial estimation of regression functions for mixing processes. *Scand. J. Statist.* **24**(2), 165–179.

Ploberger W. and Krämer W. (1992) The CUSUM test with OLS residuals. *Econometrica* **60**(2), 271–285.

Rüschendorf L. (2009) On the distributional transform, Sklar's theorem, and the empirical copula process. *J. Statist. Plann. Inference* **139**(11), 3921–3927.

Sklar M. (1959) Fonctions de répartition à *n* dimensions et leurs marges. *Publ. Inst. Statist. Univ. Paris* **8**, 229–231.

Steland A. (2004) Sequential control of time series by functionals of kernal-weighted empirical processes under local alternatives. *Metrika* **60**(3), 229–249.

Steland A. (2005) Optimal sequential kernel detection for dependent processes. *J. Statist. Plann. Inference* **132**(1-2), 131–147.

Steland A. (2007a) Monitoring procedures to detect unit roots and stationarity. *Econometric Theory* **23**(6), 1108–1135.

Steland A. (2007b) Weighted Dickey-Fuller processes for detecting stationarity. *J. Statist. Plann. Inference* **137**(12), 4011–4030.

Steland A. (2008) Sequentially updated residuals and detection of stationary errors in polynomial regression models. *Sequential Anal.* **27**(3), 304–329.

Stone C.J. (1977) Consistent nonparametric regression. *Ann. Statist.* **5**(4), 595–645. With discussion and a reply by the author.

Tsybakov A.B. (2009) *Introduction to Nonparametric Estimation*. Springer Series in Statistics. Springer, New York. Revised and extended from the 2004 French original, Translated by Vladimir Zaiats.

Vexler A. and Gurevich G. (2011) A note on optimality of hypothesis testing. *Mathematics in Engineering, Science and Aerospace* **2**(3), 243–250.

Wu W.B. and Shao X. (2006) Invariance principles for fractionally integrated nonlinear processes *Recent Developments in Nonparametric Inference and Probability* vol. 50 of *IMS Lecture Notes Monogr. Ser.* Inst. Math. Statist. Beachwood, OH pp. 20–30.

Appendix A

A.1 (Stochastic) Landau symbols

For a deterministic sequence $\{a_n : n \in \mathbb{N}\}$ we write $a_n = O(1)$, if $|a_n| \le C$ for large enough n (i.e. there is some $n_0 \in \mathbb{N}$ such that for all $n \ge n_0$ we have $|a_n| \le C$). We write $a_n = o(1)$, if $a_n \to 0$, as $n \to \infty$. Both notions carry over to a sequence of vectors or, more generally, sequences of a normed linear space. However, for real-valued sequences $\{b_n\}$, one can further introduce the notions

$$a_n = O(b_n), \qquad \text{if } |a_n| \le C|b_n|, \text{ as } n \to \infty, \text{ for some } C < \infty, \text{ i.e. } \lim\sup_{n\to\infty} \left|\frac{a_n}{b_n}\right| < \infty,$$

and

$$a_n = o(b_n), \qquad \text{if } a_n/b_n \to 0, \text{ as } n \to \infty, \text{ i.e. } \lim\sup_{n\to\infty} \left|\frac{a_n}{b_n}\right| = 0.$$

Let $\{X_n : n \in \mathbb{N}\}$ be a sequence of random variables. $\{X_n\}$ is said to be bounded in probability, denoted by $X_n = O_P(1)$, if for every $\varepsilon > 0$ there exists a constants $M > 0$ and $n_0 \in \mathbb{N}$ (which usually depend on ε), such that

$$P(|X_n| > M) \le \varepsilon \qquad \text{for all } n \ge n_0.$$

By enlarging M one can achieve that $P(|X_n| > M) \le \varepsilon$ holds true for the finitely many $n = 1, \dots, n_0$, such that the above definition is equivalent to

$$\sup_{n\in\mathbb{N}} P(|X_n| > M) \le \varepsilon.$$

Such a sequence $\{X_n\}$ is also called uniformly tight. A sufficient condition for $X_n = O_P(1)$ is that X_n converges in distribution to some nondegenerated random variable or to a point mass

Financial Statistics and Mathematical Finance: Methods, Models and Applications, First Edition. Ansgar Steland.

in some fixed $x_0 \in \mathbb{R}$. Further, we write

$$X_n = o_P(1), \qquad \text{if } X_n \xrightarrow{P} 0, \text{ as } n \to \infty.$$

Again, those two notions can be extended to random vectors and random elements taking values in a normed space. If $\{Y_n\}$ is a sequence of real-valued random variables, then

$$X_n = O_P(Y_n), \qquad \text{if } X_n = Y_n R_n \text{ with } R_n = O_P(1).$$

If $Y_n \neq 0$, then $X_n = O_P(Y_n)$, if $X_n/Y_n = O_P(1)$. Further,

$$X_n = o_P(Y_n), \qquad \text{if } \frac{X_n}{Y_n} = o_P(1), \text{ i.e. } \left|\frac{X_n}{Y_n}\right| \xrightarrow{P} 0, \text{ as } n \to \infty.$$

It is important to note that the notions $o_P(1)$ and $O_P(1)$ stand for terms with that property, but those terms can change from instance to instance, which often simplifies the calculations and proofs substantially.

The stochastic order can also be obtained by estimating moments.

(i) If $E(\|X_n\|) = O(a_n)$, then $X_n = O_P(a_n)$.

(ii) If $E(\|X_n\|^2) = O(a_n)$, then $X_n = O_P(b_n^{1/2})$.

Here is a list of rules of calculus for o_P and O_P.

(i) $o_P(1) + o_P(1) = o_P(1)$;

(ii) $O_P(1) + o_P(1) = O_P(1)$;

(iii) $o_P(1)O_P(1) = o_P(1)$;

(iv) $o_P(R_n) = R_n o_P(1)$;

(v) $O_P(R_n) = R_n O_P(1)$;

(vi) $o_P(O_P(1)) = o_P(1)$.

Here is a useful lemma that asserts that one may plug in a $o_P(1)$ random sequence in $o(\bullet)$ and $O(\bullet)$ remainder terms, particularly in Taylor expansions.

Lemma A.1.1 *Let $f : D \to \mathbb{R}$ be a function defined on $D \subset \mathbb{R}^k$ for some $k \in \mathbb{N}$, and let $\{X_n\}$ be a sequence or random variables with $X_n = o_P(1)$, which attain values in D.*

(i) If $f(x) = o(\|x\|^r)$, as $x \to 0$, then $f(X_n) = o_P(\|X_n\|^r)$.

(ii) If $f(x) = O(\|x\|^r)$, as $x \to 0$, then $f(X_n) = O_P(\|X_n\|^r)$.

For proofs of the above results see Van der Vaart (1998).

A.2 Bochner's lemma

Lemma A.2.1 *Let $\{g_n : n \geq 1\}$ be a uniformly integrable sequence of functions $g_n : \mathbb{R} \to \mathbb{R}$, $n \geq 1$, i.e.*

$$\sup_{n\in\mathbb{N}} \int |g_n(y)|\,\mathrm{d}y \leq C < \infty,$$

which converges uniformly on compact sets to a function g, i.e.

$$\lim_{n\to\infty} \sup_{x\in I} |g_n(x) - g(x)| = 0,$$

for any compact subset $I \subset \mathbb{R}$. Further, let $K : \mathbb{R} \to \mathbb{R}$ be a function with

(i) $\int K(z)\,\mathrm{d}z < \infty$ *and*

(ii) $|z||K(z)| \to 0$, *as* $|z| \to \infty$,

and h_n be a sequence with $h_n \downarrow 0$. Then the smoothing operator

$$\widetilde{g}_n(x) = K_{h_n} \star g_n(x) = \int h_n^{-1} K([x-y]/h_n) g_n(y)\,\mathrm{d}y$$

satisfies

$$\lim_{n\to\infty} \sup_{x\in I} \left| \widetilde{g}_n(x) - g(x) \int K(z)\,\mathrm{d}z \right| = 0$$

for any compact set $I \subset \mathbb{R}$.

A.3 Conditional expectation

Let $X : (\Omega, \mathcal{F}, P) \to (\mathbb{R}, \mathcal{B})$ be a random variable. First, recall that the expectation $E(X) = \int X\,\mathrm{d}P$ is said to exist or to be defined, if $\min\{E(X^+), E(X^-)\} < \infty$. Then one puts $E(X) = E(X^+) - E(X^-)$. $E(X)$ is called finite, if $E|X| < \infty$, or, equivalently, $E(X^+) < \infty$ and $E(X^-) < \infty$. Notice that due to $|x| = x^+ + x^-$ we have $|E(X)| < \infty$ ($E(X)$ is finite) if and only if $E|X| < \infty$.

Now let $\mathcal{A} \subset \mathcal{F}$ be a sub σ-field. If $X \geq 0$, the mapping $A \mapsto \int_A X\,\mathrm{d}P$, $A \in \mathcal{F}$, defines a measure ν on $\mathcal{A}$. The measure is finite, if $EX < \infty$. In any case, the Radon–Nikodym theorem ensures the existence of a $\mathcal{A}$-measureable density f such that $\nu(A) = \int_A f\,\mathrm{d}P$, $A \in \mathcal{A}$, called the conditional expectation of X given $\mathcal{A}$ and denoted by $E(X|\mathcal{A})$. If $E(X) < \infty$, then $E(X|\mathcal{A}) < \infty$. Next, suppose X takes values in $\mathbb{R}$ and satisfies $E|X| < \infty$. Decompose $X = X^+ - X^-$ with $X^+, X^- \geq 0$ and denote by f^+ and f^- the corresponding densities. Then, the conditional expectation $E(X|\mathcal{A})$ is the $\mathcal{A}$-measureable function $E(X|\mathcal{A}) = f^+ - f^-$ satisfying $\int_A E(X|\mathcal{A})\,\mathrm{d}P = \int_A X\,\mathrm{d}P$. However, the conditional expectation can also be defined when $|X|$ is not integrable, see e.g. Shiryaev (1999). Having already defined $E(X^+|\mathcal{A})$ and $E(X^-|\mathcal{A})$ as above, suppose that

$$\min\{E(X^+|\mathcal{A})(\omega), E(X^-|\mathcal{A})(\omega)\} < \infty \tag{A.1}$$

for (almost) all $\omega \in \Omega$. Then, one defines the generalized conditional expectation of X given $\mathcal{A}$ by $E(X|\mathcal{A}) = E(X^+|\mathcal{A}) - E(X^-|\mathcal{A})$, where on the null set $\{E(X^+|\mathcal{A}) = E(X^-|\mathcal{A}) = \infty\}$ an arbitrary value can be assigned, e.g. 0. If $E(|X||\mathcal{A})(\omega) < \infty$ for (almost) all $\omega \in \Omega$, then Equation (A.1) holds true and, moreover, $E(X|\mathcal{A})(\omega)$ is finite for (almost) all $\omega \in \Omega$. We say that the generalized expectation exists and is finite, a.s.

Theorem A.3.1 *A list of properties of the conditional expectation. Let $\mathcal{A} \subset \mathcal{F}$ be a sub-σ-field. Provided the expectations of all random variables exists, the following properties hold true.*

(i) *If $\mathcal{A} = \{\emptyset, \Omega\}$, then $E(X|\mathcal{A}) = E(X)$ a.s.*

(ii) *If $X = c$ a.s., then $E(X|\mathcal{A}) = c$ a.s.*

(iii) *If $X \leq Y$ a.s, then $E(X|\mathcal{A}) \leq E(Y|\mathcal{A})$, a.s.*

(iv) *$|E(X|\mathcal{A})| \leq E(|X||\mathcal{A})$ a.s.*

(v) *For constants a, b such that $aE(X) + bE(Y)$ exists, $E(aX + bY|\mathcal{A}) = aE(X|\mathcal{A}) + bE(Y|\mathcal{A})$.*

(vi) *$E(X) = E(E(X|\mathcal{A}))$.*

(vii) *If $\mathcal{A}_1 \subset \mathcal{A}_2$, then $E(E(X|\mathcal{A}_2)|\mathcal{A}_1) = E(X|\mathcal{A}_1)$ a.s. If $\mathcal{A}_2 \subset \mathcal{A}_1$, then $E(E(X|\mathcal{A}_2)|\mathcal{A}_1) = E(X|\mathcal{A}_2)$ a.s.*

(viii) *If X is independent from $\mathcal{A}$ and $E(X)$ exists, then $E(X|\mathcal{A}) = E(X)$ a.s.*

(ix) *Let X and Y be random variables such that X is $\mathcal{A}$-measureable with $E|X| < \infty$ and $E|XY| < \infty$. Then $E(XY|\mathcal{A}) = XE(Y|\mathcal{A})$ a.s.*

(x) *If $X = Y$ on a set $A \in \mathcal{A}$, then $\mathbf{1}_A E(Y|\mathcal{A}) + \mathbf{1}_{A^c} E(X|\mathcal{A})$ is a version of $E(X|\mathcal{A})$. This means, for $\omega \in A$ one may replace $E(X|\mathcal{A})(\omega)$ by $E(Y|\mathcal{A})(\omega)$.*

(xi) *If $\varphi : \mathbb{R} \to \mathbb{R}$ is convex and $E|\varphi(X)| < \infty$, then $\varphi(E(X|\mathcal{A})) \leq E(\varphi(X)|\mathcal{A})$.*

A.4 Inequalities

The following theorem summarizes some classic inequalities, which are frequently used in the book.

Theorem A.4.1 *Let X be a random variable taking values in the set $\mathcal{X}$.*

(i) *Markov's inequality: Let $g : \mathcal{X} \to \mathbb{R}$ be a measurable function such that $Eg(X) < \infty$. Then $P(X > c) \leq E(g(X))/g(c)$ for any constant $c > 0$.*

(ii) *Chebychev's inequality: If $E(X^2) < \infty$, $P(|X - E(X)| > \varepsilon) \leq \operatorname{Var}(X)/\varepsilon^2$ for any $\varepsilon > 0$.*

(iii) *Hölder's inequality: Let $X \in L_p$ and $Y \in L_q$ for $p, q \in [0, \infty]$ with $\frac{1}{p} + \frac{1}{q} = 1$. Then, $E|XY| \leq [E(X^p)]^{1/p}[E(Y^q)]^{1/q}$. In other words, $\|XY\|_{L_1} \leq \|X\|_{L_p}\|Y\|_{L_q}$.*

(iv) *Generalized Hölder inequality: Let $X_1, \ldots X_n$ be random variables with $X \in L_{p_i}$ for $i = 1, \ldots, n$, where $p_1, \ldots, p_n \in (0, \infty)$ and $r \in (0, \infty)$ satisfy*

$$\frac{1}{p_1} + \cdots + \frac{1}{p_n} = \frac{1}{r}.$$

Then,

$$\left\| \prod_{i=1}^{n} X_i \right\|_r \leq \prod_{i=1}^{n} \|X_i\|_{p_i}.$$

(v) *C_r inequality: Let X and X be random variables with $E|X|^r, E|Y|^r < \infty$. Then,*

$$E|X + Y|^r \leq C_r \left[E|X|^r + E|Y|^r \right],$$

where $C_r = 1$, if $0 < r \leq 1$, and $C_r = 2^{r-1}$, if $r > 1$.

A.5 Random series

The following theorem ensures the existence of the class of linear processes.

Theorem A.5.1 *Let $\{X_n : n \in \mathbb{N}\}$ be a L_1-bounded series of random variables, i.e. $\sup_{n\in\mathbb{N}} E|X_n| < \infty$ and let $\{a_n : n \in \mathbb{N}\} \subset \mathbb{C}$ be a summable sequence, i.e. $\sum_{n=1}^{\infty} |a_n| < \infty$. Then the series of random variables*

$$Y_n = \sum_{i=0}^{\infty} a_i X_{n-i}, \qquad n \in \mathbb{N}, \tag{A.2}$$

exists a.s. If $\{X_n\}$ is L_2-bounded, i.e. $\sup_{n\in\mathbb{N}} EX_n^2 < \infty$, then (A.2) holds in L_2. Further,

$$E\left(\sum_{i=0}^{\infty} a_i X_{n-i}\right) = \sum_{i=0}^{\infty} a_i E(X_{n-i}), \tag{A.3}$$

$$E\left(\sum_{i=0}^{\infty} a_i X_{n-i}\right) = \lim_{N\to\infty} E\left(\sum_{i=0}^{N} a_i X_{n-i}\right). \tag{A.4}$$

Obviously, the above theorem also holds true when the index set $\mathbb{N}$ is replaced by $\mathbb{Z}$.

A.6 Local martingales in discrete time

A process $\{X_t\}$ is called a **local martingale**, if there exists a sequence $\{T_n\}$ of stopping times such that $T_n \uparrow \infty$, as $n \to \infty$, a.s., and the stopped process $X^{T_n} = \{X_{\min(t,T_n)} : t \geq 0\}$, $n \geq 1$, defines a martingale, i.e. $E|X_t^{T_n}| < \infty$ and $E(X_t^{T_n}|\mathcal{F}_s) = X_s^{T_n}$, a.s., for all $s \leq t$ and $n \geq 1$.

Let $\{X_t : t \geq 0\}$ be a càdlàg process defined on a filtered probability space $(\Omega, \mathcal{F}, \{\mathcal{F}_t\}, P)$. $\{X_t\}$ is called a **semimartingale**, if it admits a decomposition

$$X_t = X_0 + A_t + M_t, \qquad t \geq 0,$$

where $\{A_t\}$ is of bounded variation and $\{M_t\}$ a local martingale.

In discrete time, we have the following characterization, cf. (Jacod and Shiryaev (2003), Proposition 1.46)

Theorem A.6.1 *The following conditions are equivalent:*

(i) $\{X_t : t \in \mathbb{N}\}$ *is a local martingale.*

(ii) $\{X_t : t \in \mathbb{N}\}$ *is a martingale transform, i.e. there exists a martingale* $\{M_t : t \in \mathbb{N}\}$ *and a predictable process* $\{H_t : t \in \mathbb{N}\}$ *such that* $X_n = X_0 + \sum_{i=1}^{n} H_{i-1}(M_i - M_{i-1})$ *for all* $n \geq 1$.

(iii) $\{X_t : t \in \mathbb{N}\}$ *is a generalized martingale, i.e.* $E(|X_t||\mathcal{F}_{t-1}) < \infty$, $t \geq 1$, *and* $E(X_t|\mathcal{F}_{t-1}) = X_{t-1}$, *a.s.*, $t \geq 1$.

Appendix B

Weak convergence and central limit theorems

B.1 Convergence in distribution

Let

$$X_n : (\Omega_n, \mathcal{F}_n, P_n) \to (\mathbb{R}, \mathcal{B}), \qquad n \in \mathbb{N},$$

be a sequence of random variables with distribution functions $F_n(x) = P(X_n \le x)$, $x \in \mathbb{R}$, $n \in \mathbb{N}$. Observe that, for what follows, each X_n may be defined on its own probability space $(\Omega_n, \mathcal{F}_n, P_n)$, $n \in \mathbb{N}$. Let F be a further distribution function and $X \sim F$. The sequence $\{F_n\}$ **converges in distribution to** F, if

$$F(x) = \lim_{n\to\infty} F_n(x)$$

for all $x \in \mathbb{R}$ that are continuity points of F, i.e. $x \in C_F = \{y \in \mathbb{R} : F(y-) = F(y)\}$. Clearly, this definition extends easily to distribution functions defined on $\mathbb{R}^d$.

The sequence of random variables and random vectors, respectively, $\{X_n\}$ **converges in distribution to** X, if the associated sequence $\{F_n\}$ converges in distribution to F, denoted by $X_n \stackrel{d}{\to} X$, as $n \to \infty$. This means,

$$P(X_n \le x) \to P(X \le x), \qquad n \to \infty,$$

holds true for all $x \in C_F$. An equivalent characterization is as follows. Let $\varphi_n(t) = E(\mathrm{e}^{itX_n})$, $t \in \mathbb{R}$, and $\varphi(t)$ denote the characteristic function (ch.f.) of X_n and X, respectively. Then, $X_n \stackrel{d}{\to} X$, as $n \to \infty$, holds, if and only if for all $t \in \mathbb{R}$

$$\varphi_n(t) \to \varphi(t),$$

Financial Statistics and Mathematical Finance: Methods, Models and Applications, First Edition. Ansgar Steland.

as $n \to \infty$. By checking pointwise convergence of the ch.f.s, one easily checks the following basic tool for establishing central limit theorems in higher dimensions.

Theorem B.1.1 (CRAMÉR–WOLD DEVICE)
A sequence $\{X, X_n\}$ of random vectors of dimension $d \in \mathbb{N}$ converges in distribution, i.e. $X_n \stackrel{d}{\to} X$, as $n \to \infty$, if and only if for any $\lambda \in \mathbb{R}^d$

$$\lambda' X_n \stackrel{d}{\to} \lambda' X,$$

as $n \to \infty$.

In particular, the Cramér–Wold technique tells us that

$$X_n \stackrel{d}{\to} N(\mu, \Sigma),$$

as $n \to \infty$, for some $\mu \in \mathbb{R}^d$ and $\Sigma \in \mathbb{R}^{d \times d}$, if and only if the univariate sequence $Y_n = Y_n(\lambda) = \sum_{j=1}^d \lambda_j X_{nj}$ satisfies a univariate central limit theorem such as Theorem B.7.3 or Theorem B.7.2, for each fixed vector $\lambda = (\lambda_1, \ldots, \lambda_d)' \in \mathbb{R}^d$. The following result ensures that convergence in distribution is invariant under continuous transformations.

Theorem B.1.2 (CONTINUOUS MAPPINGS)
Let $\{X, X_n\}$ be a sequence of d-dimensional random vectors in $\mathbb{R}^d$ jointly taking values in $\mathcal{X} \subset \mathbb{R}^d$ w.p. 1, such that

$$X_n \stackrel{d}{\to} X,$$

as $n \to \infty$. If $\varphi : \mathcal{X} \to \mathbb{R}^m$, $m \in \mathbb{N}$, is a continuous function, then

$$\varphi(X_n) \stackrel{d}{\to} \varphi(X),$$

as $n \to \infty$.

B.2 Weak convergence

The concept of weak convergence of probability measures generalizes the convergence in distribution of sequences of random vectors to sequences of stochastic processes or, more generally, sequences of random elements taking values in metric spaces. So, let us now assume that (S, d) is a metric space with metric d. Recall that $d : S \times S \to \mathbb{R}$ is called a **metric**, if the following properties hold.

(i) $d(x, y) \geq 0$ for all $x, y \in S$.

(ii) $d(x, y) = d(y, x)$ for all $x, y \in S$.

(iii) $d(x, y) = 0$ if and only if $x = y$ for all $x, y \in S$.

(iv) $d(x, z) \leq d(x, y) + d(y, z)$ for all $x, y, z \in S$.

In particular, if $(S, \| \bullet \|)$ is a normed space, then

$$d(x, y) = \|x - y\|, \qquad x, y \in S,$$

defines a metric on S.

S is equipped with the Borel-σ-field $\mathcal{S}$ induced by the open sets with respect to the topology given by d. It turns out that the theory simplifies, if the space is complete with respect to the metric, i.e. any Cauchy sequence converges. For the spaces of primary interest in mathematical finance, this can be achieved by an appropriate choice of the metric.

The most important metric spaces, which are used to model the realizations of ordered sequences of numbers, continuous function and functions with discontinuities of the first kind, are the following:

(i) **Time series**: Let S be the set of all sequences of real numbers, i.e.

$$S = \mathbb{R}^\infty = \{\{x_n\} : x_n \in \mathbb{R}, n \in \mathbb{N}\},$$

equipped with the metric

$$d(\{x_n\}, \{y_n\}) = \sum_{i=1}^{\infty} 2^{-k} d_0(x_k, y_k),$$

where

$$d_0(x, y) = \frac{|x - y|}{1 + |x - y|}, \qquad x, y \in \mathbb{R}.$$

$S = \mathbb{R}^\infty$ is a separable space, since its elements can be approximated in the metric d by those sequences that have only a finite number of non-vanishing rational elements. Clearly, a measurable mapping $X : (\Omega, \mathcal{F}, P) \to (S, \mathcal{S})$ is a univariate time series, since if π_n denotes the projection of some element $x = \{x_n\} \in S$ on the nth element, i.e. $\pi_n(x) = x_n$, then $X = \{X_n\}$ with $X_n = \pi_n \circ X$, i.e. X is a family of real-valued random variables X_n.

(ii) **Stochastic processes with continuous trajectories**: Let $S = C([0, 1]; \mathbb{R})$ be the set of functions $f : [0, 1] \to \mathbb{R}$ that are continuous with respect to the usual topologies. S is equipped with the uniform topology induced by the norm

$$\|f\| = \|f\|_\infty = \sup_{t \in [0,1]} |f(t)|, \qquad f \in C([0, 1]; \mathbb{R}).$$

(iii) **Càdlàg processes**: Let $S = D([0, 1]; \mathbb{R})$ be the set of all functions $f : [0, 1] \to \mathbb{R}$ that are right continuous with limits from the left. A random element X with values in S is called càdlàg process, cf. Definition 5.1.2. Since the elements of S may have jumps, an appropriate metric is given by the Skorohod metric

$$d'(f, g) = \inf_{\lambda \in \Lambda} \max\{\|f - g \circ \lambda\|_\infty, \|\lambda - \mathrm{id}\|_\infty\},$$

where Λ denotes the set of all continuous and strictly increasing mappings $\lambda : [0, 1] \to [0, 1]$ and id stands for the identity mapping on $[0, 1]$. An equivalent metric is given by

$$d(f, g) = \inf_{\lambda \in \Lambda} \max\{\|f - g \circ \lambda\|_\infty, \|\lambda\|^0\},$$

where for a non-decreasing function on $[0, 1]$ satisfying $\lambda(0) = 0$ and $\lambda(1) = 1$ one puts

$$\|\lambda\|^0 = \sup_{s<t} \left| \log \frac{\lambda(t) - \lambda(s)}{t - s} \right|.$$

If $\|\lambda\|^0 < \infty$, then the slopes of the chords of λ are bounded away from 0 and ∞, such that λ is continuous and strictly increasing in this case. A sequence converges with respect to the metric d if and only if it converges with respect to the metric d', but the metric d yields a complete metric space (see below).

(iv) **Càdlàg functions on** $[0, 1]$ **taking values in** $\mathbb{R}^d$, $d \geq 2$, are defined analogously to the case $d = 1$. The above notions carry over, if one interprets the absolute value $| \bullet |$ as a vector norm on $\mathbb{R}^d$. The resulting space is denoted by

$$D([0, 1]; \mathbb{R}^d) = \{f : [0, 1] \to \mathbb{R}^d \mid f \text{ is right continuous with left-hand limits}\}.$$

(v) **Càdlàg functions on** $[0, \infty)$**:** To deal with càdlàg processes with time coordinate $t \in [0, \infty)$, the following Skorohod metric is appropriate. For $f, g \in D([0, \infty); \mathbb{R})$ put

$$d(f, g) = \sum_{t=1}^{\infty} \frac{1}{2^t} \min\{1, d_t(f, g)\},$$

where $d_t(f, g)$ is defined as follows: $d_t(f, g)$ is the infimum of those $\delta > 0$, for which there exists two grids $\{t_i : i = 1, \ldots, k\}$ and $\{s_i : i = 1, \ldots, k\}$ or ordered points with $t_0 = s_0 = 0$ and $s_k, t_k \geq t$, such that $|t_i - s_i| \leq \delta$ for $i = 1, \ldots, k$, and

$$|f(t') - g(s')| \leq \delta, \qquad \text{if } t' \in [t_i, t_{i+1}) \text{ and } s' \in [s_i, s_{i+1}),$$

for $i = 0, \ldots, k - 1$, see Pollard (1984). One can show that

$$d(f_n, f) \to 0, \qquad \text{as } n \to \infty,$$

if and only if there exist continuous, strictly increasing maps $\{\lambda_n\}$ defined on $[0, \infty)$ onto itself, such that, uniformly over compact sets, with

$$\lambda_n(t) - t \to 0, \qquad \text{and} \qquad f(\lambda_n(t)) - f_n(t) \to 0,$$

as $n \to \infty$.

For the case of random variables the definition of convergence in distribution asserts that

$$F_n(x) = \int \mathbf{1}_{(-\infty,x]}(s) \mathrm{d}F_n(s) \to \int \mathbf{1}_{(-\infty,x]}(s)\, \mathrm{d}F(s) = F(x), \qquad n \to \infty,$$

for all $x \in C_F$. By approximating the indicator by a continuous function, one can show that the above condition is equivalent to

$$\int \varphi(s)\, \mathrm{d}F_n(s) \to \int \varphi(s)\, \mathrm{d}F(s), \qquad n \to \infty, \tag{B.1}$$

for all real-valued, continuous and bounded functions φ, i.e. $\varphi \in C_b(\mathbb{R}; \mathbb{R})$. But (B.1) can be interpreted as a condition for the sequence $\{\mathrm{d}F_n, \mathrm{d}F\}$ of probability measures on the real line. This allows us to extend easily the definition to general probability measures on metric spaces: A sequence $\{P, P_n\}$ of probability measures on the metric space (S, d) **converges weakly**, if

$$\int \varphi(x)\, \mathrm{d}P_n(x) \to \int \varphi(x)\, \mathrm{d}P(x), \qquad n \to \infty,$$

holds true for all $\varphi \in C_b(S; \mathbb{R})$. Many basic rules of calculation for this notion as well as many distributional limits for statistics under nonstandard conditions can be derived from the continuous mapping theorem, which asserts that the property of weak convergence is stable under continuous mappings.

Theorem B.2.1 (Continuous Mapping Theorem)
Let $\{X, X_n\}$ be a sequence of random elements taking values in some metric space (S, d) equipped with the associated Borel-σ-field. Assume that

$$X_n \Rightarrow X,$$

as $n \to \infty$. If $\varphi : S \to S'$, is a mapping into another metric space S' with metric d' that is a.s. continuous on $X(\Omega) \subset S$, then

$$\varphi(X_n) \stackrel{d}{\to} \varphi(X),$$

as $n \to \infty$.

Example B.2.2 *Here are some important continuous mappings*

(i) $\phi(f) = f(x_0)$, $f \in D([0, 1]; \mathbb{R})$, *for some fixed value* x_0.

(ii) $\phi(f) = \int g(f(s))\, \mathrm{d}s$, $f \in D([0, 1]; \mathbb{R})$, *for any continuous function* g.

(iii) $\phi(f) = \sup_{t \in A} f(t)$ *and* $\phi(f) = \sup_{t \in A} |f(t)|$, *$A$ a compact set, for* $f \in C([0, 1]; \mathbb{R})$, *as well as the related* inf*-functionals. If* $f \in D([0, 1]; \mathbb{R})$, *then* $\phi(f) = \sup_{t \leq a} |f(t)|$ *is continuous at each f such that* $f(a) = f(a-)$.

(iv) The first passage time functional $\phi(f) = \inf\{t \geq 0 : f(t) > a\}$, *$a$ a constant, for* $f \in C([0, 1]; \mathbb{R})$.

Let (S_1, d_1) and (S_2, d_2) be metric spaces equipped with the Borel-σ-field. The product $S_1 \times S_2$ is the set of all pairs (f, g) with $f \in S_1$ and $g \in S_2$. We may define a metric on $S_1 \times S_2$ via

$$d((f_1, g_1), (f_2, g_2)) = d_1(f_1, g_1) + d_2(f_2, g_2),$$

for $(f_1, g_1), (f_2, g_2) \in S_1 \times S_2$. $S_1 \times S_2$ is separable, if S_1 as well as S_2 are separable.

The product-σ-field $\mathcal{S}_1 \otimes \mathcal{S}_2$ is the σ-field induced by all (generalized) rectangles $A_1 \times A_2$, where $A_1 \in \mathcal{S}_1$ and $A_2 \in \mathcal{S}_2$. $\mathcal{B}(S_1 \times S_2)$ denotes the Borel-σ-field induced by the collection of all open sets $O \subset S_1 \times S_2$. If $S_1 \times S_2$ is separable, then

$$\mathcal{B}(S_1 \times S_2) = \mathcal{S}_1 \otimes \mathcal{S}_2.$$

In what follows, we confine ourselves to this setting.

Let $\{X, X_n\}$ and $\{Y, Y_n\}$ be two sequences of random elements taking values in (S_1, d_1) and (S_2, d_2), respectively. Suppose we already know that

$$X_n \Rightarrow X, \quad n \to \infty, \quad \text{in } (S_1, d_1), \tag{B.2}$$

and

$$Y_n \Rightarrow Y, \quad n \to \infty, \quad \text{in } (S_1, d_1). \tag{B.3}$$

The following theorem provides sufficient conditions such that Equations (B.2) and (B.3) imply the *joint* weak convergence of (X_n, Y_n) to (X, Y) in the product space.

Theorem B.2.3 (Joint Weak Convergence)
Let $\{X, X_n\}$ and $\{Y, Y_n\}$ be two sequences taking values in (S_1, d_1), respectively (S_2, d_2), such that (B.2) and (B.3) hold true. Then

$$(X_n, Y_n) \Rightarrow (X, Y),$$

as $n \to \infty$, provided at least one of the following conditions is satisfied.

(i) $Y = c \in S_2$ is a constant, i.e. nonrandom.

(ii) X_n and Y_n are independent for all n as well as X and Y are independent.

By applying the above result to the case $c = 0$ and combining it with the continuous mapping theorem, we obtain a general version of Slutzky's lemma in metric spaces.

Theorem B.2.4 (Lemma of Slutzky)
Suppose that X_n allows a decomposition

$$X_n = Y_n + R_n, \qquad n \geq 1,$$

where $R_n \Rightarrow 0$ or $R_n \stackrel{P}{\to} 0$, as $n \to \infty$. If $Y_n \Rightarrow Y$, as $n \to \infty$, for some random element Y, then

$$X_n \Rightarrow Y,$$

as $n \to \infty$.

The following useful result is often implicitly used, but rarely mentioned.

Theorem B.2.5 *If $g : \mathbb{R}^{d_1} \to \mathbb{R}^{d_2}$ is continuous, then the mapping $\phi : \mathcal{D}_\phi \to D([0,1]; \mathbb{R}^{d_2})$, where $\mathcal{D}_\phi \subset D([0,1]; \mathbb{R}^{d_1})$, given by*

$$\phi(f)(t) = g(f(t)), \qquad f \in \mathcal{D}_\phi$$

is continuous with respect to the Skorohod topology.

Remark B.2.6 *In application, that result is often applied as follows. One knows that $X_n \Rightarrow X$, as $n \to \infty$, and $P(X \in A) = 1$ for some $A \subset D([0,1]; \mathbb{R}^{d_1})$. g is well defined and continuous on A, but not necessarily on $D([0,1]; \mathbb{R}^{d_1})$. Then, one puts $\mathcal{D}_\phi = A$ in order to conclude that*

$$\phi(X_n) \Rightarrow \phi(X),$$

as $n \to \infty$. In particular, it suffices that the g is continuous except on a P_X-null set.

The representation theorem due to Skorohod, Dudley and Wichura allows us to work with equivalent versions that converge for every ω. It transforms weak convergence to (a.s.) convergence in the metric and, vice versa, allows us to establish weak convergence by proving metric convergence in probability of equivalent versions. We refer to (Shorack and Wellner, 1986, p.48 and Billingsley, 1999, Theorem 6.7).

Theorem B.2.7 (SKOROHOD/DUDLEY/WICHURA REPRESENTATION THEOREM)

(i) Let $\{P_n, P\}$ be a sequence of probability measure defined on a metric space (S, d), such that P has a separable support or (S, d) is separable. Then, there exist random elements $\{X_n, X\}$, defined on a common probability space $(\widetilde{\Omega}, \widetilde{\mathcal{F}}, \mathbb{P})$, such that $X \sim P$ and $X_n \sim P_n$ for all n, under $\mathbb{P}$, and

$$d(X_n(\omega), X(\omega)) \to 0,$$

as $n \to \infty$, for all $\omega \in \widetilde{\Omega}$.

(ii) Let $\{X_n, X\}$ be a sequence of random elements taking values in a metric space (S, d), such that P_X has a separable support or (S, d) is separable. Then, there exist equivalent versions $\{\widetilde{X}_n, \widetilde{X}\}$ defined on a common probability space $(\widetilde{\Omega}, \widetilde{\mathcal{F}}, \mathbb{P})$, such that $\widetilde{X} \sim P$ and $\widetilde{X}_n \sim P_n$ for all n, under $\mathbb{P}$, and

$$d(\widetilde{X}_n(\omega), \widetilde{X}(\omega)) \to 0,$$

as $n \to \infty$, for all $\omega \in \widetilde{\Omega}$.

(iii) The converse: let $\{X_n, X\}$ be a sequence of random elements taking values in a metric space (S, d) such that P_X has separable support or (S, d) is separable. If

$$d(\widetilde{X}_n, \widetilde{X}) \xrightarrow{\mathbb{P}} 0, \qquad \text{as } n \to \infty,$$

for equivalent versions $\widetilde{X}_n \stackrel{d}{=} X_n$ *and* $\widetilde{X} \stackrel{d}{=} X$ *on a new probability space* $(\widetilde{\Omega}, \widetilde{\mathcal{F}}, \mathbb{P})$, *then*

$$X_n \Rightarrow X, \qquad \text{as } n \to \infty.$$

B.3 Prohorov's theorem

In order to establish the weak convergence of some sequence $\{X, X_n\}$ of stochastic processes attaining values in the spaces $C([0, T]; \mathbb{R})$ or $D([0, T]; \mathbb{R})$, a basic approach is as follows. First, one establishes the convergence of the finite-dimensional distributions, called **fidi convergence**, i.e.

$$(X_n(t_1), \ldots, X_n(t_k)) \stackrel{d}{\to} (X_n(t_1), \ldots, X_n(t_k)), \qquad n \to \infty,$$

for all fixed time points $0 \le t_1 < \cdots < t_k \le T$ and all $k \in \mathbb{N}$. If X is a Gaussian process, this can usually be achieved by applying a multivariate central limit theorem in the Euclidean space $\mathbb{R}^k$. The fidi convergence is a necessary condition and, since the finite-dimensional distributions determine the distribution of a stochastic process, determines the distribution of the limit process. Unfortunately, fidi convergence is not sufficient to establish weak convergence. One has to study the latter notion in greater detail in order to derive sufficient criteria that turn out to be based on characterizations of the compact sets of the space $\mathcal{P}(S)$ of probability measures defined on $\mathcal{S}$.

Recall that a sequence $\{x_n\}$ of a metric space (S, d) satisfies the Cauchy property or is **fundamental**, if for all $\varepsilon > 0$ there exists $n_0 = n_0(\varepsilon)$ such that $d(x_n, x_m) < \varepsilon$ for all $n, m \ge n_0$. A metric space (S, d) is called **complete**, if each fundamental sequence converges. A given metric induces a topology; however, two metrics may induce the same topology, one leading to a complete metric space, whereas the other metric lacks this property. A space that is metrizable by a metric leading to a complete and separable metric space is called **Polish**. Completeness nicely characterizes compact sets by closed ones, since in a complete metric space any closed subset is compact if and only if it is totally bounded; recall that a set is called **totally bounded**, if any cover by open balls has a finite subcover.

Recall that we are given a metric space (S, d) – such as $D([0, 1]; \mathbb{R})$ – equipped with the Borel-σ-field $\mathcal{S}$ and the associated set $\mathcal{P}(S)$ of probability measures. The space $\mathcal{P}(S)$ can be metrized by the **Prohorov metric** π and the convergence with respect to the Prohorov metric is the weak convergence, i.e.

$$P_n \Rightarrow P \qquad \text{if and only if} \qquad \pi(P_n, P) \to 0.$$

In a metric space convergence can be characterized by a sequence criterion: A sequence converges if and only if any subsequence contains a further convergent subsequence. Thus, the weak convergence of a sequence $\{P_n, P\}$ of probability measures can be characterized in the following way: We have $P_n \Rightarrow P$ $(\Leftrightarrow \pi(P_n, P) \to 0)$, as $n \to \infty$, if and only if any subsequence $\{P_{n_k} : k \ge 1\}$ contains a further subsequence $\{P_{n'_k} : k \ge 1\}$ such that $P_{n'_k} \Rightarrow P$ $(\Leftrightarrow \pi(P_{n'_k}, P) \to 0)$, as $k \to \infty$, but the latter is equivalent to $P_{n'_k} \Rightarrow P$, as $k \to \infty$. Further, in any metric space a subset A is **relatively compact**, i.e. has compact closure, if every subsequence $\{P_n\} \subset A$ has a subsequence $\{P_{n_k} : k \ge 1\}$ with $P_{n_k} \to P'$, as $k \to \infty$, where the limit P' is in the closure of A. Applied to our setting this means: A subset $A \subset \mathcal{P}(S)$ of probability measures has compact closure $\overline{A}$ if and only if every sequence $\{P_n\} \subset A$ has a subsequence $\{P_{n_k}\}$ with converges weakly to some $P' \in \overline{A}$, i.e. $P_{n_k} \Rightarrow P'$, as $k \to \infty$. Here, the limit P' may depend on the subsequence. Therefore, the weak convergence $P_n \Rightarrow P$, $n \to \infty$, can be shown as follows: First, one shows that $\{P_n\}$ is relatively compact and then one verifies that all possible limits are equal to P. The latter can be achieved by showing that the fidis converge to the fidis of P, which determine the distribution of the limit P. A theorem due to Prohorov allows us to relate the compact sets of $\mathcal{P}(S)$ to the compact sets of S. This is achieved by the concept of **tightness**. A subset $A \subset \mathcal{P}(S)$ of probability measures in $\mathcal{P}(S)$ is called tight if for all $\varepsilon > 0$ there exists a compact subset $K_\varepsilon \subset S$ such that

$$P(K_\varepsilon) > 1 - \varepsilon \qquad \text{for all } P \in A.$$

Theorem B.3.1 (PROHOROV'S THEOREM)
Let (S, d) be a metric space.

(i) Any tight set of probability measures on S is relatively compact.

(ii) If S is Polish, i.e. a complete and separable metric space, then any relatively compact set is tight.

Notice that for the second half of Prohorov's theorem, the completeness is essential. Prohorov's theorem has the following corollary.

Corollary B.3.2 *Let $\{P_n\}$ be a tight sequence of probability measures on a metric space (S, d) such that the fidis converge to P. Then $P_n \Rightarrow P$, as $n \to \infty$.*

B.4 Sufficient criteria

By virtue of Prohorov's theorem, fidi convergence and tightness yields weak convergence.

For processes $X, X_n, n \geq 1$, with values in $C([0, 1]; \mathbb{R})$, one has the following sufficient criterion, cf. (Billingsley, 1968, Theorem 12.3).

Theorem B.4.1 *Suppose that fidi convergence holds true, this means*

$$(X_n(t_1), \ldots, X_n(t_k)) \stackrel{d}{\to} (X(t_1), \ldots, X(t_k)), \tag{B.4}$$

as $n \to \infty$, for all $t_1, \ldots, t_k \in [0, 1]$, $k \in \mathbb{N}$. If, in addition, there exist a nondecreasing function $F : [0, 1] \to \mathbb{R}$ and constants $\gamma \geq 0$ and $\alpha > 1$, such that for all $s, t \in [0, 1]$

$$P(|X_n(t) - X_n(s)| \geq \lambda) \leq \lambda^{-\gamma} |F(t) - F(s)|^\alpha, \tag{B.5}$$

or

$$E(|X_n(t) - X_n(s)|^\gamma) \leq |F(t) - F(s)|^\alpha, \tag{B.6}$$

for all $n \geq 1$ and all $\lambda > 0$, then

$$X_n \Rightarrow X,$$

as $n \to \infty$.

There is a similar sufficient criterion for the weak convergence of a sequence $\{X, X_n\}$ of random elements taking values in $D([0, 1]; \mathbb{R})$, which requires the following preparation. The projection $\pi_t : D([0, 1]; \mathbb{R}) \to \mathbb{R}$ is continuous in x, if $d(x_n, x) \to 0$, as $n \to \infty$, implies that $\pi_t(x_n) = x_n(t) \to x(t) = \pi_t(x)$, as $n \to \infty$. Notice the following fact.

$$\pi_t \text{ is continuous in } x, \text{ if and only if } x \text{ is continuous in } t.$$

Indeed, if x is continuous, then one can argue as follows: if $d(x_n, x) \to 0$, as $n \to \infty$, and $\{\lambda_n\}$ are transformations such that $\|\lambda_n - \text{id}\|_\infty, \|x_n - x \circ \lambda_n\|_\infty \to 0$, as $n \to \infty$, then $|x_n(t) - x(t)| \le |x_n(t) - x \circ \lambda_n(t))| + |x \circ \lambda_n(t) - x(t)| \to 0$, as $n \to \infty$, where the second term converges to 0 by continuity of x. To show the only if part, assume that x is discontinuous in t. Let λ_n be the transformation that is linear on $[0, t]$ and $[t, 1]$ with $\lambda_n(0) = 0$ and $\lambda_n(t) = t - 1/n$ and $\lambda_n(1) = 1$. Then, $x_n = x \circ \lambda_n$ converges in $D([0, 1]; \mathbb{R})$ to x, but $x_n(t) \not\to x(t)$.

For a random element X of $D([0, 1]; \mathbb{R})$ let

$$T_X = \{t \in [0, 1] : \pi_t \text{ is continuous at all } x \in D([0, 1]; \mathbb{R})\backslash N\},$$

where $N \subset D([0, 1]; \mathbb{R})$ is a P_X-null set, i.e. $P(X \in N) = 0$. By right continuity of the càdlàg functions in 0, $0 \in T_X$. However, 1 may or may not be an element of T_X. One can show that there are at most countably many t such that

$$P(X \text{ discontinuous in } t) = P(X \in J_t) > 0,$$

where J_t denote those càdlàg functions that are discontinuous in t, cf. (Billingsley, 1999, p. 138).

If $t_1, \dots, t_k \in T_X$, then the projection

$$\pi_{t_1,\dots,t_k}(X) = (X(t_1), \dots, X(t_k))$$

is continuous on a set $A \subset D([0, 1]; \mathbb{R})$ with $P(X \in A) = 1$. We are now in a position to formulate the following sufficient criterion, cf. (Billingsley, 1968, Theorem 15.6) and (Billingsley, 1999, Theorem 13.5).

Theorem B.4.2 *Suppose that, first, fidi convergence in continuity points holds in the sense that*

$$(X_n(t_1), \dots, X_n(t_k)) \Rightarrow (X(t_1), \dots, X(t_k)), \tag{B.7}$$

as $n \to \infty$, holds true for all $t_1, \dots, t_k \in T_X$, $k \in \mathbb{N}$. Secondly,

$$X_1 - X_{1-\delta} \Rightarrow 0, \tag{B.8}$$

as $\delta \downarrow 0$, or

$$P(X_1 \ne X_{1-}) = 0.$$

Lastly, assume that for all $r \le s \le t$ and $\lambda > 0$

$$P(|X_n(s) - X_n(t)| \ge \lambda, |X_n(t) - X_n(t)| \ge \lambda) \le \frac{1}{\lambda^{4\beta}}[F(t) - F(r)]^{2\alpha}, \tag{B.9}$$

or

$$E|X_n(s) - X_n(t)|^{\beta}|X_n(t) - X_n(s)|^{\beta} \leq [F(t) - F(r)]^{2\alpha}, \tag{B.10}$$

where $\beta \geq 0$ *and* $\alpha > 1/2$, *and* F *is a nondecreasing, continuous function on* $[0, 1]$. *Then*

$$X_n \Rightarrow X,$$

as $n \to \infty$.

Notice that Equation (B.8) is equivalent to

$$\lim_{\delta \downarrow 0} \lim \sup_{n \to \infty} P(|X_n(1-\delta) - X_n(1)| > \varepsilon) = 0 \tag{B.11}$$

for all $\varepsilon > 0$.

For processes with paths in the Skorohod space $D([0, \infty), \mathbb{R}^d)$, one has the following nice version, cf. (Jacod and Shiryaev, 2003, p. 319).

Theorem B.4.3 *Suppose that, in addition to fidi convergence in continuity points, Equation (B.11) holds and there exists a nondecreasing function* F *on* $(0, \infty)$ *and constants* $\gamma \geq 0$ *and* $\alpha > 1$, *such that for all* $\lambda > 0$ *and all* $s < r < t$

$$P(\|X_n(r) - X_n(s)\| \geq \lambda, \|X_n(t) - X_n(t)\| \geq \lambda) \leq \lambda^{-\gamma}[F(t) - F(r)]^{\alpha}, \tag{B.12}$$

for all $n \geq 1$. *Then*

$$X_n \Rightarrow X,$$

as $n \to \infty$, *in* $D([0, \infty), \mathbb{R}^d)$.

Lemma B.4.4 *Provided the processes* $X, X_n, n \geq 1$, *taking values in* $D([0, \infty); \mathbb{R}^d)$ *have independent increments, the fidi convergence follows at once from, provided*

(i) $X_n(t) \xrightarrow{d} X(t)$, *as* $n \to \infty$, *for all* $t > 0$ *and*

(ii) $X_n(t) - X_n(s) \xrightarrow{d} X(t) - X(s)$, *as* $n \to \infty$, *for all* $s < t$.

B.5 More on Skorohod spaces

The Skorohod metric defined above, also called the J_1 metric, has the property that jumps converge. Indeed, if $x_n \to x$, as $n \to \infty$, in the J_1 topology, then for any inner point t there exists a sequence $\{t_n\}$ such that $t_n \to t$, $x_n(t_n) \to x(t) = x(t+)$ and $x_n(t_n-) \to x(t-)$, as $n \to \infty$, which implies $x_n(t_n) - x_n(t_n-) \to x(t) - x(t-)$, as $n \to \infty$, i.e. convergence of jumps. As a consequence, the function $f_n(t)$ that vanishes for $t < 1/2 - 1/n$, equals 1 if $t > 1/2$ and interpolates linearly, otherwise, does not converge to the indicator $1_{[1/2,1]}(t)$ in the J_1 topology. This is sometimes too strong.

To obtain convergence for unmatched jumps, the M_1 metric can be used. For $f \in D([0, 1]; \mathbb{R})$ define the *completed graph* by

$$\Gamma_f = \{(z, t) \in \mathbb{R} \times [0, 1] : z = \alpha f(t-) + (1-\alpha) f(t) \text{ for some } \alpha \in [0, 1]\}.$$

Here, we use the convention $f(0-) = f(t)$. Notice that Γ_f is a connected set and particularly contains all line segments connecting $(f(t-), t)$ and $(f(t), t)$ for all jump points t. We can define a (total) order on Γ_f as follows: $(z_1, t_1) \leq (z_2, t_2)$ if $t_1 < t_2$ or $t_1 = t_2$ and $|x(t_1-) - z_1| < |x(t_2-) - z_2|$. Note that this order starts at the left end of the completed graph and ends at the right end. A parametric representation of Γ_f is a continuous non-decreasing function $\theta : [0, 1] \to \Gamma_f$ onto Γ_f. Denote $\theta = (u, t)$, i.e. $\theta(s) = (u(s), t(s))$, $s \in [0, 1]$, where $u(s)$ is the spatial coordinate and $t(s)$ the time coordinate. Let Π_f be the set of all parametric representations of Γ_f. Then the M_1 metric is defined by

$$d_{M_1}(f_1, f_2) = \inf \max(\|u_1 - u_2\|_\infty, \|t_1 - t_2\|_\infty),$$

where the infimum is taken over all $(u_1, t_1) \in \Pi_{f_1}$ and $(u_2, t_2) \in \Pi_{f_2}$. Again, for continuous f we have $\|f_n - f\|_\infty \to 0$, as $n \to \infty$, if $d_{M_1}(f_n, f) \to 0$, as $n \to \infty$. The J_1 topology is stronger, i.e. convergence in the J_1 topology implies convergence in the M_1 topology.

An even weaker notion of convergence is given by the M_2 metric, which is based on the Hausdorff distance between the completed graphs. Given two compact sets K_1 and K_2, the **Hausdorff metric** is defined by

$$d_H(K_1, K_2) = \max\left\{ \sup_{f \in K_1} d(f, K_2), \sup_{f \in K_2} d(f, K_1) \right\}.$$

Here, $d(x, A) = \inf\{d(x, y) : y \in A\}$ is the distance between the point x and the set A. The M_2 metric is now defined by

$$d_{M_2}(f, g) = d_H(\Gamma_f, \Gamma_g).$$

Convergence with respect to the J_1 or M_1 topology implies convergence in the M_2 topology.

B.6 Central limit theorems for martingale differences

Let us now discuss alternative and generalized versions of Theorem 8.3.6, the central limit theorem for martingale differences. The following result can be found in Durrett (1996).

Theorem B.6.1 (Lindeberg–Feller CLT for Martingale Difference Arrays)
Suppose $\{X_{nk} : 1 \leq k \leq n, n \geq 1\}$ is a $\mathcal{F}_{nk}$-martingale difference array such that $E(X_{nk}^2|\mathcal{F}_{nm}) < \infty$ for all k, n. Put $V_{nk} = \sum_{i=1}^k E(X_{ni}^2|\mathcal{F}_{n,i-1})$ and $S_n = \sum_{k=1}^n X_{nk}$. If

(i) $V_{nn} \to 1$ in probability and

(ii) the conditional Lindeberg condition is satisfied, i.e. for all $\varepsilon > 0$

$$\sum_{k=1}^n E\left(X_{nk}^2 \mathbf{1}(|X_{nk}| > \varepsilon)|\mathcal{F}_{n,k-1}\right) \to 0,$$

in probability,

then $S_n \xrightarrow{d} N(0, 1)$, as $n \to \infty$.

It is convenient to have a result for martingale difference sequences.

Theorem B.6.2 (LINDEBERG–FELLER CLT FOR MARTINGALE DIFFERENCE SEQUENCES) *Suppose* $\{X_n\}$ *is a* $\mathcal{F}_n$*-martingale sequence. Let* $V_n = \sum_{k=1}^{n} E(X_k^2|\mathcal{F}_{k-1})$ *and* $S_n = \sum_{k=1}^{n} X_k$*. If*

(i) $V_n/n \to \sigma^2 > 0$ *and*

(ii) the Lindeberg condition is satisfied, i.e.

$$\frac{1}{n}\sum_{k=1}^{n} E\big(X_k^2 \mathbf{1}(|X_k| > \varepsilon)\big) \to 0,$$

as $n \to \infty$*, for any* $\varepsilon > 0$*,*

then $S_n/\sqrt{n} \stackrel{d}{\to} N(0, \sigma^2)$*, as* $n \to \infty$*.*

Remark B.6.3 *The corresponding functional central limit theorems hold as well. In particular, in Theorem B.7.2 the assertion can be strengthened to* $S_{\lfloor nt \rfloor}/\sqrt{n} \Rightarrow B$*, as* $n \to \infty$*, without changing the conditions.*

Theorem B.6.4 (LINDEBERG–FELLER FCLT FOR MARTINGALE DIFFERENCE ARRAYS) *Suppose that instead of condition (i) of Theorem B.7.2 we have more generally* $V_{n,\lfloor nt \rfloor} \to t$*, in probability, for* $t \in [0, 1]$*, then the partial sum process converges weakly to Brownian motion, i.e.* $S_{n,\lfloor nt \rfloor} \Rightarrow B(t)$*, as* $n \to \infty$*.*

B.7 Functional central limit theorems

Functional central limit theorems, also called invariance principles, are mathematical theorems that form the core for the statistical inference on time series as arising on financial markets. They are also a basic tool to study statistical procedures for estimation and inference, since many statistics behind such procedures can be shown to be driven by underlying stochastic processes, particularly partial sum processes, which satisfy a functional central limit theorem whose limit in turn governs the limiting distribution of the statistic of interest.

Donsker's classic invariance principle provides the basic relationship between partial sum processes and the Brownian already discussed at the end of Chapter 4. It addresses the case that the increments are i.i.d. random variables with mean 0 and a finite second moment.

Theorem B.7.1 (DONSKER, I.I.D. CASE) *Let* $\xi_1, \xi_2, \ldots$ *be a sequence of i.i.d. random variables with* $E(\xi_1) = 0$ *and* $\sigma^2 = E(\xi_1^2) < \infty$*. Then*

$$\frac{1}{\sqrt{T}}\sum_{t=1}^{\lfloor T\bullet \rfloor} \xi_t \Rightarrow \sigma B(\bullet),$$

as $T \to \infty$*, where* B *denotes standard Brownian motion, and* $\Rightarrow$ *signifies weak convergence in the Skorohod space* $D([0, 1]; \mathbb{R})$*.*

Donsker's theorem addresses the case of i.i.d. random variables, which is too restrictive for financial problems. Generalizations to dependent increments are needed, in order to study problems arising in mathematical finance and financial markets, respectively. Here is the corresponding result for a martingale difference sequence under a Lindberg condition.

Theorem B.7.2 (LINDEBERG–FELLER FCLT FOR MARTINGALE DIFFERENCE SEQUENCES) *Suppose* $\{\xi_t\}$ *is a square-integrable* $\mathcal{F}_t$*-martingale difference sequence. Let* $V_t = \sum_{k=1}^{t} E(\xi_k^2|\mathcal{F}_{k-1})$*. If*

(i) $V_T/T \to \sigma^2 > 0$*, as* $T \to \infty$*, in probability, and*

(ii) the Lindeberg condition is satisfied, i.e.

$$\frac{1}{T}\sum_{t=1}^{T} E(\xi_t^2 \mathbf{1}(|\xi_t| > \sqrt{T}\varepsilon)) \to 0,$$

as $T \to \infty$*, for any* $\varepsilon > 0$*,*

then

$$\frac{1}{\sqrt{T}}\sum_{t=1}^{\lfloor T\bullet\rfloor} \xi_t \Rightarrow \sigma B(\bullet),$$

as $T \to \infty$*.*

The generalization to martingale difference arrays is as follows.

Theorem B.7.3 (LINDEBERG–FELLER FCLT FOR MARTINGALE DIFFERENCE ARRAYS) *Suppose* $\{\xi_{Tt} : 1 \leq t \leq T, T \geq 1\}$ *is a* $\mathcal{F}_{Tt}$*-martingale difference array such that* $E(\xi_{Tt}^2|\mathcal{F}_{T,t-1}) < \infty$ *for all* $1 \leq t \leq T$*,* $T \geq 1$*. Put*

$$V_{tk} = \sum_{i=1}^{k} E\left(\xi_{ti}^2|\mathcal{F}_{t,i-1}\right), \qquad 1 \leq t \leq T,\ T \geq 1.$$

Suppose the following two conditions are satisfied.

(i) $V_{T,\lfloor Tu\rfloor} \to u$ *in probability for all* $u \in [0, 1]$*.*

(ii) The conditional Lindeberg condition holds true, that is

$$\sum_{i=1}^{T} E\left(\xi_{Ti}^2 \mathbf{1}(|\xi_{Ti}| > \varepsilon)|\mathcal{F}_{T,i-1}\right) \to 0,$$

in probability, as $T \to \infty$*, for all* $\varepsilon > 0$*.*

Then

$$\sum_{t=1}^{\lfloor Tu\rfloor} \xi_{Tt} \Rightarrow B(u),$$

as $T \to \infty$*.*

Notice that Theorem B.7.3 covers Theorem B.7.2 as a special case.

The central limit theorem for α-mixing sequences can be generalized to the functional version leading to the following result, cf. Ibragimov (1962) and (Hall and Heyde, 1980, Corollary 5.1).

Theorem B.7.4 (FCLT FOR α-MIXING PROCESSES)
Suppose $\{X_n : n \in \mathbb{Z}\}$ is an ergodic stationary sequence with $E(X_0) = 0$ and $E|X_0|^{2+\delta} < \infty$ for some $\delta > 0$. Suppose that the α-mixing coefficients $\alpha(k)$, $k \in \mathbb{N}$, satisfy

$$\sum_{k=1}^{\infty} \alpha(k)^{\frac{\delta}{2(2+\delta)}} < \infty.$$

Then

$$\frac{1}{\sqrt{T}} \sum_{t=1}^{\lfloor Ts \rfloor} X_t \Rightarrow \sigma B_s,$$

as $T \to \infty$, where

$$\sigma^2 = E\left(X_0^2\right) + 2 \sum_{k=1}^{\infty} E(X_1 X_{1+k}).$$

For long-memory processes the following result has been shown by Taqqu (1974/75).

Theorem B.7.5 (FCLT FOR LONG-MEMORY PROCESSES)
Let $\{X_t : t \in \mathbb{N}\}$ be a stationary sequence of Gaussian random variables with mean zero and autocovariance function $\gamma(k) = E(X_1 X_{1+k})$, $k \in \mathbb{Z}$. Suppose that

$$\lim_{T\to\infty} \frac{1}{KT^{2H}L(T)} \sum_{t=1}^{T} \sum_{s=1}^{T} \gamma(t-s) = 1,$$

for some $0 < H < 1$, a constant $K > 0$ and a function L that is slowly varying at infinity. Then

$$\frac{1}{\sqrt{KT^{2H}L(T)}} \sum_{t=1}^{\lfloor Ts \rfloor} X_t \Rightarrow B_s^H,$$

where $\{B_t^H : t \in [0, 1]\}$ is a standard fractional Brownian motion.

B.8 Strong approximations

Assume now that the ξ_t and therefore S_T are defined on a common probability space $(\Omega, \mathcal{F}, P)$. The fact that S_T converges weakly to Brownian motion B motivates us to ask whether one may define a Brownian motion on $(\Omega, \mathcal{F}, P)$ in such a way that one can approximate S_T by B, preferably uniformly on $[0, 1]$ in some sense. A related question is whether the partial sums $\sum_{t=1}^{n} \xi_t$ can be approximated by $B(n)$. Provided such an approximation is strong enough, we expect that the invariance principle in the sense of Donsker's theorem also holds true. These questions have been studied extensively in the classic probabilistic literature.

The following theorem provides some well-known results on the approximation of sums by Brownian motion, which are also called *strong invariance principles*. These results are typically of the form

$$E_n = o(a_n), \qquad \text{a.s.}$$

for some random expression E_n involving Brownian motion, which means

$$P(\lim_{n\to\infty} E_n/a_n = 0) = 1.$$

Theorem B.8.1 (STRONG APPROXIMATIONS, I.I.D. CASE)
Let $\xi_1, \xi_2, \ldots$ be a sequence of i.i.d. random variables with $E(\xi_1) = 0$ and $\sigma^2 = E(\xi_1^2) < \infty$. Put $S(n) = \sum_{i=1}^{n} \xi_i$, $n \in \mathbb{N}$.

(i) One can define a standard Brownian motion $\{B(t) : 0 \le t < \infty\}$ such that

$$\max_{n\le N} |S(n) - B(n)| = o((N \log\log N)^{1/2}), \qquad a.s.$$

(ii) If $E|\xi_1|^r < \infty$ for some $r > 2$, then

$$|S(n) - B(n)| = o(n^{1/r}), \qquad a.s.$$

(iii) If $H : [0,\infty) \to [0,\infty)$ is a function such that $t^{-2}H(t)$ is nondecreasing, $t^{-3}H(t)$ nonincreasing and $EH(\xi_1) < \infty$, then

$$\left| \sum_{i=1}^{n} \xi_i - B\left(\sum_{i=1}^{n} \sigma_i^2 \right) \right| = o(H^{-1}(n)),$$

for some sequence $\{\sigma_n : n \in \mathbb{N}\}$ of real numbers with $\sigma_n \to 1$ that can be chosen in such a way that $1 - \sigma_n^2 = o(n^{-1}(H^{-1}(n))^2)$.

Related to the above strong approximations of partial sums are results leading to

$$N^{-1/2} \max_{n\le N} |S(n) - B(n)| \xrightarrow{P} 0, \tag{B.13}$$

as $N \to \infty$, which are frequently called *weak invariance principle*. Such results are strong enough to imply the invariance principle in the sense of Donsker's theorem, that is in the sense of weak convergence.

Theorem B.8.2 *Suppose $\xi_1, \xi_2, \ldots$ satisfies a weak invariance principle in the sense of Equation (B.13). Then*

$$S_T(u) = T^{-1/2} S(\lfloor Tu \rfloor) \Rightarrow B(u),$$

as $T \to \infty$.

Proof. We shall apply the Skorohod/Dudley/Wichura representation theorem. Notice that, by virtue of the scaling property (5.2.10), $\{B(u) : u \geq 0\}$ is equal in distribution to $\{T^{-1/2} B(Tu) : u \geq 0\}$ for each T. Therefore

$$\sup_{u \in [0,1]} |S_T(u) - B(u)| \stackrel{d}{=} \sup_{u \in [0,1]} \left| \frac{1}{\sqrt{T}} \sum_{t=1}^{\lfloor Tu \rfloor} \xi_t - \frac{1}{\sqrt{T}} B(Tu) \right|. \tag{B.14}$$

Therefore, we may conclude that on a new probability space,

$$\begin{aligned} \sup_{u \in [0,1]} |S_T(u) - B(u)| &= \sup_{u \in [0,1]} \frac{1}{\sqrt{T}} \left| \sum_{t=1}^{\lfloor Tu \rfloor} \xi_t - B(Tu) \right| \\ &= \frac{1}{\sqrt{T}} \max_{n \leq T} |S(n) - B(n)| \\ &\xrightarrow{P} 0, \end{aligned}$$

as $T \to \infty$. But this implies $S_T \Rightarrow B$, as $T \to \infty$, for the original processes.

References

Billingsley P. (1968) *Convergence of Probability Measures*. John Wiley & Sons Inc., New York.

Billingsley P. (1999) *Convergence of Probability Measures*. Wiley Series in Probability and Statistics: Probability and Statistics 2nd edn. John Wiley & Sons Inc., New York. A Wiley-Interscience Publication.

Durrett R. (1996) *Probability: Theory and Examples*. 2nd edn. Duxbury Press, Belmont, CA.

Hall P and Heyde C.C. (1980) *Martingale Limit Theory and its Application*. Academic Press Inc. [Harcourt Brace Jovanovich Publishers], New York. Probability and Mathematical Statistics.

Ibragimov I.A. (1962) Some limit theorems for stationary processes. *Teor. Verojatnost. i Primenen.* **7**, 361–392.

Jacod J. and Shiryaev A.N. (2003) *Limit Theorems for Stochastic Processes*. vol. 288 of *Grundlehren der Mathematischen Wissenschaften [Fundamental Principles of Mathematical Sciences]* 2nd edn. Springer-Verlag, Berlin.

Pollard D. (1984) *Convergence of Stochastic Processes*. Springer Series in Statistics. Springer-Verlag, New York.

Shiryaev A.N. (1999) *Essentials of Stochastic Finance: Facts, Models, Theory*. vol. 3 of *Advanced Series on Statistical Science & Applied Probability*. World Scientific Publishing Co. Inc., River Edge, NJ. Translated from the Russian manuscript by N. Kruzhilin.

Shorack G.R. and Wellner J.A. (1986) *Empirical Processes with Applications to Statistics*. Wiley Series in Probability and Mathematical Statistics: Probability and Mathematical Statistics. John Wiley & Sons Inc., New York.

Taqqu M.S. (1974/75) Weak convergence to fractional Brownian motion and to the Rosenblatt process. *Z. Wahrscheinlichkeitstheorie und Verw. Gebiete* **31**, 287–302.

Index

Financial Statistics and Mathematical Finance: Methods, Models and Applications, First Edition. Ansgar Steland.